Selection of Writings of
MIGUEL IRADIER

Volume I

PHILOSOPHY, SCIENCE and CULTURE

(2004 - 2022)

Edited by *Hurqualya*

www.hurqualya.net
Cover: Adelaida Rondán

Índex

Selection of Writtings of
MIGUEL IRADIER
Volume I
PHILOSOPHY, SCIENCE and CULTURE
(2004 - 2020)

Selection of Writings of
MIGUEL IRADIER

Volume I

PHILOSOPHY, SCIENCE and CULTURE

(2004 - 2022)

Edited by *Hurqualya*

SAMKHYA AND GOLDEN MEAN (The space of Analysis in the age of revelation of form)

May, 2004

SINOPSIS

This book is basically about philosophy. Samkhya, which is the oldest Indian philosophy, had a decisive influence on the development of Buddhism. We will use its viewpoints to tackle a wide-ranging physical and mathematical question, related to the asymmetric principle of measurement established and generalised by Alexey Stakhov, founded on the mathematical principles of the Golden Mean or Continuous Proportion, which is commonly designated by the letter *phi (Φ)*.

The Samkhya is an exact, mathematical philosophy, which studies that waves and fluctuations of any kind of motion, with the mind considered as merely another form or order of motion in general. Samkhya means *proportion, measure* or *analysis*. Usually, the purpose of Samkhya is self- observation or the direct perception of the fluctuations of one's own mind, with a view to achieving balance and an eventual end to its motion. Samkhya is thus the philosophical base that underlies yoga. However, although Samkhya generally focuses on direct perception or discrimination on the mind by attention or awareness, it is a universal theory or framework for the generality of the nature of motion, based on the concept of *modification* or *fluctuation*. Any modification of regular motion exists in three modalities or *gunas*: *Sattwa*, *Rajas* and *Tamas*. With *Tamas* as inertia, *Rajas* as action or mutation and *Sattwa* as the relative balance or equilibrium between the former two, or the joint sensitivity of the system, all three modalities can be matched perfectly with Newton's three principles of mechanics, which define rest or inertia, force and action–reaction. However, Newton's three principles oust the modality, as we can see particularly clearly in the third principle or law, which, being immediate, defines no *medium* for action–reaction, thereby making the *real* or specific transmission of communication or information impossible. Indeed, information is produced when the reaction of a system is *not equal* to the action received, i.e. when there is asymmetry and also a lack of temporal synchronicity. Newton's absolute time is an implicit principle of *global synchronisation*, which has persisted in the theory of relativity and quantum mechanics. A summary of this topic can be found in Koichiro Matsuno's online paper *Information: Resurrection of the Cartesian Physics*. Global synchronisation or absolute time is at best a metaphysical statement, although it has made our idea of physics and its constants possible. This absolute time means exactly the same in Newton'*s* three laws, albeit in a different order, as the absolute awareness or *Purusha* of Samkhya with respect to the three modalities or *gunas*. But study of asymmetry or asynchrony between action and reaction is an ideal subject for the application of the Golden Mean or Continuous Proportion.

Samkhya may coincide with mechanics in its idea of the oscillator. The only empirical science derived from Samkhya that we are aware of is pulse-based medical diagnosis or *nadi vigyan*, where the pulse is a non-linear or non-analytical biological oscillator. Samkhya, however, is non-linear analysis *par excellence.*

The three *gunas* or modalities of Samkhya, when applied to the pulse oscillator, become the three *doshas* or humours of *ayurveda* and *nadi vigyan*: *Pitta*, *Vata* and *Kapha*. Since *Pitta* is equivalent to action or force, and *Kapha* to inertia or retention, *Vata*, as a principle of action and reaction, is the sensitivity of the balance between both. This strict correspondence is appropriate in relation to the moments of dynamics, although we have performed a transposition to a non-linear order that cannot be quantified as precisely, given the interdependence of the terms.

Pulsology was the first semiology to achieve both internal and experimental consistency, in a way that numerous modern attempt have been unable to equal. Pulsology, as morphology for diagnostic purposes, identifies between five essential properties of the pulse: rhythm, frequency, intensity, amplitude and *form*. The first four mean much the same as they do in harmonic or wave analysis, but the fifth has a unique meaning that is only found in morphology. It is a genuine, elusive *quintessence* that cannot be reduced to analytical interaction between the other four, and best expresses the non-linearity of the system, including an synthesising the full complexity of all the potential pairings between miscellaneous cycles that could affect this apparently isolated pulse. The *form* of the pulse is *not* just the irregularity of its wave compared with that of a harmonic oscillator; it is also a *thickness* characteristic of its plot or signature that can be translated into a buffering or lack of determination that is a *characteristic* of the *form* of the pulse, rather than a mere lack of determination considered as the incompleteness of the information available to us in this regard. Until recently, the *form* of the pulse could not be recorded by any mechanical detector, but could only be perceived by the direct sensitivity of the fingers of the *vadya* or examining doctor. Now, however, it can be captured properly by small laser diodes and also studied acoustically. Thus, it is now possible to study and analyse this *form* in real time under variable conditions.

We consider this *form* to be a universal characteristic of *all* motion, not only the pulse; all that has happened is that mechanics and analysis have ignored on a matter of pure principle. We may therefore be dealing with something that is of great importance that demands its own independent method.

The *preliminary* measurement of the three modalities tends to be rather imprecise because of the interdependence of terms mentioned above; they can only be considered as statistical proportions or weights of fuzzy sets within an expert system that can, at the same time, be considered as a control system. These fuzzy sets may be related to the Golden Mean at the intersection of sets — with the Golden Mean as a ratio of ratios, it must naturally correspond to these intersections or overlaps of fuzzy sets resulting from areas of inference applied to the curve of a graph. Several studies have already been done of the relationship between fuzzy sets and the Golden Mean. Nevertheless, this remains a purely heuristic, empirical approach, lacking any well-defined theoretical concept, although it may lead us towards some interesting ideas. In any case, it is possible to define a genuine theoretical model.

We do not believe that a force applied independent of time can exist in nature. In any case, the **impulse** seems more real to use as a fundamental unit. The impulse, conceived as force by time, as well as a variation of the moment, fully matches the notion of modification in Samkhya. A balance between impulses can, in turn, be conceived as an impulse with its own temporal dimension, if we fit into our imaginary balance an adjustment time between successive operations of removing and adding weights — a time to which we can also ascribe variability. Thus, we can conceive motion and velocity as balance between impulses in different or opposite directions, equalling not only the inertia and force in the wave of the pulse, but also the third principle of action and reaction, in the actual modality of the sensitivity of the balance. This estimate of the time interval between discrete operations now gives us a measure of the inertiality of the balance as a medium or instrument, or, if one prefers, its sensitivity. The three modes now share a time element, whereby both the mass and the force can be equalled with the impulse by means of the sensitivity of the balance. It should be noted that since the sensitivity of the balance can be computed as inertiality (although not immediately), we obtain a definition of the recursive or self-recurrent impulse, of the same nature as the Golden Mean itself. There are also certain clearly defined logical rules for the sequentiality of this self-recurrence, which are touched upon in the paper. Time itself can be defined as the actual and potential supply of operations within an interval; the possible operations are limited by the current loads and the available weights, which may differ in their magnitude and proportion, and which become optimal in generalisations of the classical problem of Fibonacci'*s* weights. But it is the *form* of the pulse, the thickness or plot of its signature, that fits the possible solutions and *defines the relevant range of precision.* This is the key point, since there is no exact motion outside differential calculus, and all motion —classical or quantum— must have its own degree or plot of precision, inasmuch as none is purely linear or ideal.

In physics there are no *natural* units of energy, force or even mass, if we assume that pure inertia cannot be computed. Moreover, nor can we give any content to the notion of the cycle, which is reduced to the mere formality of the circle, and since Planck'*s* action quantum is a unit of energy per cycle per second, we have no idea of what kind of recurrence this cycle corresponds to, which is overlooked when the unit is defined. Obviously, a cycle with no content or a purely formal cycle would only exist in the absolute absence of mass, which is not the case with any particle in motion. Owing to the recursive definition of the impulse, having been able to find an equivalence with mass or inertia, force and specific action–reaction, we can do the same with other magnitudes or notions, such as pressure, tension and many other existing or totally new ones. We will also find contents for the notion of the cycle as a recurrence. Why? For the same reason that we can make new definitions for numbers with their corresponding number theories within the golden system of numbering as conceived by Bergman and generalised by Stakhov. This seems barely credible. We have a universal translator of notions and units; and not only for known ones but also for any others we might come up with. This covers a wide range of different spheres that would not appear to be in contact with one another, from physics to psychology, via biology, economics and many others.

What is more, this physical principle of translation by material implication is inherently linguistic, and inevitable also covers all the possibilities of human languages. The three modalities can readily be associated with the subject, verb and predicate of our sentences. The same Continuous Proportion and modalities exist as connective and disjunctive functions, with an intermediate term in the *–and*-and *–or*-to allow for mixing and separation. In this paper the only non-rhetorical characterisation of nature is considered to lie in "joining and splitting", which only makes sense in relation to inequality, rather than for the equivalences and identities predicted by equations. Yet we also have Fibonacci'*s* "words" and "strings", as logical sequences, which allow the transfer between content and form in the maximum possible number of orders of construction. "String" or "link" is one of the primary meanings of the word *guna*; meanwhile, the theory of golden numbers and the principle of asymmetric measurement closely related to them is the most efficient analogue–digital converter that we have found. Thus, we can linguistically access the pulse or any other motion with as much property as can be achieved through mathematics, which in this sense is considered as just another language, rather than a privileged one. This assertion in no way an exaggeration, and it may even be possible to demonstrate it, although we have neither the space nor the skill to deal with it in depth. The languages of nature, and their motion, are not just a metaphor, although they exist in metaphor: mathematics also has its own transformation space. Yet the density of a language hinges on its material implication, in the same way that mathematics only seems to become transcendent through its relationship to physical reality. However, all this is much too noble and serious to be turned into a simple formal game. We have no wish to add more storeys onto the Tower of Babel nor deepen the moat at its feet. Moreover, all this leads in the opposite direction to that of formal artifices, and Patanjali himself, the supreme summariser of Samkhya, pointed this out quite adequately: the sense of any language exists in its own right in the gap that separates the three moments, thus confirming that Samkhya is the true analysis, with mathematical analysis taking the opposite direction — that of composition. It is worth making the effort to comprehend this fully.

The true root of translation is motion itself, the *own form* of the motion, which has not until now been taken into account. Form and motion are the same thing: form cannot be conceived as being external or unspecific. If the phrase "the medium is the message" became famous for its ambiguity, we can be equally sure that with respect to motion *form is content,* and the more layers of ambiguity we remove, the truer this becomes. These layers are hidden in the very idea of exactitude that we apply to differential analysis, which does nothing but export problems to further spaces for abstraction — in order to remove ambiguity, precisely. But the fit, whether it is ambiguous or not, belongs to the form itself, and in this property ambiguity would just be irrelevant. Since our idea of "limit" and "function" refers to an irreducible morphological component, we can assume the unity of design and intent, i.e. understand through the motion itself the impossibility of fully separating efficient causes from final ones, or material from formal ones. Without this, there can be no theory of real-time evolution, but just an imaginary appeal to statistics and possible or past worlds. **The question boils down to this: we are now in a position to extract more, better information from the form of motion than of structures, with the justification for the latter being the shape of the motion itself. Furthermore, this allows the same information to be folded or synthesised. This remains valid at all levels.** By "structures" we mean whatever is intended to be free of conditioned motion and fixed by the resolution of the analysis, whether we are dealing with genes, conventional non-analytical curves or normal differential equations.

I am not a physicist or a mathematician, which is why I am raising this problem for anyone who is interested in it or capable of addressing it. The problem would seem to be related to the perturbation theory applied to an oscillator — but here we have a recursive or self-referencing definition, as well as a level of adjustment for precision or lack of determination in optimal terms of maximums and minimums within the actual *form* of the motion. This has never been suggested, and if it is understood, it could lead us to a completely new space for analysis. This is why I have subtitled this paper "The analysis space in the age of the discovery of form".

Naturally, we cannot expect to find any analytical solutions here, by positioning ourselves beyond the limits of differential calculus. The usefulness of all this is that non-analytical solutions for such systems take us to a new level of description to allow *translation between different systems*, as well as any possible redefinitions within a given system. The scope for this is too great for us to determine even within formal systems, but it can also transcend them via its capacity to redefine our actual concept of unity. What we lose in the general representation of nature, a mere surplus, we can gain in the capacity for real-time integration at all levels, including computing and work time. The new organisational power that resides here cannot be missed. In all our models representation it is either or question or it is ineffective — a mere obstacle. That said, all the excesses of abstraction to the point of unrecognisability are a consequence of this representational component.

We think that all the problems of contemporary physics related to the scalar component of fields pose the same difficulties as what we are dealing with here: the Higgs field for the masses of particles, the *lambda* tension constant or cosmological constant for the "weight" of space closely related, and even "dark mass" or probably inexistent dark matter. All of this leads to a break in the symmetry of our equations and the conditionality of their structure. We now find that that vector forces independent of local time have nowhere to rest and nothing to which they can be applied, and they are thus revealed to be largely fictitious. This could shatter the idea of the physical constant. Curiously, this scalar component of fields is also designated by the letter *phi (Φ)*, and in a context of adjusting functions with optimal local values. Meanwhile, some physicists, such as El Naschie have proposed focusing and unifying all physics on the constant *phi*; such attempts would appear to us to be far too speculative, since they are based on transfinite Cantorian spaces that make constructive procedures impossible. The mathematics of harmony driven by Stakhov is based on living priority to constructive definitions and limiting the inflation of imaginary, abstract spaces. This direction is one that merits further study, since too much emphasis now tends to be placed on complex space, with little understanding of real space and its degrees of compossibility.

But our problem is much more basic and general than the ones being addressed at the cutting edge of physics, which are the end result of extreme specialisation. One may think that we could not have chosen a more fragile motif than our balance to point out the stability of processes and things. But is stability guaranteed anywhere? We do not think it is, and the misunderstanding is based on both prejudice in favour of approaches with differential mathematical solutions and our idea of the mechanical, which always refers to inertia and rest as pure nothingness. The principle of minimal action may be the only reference for stability; but it is precisely differential calculus that has deprived this principle of the attention that it is due. There is no stability — only degrees of equilibrium. The contrary would only be possible if the mass and inertia were identical, in which case the mass would be zero and equivalent to rest; but this makes no sense at all, especially for physics, because for dynamics the mass is the point of reference and reality, having removed all properties of reference from rest. This sums up our concept of mechanics, and reductionism itself. Thus, mass can only be described in terms of non-linear motion, i.e. motion with its own *form*. This is the only way to be fully consistent with mechanicism; anything else is running away from the problem. Everything therefore depends on how the conditions are arranged: Samkhya is the doctrine of pure conditionality in nature, often expressed as the claim that "nothing exists on its own", but only in terms of its relationships and contrasts with other things.

As history has shown us, nature and the form of motion is one of the most general topics dealt with in philosophy. It can properly include both the mind and beauty, because beauty, first and foremost, is also motion, at times evidently so and at others only in deeply hidden form. Meanwhile, *phi*, the simplest ratio of recursivity or self-reference, leads us — thanks to its adherence to the singularity of things — to the ultimate problem at hand, the frame of reference, the most inexcusable and esoteric of all. Where is it, if intellect is, like *phi*, pure adherence? Although we have always dealt with reality, we have tended to confuse it compulsively with our arbitrary demarcations and definitions. This is the final link that is being dissolved.

The frame of reference and its definition are now in our hands but, more than ever before, all will depend on our intention to insert into reality.

SAMKHYA AND GOLDEN MEAN
(The space of Analysis in the age of revelation of form)

§ 1 §

Why we say about something that is "simply mechanic"? A simple harmonic oscillator, a pendulum without friction, is something easy to understand. To construct real pendulums fit to that ideal entails increasing degrees of difficulty and complexity. The simplicity of mechanics resides properly in its idea, more than in some manifest property of the things. An ideal mechanical system is only a limit for real systems; to turn ourselves into an inert system subject to mechanical forces involves the idea of the death. Besides and by definition we do not know any force at all that could be not mechanical nor subject that could not be inert at the moment of receiving it. No later extension of the physics has altered this fundamental conception, nor could at least imagine how to do it. The invisible nexus of the causality can be stumped, but it cannot be cut: otherwise we would be already outside of mechanics. Therefore, the idea of mechanics would be identical with the one of the ideal causality, that occurs in a single and same plane; in this item the program of pure reductionism would be condensed implying nothing else than another ideal, another limit for the last remission of the phenomena, no matter how varied could be the levels or emergencies of events. That unique and same plane in which all the forces, their actions and reactions would be developed, is not other one that the ideal plane by which it rolls perfectly a smooth ball with total monotony that generates the inertia definition in Galileo and in Newton later. Like first of its variants, also the gravity was conceived originally like circular motion around the centre of grave bodies, now accelerated, in the famous experiment of the inclined plane.

This it is the beginning of the mechanics, also called dynamics, and of course that seems something too fundamental. Anyway, we know that in the meantime the uniform motion has been equaled with rest, and because of this, any characterization of a point in rest in the real space has been made impossible in other terms that those of dynamics or of the motion itself. From they arise here, then, all the difficulties around the reference systems that have been constituted in the problem par excellence of physics, with questions such as the weight of the vacuum and the generation of the mass of particles. I will not solve it, although one can only think that it would be more consequent with dynamics to assume that real space does not exist anymore, and that this space is not but a group of parameters for the adjustment of a temporary function. To assume this would surely require watching towards the sea that

breaks against the border of the differential calculus; but mainly, it would require watching towards uniform motion more like a destination than a point to begin with.

§ 2 §

To contemplate uniform motion or even the rest as destination would require seeing it from within. But it not exist a defined point where the external begins to be internal. Let us take the definition that gives Patanjali of the mental substrate during the gradual concentration in the meditation:

Tatah punah santoditau tulia-pratiaiau cittasiaikagrata-parinamah, Which can be translated: "the concentration takes place when the mental waves that rise and fall at two different moments are exactly equal." This is the definition of a simple oscillator, or an inert ball that rolls on an ideal plane. What for us, seen from outside, has the full aspect of inanity, what makes think us irremediably about a flat encephalogram, it is for Samkhya the beginning and the condition of the most perfect life in awareness, and therefore, of the most perfect life also. For nobody escapes that a maintained attention is required to maintain that state; an attention that is more and more difficult to obtain for us nowadays. Who said inertia? What was about something "simply mechanic"?

§ 3 §

Samkhya, attributed to *rishi* Kapila, and exposed with insurmountable concision by Patanjali, is the philosophy that underlies and defines *yoga*: yoga is the cessation of the fluctuations of the mind. Or, in a more progressive form, the whole transit that leads to the arrested state. Therefore, it simply includes everything, since mind is also a mere object and a mere object of knowledge like all the others, although less graspable in appearance. To understand it: the objects or are not only pervaded or surrounded by our mind, but, being our mind only another kind of object, objects themselves have to exhibit mental properties, being knowables. The mind is the order of the fluctuations, and everything with fluctuations will participate necessarily in it. The consciousness or pure awareness by definition is not an object and therefore it does not show qualities. Nor it is an attribute of life, and for that reason it is beyond life or death; of inertia, of the forces or their reactions or fluctuations. Only assuming this we are free from absurd, anthropic principles and other perplexities. For the modern scientific mentality, dominating over any other one, the problem is how the things have begun to move from a rest that is excluded by definition and it becomes a non-being; for the perspective of Samkhya, that is not alien to the traditional and pre-scientific perspective in general, the things and beings have been moving always —for that reason they are actually beings —, and the only thing that is possible to contemplate is their possibilities of return to non-being and to the rest beyond the fluctuations of the time. For one and another, motion is the problem and the solution is rest.

§ 4 §

Samkhya, as the study of the fluctuations of the mind that possibly can be lead to the rest, is an oscillator theory. But the qualities or moments that Samkhya detects in the fluctuations cannot be purely extrinsic, but internal or inherent, the discernment that only characterize them is realized inasmuch the mind is contemplated or testified by the consciousness with no qualities. There is not contradiction here, but continuity, since the motion characteristic of mind in its steps to acquisition of knowledge results from the same number of steps towards rest and therefore relative or tangential contact with the same knowledge as source. We will see that this break in continuity is the subject par excellence of Samkhya in any possible approach.

§ 5 §

Between the varied meanings and possible translations of the term samkhya we have those of *analysis, detailed scrutiny, number, measurement and proportion.* The analysis that practices Samkhya is purely discriminative and intellectual, and in such sense it would be necessary to consider it objective. On the other hand the discrimination or *viveka* is made basically on an internal object, the mind, without another contour or profile that the one that conscience or consciousness can give it. Again, it is necessary here to give some explanations. For us, an object without external profile is only subjective, although only by the deformations that we have imposed on such conventional terms. Even so, this would not be a problem for Samkhya, which defines the supreme state of knowledge like the suppression of the same object and the "objective consciousness" in the sense that commonly we give him. This is remarkable. In the continuity wake that leads to this knowledge, the data and the apperception in general must be of a more and more internal order. Well, even so, throughout the whole scale of the conceivable objects, from most external to the most internal, which is irreducible at the end of the analysis of the fluctuations are the three *gunas*, ultimate qualities or modalities of the conditioned nature, constituent and inherent to such fluctuations. We can understand by conditioned nature all nature now considered as manifestation —which already is inherent and coordinated as *gunas*—, now considered material or not (what it would be just an opinion), now perceivable or not. The conditioned nature is the order or inherent disorder to those fluctuations; when the three modalities reach the total equilibrium also cease the fluctuations and the order or disorder of manifestation.

§ 6 §

Gunas, the moments or modalities of nature, can admit an extensive or an intensive breakdown, they can be explicit or implicit, but in no case they leave their inherency, what is to say that they are not susceptible of later analysis. They mark the end of the analysis, and therefore any consideration or contemplation on them begins where the analysis finishes. This has to be worth for which now we understand like mathematical analysis and also for everything what studies physics like intrinsic of the differential calculus. The break or the continuity is not marked by one theory or another one, but the same application of the calculation when it lets find solutions or these are multiplied becoming irrelevant; it is well-known that is enough three variables or three independent bodies in a problem to throw to us fully in non-linear dynamics. The non-linear thing is the non-proportional thing; the sense of the proportion in Samkhya aims towards that apparent lack of proportionality. The three *gunas* are different, they are the different thing par excellence for the discernment, and nevertheless they only can exist by its mutual correlation; in the same way in the non-linear analysis the difference between dependent and independent variables begins to become semantic and tends to become an open question in dependency of the posing and the solutions. If we took the famous three-bodies-problem, only by the initial posing we could say that those bodies are independent; what decides the question is the form to treat it, the perturbation calculus, that only can go on sequentially from an adjustment to another of the successive bodies, or to draw up a surrounding approach for the limits of general resolution of the problem. Of a significantly analogous way for Samkhya, mind, that is sequential in its own nature and it cannot perceive two things simultaneously, is with more reason incapable to contemplate simultaneously the distinctive character of *gunas*; in fact mind cannot perceive itself directly, so it can perceive itself or *gunas* by inference, received knowledge or impression of consciousness that has no memory.

§ 7 §

Sattwa, *Rajas* and *Tamas* are, in their primary form, sensitivity, activity and inertia. Or, if we prefer, (relative) balance, mutability and potentiality. It is worth to notice that some circumstantial divisions such as *Grahita, Grahya* and *Grahana*, or the knower, the knowable and the means of knowledge, are not equally primary, because they are not exclusively referred to the conditioned nature, since we can understand the knower ambiguously either like

Purusha or pure awareness, either like the intellect of an empirical subject that belongs to a material modality, *Sattwa* in this case.

On the matter of mind, we can say that it is not an exclusive attribute of *Sattwa* or the sensitivity, but that mind itself, although more clearly manifested in the state of *Sattwa*, contains mutability and inertia like inherently associated qualities. *The three* gunas *are included and present in any object or subject that we could separate: therefore the* gunas *or primary qualities are those that cannot be separate or isolate in any way, except by the momentary or relative dominion that one of the three always exerts on the others.* In fact, if we considered the gear of the three like an internal or proper clock "isolated" from any other, we would reach by definition to the conclusion that does not exist any event or moment if some alteration or mutation between *gunas* has not taken place, no matter the slight this could be. *Gunas* themselves are temporary propensities and the changes of their disposition mark the "internal clock" or the proper time of the entities and processes. This does not happen because the entities could be isolated from environment in any special way, which is not the case, but simply by the fact that the change in anyone of the three qualities only can be detected in relation to the others.

§ 8 §

Then, we could get a clock for proper time. Let us remember that the inertia principle of Galileo leads, in the interpretation of the correlative motion between systems, to the principle of Galilean relativity that tell us that no system can measure its motion from itself, but in relation to external others. The later theories of relativity, special and general, are extensions of the same case taking the speed of light as reference. We continued counting the time with an absolute clock like the Newtonian one because we are incapable to find a clock of proper or internal time. The quantum theory of radiation and matter does not change this because it is postulated like invariant with respect to the time, the space and the direction. Peculiarly, often it is said that their events do not depend on the absolute time, only trying to mean that they do not depend on a "before" or a "later" within the universal synchronous time. In fact, such invariants are postulated to maintain the principles of conservation of energy and moment, in last instance dependents of the inertia principle, facilitating therefore the calculations. Certainly, there is not another way, because even if variations were detected, *it does not exist any other clock* to measure them or to compare them. They would be always random events, which always will have a place in the barely outlined frame of probability amplitude. What is decisive here is that within the frame of the mechanics *it does not fit nor can conceive another clock: if we could have it, we would not know what to measure with it.* The Universal, Global Synchronization Clock ,always forces to us to measure the things in relation to external events, appealing if necessary to the Zero Time and the Big Bang itself as the release of the spring; but it is not possible another idea of the clock from this perspective, the clock is already the perfect and absolute synchrony of the events –an apparent a priori that nevertheless is derived from the inertia principle and its ideal plane. In the sense of the global or universal synchronization, the key of every other, it is not true at all that the theory of relativity has overflowed the kantian frame, because it has not overflowed the one of Newton nor the one of Galileo either; the fundamental statement has not changed and it has been merely enriched with contents. The same is valid, although not in the same proportion, for the quantum mechanics.

§ 9 §

It works in the same way when we consider the so called "internal clocks", for which Biology is the privileged field. Such clocks, now molecules now groups of cells, do not play another role that this one of synchronize itself to each other and altogether to synchronize themselves with the outside. ADN itself is a timer with very delicate dependencies in its modulation. Of course, pieces for this mechanism exist, and they must exist; the question is where the knot of its correspondence is. In other words, we have to know if there is a mechanism in addition to the pieces. As both things are assumed, the "mechanism" must be synchronous with the exterior, and this is already what defines the mechanisms in the way we have used to understand them. From this point of view with no possible escape, it is absolutely inconceivable that something not mechanical could exist, and the only that make some differences are the degrees of determinism and predictability. For that reason any variety of biological determinism has already given a form in which it must fit materials and data; this form is not so much any content of chemistry or physics, for example, but the forced reference to the outside without specifics breaks of continuity.

§ 10 §

Gunas or primary properties of the Samkhya, as the last discernible by the intellect, and only if it becomes present to the awareness, would only have to be the objective par excellence, and this is precisely why, whereas immediate knowledge, is absolutely unattainable for any attempt of direct measurement. This is excluded by definition: whoever that want to perceive directly *gunas*, already knows what must do. For that Patanjali wrote the *Yoga-sutras*. To perceive differently *gunas* implies already to perceive them free, free of all mixture, and therefore, immovable – not nonexistent, but incapable to manifest anything. To this we can refer the state of *Kaivalia*, the state free of qualities and released from the bonds. As the bonds are *gunas* for themselves, the state of *Kaivalia* or exemption can be preached with so much more property of the conditioned nature that of the same consciousness, that already is postulated like unaffected. If the identity that underlies in this is denied in a separation gesture, is for avoiding the same compulsion of identity like synonymous of the confusion that is worth to avoid; beyond this extreme consequence there is not the least reason to consider the Samkhya like a dualist system. The more we could say is that, if liberation exists, in that liberation the same nature concurs. This would have been a maximum motif for the philosophy if the theology and the dogma would not buried it.

§ 11 §

Returning to the subject of the measurability of *gunas*, since they are only exempt at the moment of their cessation, it could be assume that they have some sort of mixture or confusion while they exist fluctuating, or simply exist. But this it is a very delicate point since the own nature of *gunas* is the distinction, and in addition that distinction is the constituent of the moments in a "proper time" that does not need an external reference. Therefore, it is always preferable to speak of an interaction, marked always by the dominion of one of the qualities over the other two. It could be asked if this interaction is temporary or strictly simultaneous, that is to say, if the reconfigurations of *gunas* obey to a causal struggle between them which entails an internal time, or if they are reshaped without a factor of time in pure simultaneity without internal cause at all, that in any case would be necessary to look for outwards. It is easy to see that this takes to a new version of the everlasting antinomies of motion, the continuity, the time and others. Since we do not try to find solution them with the mere thought, we have to look for experimental models with formal validity also, that is to say, adapted to our methods of measurement. The possibility of such methods was the departure question, that is to say, how to measure what it assumes immediate. The definitive answer that can give the Samkhya to us agrees in this with the one the Vedanta, the classical non-dualism could give: it does not exist nor it can exist a last and irreducible difference between mediate and the immediate things, continuous and the discontinuous things, the motion and the rest or the interior and the exterior. We could not speak about anything if such things could exist, and if we did not speak, hardly they would exist more.

Therefore, the less than we can do is to leave us by the tangent of similar conventions created by and for the thought, and to see what the tangent draws between the real things and the term of cease of thoughts. Which will be constituted as well by new thoughts, now perhaps guided by a better understanding and fortune.

§ 12 §

Samkhya conceived a purely empirical science that hardly could have some value for the eyes of the moderns. It was *the Ayurveda*, the science of the care of the life, not very distant in the spirit of other old medical arts, like the hippocratic medicine or the Chinese one. What distinguishes the ayurvedic frame in its humoral theory is to be an application or contemplation of the frame of *gunas*. Three humors or *doshas*, *vata*, *pitta* and *kapha*, or wind, bile and phlegm, are not but reactive forms of the inherent properties of sentience, activity and inertia. By reactive it has to be understood that its predominance on the other properties is able to generate imbalances and ailments throughout the time; but also that they are secondary or derived form from more original and less privative modalities in the constitution of the subject: *prana, tejas* and *ojas*, the vital breath, the brilliance of the combustion and the radical oil that pervades and nourish the body. Now we treat with a much more material qualification of *gunas*, but exactly in the same way subject to the same fundamental logic.

It has been noticed sufficiently by the scholars the influence of the ayurvedic concepts in authors like Plato, in *Thimeus* particularly, where he wrote openly about the "air, phlegm and the bile", besides to mention "the established periods of life... in conformity with the triangles of the nature of each one" .It is really peculiar to observe how this book of mysteric root and pole of the enlightened hermetic philosophy has provided a rational inspiration for varied readers of scientific die, from Kepler and Galileo to Heisenberg. Another one of the platonic divisions of the soul in vegetative, sensitive and rational, is also of net Indian origin, or in any case it at least and admits a strictly congruent superposition with *kapha*, *pitta* and *vata*, and the corresponding functions of vegetative plasticity, irritability, and sensitivity.

Even the traditional contemplation of the man like body, soul and spirit keeps a strict correspondence, if not in extension or expression, yes in the most intimate nature of its correlations; being this consideration the only thing that matters. We will not mention the numerous triads of identical nature that the reader can easily imagine.

§ 13 §

In the beginning, any attempt to define better *gunas* or the correspondent *doshas* exhausts in mere epithets and paraphrases. We can use some semantic approaches, but trying to grant to the semantics the minimum of respect that deserves. After all, our investigation can take us to the conclusion that it does not exist a way to eliminate the "semantic questions", and not in last instance but throughout all its formulation, since we would be trying to show in a convincing and useful way the non-separation principle or property.

The three *gunas* also can comfortably be classified like centripetal, centrifugal and orbiting propensities. In respect to the *doshas*, that aspect relatively more material of the primary qualities, *pitta* makes reference to the heat and the energy, *kapha* to the plasticity and solidity, *vata* to all the phenomena of circulation from which the form is derived –of the more isolable or pure aspect of form insofar as independent of the plastic and the dynamic moments. It could be objected that this last one is a mere entelechy; we will try to see that it is not so simple.

Contemplating a tree we can see and touch characteristics such as colour, forms and texture. To say that the intrinsic combination of those external characteristics is the result or the appearance of primary qualities like the growth, condensation and formation can sound too unspecific and arbitrary. But if we would burn the tree, we could distinguish perfectly between something that actually burns, something that is reduced to ashes, and something that rejects both things and is freed in the form of steam and smoke. *The form* is what it evaporates and disappears, speaking in the superficial sense or in the deepest. Steam and smoke not is simply the result of that which burns, but very contrary is what escapes of burning, as well as, on the other hand, escapes momentarily from the gravity. Although the scrutiny of this can take much more far, is sufficient to make see that, of the most violent form, while these qualities still maintain some relation to each other, they have unmistakable behaviours. The form is not absolutely an abstraction, unless we could understand by "abstraction" the very distillation or circulation made by the same tree and nature itself.

§ 14 §

Within this ayurvedic frame of *doshas*, there exists a clinical procedure par excellence: *nadi vigyan*, the auscultation of the pulse. Those who practice it with unequal fortune are often called vadyas, seers or performers, according to the literal translation. The auscultator successively feels the radial artery of the subject with the three middle fingers of the hand, alternating the pressure, in an analogous way as playing a string or wind instrument. The auscultation becomes after a sufficient interval of rest.

The separated or "independent" characteristics of the pulse are recognized from always like five: rhythm or rate, frequency, intensity, amplitude and form.

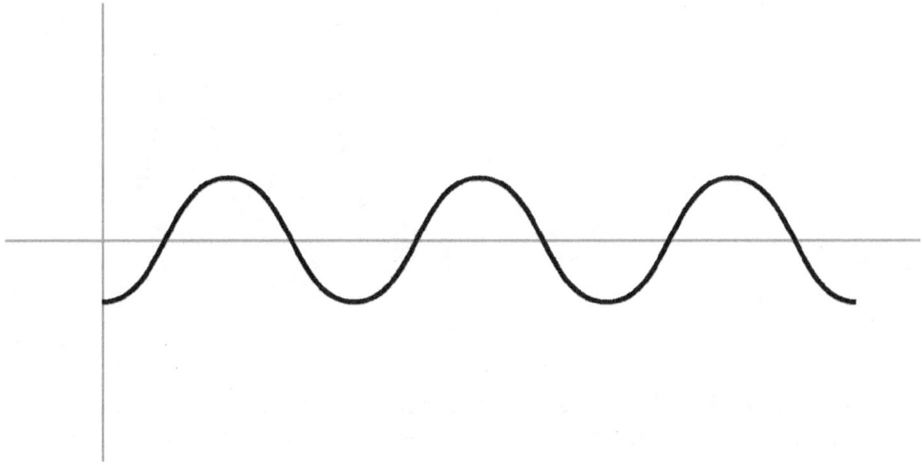

Fig. 1

We have now a biological oscillator. We can perfectly understand the first four of them in the analytical sense of an oscillator. The basic frequency is the number of beats or cycles in a sufficient period of time, a minute, for example. The rate or rhythm is the measurement of the regularity or irregularity of the cycles throughout the greater period. The intensity is the peak altitude reached about the tops of the wave of the pulse. Finally, the amplitude registers the total difference between those maximums and the minimums of the bottom or valley of the wave. Of these four, the obtained frequency is constituted apparently in rigid base for the possible variations of the others: the rate or regularity not only affects to the relations between cycles, also it is necessary to consider the compression or stretching of their own duration. The intensity as the amplitude can be variable also, based on the pathologies. The basic frequency can be considered rather rigid because in normal conditions the variations of the other elements would not have to relapse into it; but in a more comprehensive plane also the frequency is determined by the others. In addition to these four elements that we can treat analytically, we have *the form* of the pulse. During a long time, the mechanical sphygmographs or pulse-graphs could not detect with sufficient acuity this true and evasive *quinta essentia*; it has been necessary to wait to the small laser diodes and other technologies to catch his profile suitably. These technologies do not register anything essentially different from that which an expert auscultator catches with its fingers, the only thing that do is to shape them graphically and to offer them as a challenge to the analysis.

Is the form of the pulse a strict result of the other four analytical components, or contains something irreducible? But we have still not specified what we have to understand in the pulse as *form*. The form is not only the set of variations of height inside the phase or cycle,

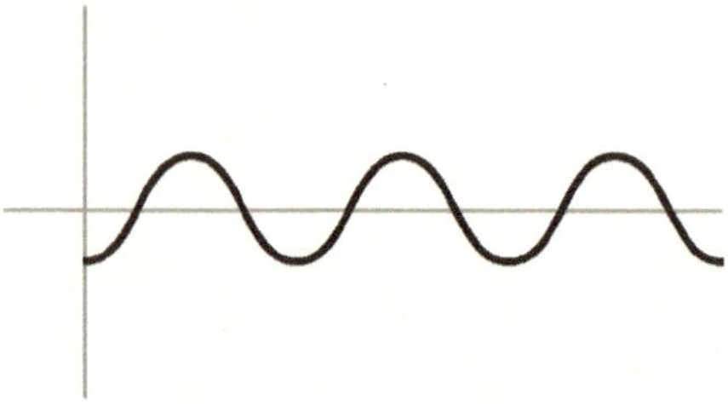

Fig. 2

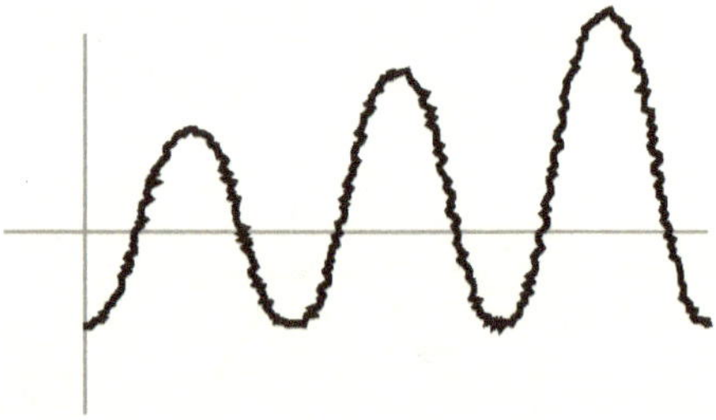

Fig. 3

as it is appraised in figure 3 with respect to figure 2; it also incorporates in itself the thickness of the feature as a whole, that can be more or less homogenous, or to have very important differences (Figures 4, 5 and 6),

Fig. 4

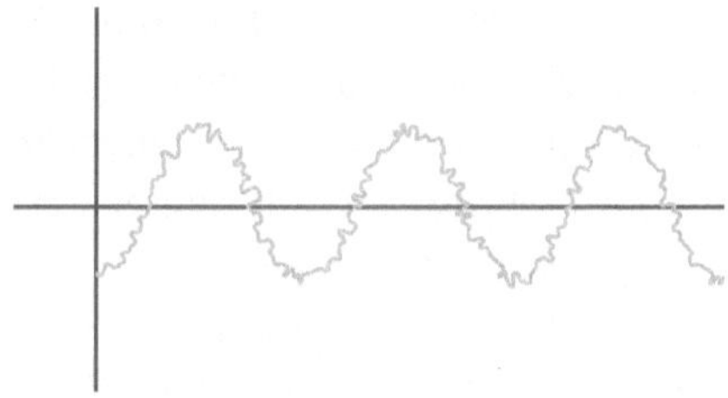

Fig. 5

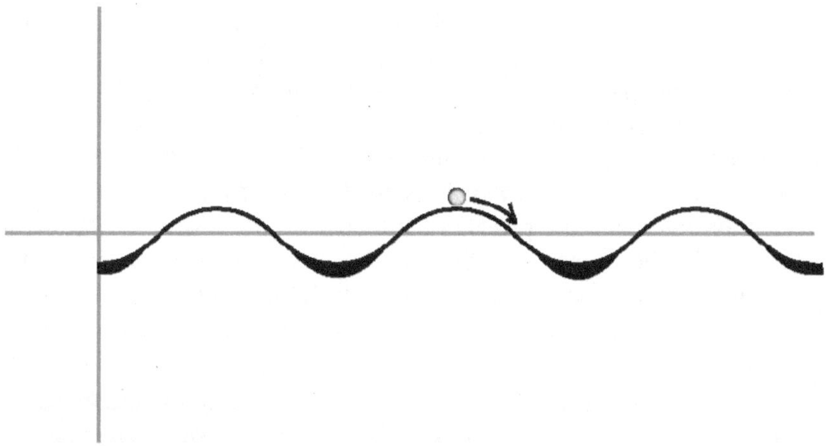

Fig. 6

either comparatively among them, or inside each cycle for the same pulse, or, even, between different pulsations or beats from a same auscultation, that would be resulting of a great imbalance or of an internal or external removal. The Chinese pulsology, in essential agreement with India, has characterized with epithets such as "silk worm having nibbled a mulberry leaf", or "small ball sliding in a porcelain cup" to pulses similar to those of figures 5 and 6. It is easy to understand that such shades have escaped for a long time to mechanical detectors. Now that problem no longer exists and all this can be caught by minimum devices; in addition, in this way all type of dynamic sequences in motion situations or effort can be obtained, something impossible in the traditional procedure. The changes of the ambient temperature also introduce revealing modifications. To this another treatment of great interest can be superposed, the sound recording and its harmonic analysis. Not we forget that the main meaning of word *nadi* is channel or hollow tube, and that in the Hindu conception is the relative "vacuum" what resonates, more than what is filling it. But we return again to the subject of the form.

It is evident that this form of pulse recording, considered separately, admits an analytical treatment, since the reader connected to the detector is integrating the results according to the rate of the samples; on the other hand, also the thickness of the plot or feature admits a differential by its own fluctuation in time. But here finishes all the similarity: what we cannot integrate is the five characteristics with its different ranks of variations in a formula simpler that the obtained reading pulse itself. For that reason it is a non-analytical system. This seems to be largely studied by the different forms of modern cardiography, which is experiencing an extraordinary development now. It is indeed by this technological development, and, peculiarly, by the non-analytical character of the data, that the cardiology and the medicine of the biomedical model ignore completely, or at least as much as it is possible, the intrinsically semiotic character of the pulse in the traditional conception. That is to say, being a signal not of analytical character, experts tend to fit in it everything what effectively is possible to analyze or, at least, to separate in parts, which results more in the detailed pursuit of the implied corporal structures. However, well understood, this motion would not be absolutely in opposition to the traditional posing, which matters very little that pulse as "system" is not analytical, since, on the contrary, which looks for in it is *the more irreducible signature and the more able to synthesize the rest of the data, known or to be know.* One assumes that the profile of the pulse is the result, definitive and unquestionable, in the same way that in the stock market, the market of the speculation, the same speculations in the whole possible set of orders and powers always refers to the present or current value –indeed, the reason of speaking about "to take the pulse" to the market.

Is to be present that also the three *doshas* or implied reactive humours in the pulse can easily be separate in three factors or precise mechanical moments: in most general of the terms *pitta* corresponds with the contractile impulse of the heart, *vata* with the state of the walls of the vessel and its own conduction of the contractile wave, *kapha* with the fluidity or viscosity of the blood. Even like independent terms of a problem of mechanics of fluids, these three factors already are sufficient to create a non-linear system without stable values in the solutions; but in fact what happen is that there is an intrinsic dependency and feed-back among them. Still more, the great relative stability that only throws the results can be conceived by its mutual reaction, which often is denominated, with unlucky term, *self-organization.* To understand to what extent is not necessary that "self-organization", is to go further on as much of the mechanics as of the theory of the ignorance that we call complexity theory.

§ 15 §

About the comprehensive and predictive value of the coordinates of *tridosha* in the medical exercise of the diagnosis and the prognosis, the subject is too open for hasty conclusions, ignoring those who consider such marks as empirical rudiments or even superstitions. And it is opened by several fronts and reasons. First of all, we can be sure that in this medical art, like in any other, are many those that practice it and well few those that have some mastery over it. But it is not enough to say that the subject is as vast as the same health or the life, that is difficult and complicated. It could be really, specially if it is lacked necessary sensitivity, or of the necessary education of sensitivity. Certainly the term complexity deserves often an indicated parenthesis. Operations and the diverse manipulations could be "difficult" or "complicated"; listening or the understanding does not try to thread or to untie the knots, but to follow them in which they can give of themselves.

Although these deficiencies of sensitivity that all we suffer do not have quick solution, still turns out more devilish paradoxical to want to solve the subject giving it parameters. We can, is certain, to assign values or relative weights to the three *doshas*; but this only has an approaching value, to which it is not possible to give a scale very fit scale of precision. Nobody, that we could know, has measured the relative weight of the three tendencies in centesimal values, which even seems absurd, unless we were filling up a test by points. Not even we know if there is a degree of relevant precision, aside from more fundamental questions like the nature and sequence of the reactions. Soon we warn ourselves of the unnatural thing that is to want to turn a procedure into a method. By very valid this could be, it must be recognized that was not developed thinking about numbers, even if samkhya means *number*. And nevertheless, the nucleus or subject of all this continues being something as simple as the proportion and the balance: and not of an immensity of factors, but only three...

We can return to ask to us if the pulse really contains as much information or if it is not rather an illustration of the knowledge acquired by the doctor with other symptoms. But it is not so; it is more probable that the aspect of a patient could be illustrative of the pulse that in the contrary way, if the doctor has enough aptitude; although both enrich and complement the judgment. On the other hand we can also try to reframe to us if the frame of the three tendencies is really necessary and sufficient. Naturally, *vadya* will say yes to us, and the only thing that we can do it is to accept or not to accept his rules of game. After all, for Samkhya the only form to verify if *gunas* exists or not is following them until the end. Samkhya is the procedure of that verification, not a theory or method.

§ 16 §

Then it will be necessary to make use of our immense arsenal of applied mathematics to headless problems. An ideal candidate would be the fuzzy logic and its sets. Ideal, because the very high grades of precision do not seem pertinent; in addition, the fuzzy logic is a suitable treatment for control systems and expert systems. Both things is what represents the pulse if is considered alternatively it from within and from outside.

In order to create an expert system we must transfer the knowledge of the expert, in this case the *vadya*, turning it into quantitative estimations and rules of inference. It will be needed a minimum several hundreds of samples, that can ascend to a thousand or several thousands. The estimation of the proportions of the agents or tendencies becomes as precise as it seems opportune for the *vadya*. In no case such estimations for the input of data will be of a very high precision; probably a scale from 0 to 10, or perhaps only from 0 to 6. For example, 4 of *kapha*, 1 of *vata* and 1 of *pitta*. Which already is telling us that such estimation is more than nothing indicative and that what matters in this case is the dominant one. There could occur cases without dominant too (a third of each one), and those pulses will be most regular and less not-linear. If this scale of entrance is not even very precise, the sum of cases and its averages by the expert system would tend in the long run to increase too much the precision, if this one really were relevant. It could move perhaps in a rank of hundredth, or thousandth even. Have we gained something with this? Are not we deforming the original data yet? To this it is necessary to say that the expert person, in the first place, *does not realize everything what he knows.* That is to say, that his knowledge is not explicit except in a small degree that permit him to communicate, to learn and to teach. Surely he has not thought at all about quantitative or numerical terms, and he will continue without needing it. His sensitivity gives another kind of data to him, and among them, a very important one: the tendency within the tendency.

§ 17 §

The *Vadya* discern between *prakriti* and *vikriti*, the biological and the biographical thing, the original or innate constitution of individual and its tendencies acquired throughout the life. By definition, any later state in the health of the subject is dependent and it is refer to the constitutional tendencies; so that the habits or alterations produced in the meantime are, on one hand, already assimilations of the eventualities to an original conformation, and, on the other hand, that same conformation that has its slopes of reaction respect to the accidental conditions or *vikriti*. If *prakriti* and *vikriti* come to be like substance and accidents for a given pulse, for that reason it is not necessary to forget that in the long run those accidents or way of life can get to be most decisive for the conservation or destruction of the balance that represents the health. But not in vain it is spoken then about the way of life as the specific form of life. In an essential way, the form only is modified by its own form of wind itself in time, and that thread has to be perceived by *vadya* with right sensitivity.

However we consider it, it does not stop being an extraordinary wonder the fact of that fundamental persistence of the form of the pulse throughout all a life, if we think about the immense variety of circumstances that can surround it and alter it, and how slight is its signal. If we think about that this little thread of time is going to maintain something of its own in the most adverse circumstances, beyond the successive losses of identity of the memory, and that not even a violent death can rob the right to him to disappear according to its own and intimate law.

§ 18 §

This distinction of the tendency within a tendency is so fundamental for our study as it was it the distinction between speed and acceleration for the birth of dynamics through the concept of force. The hydrodynamic systems or fluids mechanics -a circuit of pipes with water to pressure- cannot establish this type of differences because the components are not connected from the origin, and all the differentials we could measure later depend on the arbitrary variation of parameters. A primary and original reference does not exist. It is well known on the other hand the dependency or sensitivity of the not-linear systems with respect to the initial data, considered here in another order. Therefore, the initial vagueness that results from the qualitative study of the pulse also has a very important counterpart that is unthinkable in the traditional analysis. Only by this already would be worth to study thorough *nadi vigyan*. But it is that in addition this original qualification to *doshas* would have to give us the key to measure that *proper time* of beings and the entities that is not possible to conceive in the analytic frame.

In another order of things, and ignoring which could be its relation, it is worth to remember that still it is object of discussion, study and modelling the problem of the forms of flow of the sanguineous torrent. It is conjectured if this flow, that of course is not laminate but with very varied turbulences, can follow helical or spiral guidelines, which doubtless would optimize its diffusion. I am convinced of the rightness of this hypothesis, and is surprising in any case that such elemental unknown factors would persist for processes in our immediate interior and within canals of millimeters or centimeters of span, when it has been possible to determine with great precision particle interactions in distances fifteenth magnitude orders smaller. This gives us a certain idea of the existing abyss in the analysis for different disciplines. And we can remember that until the year 2000 has not been generally recognized that the fibers of the heart form a spiral, a helix that simply stretches and it shrinks. And in the meantime there have been thousand and thousands of open-heart surgeries, and almost five centuries of detailed anatomy.

§ 19 §

Then, it is worth to give values to the *doshas* like fuzzy sets capable of relation. The expert system will be able to adjust more those values thanks to the average from many samples as well as thanks to the inferences that we allow him to generate. There are also simple rules of inference and others of superior orders. A simple one is, we take as example, "if the intensity increases, *pitta* also". The quantitative correlations by diverse means can settle down soon. More complex it would be: "if the intensity increases but the frequency diminishes...", etcetera. It is a simple work of exhausting, of exhaustion of the real possibilities. Anyway, the canonical classification of the fundamental pulses in twenty-odd types already facilitates sensibly the ordering of the criteria search. One of the advantages of the fuzzy expert systems is that they allow to make finite the number of inference rules to use, which is not the case for other models of probability. To make them finite is still far from making always the number of rules necessary to cover any system manageable, but it is no doubt that for a low not-linear system as this one there are no great difficulties of principle. The fuzzy or diffuse logic often has been erased to be simple logic of probabilities disguised. This can seem certain if it is wanted to see from outside, but the case is that the traditional logic of probabilities in its different variants cannot make finite its rules for the great majority of problems, so something important can be out of them. In other time, J. M. Keynes, the famous economist, tried to accede without success to an suitable formulation of the principle of "variety of limited logical independence", that at least was proposed. Today we see that this is impossible and inadmissible for the general context of the analysis, virtually infinitesimal. Nevertheless, that to which Keynes talks about in his approach to the logic of the induction is something so reasonable because the limited cases cannot depend on a limitless number of rules. But so that this happens, rules are needed that are not perfectly analytical, that is to say, different rules that often occupy such cases in the same space of probability. In fact, the logic either is vague from the beginning, and crucial depends on the fortune of the applications that this is being notice later or earlier. The fuzzy logic turns over that deficiency as far as possible and can determine the weight of the redundancy of different rules. One of the initial theorems of the fuzzy logic demonstrates that the set of the real solutions is minor than its possible, statistical parts; this way is necessary to begin, since for the traditional probability the set is the sum of the whole separated possibilities.

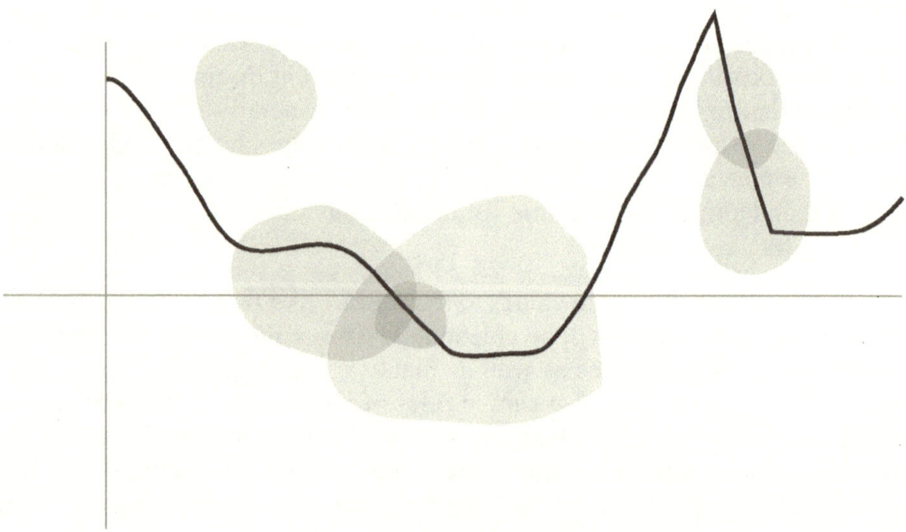

Fig. 7

§ 20 §

"The weight of the redundancy of different rules" is not but the fact that a position is confirmed from different angles or for different reasons; this overdetermination is the counterpart of internal precision to the external imprecision that offers the fuzzy logic in its rules and values. One begins to suspect that this logic is not a mere approach of force to a problem that we ignored, but that even can help us to conceive that paradoxical "precision of the lack of definition" that with so mastery and indifference exhibits nature. This that we see like mere redundancy in the space of probability occupation, can be authentic over-determinacy in the "internal space" of the system or process, or better, in its "internal time", and to say it definitively, in its *proper time*.

Thus, at least, we could begin to conceive something that as a start and coming from the analysis seemed to us inconceivable.

§ 21 §

Being *gunas* authentic and genuine modalities, it is possible to ask ourselves about their relation with the modal logic, the logic of possibility, necessity and contingency. But these are so vast categories or modes that it turns out pretentious to speak of this logic like a real formal science. Surely, insofar as can follow the nature, it could be refined and be outlined in the Samkhya, instead of trying to give form to it, which would be absurd and in last instance impossible. The modal logic follows the modes, so the modes exist more than the modal logic. This justifies it and gives the limits to it.

§ 22 §

We return to the question of why such a simple and slight signature as this one of the pulse is able to represent so much variety of information. Only if the time has its own density, that is to say, if it admits different densities and thicknesses, we can know of what is reduction which we are perceiving like limit. The same signal already is the perfectly fit limit, inverting the perspective of the analysis. What before we called *form* of the pulse, as much in its particularity as in its surrounding, is as much the form of the limit and its formation.

§ 23 §

The monitoring of the temporary evolution of the pulse in dynamic sequences, that is to say, during the motion of the subject in situations of action and passion, of pleasure and pain, allows the extension of the reference field that the three *gunas* constitutes. Since *Tamas* already is the inert receiver the passions, *Rajas* is the removal by the action of the pain that as passion is its means, and *Sattwa* is the relative safe-placing as much of the pain as the action, which altogether we consider as pleasure. This way we can catch to what extent biological and the biographical thing, *prakriti* and *vikriti*, are not heterogeneous, but mutually continued. This would not be possible if the *gunas*, in any given constitution, did not incorporate in themselves their relation with the environment, of which the temporary evolution already is explanation, untacklement. In the total unity of measurement and mediation it resides all the excellence and the apparent difficulty of this procedure.

§ 24 §

Making use of this principle of over-determination or superimposition of the space occupied by the inferences, that refers to the weight of the things and to its cause in the sense of sufficient reason, we could verify that the same system or its equivalent graph tends to open, to close and to transpose spaces, or what is equal, it gives back guidelines of induction, deduction and hypothesis to us that we can try to complete in case the relation between values of the parameters and rules of inference has become stabilized or converged. Soon it can be judged to what extent and of what specific ways the system is breaking the moulds of the differential equations that we attributed to it. This has fundamental relations not only with algorithms and the study of the steps to describe a problem, but also with the economy of operations of the knowledge in general, and its composibility. If we can describe sufficiently a process without using differentials equations and analytical terms, nor there is an analytical form to decide which is the minimum sequence of inferences or its order. Although we handle with someone always.

§ 25 §

This seems inconceivable; more inconceivable than any principle of uncertainty. The smoothing of the roughness or complexity of a problem, of the joint of a function, seems to depend on an optimal order in the sequence; but that minimum requires for its definition a maximum of super- determination, of superior modes. But these also tends to yield and to be fused if they are really contained in the profile or *fit* form of the function. Now it can see that we are playing with a parallelism between the "expert system" and the "control system", between the cognitive approach and the natural process, as if they concurred in an identity. Both are needed for "working", although its independence is presupposed. Samkhya affirms that no absolute limit for the knowledge as for the perception does exist. Also affirms that all the set of the thought or knowledge subject to thought is, whereas mediated, least in relation to the being or the consciousness. If, following with the same logic, we wanted to try to understand how the paradox on the dimension of the knowledge is possible, its non-decidable compressibility or non- compressibility, we would also have to ask if the nature, the conscience, or being, is with respect to us an "expert system" or a "control system".

§ 26 §

But we return to the small things. If the pulse itself is one precise waterline for the set of the activities of a greater system, the body, manifested or not, we would have to be able to detect its temporary correlation with other subsystems. This correlation does not have why to be very direct since the pulse would be a luck of indefinite integral with respect to the rest; the pulse would be a diffuse and fuzzy set in the other systems, and vice versa. In addition, the signals that other systems offer us not always give specific curves of the same type than the pulse. In this respect we must remember that for the medical technology and clinical all these indicators, whereas signals, are considered as *non specific*, in the sense that they are not sufficiently revealing of the functions, structures and mechanisms; from our point of view, we considered *specific* everything what he is inherently revealing of *gunas* or *doshas*, that is to say, what it express with sufficiency the modes.

The electrical signals, like that of electroencephalogram or the galvanic resistance of the skin, are less directly expressive than those purely mechanical of the pulse or the one of a neumograph adapted for the breathing. In electroencephalogram, for example, it would be necessary to add different frequencies or simultaneous amplitude to obtain equivalent from the thickness of the signature in the form of the wave, being blurred perhaps the form of the phase. This offers interesting questions of conversion of signals which now we cannot treat, related to the same nucleus of the quantum mechanics. In any case, in this type of measures we are far from the uncertainty threshold.

Finally, electrical or mechanical pulsating and oscillating rates cross the different systems until arriving at the same cells and their division. The important thing is in what degree we can catch the integral phenomenon of the form with the sufficiency that we found in the pulse. Then if there is something able to connect all these oscillators heterogeneous in many ways, that something is the form.

§ 27 §

The fuzzy logic has contributed to us something more than instruments at the time of finding the relevant qualities of the pulse; it has also give us a way to glimpse to what extent the nature can act independently of all our analytical logic and our equations, increasing on the contrary the thickness of his internal or own consistency. This already is much, and is possible to ask why the subject has not been investigated with more zeal. Perhaps the method is too strange to our concepts of formalization, and its apparent procedure towards the dark by darkest can produce terror. And nevertheless, many systems can adjust this way with a maximum economy.

Speaking about the form we have arrived at layers more and more unformed; and this could be the best signal that we have obtained a more real approach to the nature. But this absence of form disgusts and moves away to the mathematician, the only people who could help us here. They need an acceptable link to cross to this side.

Irremediably Galileo comes to the memory. It is well known that in the famous experiments with inclined planes he had to resort to the beats of his own pulsation like measurement of the time, in absence of a better one. To make matters worse, when he began to study the oscillators by means of improved and primitive pendulums, he could not arrive more far by the tragicomic circumstance that he needed a good pendulum clock to measure the time intervals accurately.

As for the rest, our modern notion of the oscillator does not go beyond knowing how round is a circle. One does not know what circulates because the reestablishment of the symmetry and the balance are already assumed. Even when it is spoken of a symmetry rupture is for justifying the supposed symmetry, that already it is given, outside there . It is not to be strange then that seems to us strange the notion of many cultures of a circular time. Why they want to go nowhere with as much delivery? When it is so easy to go outside there, where already is everything. And giving rounds, in addition.

But it is not difficult to see that the only thing that moves is the imbalance, and that is the balance the precarious, problematic and the delicate thing. Seeing thus, the science of dynamics could treat more on any thing than on the motion.

Very little of this is rhetorical if we return to the unavoidable surprise and difficulty that a simple phenomenon as pulse according to our exposition provokes. It is necessary to see first the range of our ignorance when we spoke of "self-organization". We do not really understand that although nature would like to organize itself in its free moments, it results that it has not free moments because we have already given to it a permanent and full time occupation in the space of dynamics. Still we even look for new spaces and places to work more! An infinity of spaces, without the smaller temporary density. And that this is a real problem for the analysis, is evidenced in the fact that the more recent theoretical physics, after its extraordinary deployment in eleven dimensions, still has had to resort to objects of dimension zero and non-commutative geometry in which a point no longer is a point but all a matrix of possible values, discountable values, we suppose. It seems that we want to catch something already excluded in the definition. Our strange means of inference, however, have been nets to give a form to that without form, more than to capture this slippery and mysterious fish. Leaving this preliminary squids fishing apart, it would be very easy to say that we are treating completely different problems, but we do not believe it. No analysis worth of this name tolerates completely different problems.

§ 28 §

Everything would be simpler if our measurement of the time were already a measurement of the asymmetry. But we can ask if such thing exists and in what respect. Modern cosmology already parts of events of balance rupture, from which data and observable things would be the result. We, on the contrary, and without opposition spirit, give by fact that are the imbalances those that exist from always and those that want to be perpetuated always, which could not be for a moment nor without pacts with the balance or even series of pacts, that constitute them and maintain. As much the act to persist in its being as the one to agree to the balance necessities falls under the concept of reaction of *gunas*, and surely is impossible to discern if in last instance they are not identical.

But we need some departure point or reference. The physics uses rigid, external and absolute values of reference: the speed of light, the action quantum, the intensity of gravity. Other absolutely necessary values of reference exist, such as the specific weight of the vacuum and its relation with the observed minimum masses, those of different particles; but in this case no longer we can know if there are rigid values, neither external, nor absolute. In any case the previous constants give us the meter and the clock with which to measure and to compare all type of events and things no matter how different they were. In the same way which we ignored the relation of these three constants with the indefinite fields of reference, we cannot either know if these three constants are so at least congruous, that is to say, if they have some real possibility at least to converge; as we either do not know in the other hand if the duration, the mass and the distance are homogenous or non- homogeneous properties to each other, or in other words, if its relation can be closed and be defined lawfully in its own terms without another external reference. Such case, the external would have become internal and we could enjoy a lasting transparency. But I suppose that no physicist tries to take the things so far.

Our reference point can be neither rigid, nor absolute, nor external. Nor internal at least. It must be proportional, be based intrinsically on the proportionality. But the proportion is based on a meter or scale, is said.

§ 29 §

Often the things show us their own proportions. For example, the average distances between the successive bifurcations in the branches of a tree, or in the sequence of lengths of our same members; those proportions do not require an external reference, since they are independent of the meter whereupon we measure them. We even can compare between them the specific proportions of different examples of trees of a same species or different species, and they do not lose anything of his intrinsic character. For the analysis, all this accumulation of varied relations conforms the circumstantial, the descriptive thing, *the unspecific* of an object, even when could be the most significant for us. Since the analysis has convert the time into space without admitting appeal, the only worth to do in this case is to reveal the temporary element that there is behind these processes or developments that occupy a space. The "external" form of a tree is already the form of its development, in first plane and over any other consideration. That first plane surrounds to all the others. The form is not something secondary, adjective or circumstantial. In fact, it does not exist at all something like an isolated external form, exists an outer contour of which it is broken through, indeed, by the form of action, and that is intrinsically temporary, and therefore, neither internal. Time has nothing to do with the "inside" or the "outside".

§ 30 §

In the last decades of the XX century was developed an increasing interest between the mathematicians on the generative relations that proliferate around the *Phi* number (Φ), also known like Golden Mean or Golden Section, and other epithets. As it is well known, the golden mean is the relation given by the cut in a point C of the segment unit AB of such form that the smaller segment AC has a measurement with respect to the greater CB equal to which this one has with original segment AB

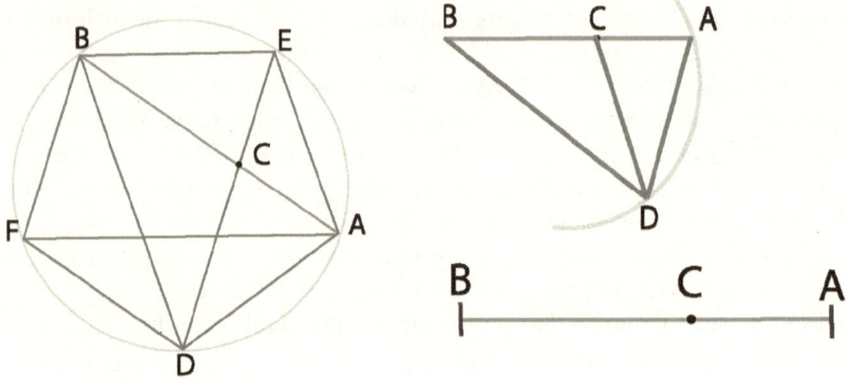

Fig. 8

The value of this relation can be expressed indifferently like 0.618... or 1.618... (1+V5)/2, according to if we fold or we unfold the segment towards the interior or the outside. This continuity of the inequality, or this discontinuity of the equality, as much by the operation as by the proportion, and as much towards the interior as towards the exterior of the indifferent unit of reference, is deserved of being considered like *the Continuous Proportion* par excellence, and thus we will call it onwards.

If we looked for proportions between the different measures from an object, as well as of the different reasons among them (being the same reason than proportion), the concatenation of those reasons and the reasons of reasons in indefinite number always would have to refer us to some approach to Φ like the simple possible reason. This, by definition. It is for that reason that the value of use of the Continuous Proportion and anyone of its approaches does not have in itself anything of arithmological -as somebody could believe-, inasmuch it always refers to the simplest reason. What is subject of doubt here, what is problematic is the application, the pertinence: but this one is the problem that defines *phi* itself, or rather, the reason of its application. If we remember the problem of over-determination in our previous sketch of the pulse like a fuzzy system, we can see that it is the same in essence. The Continuous Proportion, *phi*, is as much the meter or the measurement as the set of the measurement modes. It is measurement and mediation in a single and same act, which give it the character in all the measures and of all the systems of measurement. It could only be alien to those systems of measurement that had demonstrated the internal self- determination of their constants, which would constitute his absolute character. At the moment, this is not the case for physics, and it would not be little know how to formulate the question of if this is possible in general. Or in other words, the whole exercise of physics is that question without its corresponding formulation.

§ 31 §

It has been said that the Golden Measurement is the top of a pyramid of which we do not know everything, since nothing we know of its height or its base. For Kepler, it was a treasure only comparable with the formulation of the Pythagorean theorem, that as well it is known was known and diversely demonstrated by other previous cultures. Also the Continuous Proportion, raised for annals by Euclides, could be well well-known by other cultures even if we only can conjecture on the matter. In any case it seems evident that the same Greeks knew it before Euclides, and, in fact, we can superpose the Pythagorean theorem and the Continuous Proportion in a same problem in one of his simpler expositions, and in simplest and equidistant of the problem of the equality and inequality, the mathematical rationality and the irrationality, that inaugurates all the possible slopes of thought. Thus it shows before us this Proportion its irreducible nature, its depth and its sublimity.

§ 32 §

We know inherently in which slope the Pythagorean theorem began to roll as a ball, by the same fact that all the fundamental concepts developed later, from the differential calculus to the uncertainty principle, are refer in its irreducible form to the mentioned theorem. Thus, it is possible to contemplate in front of us the whole panorama of the Analysis. That we have forgotten almost completely that must exist another slope, the chance whispers maliciously to us that the mentioned theorem takes the name of a Greek philosopher and not the one of a Babylonian or Indian mathematician. It exists, in addition, the extremely remarkable fact that the innumerable developments that are receiving the Continuous Proportion and that unfold in all the imaginable directions are not mixed in minimum with the developments and key points of the Analysis, to which are not of any aid. Perhaps by this it has gotten to be spoken of his mutual complementariness, which actually means that mathematical of the Balance or the Harmony is complementary of the Analysis, whereas the Analysis is completely self- sufficient and it does not need complements.

This situation could change.

§ 33 §

Therefore, and for whom that knows to see it, the subtle winding of the thread of the Continuous Reason in the recent years offers a moving spectacle. Treating about the non-analytical thing, it is as much the product of the Analysis as what it is the vomited by it. Its empire is the nobody-land that grows everywhere between the different disciplines as these are branched off and diverged. But whereas Continuous Reason, has its own winding rhythm and familiar forms to which it does not resign whatever it is the treatment: like a brilliant living serpent appearing from a machine for sausages.

§ 34 §

Leonardo Pisano, Fibonacci, gave with the elementary sequence of numbers that tends with increasing exactitude to the irrational value of Φ: 1, 1, 2, 3, 5, 8, 13, 21, 34, 55, 89...The numbers are generated adding both previous terms, and the reason between adjacent terms give us the successive degrees of approach or precision. Fibonacci deduced the sequence as solution of two problems without the least apparent connection: that famous of the reproduction of a pair of rabbits in ideal conditions and the other, less known but equally revealing, of the optimal weights in a balance to make the smaller number of operations or manipulations. We can not to know until what conscience arrived him about the relation with the Continuous Proportion, as we either do not know the one that could have Al-Khwarizmi or Abu-Kamil, of whose problems extracted its wonderful generalization; but it is inevitable to see that only the development of the analysis could give to the subject its own relief and its contrast. And in effect, the generalized forms must mainly wait for until Binet and Lucas, until the development and maturity of the same concept of function.

Without its generative and self-recursive capacity of approach, *phi* would be to us of so little utility as the value of *pi* for a circle that does not admit the division in degrees; in both cases that intrinsic property makes possible algebraic generalizations and the elementary and superior functions. As on the other hand the connections of the Continuous Proportion with *pi* and *e*, the second subordinated constant of the analysis, is always weak and accidental, it has not been possible to arrive very far in the marriage between this one and what A. Stakhov has called, with enough justification, harmony mathematics.

§ 35 §

Somebody said, with great sensible choice, that if we know to truss the framework for the relations between *pi, phi* and *épsilon*, in addition to its contacts with music and geometry in general, we would have reached the simplest and optimal understanding of the nature that within the formalisms we could dream. It is not a small program. *Epsilon*, the measurement of the non-linearity, in its varieties or constants, already constitutes in itself a sufficient labyrinth for anyone, and in any case it seems the knot that ties a whole package of problems, well closed for us. But it is to note that here nobody talks about the *e* constant. We amply know its essential paper in the analysis, their inherent character without question; soon is here a sovereign indication. The number *e*, synonymous of the compound interest and number of the accumulation par excellence, mirror of the proportionality in the functions, include in itself and like limit the series of all the fractions. Like exponential number, it reigns in the probabilistic space of the values and its products; and it is not necessary to say that almost all the values of the analysis already are products, result of the multiplication of factors. As well as the generativity is expressed through sum of terms and their limits, it is possible to say that the values product of a multiplication of terms already express a jump in the emptiness that is impossible to quantify from the result because it depends on the nature of those values. Let us think for example about the mass or the energy of a system, for which we do not know its natural units and we must be contented with equivalences, resultants as well of other products. The number *e*, like operator in the corresponding functions, has become to represent that necessary bridge between sum of terms and their products, but in decided please of these last ones, which are those that conform the elementary values. It is no wonder this expansion of empty spaces has provoked the maximum exigencies of rigor since Gauss and Cauchy. But in physics no rigor can affect the base of this process: he legitimacy of the conversion of sum in products will always depend on that exist pure units, unaffected and free of factors. To want to look for something thus in physics would be today inappropriate and absurd.

§ 36 §

And nevertheless, we remain feeling that in this way something very important is ignored or is harmed in the integrity of the processes, and the things. The statistical spaces, the measures of order or disorder, the entropy as logarithm of the probability, are only unquestionable in pure discreet mathematics; in *the real* world we only have correlations whose factors must be affected in last instance and necessarily. We can think about the three constants of the physics before alluded to, or in the three more fundamental magnitudes.

So that, not without audacity, we can venture that the unavoidable presence of the *e* number in the analysis is a good indicator of all that we attribute to chance and the ignorance in the real processes, whereas the presence of *phi* indicates perhaps an endless search of the generative sufficient reasons. The numbers *e* and *phi* would express antagonistic tendencies, more than complementary, although fortunately nothing prevents us to work whatever we want with both.

As much *pi* as *phi* are “highly natural” ratios, within the context of mathematical formalization, and could be known by different peoples from the antiquity. The number *e*, nevertheless, is an exclusive concept of the later Christian West. We do not know for how long will last its reign.

§ 37 §

With too much frequency one says that the man fears at chance, and that he try to take refuge from it with systems or theories. But in many aspects we need it so much as to breathe, and nothing would frighten so much to us as to suspect that the things could be ordered until the last degree, and mainly if that last degree depended on us, who completely we ignored it. The chance is ignorance and the ignorance a positive support for our acts and also for the same knowledge. It is not conceivable that the chance could be in the existence, but for the existence, of which it is deduced that many have conceived little both existence and chance. Much more that the chance, to the man has worried to him and it would have to worry to him not to ignore or to be in opposition to the laws that make their existence possible, those laws or law which never is necessary to give form. The reasons that the Continuous Proportion looks for — or that resides in it—, even starting off very substantively of the form, lay for that direction, and in that same measure it can be save of the infinite search and the delirium. If we did not request too much the things are enough. The Analysis, with their emphasis in the precision or exactitude, is already the ideal frame to isolate the systems of their context, considered in form of strange influences to which it is wanted to consider. Even it surprises that this were not more fragile, if one forgets that this has been the means of its sensitivity.

§ 38 §

In the same way that the number *e* can generate all the frequencies, the Fourier transformation can "analyse" any package of waves in the resulting spectrum. Since it may be infinite different packages for each spectrum, but a single spectrum for each package, is evident that there is loss of information, in particular of the phase of the waves. The so call principle of non- determination or uncertainty is derived from these inherent limitations of the mathematical method, and is difficult to see why would have to be a general principle of nature. The same quantum of action is not simply an amount of work or energy per second, but per cycle/second, omission that tends to forget the phenomenon of frequency like evident, when it is all the opposite. By the same way we arrived at the superior generalizations of the gauge fields, that takes care of conservations of symmetry –local or global-for the changes of phase, and not for the same phases, which in this context no longer seems to have any sense.

The space of waves of the physics, more synthetic than analytical, is basically a one dimension, linear and abstract space that admits the most ideal of the extrapolations to the ordinary dimension of the events. But —and this is the important thing here— that idealization does not have why to absolutely mean a simplification of the facts, but it also adds an imaginary complexity to them that would only occur in the case that waves did not have their proper binds and their own modes: those that on the other hand could not find place in the analysis, except by certain signatures and signs. That is to say, by nothing proper or specific. Here is the difficulty and the general incredulity when someone speaks of such hypothetical proper modes, that would have to be everywhere.

§ 39 §

In the meantime many theoreticians and dissident experimenters, or simply freer of commitments, give to tests with water in the bathtub or basic simulations in a computer. And certainly their unrest honour them, and is quixotic their persistence, in the same measure which their devices cannot finally elude the margin and the context of the analysis, in which sink all the attempts until being diluted. And it is not that there are not here very original and deep penetrations, although of different magnitude and draught, but which follows forbidden is the auto-definition, not being sufficient consequence of the analysis. But of what analysis?

§ 40 §

The waves that fluctuate with the proportion of *Phi* (Φ) can be added and simultaneously multiplied of non-destructive form, well-known property by those who generates radio frequencies for the amplitude modulation. In addition, *Phi* generates its series of powers automatically. This way, we see that the Continuous Proportion fills the real space of the waves in all the possible senses, whereas the waves of the analysis deduce real solutions of the complex plane. Thus, the Continuous Proportion is the true sea in which they sail, emerge and sink the real solutions of the analysis, the phase differences considered either as waves either as particles; but on the other hand the complex plane that defines the analysis contains these proportions like parts they are of the real numbers. We can ask how it is possible this.

Bergman created for the first time a system of numbers with irrational base, indeed the value of *Phi*, by virtue of its algebraic properties. A. Stakhov has generalized what it seemed a simple case and he has put on relief essential properties of the system. All the integer numbers can be generated with the set of *phi* and its powers, admitting a constructive and algorithmic treatment. This already is something extraordinary. In addition, new definitions of numbers for each one are generated of which a theory of numbers exists, being then possible a infinity of "theories of numbers"; the same classic or Euclidean system of definition becomes a "degenerated case" of the general theory. Therefore, we have an unsuspected new dimension in the density of the real numbers. The "all is number" of Pythagoras well would admit its reformulation in "all and everything is Golden Proportion".

The itself, the exact and generator number of Euclid, the Principle or Unit, is reformulated like *Phi* in this ampler and comprehensive meaning.

Between the new advantageous properties of the numbers of Stakhov, we have its possibility of ternary formulation without losing the favourable conditions of the binary notation, as well as a mirror property that allows to the test of the system itself thanks to the reflection of the generalized golden numbers and its powers in corresponding his counterpart of Fibonacci numbers: the Z-property. This reflection between discreet and continuous numbers or processes makes a entirely new *asymmetric theory of the measurement*, completely algorithmic, that can lead to us very far. As it can be seen, we are in front of an extraordinarily robust model whose implications now we began to glimpse.

This new order of the Mathematics also includes its own functions, matrix and a long et cetera. How it is then possible that its own density is not mixed with the one of the Analysis? If the complex plane is like the sea for the solutions and waves of the analysis, the measures of the Harmony Mathematics are like drops of oil that can possibly be united and they extend in the surface. The so called "harmonic analysis" of Fourier would have to be called without sarcasm “compound analysis”; the true harmonic analysis will depend on the modality.

The functions of the analysis trim their real values against the *sui generis* continuity of the complex plane, that properly never is outside; if, beyond the modality, and taking of the hand to Hegel and Gauss, we would define this in the old terms of dialectics, we would say that the being stands out against a background of non-being that at no moment would have to be alienated of its definition. That is indeed what happens. This “non-being” to be generated by the same functions it is outside them. Or more simply speaking, which is outside is the time, that is left into the hands of the logic. What the analysis cannot get up is the temporary asymmetry; however for the harmony mathematics this is something inevitable, since the Golden Section is the measurement of a in the habitual sense non-generalizable own asymmetry , but to which is possible to refer all the external or internal asymmetries like members. These are not only heterogeneous to each other, but that would have to be, by definition, heterogeneous with itself. Since *phi* can translate different patterns of measurement to each other, responding to the old problem already created by Eudoxus, the *Continuous Proportion* can be considered like the original module of integration. Like asymmetry already in act and with no need of a symmetry of reference, the Continuous Proportion admits the real time and what is irreversible in the mediation and the communication. But to take more far its potentiality we need a proper setting of real problems.

§ 41 §

Naturally, the Continuous Proportion and its closer ratios have been found in the pulse, the temporary intervals of systole and diastole, as well as in the maximums and optimal minimums of the sanguineous pressure. Also of very significant form in electroencephalogram, as well as in other miscellaneous rhythms of the body studied with different means and degrees of dedication. Its relation with the development, the three-lobular segmentation of the body and its gauges, and even with the asymmetry in the cellular division is well-known on the other hand; not to mention its roll still somewhat ignored in the modulation of the DNA. Aside from this powerful ubiquity, is impossible to value suitably the real importance of this module, since as soon as it is not said to us in relation to what it operates. It remains suspended in the air without hardly relating to another thing that to itself. However, which is absolutely evident is its aptitude for the optimization, its determining capacity of maximums and minimums *whenever they are referred to itself.* This can be a vicious tendency, but in fact it is already in this context that has become its self-reference evident. And in such sense, all the values related to *phi* must be considered like *self-values*, index of themselves. But this is not at all in discord with the spirit of the harmony mathematics. If on the other hand it has been wanted to appeal to the natural selection like sufficient reason of the surprising insistence in the appearance of these optimal values, is simply because it is not possible to go to another more ambiguous and general external reason. It always lacks a background. Even, instead of adaptation, it is possible to speak of the virtue of *adherence* of *phi* to the irregularities of the objects, of his fidelity to them. This is already the maximum criterion of selection, that other principles would try to imitate. However, From where it is made? Any theory that leans in the distinction between the interior and the exterior is as primitive as to explain everything.

Either we have the great generalization of the Golden Section for the number theory, and it is time to advance a generalization to define its inclusion in the real time of the things and the entities. For want of experimental and formal development for this intention, we will be contented with the philosophical and qualitative definition of the subject.

§ 42 §

Let us think again about that great set of fluctuating biological rhythms that occur in our own body: the pulse, the electrical activity of the brain, the one of the skin, the breathing, the pulsations and divisions of the cells, the cyclical emission of determined bio-molecules, and a long et cetera. That all these biological oscillators have a concatenation and connection, is a fact as doubtless as the manifestation of life itself. However, the inter-mediation that is supposed between all these rhythms is so complex and detailed that hardly we would be able to understand it without loosing ourselves while we disentangled the maze. The question is if a scale exists that crosses all the scales longitudinally and in cross section, a generalized rate of the rates. Of course, that has seemed until now a chimera, in spite of the reasonable of the supposition. We have already seen that the Continuous Proportion gives an ideal mathematical frame us to make this correlation of measures, although by itself it is revealed insufficient. The Continuous Proportion is a module of integration of asymmetries like proper measures; it speaks to us of the measurement of any possible connection. We need a selection index to give us real connections. But, on the other hand, it turns out inadmissible to refer *Phi* to another thing that to itself, because it would supposedly cut the virtuous circle of its self-recursion. And speaking of circles, also we have seen that the connection of *Pi* and *Phi* is always fleeting and miscellaneous: what it already seems to say to us how mysterious and subtle must be the nature of this self- reference. However, we spoke before of if there is something able to unite the rates of the different fluctuations, that something was the phenomenon of the form, the ungraspable and specious form. And of which the minimum signature for any specific modification of the form was the conjunction of the three modalities of the nature according to the Samkhya; what we applied to an oscillator like the pulse, the most primary and generic of which we can find in the body. Being the Continuous Proportion a purely formal reason, and giving us the Samkhya the more purely empirical and modal reason of the form like fluctuation, and being both inherently ternary, necessarily has to coincide in their own sphere, that for more abundance, is the generic circle of the oscillator.

§ 43 §

For the Samkhya the *gunas* are the last reason of the fluctuations and, therefore, its concurrence is necessary and of universal character. The only thing that changes from a phenomenon to another one are the proportion and the magnitudes, being in last instance the magnitudes also an effect of the proportion. What this would imply is that if we know the factors of the modifications in the pulse of the man, also we can know them in any other animal, and any other system that keeps analogous minimum conditions: the Sun is a great example for this class of analysis, even though to anybody escapes the abysmal difference of the conditions.

§ 44 §

All this would seem absurd if it did not find his natural hollow and its correspondence in dynamics. Matsuno and others have treated the commonly avoided fact that the third law of Newton referring to the action and reaction does not give us a reason nor mechanism for its fulfilment. The third principle would have to be the passage of the most general and abstract principles -the establishment of the rest and inertia, and the definition of force with respect to that-to the real world of the innumerable interactions. On the contrary, not giving it any material plausibility, it turns out to be the distinctive characteristic of his perfect isolation, simultaneously that guarantee and safeguard of the operation of the law of the gravity in the context of the game of the forces. What the third principle wants to establish and to consolidate is the dominion of an absolute time of global synchronization that necessarily evacuates the problematic of the signalling or the local communication between bodies or agents, really, the question of the information or mediation. That global synchronization, refractory to any arbitration or measurement, is completely conserved in the theories of special and general relativity by means of the transformation of Lorenz and the notion of covariance. The quantum mechanics is not either made question of it, fitting its cancellations so that they agree with the exigencies of the conservation principles. So that even if we want to describe in detail the action and reaction between two billiard balls that hit, the interface between two kinetic energies of input and output, the statistical or thermodynamic mechanics is included and explained by a quantum mechanics that ignores those problems to give back without touching it the balance of accounts to the classic mechanics. This way the principle of irreversibility of the thermodynamics, unique that can give account of the real interchanges of communication, is sent of return towards no part. Indeed the best definition of information, not to say the only one, is the one of inequality between an action and a reaction. **Local asynchrony is the only proper measurement of the communication or mediation.** Any other measurement is a remission to a composed analytical space; any interaction of waves or particles is a mediate attempt of detaching a principle that already has excluded the mediation. What we looked for, on the contrary, is the immediate and general form of the mediation possibility. And although we spoke in terms of the modality, is evident that the three principles of Newton are there for demarcating the field of necessary and the possible thing, without never entering the ground of contingency. A third principle never was more to the service of the other two, neither had less mediation capacity. Here is the strength and the independent expansion of the physics like discipline.

§ 45 §

If now we return to our fundamental oscillator, we can include in the game of the ascent and reduction of the waves the inertia and the force with the greater facility, and say in the same way that the principle of *action* and reaction of both comes defined by the perfect continuity from both, so that it does not admit nor it demands a specific location. For the ideal case of the ball rolling subject to a force in absence of friction, mass and force become indiscernible by definition.

The formulation that Newton makes for his third principle is extremely careful: "For all action there is always an opposite and equal reaction, or the mutual action of two bodies the one on the other always equal and is directed to opposite parts". The terms have been left free so that the term action can mean indifferently as much the transmitted force as the body with its mass; and when it is spoken directly of the bodies which among them it gives is not another thing again that its "mutual action". We do not know if the action-reaction really takes place between bodies or their actions and respective reactions. The action and the reaction include the mass and the force and can occur without another relation that the one of "action and reaction", indeed. In comparison, the most modern equivalence of mass and energy not even have been tried to explain, since it emerges already excused from the pure formulation.

§ 46 §

We know that no natural unit of mass nor of energy does exist at the moment; and in truth it is completely improbable that such thing could be found out of the complex correlations, in view of the abstract character of such magnitudes. The force notion can *be made up* or compose sufficiently within the analytical framework, but also lacks natural units. We have already seen that the concept of "quantum of action" introduced by the quantum mechanics is valid only like measurement of energy-by cycle-second, being the unit of time arbitrary but irreducible the one of frequency. Of all this it seems to follow an important lack of harmony and correspondence between the different forms from expression of the magnitudes, arisen from the conveniences of measurement in the historical bifurcations of different problems.

In repeated occasions like fundamental unit of the mechanics has been proposed. Domingo Acosta, for example, gives us a definition of the mass like a set of impulses in all the directions and the speed like a balance between impulses in different directions: the impulse is always a unit of force applied during a time. Acosta is wondered if any force can occur or be applied independently of time, and reaches the conclusion that in the nature nothing similar can exist; we are of identical opinion. Grejzdelsky approaches of another form the same problem and points the convenience of replacing the ideal notion of absolute rest by the one of initial imbalance with a minimum value quantified by the constant *phi (Φ)*.

§ 47 §

Let us try to lead back these valuable indications to the subject of our oscillator. Of course, nothing forces to us to consider the impulse, as it is defined, like a natural unit; we simply considered it like a more suitable unit to our interest. We part of the principle that a simple harmonic oscillator is not indicative at all of a proper time, but of a reversible and completely unspecific time. Nevertheless, this one must serve to us as reference, or otherwise we would not know how nor by where beginning to measure the inequality in the balance or asymmetry in a given complex oscillator. But this does not mean that we leave from that ideal oscillator and we apply continuous deformations to it in the way of figure 9,

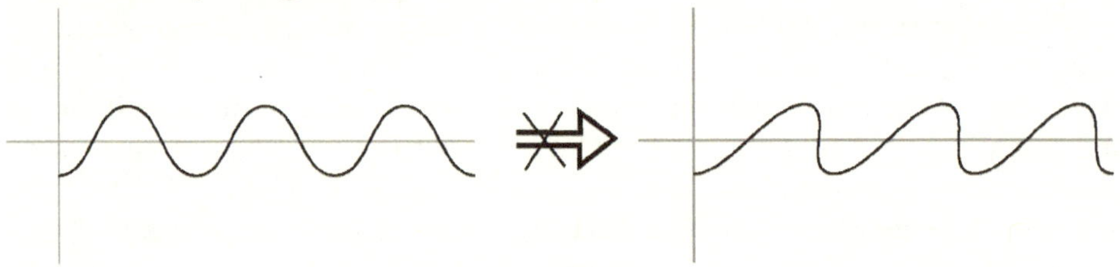

Fig. 9

then in fact, our assumptions go in opposite direction, although no definition of the impulse as force per time oblige us with respect to the cycles.

We can interpose with perfect arbitrariness a regular or linear oscillator in the middle of a non-linear oscillator.

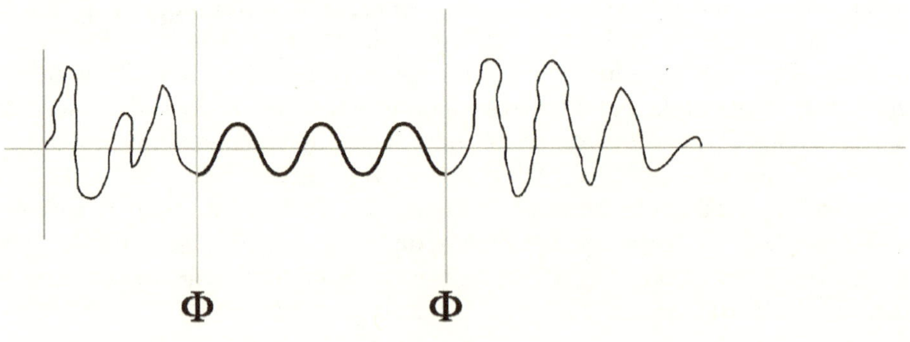

Fig. 10

We know that things thus do not only not occur in the nature, that cannot either take place such steep cuts nor in the best conditions of control. Aside from our manifest impossibility to imagine it, that denominated phase transition exists; that transition of phase includes as well diverse factors like the damping. Nevertheless, and beyond all it, subsists the fact demonstrated by diverse theorems that *Phi (Φ)* is the last ratio between frequencies and amplitude that its stability, although it only go by consideration to its "specific density" with respect to the whole and real numbers. So that the Continuous Proportion really is not in one or another part of the oscillator, nor in one or another of its phases, but along the phase itself and in anyone of its cuts, that which has come to be called edge of chaos. But in the same way in which the pulse allows us to accede to an inner section of the organism without carrying out the smaller invasion or cuts, the Continuous Section is inserted in the reasons of a process without it has sense to speak of inside or outside. It is already truly eloquent that we begin to notice so subtle thing.

§ 48 §

A balance between impulses can be conceived as well as an impulse with its proper time span, if we equipped to our imaginary balance with a time of adjustment between the successive operations of add and remove "weights"; time to which also we can grant variability. That is to say, we located ourselves completely in the asymmetric theory of the algorithmic measurement that Stakhov proposes as generalization of the classic case of Fibonacci. That estimation of the time interval between discreet operations gives us a measurement of the inertia of the same balance, or if it is preferred, of its sensitivity. The inclusion of this third element is necessary to obtain an exact correspondence. Since the moment in which the three share now an input of time, as much the mass as the force can be equaled with the impulse by means of the sensitivity of the balance.

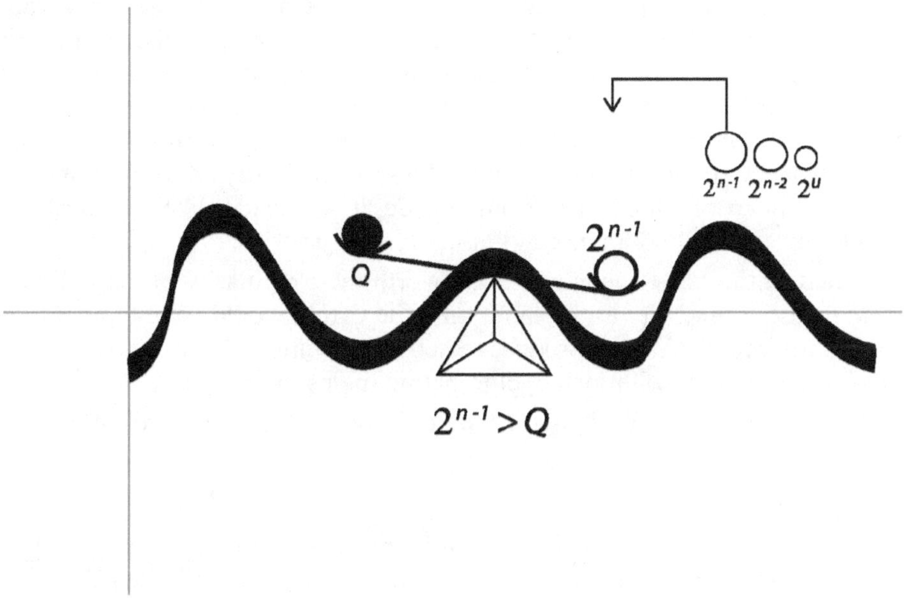

Fig.11

This disposition allows a more fluid and also operative conception of the problem of the motion in an oscillator, with no need to be refer to the sources of the impulse; in fact, which allows us is an absolutely general conception of the motion. The own time could then also be defined as the real and potential input of operations in an interval. The possible operations already come limited by the present loads in course and the weights available, that can be of different magnitude and proportion. This is, cannot conduct any operation in any amount of time. By incredible that seems, the physics does not take care in general of restrictions of this type, but of general much more relative to the conservative principles and constants, that does not touch the case. Without these restrictions, it is not possible to find a specific form in the processes. In addition, without these restrictions the conservation laws remain unfounded as incomplete, and always more supposed than truly reasonable. Also it is doubtful to speak of the absolute energy of a system, notion that is surely indemonstrable by the absence of a fixed value of reference; but to deal with the energy available, only the real one, is used the absolute energy like reference. Therefore, a more conspicuous measurement of the energy must be possible. The notion of impulse and its adjustment in the temporary balance would have to approach us to it.

§ 49 §

The relation of Φ and Π in an oscillator is too ambiguous and open in absence of *proper weights*. In principle it seems too evident that the primary reason of a regular cycle is the maintenance of a stability, until the point of which cycle, regularity and stability practically sound to us like synonymous. The more relevant property of the Golden Mean seems to be the conservation or maintenance of the asymmetry or inequality at different levels, disappearing in the infinitesimal continuous to reappear by itself in ampler intervals, without another solution of continuity that the one that indicates itself like Continuous Proportion. Or in other words, the Continuous Proportion expresses the balance in the same imbalance, the self-sensibility as form of stability. This is the heart of which, seen from outside, it is consider like the problem of the optimization of maximums and minimums to us.

The Continuous Reason makes be worth the irregularity of the individual throughout the time, but at the same time the Circle would have to tend to annul the differences. Which produces friction on which? Both ratios do it and both can mutually be eliminated, eliminating themselves with it.

Little more we could know if they did not interact in a third ratio or property.

§ 50 §

Instead of loading our balance with impulses we can load it with the fuzzy weights attributable to the three *gunas*, or if we followed with the pulse motif, to the three *doshas*; the mass or inertia in terms of *Tamas*, the force of *Rajas*, the temporary sensitivity of the balance, an inertiality of superior order, like

Sattwa. Naturally, the three appear already in a whole undivided set that as in the previous example tends itself to be not perfectly located and to assume layers of time span. The specific weights of the three *gunas* with their mutability would give us the values of *Epsilon* as the generic frame of not- linearity. Now studies exist on the presence of *Phi* in the fuzzy sets, and the relation between both is quite natural if we consider the mentioned "density" of the golden numbers and the problem of the over-determination or intersection of areas and spheres of inference. This would be the optimal experimental and formal frame for a definitive advance in the subject, and perhaps, to go beyond it.

§ 51 §

The temporary sensitivity of the balance can be graduated in diverse ways, and we will obtain different types of links of series or sequences; but more even than the type of sequence it concerns to us the possibility of expressing and of modelling suitably the phenomenon of the thickness of the signature in the characteristic *form* of the pulse. One assumes, and also this is essential, that this graduation of temporary sensitivity must also give us the relevant degrees of precision for each case. The "semiosis" problem, or infinite remission from terms to others, in mathematics the impossibility of solutions, is avoided because an immediate location does not exist, but rather an immediacy not absolutely located, that is an optimal adjustment or tends already to an optimal adjustment of terms, which sets out like intrinsic nature of the same motion. This also supposes intrinsic or proper limitations within the interval. The infinite remission of terms throughout the optimal adjustment tends to be stumped or to cancel itself throughout the optimal adjustment; cancellation that never can be total and absolute, but more or less suitable or simply sufficient. From a technical point of view, this would be a new frame for the probability spaces.

It is evident that in systems under control the balance cannot "become crazy" at any moment, safe by the gradual or sudden destruction of that "control"; the biological organisms and the human being are part of this set of systems. Here the problem is not as much to find stable solutions, as appropriately to graduate the spaces from which the instability is pending.

Without doubt, we could not have found nothing more fragile and precarious to give account of the stability of things. We move in the tightrope of the pure conditionality. If this conditionality were not naturally softened in some kind of buffer, nothing would be possible.

§ 52 §

Almost without trying it, we have made a somersault. But we have still not landed nowhere, nor we know if this first twist is not only the first of an indefinite number. The used terms have corresponded of inevitable and natural form. Our tripartite definition of the impulse is, like the Continuous Proportion self-recursive. Also like in Samkhya: the inertiality can be computed like sensitivity, but not of immediate form.

It happens that if we can define the mass or inertia, the force and the action-reaction within the self-referent ratio for the impulse, also we would have to be able to make the same with other magnitudes or notions, like the tension, the pressure and so many others. Why? By the same reason we can make new definitions of numbers with its corresponding number theories, in infinite number, within the system generalized by Stakhov: that it is, in addition, an algorithmic and constructive theory of measurement, with the amplest and precise criterion of analogical-digital convertibility. That is to say, we can redefine the magnitudes, the conditions or parameters, the constants and all the notions as we consider suitable. This is a luck of universal physical translator with its own rules of adjustment, their required logic and all the constructive means. With complete autonomy, and of such an amplitude as we do not find it in any other experimental discipline. If after this we talk about subjects like the compatibility with other metric and disciplines, it is simply because still we have not begun to go deep in which there is here. All the later references to other sciences and disciplines hardly are another thing than imaginary excursions to attract the attention on which it is not here revealed, in this new frame.

§ 53 §

Like the dialectics of the old times, the modalities of Samkhya do not express another thing that that of which nothing exists by itself. When we reduce this to antinomies or polarities, like the one of the cold and the heat, we cannot leave triviality; when we introduce a third element, which normally we cannot leave is of the arbitrariness, that allows to affirm and to deny everything indifferently. Even in the logic of Aristotle there are triangular contradictions of impossible solution, as Giorgio Colli underlined in his time. But these contradictions of qualitative order have always occurred by the incapacity to establish a bond between the qualities and the proportion, which on the other hand is something unavoidable and natural, and conditional also. If we managed to express that conditionality suitably, which we have is the key of that old and unattainable philosophical reason, the conversion of the quantitative thing in qualitative, and vice versa.

This cannot either get rid of the paradoxes, since the formalization, as the own form or the motion, never can be fixed, but be managed or arbitrated at the most. But also it can transform itself with a defined rigor, and here we come into a new space of transformations.

§ 54 §

With these three components we have defined the moments, modes or fundamental modalities of all motion, although we left the ends of that definition *necessarily* loose here, what would require a special development. Nobody will be surprised of which here there are not analytical solutions, since which we try to go deep in what necessarily escapes to the analysis. The non-linear balance would be the more simplified analysis of the non-analytical in general. The principle of least action is present here fundamentally in the number of operations for a partial balance from moment to moment, not from instant to instant. This principle also may be in the number of weights of both "plates", as well as in the time of reaction of the balance that also can be comparable with the others. The motion, the resultant of these three ways, is seen necessarily surrounded by this third mode or sensitivity, to which in fact we can consider indifferently first. Therefore, which seems to surround to the motion is just the same that undresses it, and its lack of definition, its authentic proper definition, moving away of the ghost and purely abstract motion of the analysis.

A smaller time of adjustment of the balance, a finer and defined outline of the curve, necessarily does not mean a better or more optimal adjustment to the property of the motion, since also it can mean the opposite, a fragility, an excess of instability.

The question of the possible couplings between coordinated motions would seem to be indicated by the time interval that quantifies the sensitivity or reaction of the balance, being able to consider these intervals then as bars of balances on which to put other balances, forming indefinites series. It is always the mean part the one that evolves, the one that surrounds and fuse in its limit the past and the future, the interior and the exterior, those predicates whose relation we excluded for the homogenous space of the analysis. These couplings or transitions of phase would have to occur even within the apparently simple systems, like the pulse or the undulations in the water of the sea, and have to be related to the generalized hyperbolic functions of Lucas and Fibonacci.

In all the modes of each proper motion we see that different possibilities of purity with respect to the form exist, understanding by purity the greater homogeneity within the greater definition or clearness. This notion of purity is more improper when it settles down by comparisons between different entities from the same sort, and more proper when it is considered from the scope of possible evolution of a single entity. Anyway, it is not possible to be denied that intrinsic degrees of purity, definition or perfection in any properly natural motif exist, either we think about the ribbing of a leaf, that has supported degrees in its development, in the spots of a leopard or whatever we consider. Intrinsic means that they can be appreciated in his own context, to which we simply we are giving a frame. We took to intention visual examples to make see that this same principle is applied to things and orders that little have to do with the visual appearance. We could consider it in the plane of the sound or any other.

§ 55 §

It is to be present that we are not taking care here of the general question of the compatibility of this model of the motion with the analytical space, thus we say, the one that we applied in a billiards table, which we left into the hands of the competent specialists. What demands our interests is to be able to define specifically the form of the motion in real time, leaving at the moment to a side the possible metrics of the space. It would have to be a fact that this form of the motion manifest in the most external appearance of a process or a living entity; and if this is true, the metric of which we considered outer space, a pure construction, would have to adopt other modules.

It is a serious error that still persists to consider the motion like something perfectly explained by the modern kinematics, also developed in the baroque age. Paradoxes as those of Zeno have been tried to explain by sum of infinite series of terms, something that surely nothing has to do neither with the motion nor with reality. Alexis Jardines remembers to us that exactly the intention of these paradoxes was justly to demonstrate the impossibility of represent the motion in the space, so that all these "demonstrations" and their acceptance continue being mainly ridiculous, simply a form of not wanting to see something. Jardines jointly subdue quantum kinematics with their limit of uncertainty and the relativist kinematics with its ambiguity or relativity of the simultaneity within the same classic case for the motion of Zeno, coming to say that both are the extreme illustrations of the same thing, motion itself, although do not try to give an operative model for this motion. We will say the same with respect to our model? We agreed with Gardens, and this is the fundamental thing, in that a single motion exists, independently of the scopes, or otherwise it would be vain to try any unification. But Jardines, with extreme consequence, radicalize a position that no science can assume, nor to digest: on the one hand, dynamics only describes the rest, and, on the other hand, a same object or process from different positions no longer is a single object, but different objects. In the middle of everything, the motion is intrinsically indeterminate, since the same time is a wave. The arbitration of a theoretical enigma already is a existential problem, and in this case, the solution is too threatening for any scientific notion of the reality. As far as which our model can locate in means between quantum uncertainty and the relativity of the simultaneity, or the ambiguity in the intervals, it can be said that in the first place we do not know the rank of validity for relativity, but mainly, which we do not know is the general compatibility of this temporary model with respect to the metrics of the analytical space. No of these things constitutes the immediate nor the fundamental thing of our problem.

§ 56 §

As we have characterized it, the motion has a surrounding aspect with respect to its moments, no matter how hard this surroundment is the own undressing of their form. In a identical way, also the mind adopts this surrounding character, and then **it is possible to ask if the mind is not another thing that the same motion**, motion that is inseparable of its form. Not only we create this, but we ask to anybody what another thing can be thought. It is not possible to think the mind but like motion, and if we resisted to it, already is by the own inherent motion of thought, reluctant to the identification or fixation; if in this respect we are forced to believe in other things, is by the absolute deficit that any representation supposes for the motion. It is more, we will say that to think that the mind is another thing that the motion, that already has all the necessary complexity, it is simple fraud. Soon the same space of the analysis pushes compulsively towards the vagueness and the inconsequence to us.

For Samkhya or Yoga this vagueness does not have sense. Try anyone to stop the mind and will see that one does not face another task that to stop the motion by the same motion, with its inertia, its impulse and its sensitivity. Here the things are recognized of first hand, with no need of any representation. But to all this respect, think about the spectacle offered by the western philosophy speaking of pan-psychism and the universe of *qualia*, and the *mind-body problem*, still following the Galilean scheme of the secondary and primary properties and completely ignoring that as much on the qualities as on the same motion the Indian philosophy had genuinely resolved the problem already thousands of years ago. This was only possible remaining naked in the problem and carefully avoiding all the trifles of the representation, that after all, are only other forms and other clothes for the motion, and successively far away of the question.

Therefore, it would have to be the easiest thing in the world to find mental "correlates", since they exist in all whatever moves, thus is a piece of rock. Another different thing is to want to fill up the content of those correlates, which is not allowed to us nor even with our own mind. For example: if we want to carry out feedback with the waves of ours encephalogram in real time, rendering attention to the modulation of the waves flock any other possible content, no matter how hard these exist in the same motion with the form of inertia or other qualities. This is the basic deficiency of this type of instrumentation, simultaneously that simple illustration that the mind is the mere form of the motion. Therefore, it is not necessary to look for the mental correlates -and we do not say, sancta simplicitas, the consciousness- in determined groups of cells of the brain, no matter how hard also they are there: the only real location of those correlates exists in the time and the motion, its asynchrony or its synchrony, and this extends from those same nervous cells to the pulse, the breathing and the most external corporal motions. The same spatial, taxonomic location, has to depend in last instance of the pure temporality of the development, of which each act is a landmark.

If we consider the necessary couplings between the form of the motions that occur in our own body, we will have a much more resolving frame for the question of the interdependence of the body and the mind. Because if by one hand it is idle to think about the mind without conditions of motion, is not the same with respect to the materiality, that already in any case is conditional, and conditional by the motion, indeed. I don't want at all to enter in the subject of the independence of the mind of a given body, although in fact would have to be easy to see that this is not a real problem, being the problem the conditions of the form: for this it would be enough to see, far from any rhetoric, the binds and the unbinds that take place in the modes of the water of a simple undulation in the sea, as we are considering. As we said, the content of those correlates is something that is only incumbent on to that mind from its same condition, and in no case from outside, unless we have desire to waste the time.

§ 57 §

If the estimated weights of the three *gunas* become equal and they balance, momentarily and by definition there is no action nor reaction possible. This is, the *gunas* naturally conform an equilateral triangle integrated in the unit circle of a regular oscillator. But like the *gunas* can only exist really concatenated by the action and reaction of its difference or inequality, is clear that we are leaving outside the process of its proportional equalization, the optimal way that takes until there. We are leaving outside the Continuous Proportion. Relying us on our own categories, a circle and a triangle such could only exist outside the time or like an unstable moment without proper temporary duration, that is to say, they would not be able to manifest themselves.

§ 58 §

Nevertheless, we can see that the three *gunas* tolerate so much the character of cycle or circle as the self-recurrence with specific proper values. We have already said that *gunas* are anywhere as much inside as outside and that we can observe them if we learn to analyze what we see; that there is not another thing to know in last instance.

If the existence of *gunas* is based on its inequality, on the other hand it cannot modification or fluctuation without the dominion of one of *gunas* on the other two, which makes possible that that modification can be notice properly as a moment. That dominion has the form of a modification of the weight or value of one *guna* that reverts on the weight or value of the other two. This admits and demands characteristic sequences that, beyond the abstraction, constitute the only form of causality which we can know. That is to say, there is no another one, and any other apparent form of causality will be revealed as devoid of form, like a correlation that must by necessity to be refer to the only form with its own modes; we can then call them degenerated cases of causality. All whatever that can become intuited is here.

One *guna* or modality cannot exist separately, so that a state dominated by *Sattwa*, we take as example, already includes the presence and subordination of the others, as much for the most stable cases like for the modifications triggered by the fluctuation. The same it is valid for the other *gunas*. We can say that there is no action without potential energy, neither sensation or sensitivity without action, but in no case we can reduce everything to the potential energy, nor to the action, nor to the sensation, since thus separated we would lose the reference or the mode. It must be clear that these ways are only applicable to the real cases, not to imaginary entities without comparison terms; it is by this that *gunas* are considered like genuine material causes of the mutations of the real, conditioned things. Patanjali distinguishes between diversified, undivided or monadic, only indicative, and non-indicative states of *gunas*. The same empirical or mutable ego is an undivided and indivisible state actually, no matter how hard so many insist uselessly on their theoretical dissolution; the pure sense of I, the existence like mere attention, is an only indicative state, *Linga-Matra*, reason why the non- indicative state or *Alinga* properly is unmanifested, rather than non-existent, if we want to fit expressions that not always we used with much property. This is, the state of balance of *gunas*, whether it is considered or not "metaphysical" term (which would have to include all the well-known reversible physics), not is attributable to it neither the existence nor the not-existence, and the only thing that is possible to say of it is that it does not generate indications nor references, nor terms like the previous ones that always have a fictitious element in their framework.

The fundamental states or sections of *gunas*, in their divisions in groups of six or five, have an extraordinary importance if we want to deepen in this dominion, although now we cannot get into it.

§ 59 §

If the balance of the three *gunas* constitutes its non-manifest state, it is because there is no place in them for the dissolution, growth nor decrease, since they have no further material cause. If we follow sufficiently the growth or manifest decrease of *gunas* in an entity or system it is inevitable to verify how values that seem to be diluted can emerge from their latency, and on the contrary also, which is in substantial harmony with the conversion of continuous and discreet values in the development of the Continuous Proportion. But a value cannot sink in the latency, relatively non-manifested, without passing through the modality of the Latency or Retention, that is to say, *Tamas*, in the same way that cannot emerge without action or *Rajas*, nor acquiring certain balance without *Sattwa*. Then, there is an eloquent continuity between the content of his values and the form in which they are manifested.

§ 60 §

Even accepting that *gunas* themselves neither increases nor diminishes, any action or predominance of one on the others in a empirical or temporary entity has to be seen in such terms for the understanding of the sequences. A sequence begins by necessity in the *guna* that then is already dominant. But this already produces a redundancy that can be detached indifferently of several ways, since what is increase in one always is a correlative diminution in another one. We can detach this of several indifferent ways, one of which would be:

1°. If *Sattwa* or sensitivity "increases", it entails a correlative diminution of the activity or *Rajas*, and never it increase. (This does not seem nothing evident and there are many experiences that at first sight seems to contradict it). It can produce neither "increase" nor "diminution" of the principle of retentiveness or *Tamas*.

2°. If *Sattwa* diminishes, it can only happen by a correlative increase of *Rajas*, and never by its diminution. It is why the term *Rajas* is synonymous of contamination or dust. *Tamas*, as in the previous case, it is not affected directly by the modification of *Sattwa*, but already the increase of *Rajas* operates its diminution.

3°. The diminution of *Rajas* also corresponds to the increase of *Tamas* or inertia.

Thus they are included the four possible direct motions of *Rajas* or activity, and both of *Sattwa* and *Tamas*. *Rajas* is in the middle of the three *gunas*, and is the only possible connection between the other two. A defined order is conserved although what it seems beginning of a sequence can be seen from another modality like consequence, and the increase, like corresponding diminution, and vice versa. We move already in the circularity. Of course that we are not going here to advance the formalization and quantification of a system that seems insignificant and that in fact it is possible to be complicated of indefinite form, being able to deal it any logical with more pleasure and competence. Certain that it would be possible here to speak about disjunctive and conective functions, but we would prefer to adhere to which the nature is able to unite and to separate, which has a very different reach; we will return more ahead on the subject. Also elementary figures and diagrams with indefinite combinations can be obtained. Now it is enough with seeing that the order of *gunas* is simply a scale or ladder that simultaneously allows to the union by degrees and the separation or exclusion of certain immediate motions. That is to say, it is an elementary model of the mediation, that in itself admits almost infinite variations.

§ 61 §

Until now it never has been possible to elaborate a quantifyable model of the mediation operative in the empirical dominion, that can take us to some checkable knowledge. If we want to ask what makes it possible now, the answer is in all the previous one. Or if it is wanted to summarize, in *Pi, Phi* and *Epsilon*. It is necessary to say that we move here to look for, not the most exhaustive and powerful model, nor the most comprehensive and general, but the optimal one, that in the end could be most exhaustive and general.

We have then in this model an oscillator with purely quantitative elements, a scheme of the modalities that in their quantification even refer to the purely qualitative element, and a purely optimal element that even operating in contexts of maximums and minimums never can be considered neither quantitative nor qualitative. So that we have the elements for that balanced search of the problem of the balance in the plane of the discovery of the form.

§ 62 §

In this form, we can see how the three *gunas* correspond to a cycle or circle through their "continuous" mediations, whereas they demarcate "discreet" sections defined by the exclusion of determined immediate reactions: the interaction of both throughout the time for the real cases will demand the inclusion of an optimization term whose more generic reason will be always the Continuous Proportion.

§ 63 §

The observation and the self-observation with respect to our context within *gunas* always would have to tolerate more lucidity than mere illustration with "examples", which would not have to be there to add themselves simply to the general case.

In any case, in relation to the sequence of *gunas* we can think that we can become from a state of lucid attention to another one of estupefaction, forgetfulness or inadvertence without continuity solution; but it is this type of sudden conversions the one that makes impossible the same nature of *gunas*. Without a doubt it must have had emergency of other thoughts or impressions before whereas mutations, although that modification whereas thought has been able to arise from the loss of the balance and whereas impression could have been a loss of inertia in form of released latency. This could as well trigger a series of reactions amplified until a new modality becomes dominant, in this case a certain unconsciousness. In the same way, the increase of activity cannot be in itself the cause of the pleasure, but by the momentary liberation of a determined weight of conscience or unconsciousness, giving passage to diverse sequences. The fundamental nature of *gunas* dominates on what is accidental of the reactions, but that is not so easy to see without considering the specific weights of the situations, that, in last instance, refer to a native constitution generated by alien forces and an apparent cut.

In fact, from the possible sequences of *gunas* and all its impossible mirages we could derive all fables of the myth and the cascade of the literary forms and its contents.

§ 64 §

Always there is to be possible the distinction between the momentary balance and the constituent, that goes progressively being altered by the first, of entirely analogous way to as *vadya* can clearly distinguish between *prakriti* and *vikriti*. Without this previous landmark we lack the valid reference for later inferences. But both factors already are undivided in the form subject of gauge, in this case the pulse, reason why they are considered by its mutual reference; what implies as well an active balance whose value is given by the same motion within its own intervals.

§ 65 §

By opposite that can be to the perspective contemplated by the thermodynamics, as well as to a discreet mechanics that cannot solve entirely its foundations, the temporary irreversibility in non-ideal processes exists because the complete loss of the information of the previous states does not occur; that is to say, because it is not possible to be eliminated or to be suppressed. If it is considered in diffusion terms, any loss or liberation to the "outside" of the system will correspond with the consequent reactions in the interior, and to this it is not necessary to give many rounds about it: rounds that we give it in fact taking advantage of the virtues all type of cycles to eliminate the signs. From the quantum of action (whose cycle is omitted), to the most virtuous biological cycles erasers of the information -the only one that there is, the one of the previous states. The impulse notion that before we exposed demanded a temporary interval, interval that can be spread and be emerged based on different potentials. But it seems already lamentable that we must resort to mixed terms with those of the analysis to make reasonable what by itself it would have to be evident. We arrived at the situation in which the most demanding reductionism is incapable to do nothing with a mechanism as obvious as the circle, because it is incapable to apply it itself. The determinism tends to be applied so on nature in a way that we remain undetermined, and what at first it could be a Jew-Christian prejudice has gotten to be with time the decided search of a situation of maximum advantage. The India philosophy, on the contrary, part already of that situation in which to find advantage is an impossibility, and in addition is undesirable.

§ 66 §

Perhaps the principle of minimum action is most general and including in nature. It has been said that this principle cannot acquire relief precisely by the differential context of the analysis, but this cannot be but part of the truth. The analysis seems to benefit enormously from this unexplained simplicity of the nature, simultaneously that seems to adapt it; what it cannot happen without something very fundamental has been relegated. The optical ways of the light, in the refraction of a ray in the water, for example, are understood like minimum routes because they run by the way that spends less time. Nevertheless, it is more than doubtful that a natural phenomenon can be directed by a temporary principle of economy, only imputable to certain subjects of the men; at the most, this could be an indirect effect and collateral of another disposition of the motion in time. In other words, if we considered the existence of "atoms" or waves of time, these would not have where to go: so it is the concept of modification within a sequence for the Samkhya. Concept that besides is not incompatible with the quantum mechanics, although this one cannot contemplate it because it does not leave site him. **The "action" defined by an impulse tends to last in the time in the maximum extension of the sequence or sequences.** Yes, this seems to us an intrinsic property in the nature in addition to harmonic with the perspective which we are contemplating. By the way which it is not necessary to make a great exercise of etymology or *nirukta* to notice the relation between the word *krama*, sequence, and *karma* or action: *Karma* or action is not but the round circular figure within the sequence, also in harmony with the nature of the operations before alluded.

§ 67 §

Thus, the principle of minimum action or the minimum, or optimal ways, acquires a relief very different from which allows the differential calculus to it. But it happens that what is contemplated in the analysis of waves is the change of phase and not the same phase. Contrary to which often one assumes, the uncertainty principle does not prevent that we could to register changes throughout the phase; the limitation only exists for the spectroscope and the Fourier analysis. Numerous types of sensitive detectors to the phase exist, some of which are even used in colour televisions; the only limitation for its sensitivity is the capacity to eliminate the noise. Naturally, this does not contradict the uncertainty principle, but it allows us to obtain another type of information that can be more relevant in other contexts, like that we propose. On the other hand, the quantum theory does not contemplate things such as a component of acceleration for photons, nor modifications that can be equivalent in another context. For regular oscillator of this type, that do not admit real fluctuations, neither the non-linearity nor the other meaning of the principle of least action have nothing to say.

§ 68 §

If, as limit case, we could completely make external one to each other the five elements of the pulse, we would have a pentagonal symmetry like which so very insistently we have in living beings. We could to use fuzzy graphs, but these elements and its values always will overlap, never will appear exempt, as they do it in the more patent way in the external form, that for the rest, if we followed it in its order of generation, takes us to the Continuous Proportion. The pentagonal symmetry has been in all the cultures incarnated symbol of the microcosm, of which it acquires differentiated laws amplifying a very small or infinitesimal difference, but never unsubstantial. The hexagonal symmetry, however, has been always symbol of the balance of the macrocosm: not of its closing, but of the ideal interpenetration of the space and the time in its three respective dimensions. We cannot then oppose both symmetries as they oppose the organic and the inorganic; because for physics, that deals with the conformation of the forces in the inorganic in general, these three dimensions of time do not exist properly like slopes or tends, but like unspecific and arbitrary distinctions. By the same reason the dimensions of the space are made a whole undivided that becomes mere support of the infinite directions of the vectors, without on the other hand it finds specific principles of direction.

The hexagonal symmetry is then epitome of a principle of least action more inaccessible to us than we think, and the diverse levels of the pentagonal symmetry exhibit the laws of persistence of the differences. Both seem fundamental to be able to understand the relations between the singularity of an object and its insertion in a vaster scene. The theory of regular solids will hide for always so many contents as those that we are able to formalize, and that why a mathematician as Felix Klein considered the regular icosahedron like the mathematical object with more possible connections.

§ 69 §

It seems also evident that the most intimate aspects of the sound, related to the harmonics, the material implication of the timbre and the resonances, as well as music and its sequences, can be deal with more explicit and different form, without for a moment leaving the scope of this extended classic mechanics. The same stairs of the modality, the conditionality, are a musical scale, even if eludes our representation and our ear.

For the usual classic mechanics, the sound is a mechanical wave that needs means with a density for its transmission; for the India conception, the space and the possibility of being crossed by the sound are the same, soon are not needed means transmission and the problem of the emptiness or the ether is indifferent in such sense. What defines the problem is the aptitude of the receiver, its sensitivity. The receivers adapted to the resonances in a planetary system would be the own planets, and if we do not know to model the conditions that make them sensible are because there is no space for it in the mechanics to the use. Perhaps the appropriate rank of precision or definition for the mechanics of this system is inseparable of the radii or diameters of the same planets, whose average orbits would represent whole numbers, although to consider this without another method of analysis is completely arbitrary. As for the rest, something as well as "pure dynamic resonances" cannot exist, which is equivalent to a pure fiction; if there exist resonances they have to have an acoustic and net mechanical character. The Pythagorean music of the spheres would not be a metaphor.

§ 70 §

If this model of motion and fluctuation fructifies, we would have to be able to apply it of an appropriate form to all the things that move and fluctuate. Departing from that, we could posse necessarily the form in which the different motions can make contact, friction or coupling. This is, can be considered in vertical different changes or phase transitions, no matter how hard these are not simply affection by the verticality or hierarchy, and since they have a certain diffuse aptitude to reflect a complete and self-sufficient set. Supposed the universality of the form in the motion, this is applied without rank distinction, whenever they are satisfied minimum conditions of description. So that the connection between the different subsystems from a given system could not only be study, as it is the case of the biological oscillator, but that also we would have the way to appropriately compare the most dissimilar and separated things by its characteristic forms of motion through the general constancy of the form.

§ 71 §

It does not seem that could be made use of this proverbial Gold Cane to measure the things with inadvertence of which those measures mean and involve, which provokes the doubt that as well really their use is given to us. Here we would enter for the first time a dominion that surpasses our notion of the analysis as exhaustive study of the alien, separated things like objects. We are not customary to it and most probable it is than we try at all costs to formalize it like something without direct implications. But also it is certain that it has been an increasing process of formalization which has taken to us up to here.

Let us try to imagine what the application of these categories would suppose for the economy, for which has been called "economic analysis". Everybody admits, beginning with the same economists, the technical impropriety of the tools, not to mention its incompetence; in spite of which they are continued applying with the great power of those who pretend an absolute technical superiority. The reality says to us that that superiority is based exclusively in capitalizing what it is made suppose that is the only measurable and the observable, that is to say, in the imposition on great scale of the same categories whereupon the analysis takes place. In spite of everything, it is necessary to recognize that it is the whole genealogy of the analysis, and not only the economic one, what has lead to that state of things in which already are certain privileged criteria.

What is what will fit those who tried to fit everything? The extended motif of the balance already is the most generic form of transaction, simultaneously than its more acute indicator; from now on also we can use it like symbolic interpreter, without it lets depend on us the purity or fits of the interpretations. A strict and consequent analysis of the economic curves would keep awake, in the first place, that the denaturation of the economy begins by the intervention and distortion of the cycle of the cycles, the currency. From here all the consequences for the theories of the balance can be followed in cascade for the general distribution and balance theories, until now little more than dependent macroeconomic notes of those control "constants" of the employment, the interest and the money. The consequences of the application of the model of partial balances would be often so strange, that it would be necessary a good time to purify the concepts and to undertake actions. By the same reason, the application of the model to the coupling of the different sorts of activity would be essential.

§ 72 §

It is not necessary to look for examples or enumerations of the fields to which it could affect this general vision, because what we are treating is the overflowing of the classical methods of the analysis and his fulfilment in the particular.

As we said, it would be comfortable to maintain all this in the tolerable margin of the theories of the complexity, but the fact is that affects the same hearth of the analysis, without meaning of objects. We can then ask which is the relation of our idea of the analysis with the physics, that great symbol of our commerce with the reality. The answer, that already we have advanced, is very simple: if the physics takes care of forces that do not fluctuate, in that same measure our concept of analysis does not have anything to say. The Samkhya is the specific analysis of the fluctuations, whatever they were. Therefore, which we have to ask is which is the relation of the physics with the fluctuations. This relation is expressed in the variables and it is limited by the constants, that can be of very diverse ranks. But the changes that express those variables are in principle and by definition continuous, even though they come from discreet changes, reason why also the relation with the fluctuations is weak and unspecific, and accidental in any case. Yet, it is the constant notion that seems definitively to limit the dominion of the fluctuations. The most fundamental constants talk about the forces, the speed and to the definition of the least action; the non-linearity is of side although in fact it can be in the middle of the others of multiple non-explicit ways, deciding its relation.

We have already seen the character more limitative than complete of the definition of the quantum of action. As far as the speed of the light, often its variability in the context of the space-time sets out, contemplating, for example, also the possible variation of the fine structure constant of electromagnetic coupling. With respect to the gravity, the first great constant in order of appearance, any mention of its variability is always seen with incredulity, and its interaction or fluctuation with other forces is so weak that hardly could conceive something about it.

§ 73 §

But in fact it is easy to see that the physics has needed the constants for its constitution. If the gravity, for example, had a variable intensity, the calculations and extrapolations would not be made difficult, but practically impossible, at least within contexts like the present one. In addition, this could affect the rest of the constants. Often remarkable variations, much greater than the factor of relativist correction, have been detected in systematic experiments with graves objects in the same Earth and even in a same place; not to mention the estimation of the masses of some planets of our small Solar System. But, in absence of a correlation factor that has been considered acceptable, such variations are refer to the little precision in the estimation of the mass of the Earth -what it is surprising enough-or to any miscellaneous factor that are possible to suppose more than to find. It would be absurd to hope that the physicists could renounce to their precious accumulated capital if there is no clear possibility to still win more; not in another way the scientific knowledge is managed, and nobody is to be blamed for that reason. There is no doubt that such fluctuations would stop being ghost if could be worked firmly with them; but this does not talk about the *what* or the of the implied factor, to physical causes, as to the whole form to practice the analysis, to its procedures or methods. What, for many, it is still worse.

§ 74 §

Someday in the future it will seem to us incredible that our ancestors clung to as improbable beliefs as the universal constancy of the gravity and other forces; and until we can be asked by virtue of what means it could not seem improbable already to the same Newton.

Before answering this, it is necessary to recognize that the present attitude of the physics already invites to pose the problem. The hollows generated by the idea of the constancy begin to occupy too much space, although, perhaps in a comical way, is wanted to fill with new material causes of the most improbable materiality: thus we have the dark matter, the dark energy of the "cosmological constant", or another version of that constant in form of multi-compound scalar field to cause the masses of particles. These three problems, leaving of side other many existing ones, are referred to the scalar component additional of the vector magnitudes or forces, leaving at the moment unaffected to these last ones; what has the appearance on a desperate attempt to preserve the invariance more suspended every day. The scalar fields arose in the middle of the XIX century for problems, so alien to the classic physics, of the thermodynamics and the mechanics of fluids, and is also improbable that they can be applied to a greater resolution without forcing by counterpart to a dissolution of the rank of constancy or action. That is to say, the scalar fields, also usually denoted by the letter Φ, the only thing that can obtain is a *local adjustment* of the functions, which already puts in interdict its constancy in the space and the time. If this is applied in addition to the mass as to the gravity as much, the problem of the stability of the solutions as well as the absence of really applicable minimum units of measurement are elements that mutually tend to scuttle. This is equivalent to say as much that we are in a transition of phase for the same form to make physics, as for its contents. And the test is that nobody has the smaller idea of how fitting the components of these fields. The pretended constancy of the forces and the masses to which they are applied is more and more at the mercy of the fluctuations of fields that, like in the hydrodynamics, hardly could be modelled. The situation remembers vividly to the one of a small paper-ship in the sea.

§ 75 §

In addition to all this, and in intimate relation, are questions like the one of the hierarchy of forces and constants that makes possible the "fine adjustment" of values as dissimilar as the gravity, the masses of particles and others so that they make a universe so relatively stable as this in we live ours. But this cosmological perplexity, whether related or not with that "anthropic principle", it is not but an indicator of the extreme conditioning which that initial exposition is subject; aside from which already it is more than doubtful that we could to calculate the improbability of this universe lacking the suitable units for it.

In the same way, one have more than reasonable right to doubt that matter has identical characteristics and fundamental values wide and long of the whole observable universe. This is the same class of assumption that with the constants. Of course, something we know on the limitless capacity of variety in the nature as to believe without remorse in a monotony introduced only for our convenience. Certain that there is monotony in nature, but how different from the uniformed normalization of the analysis.

Diverse variations have been proposed, like for example, the one of the length of the radius of Bohr of atoms, using this one like reference and cause of the relativist deformations, in addition as local clock. But if we deepened more in these and other alternatives, we would always arrive at the consideration of which only the local variations can give reference frames to us, instead of considering the global uniformity (cosmological principle) like guarantor of the constancy and equivalences. Briefly: we cannot decree the universality, that from that moment would be empty, just as the rest of the so called "democratic principles". **There is no another universality that the one that can guarantee a stability of the local conditions.** Thus it has been always, it is, and it will be. In the meantime, it is necessary to remember that does not exist any theory that guarantees the stability of the matter and the atoms beyond the exclusion principle, that the same physicists tend to consider like secondary and somewhat phenomenological, although that depend on the context in which we could to frame it. This context is not clear either for physics.

In such conditions, one does not know why we allow ourselves to speak with as much familiarity of the matter, that, by definition, would have par excellence to be the locally conditional and most contingent thing. This is, would have to be able to be treated like the most faithful indicator of our contingency with respect to any other possible states, whose partial occupation is surely one of the best reasons of the existence of which we called universe.

But, leaving to a side other things, the same atomic theory has sectioned whole parts of the behaviour of the matter that can be observed and measured, and that often and incidentally remerge in the context of the "exotic states of the matter", virtually infinite. One of the greater limitations of the atomic theory is that it only contemplates extremely short times of reaction, escaping in this way many transitions that are equally relegated to the phenomenological and miscellaneous. The stability of the matter can only be determined comparatively by the form of the reaction curves before similar stimuli, and the criterion is the same one that unites and separates *gunas*: in what extent diminution of the own action is defined by the retentiveness or sensitivity gained, and to what extent the increase of an own action generates or is generated by one of the other modalities. In order to define such own actions the logical thing is to begin by sensitivity to the stimuli.

§ 76 §

The discovery of the inequality in the cosmos, its variety, will cause a impact much more deep than which caused in its day the expansion of the spatial and temporary perspective. Watched well, such expansion had more of theoretical movement of generalization that of real inclusion, and therefore, affected more to the imaginary than to the possible action. The change that is being experienced now is the one of the local insertion of the phenomena, just at the moment at which everything seemed forced to uniform itself by the requirements of the global definition.

§ 77 §

At no moment its universal generalization of the gravity could seem to Newton a boldness: in the context of the analysis, the gravitation became general by itself and by necessity. The organic reintegration of the law of inertia within the superior framework of the three laws or principles of the mechanics, without doubt the greater profit of the absolutist physics of the baroque and the physics in general, was enough. One would say that that great speculative moment, so clear in the philosophy of the age, has been perpetuated of unsuspected way in the physics until beginnings of century XXI without getting to become totally aware of itself. In the three laws or principles the notion of absolute time or global synchronization is already implicit, but only outside them it has his place to become explicit; just in a parallel way and without contacts as like the Samkhya defines the situation of the modalities or *gunas* with respect to the absolute awareness or witness,

Purusha. But if after all Newton arrived at the total trust in the exactitude without limit of rank of the law of the gravity, as well as the absolute time, being both as unjustified proposals as Metaphysical, it exclusively had to the intrinsic nature of the analysis or infinitesimal calculation, and to no other thing. The absolutist prestige of the universal gravitation must *to the space of the analysis*, the only form that can give expression to it. This means that beyond this space the universal gravitation does not exist. Some conscience of it could have Newton himself obstinating so absurdly with the subject of the authory of the differential calculus.

The historical fact is that Leibniz was the first in making public a mature concept of the calculation, and was not mere chance. While Newton could handle it before, it is well clear that what was worried to him was the own defined solution of his problem, created in net geometric terms. The ease to generalize the calculus, that the same form of the annotation exposes, demanded sufficient independence with respect to the materiality of the problems, which was not given to Newton. Leibniz perceived in addition in the potentiality of the calculus a whole orb of problems and a perspective that no but he himself could suspect, and that still today wait to be correctly raised. Therefore, the autonomy of the space of the analysis is born with him and no other, indeed because Newton became their maximum applicator, and only in a slope or direction of the problems that this space raised.

The space of the analysis "has invented" the universal gravitation, not vice versa, and this explains the irritation and the mutual misunderstanding in this extraordinary knot or controversy. No distortion will erase neither the truth nor the weight of this fact.

§ 78 §

Already it has been spoken too much, although never with sufficient sharpness, of the perfect antithesis that composes these two counter-figures of the baroque. For example, with respect to the election of the principles, that in Leibniz shell in a plurality that tends to be annulled by its competition, and that in Newton are organized with such an economy that even today our incompetence to penetrate them is revealed. Leibniz wanted to found his whole philosophy on the modality, whereas the principles of Newton exist exactly to evacuate it. The mechanics of Newton has a fully plastic appearance and is so convincing for that reason, although as we have seen, this is exactly appearance. The dynamics of Leibniz or its monadology continues being interpreted indeed being so badly seen like plastic, that is to say, under the glance of Newton, when it is evidently a musical order par excellence, a thought order to cross the appearances. Most of the philosophical subjects of Leibniz continues being interpreted of very pedestrian way by the aforesaid reasons, and is for that reason that the most advisable form to approach them would be indirectly, through the empty space left by the same Newton.

§ 79 §

The pre-established harmony of Leibniz, for example, that seems most of us the *summum* of the gratuity, if not of the arbitrary, acquires relevance in the same and exact measure in which we realize of which the absolute time of Newton, the global synchronization, is not less metaphysical at all. And not only that: if we perceived the net metaphysician character of this last one, we began to understand or to at least suspect that to which Leibniz is alluding with the question of that pre-established harmony is exactly to the possibility of synchronizing entities and processes that already are asynchronous, since it is of themselves who derive their own law. But on the other hand the monads not even exist in the space, but that are location principles, as much with respect to the space as with respect to other many substances or planes. In the same way that only for asynchronous processes can have something to synchronize, exists the non-located for the locality, of a form that in the space only partially can have tracks. It was the same Leibniz who defined the same point like pure modality.

§ 80 §

The monad in Leibniz is already the best figure of which he understood ideally by integration; more than the defined integral, the indefinite sum of terms. Also this distinction lacks any value for the modern context of application of the analysis, but not for the context of the monad, neither in its affection, nor in its modality. In fact, the monad is the space of integration par excellence of all that that now we called non-integrable, no matter how hard the question of the not-linearity could not be considered then. For that reason Leibniz introduces the principle of least action and the optimal ways, in a context and with a reach that little has to do with which in optics could notice Fermat. The characteristic calculus did not aim either towards the rudiments of a logic of propositions, but rather towards the revealing of the non-divisibility through the ways in the individuals. We could extend with other reasons like the one for the discernibility, the necessary fuzzyness of the perception, the fracture of the real sequences in our understanding, the idea of the representative like sufficient index and not like representation space, the mere incompetence of the mathematics to choose between different descriptions of phenomena, the correlative motion and many others; it is difficult not to see the full unit that surrounds to these motifs, that, if remained in fragmentary notes, it was as much by the technical incompetence of the time as by the absence of minimum required sensitivity. The moment of all this was very far, and the mechanics already gave problems enough to work.

§ 81 §

The idea of the constant is a superstition generated by the effectiveness of the infinitesimal calculus, and the later developments have only contributed to prop up it. Of course, also it has been an enormous utility, in addition to indispensable work hypothesis. The restricted frame of the analysis has prevented any serious effort to resist this idea or to handle itself with its variability; on the contrary there have been made all kind of efforts to isolate these values of "disturbing influences", rejecting with it an essential part of which it constitutes them. There are then good reasons to consider these constants of so many decimal numbers of precision as mere statistics that fluctuate with respect to something unrecognised or unknown.

When these values so fantastically precise occur us, we can only ask how many things have had to be left of side; without a doubt this bonds for all these displays of precision in the analysis.

§ 82 §

If physics is an indicator of our commerce with the reality with secondary repercussions in our imagination, the theory of the evolution is as an indicator of our commerce with the imaginary, and even the fantastic , with not at all despicable consequences on the reality. Both are needed as the left hand and the right, but not being able the physics to conceive another thing that limits, and being the evolutionary theory incapable to conceive the limits of anything, the joint impotence of both to define any thing that really concerns to us is manifest.

If something has been misunderstood and even ridiculed within the philosophy of Leibniz, that has been the call principle of the best, than often it is reduced to that on which we lived in the best of the possible worlds. Since we have had neither balance nor weights to resist that affirmation, we cannot either know not even to what is talking about. As it already noticed Ortega to this intention, the famous affirmation does not mean in fact but that our world is the less bad of the possible ones. What concerns then such proposition are the intrinsic limitations of the things, of the set and of each one of them: its own imperfections. The only possible teleology, the only legitimate study of final causes, only can talk about *an empirical disteleology*, that is to say, to the study of the imperfection in the nature, of its inherent limitation, non extrinsic. They are inconceivable the one without the other, and both must find their sufficient expression (that not their sufficient reason) in the form of the motion.

We excuse to say that this has been completely unapproachable for the evolutionary theory; and although this one tries to defend itself of the teleologism and of the context of the final causes, the certain thing is that without this it becomes a pure episode, anecdotal and irrelevant. To whom they really interest last million or thousands of million years if they do not have no relation with which from now on it can happen? It would already for that reason only lose the relation with the present. The theory of the evolution cannot be allowed to be transparent to itself.

§ 83 §

And the same is possible to say of physics, that today is constituted through cosmology with evolutionary arguments, when its ideality and sustantivity are based exclusively in which it can be defined of a non-temporal way. Between both great branches does not exist the smaller real contact, but, simply, what it is necessary to suppose to us.

The physics can not absorb in itself neither the reference nor the meaning. We have seen yet that the three principles of Newton exist to exclude absolutely the modality, to evacuate it, although that is virtually impossible since the necessity is a modality. The plane of ideality of the principles of the mechanics corresponds, in the scope of the sections of *gunas*, with the exclusively indicator plane (*Linga-Matra*), that is privative of the pure sense of I or *Mahat*, to differentiate it from the mutable and empirical ego. This plane is the one that, directed towards the more differentiated sections, behaves like generator of the fiction, whereas its fictitious character only is reduce before and in the direction of the own un-manifested balance of *gunas* or modalities, the only state that it does not indicate nothing nor needs it. The balance only can be reference for the imbalance like its own internal space. Repeating it once again, Newton mechanics has already excluded the balance from its internal scope, and thus they have continued it making all the later developments. On the other hand, it is to note that the metempiric ego to that we alluded here continues being simply a state more of nature, and thus we must see it inasmuch the consciousness without qualities is omitted. This implies as well most intangible but inevitable of the questions: that a state or modification is defined upwards or downwards, towards its qualification or disqualification, can depend as much on the conscience as immediate instance and for that reason unavailable-as of the same mediation of the qualities in its own circularity. Naturally, without some class of immediate knowledge the knowledge of the mediation is impossible. This it is the inescapable context of any possible evolution, including this one of science.

§ 84 §

Let us repeat once again that without a mediate principle of action and reaction our world slides magically in an absolute vacuum or nothing.

The third principle of the mechanics eliminated the friction definitively. Do we realize what this means? The thermodynamics tries to be explained by a reversible quantum mechanics that again fits its balance appealing to the classic mechanics, another reversible system. The mutual incompatibility of the sequence jumps at sight. The friction magically is erased again, being transferred of an unreal scope to another one, in an endless process that cannot never be filled. The one of Newton also was a somersault, that conceived like a landing, with the known consequences that still today good heads being are wrecking their necks. Considering it a landing, little conscience could have of the nature of the jump, that was imposed to him for the preservation of an acquisition, the discovery of a law. We have already seen that this one is insignificant in relation to the rank of the principles, so that it was a bad investment justified by the availability of mathematical assets. What now it has been exhausted is that availability by the expenses that it itself has created, so that it is not left more remedy than to return to the principles disinvesting as much as necessary. The problems with the scalar magnitudes are thermodynamic problems, of imbalance or irreversibility only because this is what it has been evacuated; in other conditions and with other principles the same irreversibility would adopt a completely different aspect, because it would not be dissonant. The tri-modal analysis is also applied to these three great moments of the physics, where unfortunately the physicists are obstinate in seeing only two. But here it is idle to do juggling between branches to match that already have become different by pure exclusion and convenience: the only worth is to give with a undifferentiated object of the suitable rank and to adhere firmly to its possibilities.

§ 85 §

The same it is worth for an evolutionary theory, that as Peirce saw, can be reduced to the three moments of chance, statistical selection and habit or conservation. This last one was translated in XX century like genetic inheritance, a reduction that the time is showing like too insufficient: the gene concept has been evaporated and there are no simple forms to explain how the genetic inheritance is modulated. More than programs, the genes begin to seem what really they are, considerably integrated construction materials.

In the same way that in physics, but in an incomparably greater degree taking advantage of the inherent complexity of life and Biology, there are an infinity of appeals to one or another contiguous instance -chance, selection, inheritance-without respecting the minimum logical order of the sequences, so difficult to arbitrate here, as for the rest. But it is the same, because in fact any mechanist explanation of the specification of the biological forms, when the smaller principle for the same forms does not exist, only can be a nonsense and an absurd, lacking of any sense of the property and the elegance. And indeed, the theory of the natural selection is only an appropriation of which it is not understood by the most inappropriate means. Besides, it is necessary to say that not even the most complete theory of the motion and the form will have never too much to say on the origin of the species, an imaginary space that is not incumbent on to us and that never can be fulfilled. The evolution in real time of the systems and living forms is thousand times more instructive and beneficial for us.

§ 86 §

The Samkhya little must say on the consciousness, since it is the reason for its development as much as it goal; in any case it seems the summit of ingenuity to limit the awareness to a mere subjective condition, when these already is for us pure possibilities whose limits nothing we know. By the same it is desirable to avoid to its respect all objective association. Then it is not necessary to look for where is the consciousness with respect to us, which already it is impertinence, but to see by it where we are. With the little effect that is wanted, this is what it has been said from always.

§ 87 §

As much the impotence of science as their value is show to us through the reaction that provokes the *hollow Earth* fable, that perverse and of course grotesque new edition of the cavern platonic fable. If somebody says to a scientific, instructed spirit, that the Earth in which we live is the convex surface of a bubble within an infinite rock ocean, we can believe to us with reasons to laugh at the extravagance; but if, as we argued descriptions, observations and data, we verified that our interlocutor is able to pleasing invert all those arguments in detail, the scientific, instructed spirit undergoes invariably through the foreseeable sequence of sarcasm, irritation, indignation and exasperation. A fatal moment arrives at which it must argue the details in the motion of the poles, and if the contradictor spirit also is able to invert this argument, it seems that it is not left another instance of appeal that to bore the Earth to sufficient depth, which at the moment is far from being possible; although not even we know here, would say our ill-fated contradictor, if, in case that formidable sounding did not find where to stop, new theories would be invented on a space-temporary expansion of the rock.

We don't need to take this very seriously seeing that the mathematical laws and descriptions are always susceptible of arbitrary inversion; that the experimental data are it much less, but than, if in last instance they come defined by those mathematical descriptions, also will get to be inverted of one or another form. The hollow and convex Earth only reflects the autism of science, its incapacity to talk about suitably of another thing that itself, and that is denied to it to recognize by its lack of own reflexivity, of a space for it. Fortunately almost nobody, nor the same scientists, have hoped that science served the man to as reference. And nevertheless, what drags and what rhetorical.

Until now, science has lacked a self-reflexive space, no matter how hard it forces reflection as much to us if we located ourselves contemplating it from the outside as if we were applied in it: both moments become irrelevant for this because they are mutually self-excluded, which on the other hand will recognize any scientist given to make a little of philosophy in his free moments. If the knowledge of science cannot be valued by the same knowledge and has to be sent to the applications and to the horizons for the action, we always have science without conscience that uselessly appeals to the external ethical considerations. There is more conscience or unconsciousness than the one of the same form of the procedures, and is too vain to appeal to the conscience of the society for the scope of the applications or the mere consume if there has been no conscience when with the greater care they have been instructed and plot the protocols of the investigation.

§ 88 §

Day to day presents to us more urgently the question of what we can do with our knowledge and the why of our dependency of it. More than to be able, this class of knowledge begins to seem to us shipwreck and misery. For that knowledge, the miserable dependency and the lack of self-esteem come to be confused in the same, and thus they let it know to the knower when it exerts like so and not as poor instrument.

The same scientific knowledge develops an addictive circle: once we assumed anyone of its marks, we felt compelled to make all type of questions, when on the other hand we never would have made the questions that already arrive to us like consequences. We must respond to questions that never we had raised, and this drags to us. We, who so superior we can feel before a farmhand that did not know that the Earth is round or turns around the Sun, had never discovered this by we ourselves, nor so at least us we would have asked it. In itself, our knowledge is of a fragility so, that it would have to induce us more to the fear than to the trust, not to mention the security. Any mediocre man can dominate and harvest the lights of all the passed enlightenments, but at the cost of the freedom to make his own questions. The history of science is the madness to accept a bet that cannot be won; at the moment this madness has not been instructive to us.

In addition, the exuberant growth of patterns that we believe to see in nature has resulted in the proportional impoverishment of our sensitivity to perceive them, until the point that to speak of any direct perception it has to sound to us unavoidably like a fable, or to clumsy translation of some underlying mechanism or highly refined pattern. Our imagined balance it would have to teach to us why this does not have to be so, and that we can well have forgotten to balance the weights. Refinement and dullness can go perfectly of the hand.

§ 89 §

In these circumstances, one can consider seriously if, more than an increase of knowledge, on the other hand inevitable, we rather did not need a way that guarantees its correct elimination. In opposite sense, the compression of information, that exists in addition to limited it is self-defeating. But Compression of the knowledge? Perhaps one did not say whenever "knowledge don't need place"?

That knowledge is based in the possibility of forgetting the object of knowledge, in an independence with respect to it. The elimination is the guarantee of the assimilation. The knowledge cannot be compressed, can be condensed, thicken, only leave sediment, by virtue of the process of independence of the object, and this not only like consequence, but from the same beginning. It would be possible also to ask for the retroactive effects.

An inverse relation between formal knowledge and the knowledge of the own form of an object can exist, so that one increases by the diminution of the other, and vice versa. The problem of our balance is located already in the middle and in the edge of that question. What would be to wish is that the problem as a whole simultaneously admits to be expressed with the maximum formal pertinence that leaves in the middle a continuous hollow in the middle and with degrees for its de-formalization, for the pertinence of the proper form, for the most intuitive and direct perception of the object and of we ourselves. Only in such case we could accede to an optimal and own form for the condensation of the knowledge, condensation that as for the rest occurs continuously in fact even hardly with some conscience or direction.

That the synthesis of the formal knowledge is often possible is demonstrated along the centuries; but whenever that has become reality, increasing the abstraction, has not made but increase as well to the possibility of dissipation or dispersion. They do not seem to be then very stable syntheses. If the possibility did not exist of leading back this towards the immediate or intuitive knowledge yet we would be lost; but until now that form of renewal has amounted in interfaces and technological products, with the proliferation of media and their masking, hindering the intrinsic and inexcusable of our own mediator character.

§ 90 §

The balance with series of adjustable and self-adjustable weights and temporary sensitivity are already the expression and the possible simplest symbol of this internal convertibility of mathematical structures and sensitivity. This model cannot be surpassed without entering completely in the absolute absence of form, of which little can be said; so that it is given to us like mediator. This can be known with no need to enter in the scopes of solutions of the problem. The simplest and intuitive form of any real problem, in economy, psychology or physics would have to be expressed and including in it. It is possible to ask if all the real solutions of functions constructed in the complex plane would not be degenerated cases of this type of real function, whose proper density also we have seen that it differs from the one of the analysis.

But this is secondary things. If this model does not serve so much to educate the sensibility as to make calculations, it would be necessary to judge it of doubtful or null utility.

§ 91 §

Here we are only presuming the existence of solutions for this type of functions basing not only on its majority, but also on the immediacy that the given real case with respect to the possible solutions represents. These must admit different sets of weights based on the definition of the time interval. As in the minimum non-analytic interval there are equivalent sets, *the diffusion* or expansion of the interval will define as well a function for the adjustment; and thus we see as the non-linearity entails its own non-locality. This is not only one limitation, also is the proper setting as much of the problem as of the solutions, whatever it is the form that they admit. The touchstone continues being if there is some combination of the general problem of the balance able to describe the complete *form* of the wave of the pulse.

In the context of impulses that we have given, one would say that any process is partially non-located, and that the tendency to be located or to find its own or proper location is what conforms its existence. The location of the fluctuations is always a temporary phenomenon, no matter how manifold are the external factors that affect the system.

§ 92 §

This model of the balance, formally based on the asymmetric principle of measurement discovered by A. Stakhov, and materially concerned by the weights of the modalities, would have to be able of being expressed with such simplicity that a child of ten or twelve years did not have the smaller difficulty to understand the rudiments and perceive the implied principle or principles. And although coming from the world of the formalised objects that it can now seem easier to simulate in a computer, has to be inevitable to catch the very intrinsic of the motion, so that it is possible to be perceived more naturally and outside of control spaces, of which always is to leave. In the direction of that exit it would have to be always education.

§ 93 §

As well as the last nature of *gunas* only can be elucidated assuming the perspective of Samkhya, only from the mathematics we can get to conceive the possibility that the nature and the world necessarily is not written in mathematical language, releasing us from that perspective in which we have confined ourselves from Galileo. In fact we know that it can be indefinite descriptions adapted to a section of the phenomena that can be mutually non- consistence without that affects for anything the march of such phenomena. We have already seen that a fuzzy system of inference can entirely describe a complex process with no need of differential equations. In this respect it is necessary to say that the applied theories of measurement in these cases can be different also and generate contradictions between them. The principle of asymmetric measurement seems to release of the arbitration or election problems of to us.

Perhaps it is possible to advance in the understanding of the independence of the reality with respect to the mathematics turning the formal necessities into necessities of material implication, and although this already implies formal operations, the important thing is the possibility of convergence in front of the constant bifurcation of branches of mathematics. It is more than doubtful that the mathematics ever had a central problem, but rather a fluctuation between reasons. Perhaps it cannot be of another form. The central problem that cannot be posed is its relation with the existence; the relation with the physical and measurable reality is only the end that leads without a predefined continuity to that other greater question. We don't need to say that for the India philosophy the being is independent of any formal superposition. That independence is the centre of convergence for the mathematics, the point from which radiates its necessity or non-necessity. However, this that seems a ultimate perspective has all its material implication in the problem of the balance, of which surely it is possible to demonstrate that it has decisive connections with all the important branches of the mathematics, and, in the most general plane, with the moments or interactions of the arithmetic, geometry and algebra. Therefore, it is possible to propose this problem with all its implications like the subject of the convergence of the mathematics. The questions of consistency raised by the logic or axiomatic remains like exterior and more purely formal to this one, if it is that the inconsistency is only the external aspect of the impropriety. As for the rest, when the logic assumes proper temporal modes enters conditions of material implication. This would also open a breach in the rampant abstraction of the algebraic forms.

In general, and fortunately for the rank of the mathematical task, the convergence of the mathematical forms depends much less on formalization problems that of the problem of the appropriation of the forms that we call understanding.

§ 94 §

Also it would be to hope that the problem of the balance throw light on the question, now entirely fundamental, of a more complete definition of the always slippery concept of potential, with the implications that this has for the physics and the topology. What it is not known is to what extent a sufficient definition would be compatible with the now habitual concepts.

§ 95 §

The computation has become "the natural" scenario to fight with the complexity, but at the cost of increasing the complexity and the autonomous sensitivity of the other scenario, it is not known if virtual or real, but in any case in motion. Without a doubt there are many areas of computer science already sufficiently developed that have not reached a greater integration by lack of development of the suitable object of reference, everything which is intimately united to the privileges granted by a conception of the analysis as narrow as little restrictive. The notion of modelling top-down and from the effects to the causes, as well as the programming directed to objects, tends to invest this tendency, but with too ample and still general approaches, caused more by the convenience of the procedures that by the direct light which they receive. That is to say, they are still blind tunnels, of which often nor one hopes that they see the light. This would have to radically change with the new class of object and process that we are redefining like the objects and processes par excellence of the analysis. Around this definition it orbits a variable constellation of elements that by themselves have never reached a critical mass, but whose set not only must exceed it, but also to generate another very different light. Between these elements we can include of more or less miscellaneous form the fuzzy logic, the algorithmic asymmetric theory of the measurement, the ternary principles of computation, the new kinds of analogical-digital converters, the variable architecture and re-programmable circuits, the relation between computation in parallel and asynchronous circuits without clock-cycle unit, the languages between objects, and the progressive thickening of levels in almost any interface. As much inside as outside are reasons enough to put in motion this orgy of complexity, but without a reference that can be simultaneously physical and formal they can only proliferate outside situation. The funnel of convection for all this storm, the so call "artificial intelligence", cannot be but another proliferation without limit of formalisms looking for to find its own frame. To call it "mechanical intelligence" would be an humorous term; nevertheless, without a mechanics that contemplates the degrees of sensitivity in its interior and knows to take advantage of them congruously, we are ourselves led to the proliferation of intelligent agents in the less intelligent of the perspectives.

But in no way agrees to make a joke of our increasing dependency and symbiosis in this unforeseeable and fuzzy scenario. The dense networks of the information and the tendencies in their configuration acquire of increasing way the function of destiny, Nemesis or necessity for that other social function that it is suppose that have to model them. On the other hand, artificial intelligence, that more than a defined object is the process that would have to fill and to crown this interaction, is purely synonymous of autonomy, the only concept that confers an own right and weight on it. So that we seem to be looking for the autonomy of the whole process of interaction absorbing all reference. To this almost indescribable situation it tends to respond our exposition, knowing full well that by the same form will come given the effects. The attempts to avoid the physical supports derived from this persistent tendency are deceptive and destructive; self-destructive, also.

§ 96 §

Known it is that two superposed glass sheets throw a number of possible trajectories in the reflection that adjusts progressively to the succession of Fibonacci. The most powerful means of calculation whereupon can be dreamed, the quantum computation, are only disguised analogical computation that looks for a digital yield, in the same way that the Feynman path integral is only the principle of diffusion of the light of Huygens -of all the points to all the points by all the points-with a finite speed for the light. Therefore, and ignoring incompleteness of the quantum mechanics and the analytical undefinition of the motion, the quantum devices will depend on a crucial way for their usable yield as well as for the discovery on relevant algorithms on the questions on conversion analogical-digitalis raised by Stakhov with regard to the Golden Section, in addition to the physical ones raised by us with regard to the motion. It is not either necessary to extend on the absurd of the interpretation of the quantum mechanics like the updating of all possible worlds, another form to evacuate the uncompossible to imaginary worlds. In fact all interesting and genuine of the quantum computation falls of the opposite side, in the evaluation of uncompossibles, of the mutual incompatibility in general. It becomes evident the inertia of the present analytical notion of the calculus. This makes also think to what sophistication levels it is necessary to arrive sometimes to return to deal with the most sensible, immediate things.

§ 97 §

To what extent the confusion with respect to science, and to contemporary science in particular arrives, it is illustrated by the topic so frequently repeated that it comes to say that the Newtonian science was "hierarchic and theological", whereas the present one "destroys hierarchies and finds order in the chaos". The opposite is almost the certain thing, although, not to simplify in excess, it is possible to detach: "Newtonian" science is the less hierarchic than it exists, and in that exactly its absolutism is based. We have already seen for what the three principles of the mechanics exist. The modern science to which topical refers, the heterogeneous disciplines of the emergency, the evolution and the complexity, which does indeed is to create hierarchies, levels and meta-levels where more likely there are none, and everything because they are incapable to accede at a level apparently simple constituent level like the Newtonian one; as far as of the order of the chaos, still we are hoping to see it, if it is worth the trouble to speak of it.

But what is certain is that as much one posing as the others run like chickens without head. It is necessary to leave the soft fantasy of which both methods are complemented, since they would do the classicism with the romanticism. No, because a single meeting point does not exist, but lines that are superposed to different levels. Let us see, For what we want so much "fangs and claws" if the third law of Newton not even allows the friction? A mechanist *explanation* of the life, they say. What we are playing? It is simply impossible to maintain the least intellectual respect by this arrangement of things, excused product of the collusion in the specialization. We began to verify the power of the disjunction.

Without a doubt the evolutionism and all the phases of romantic science have benefited enormously from the credulity generated by the amplest emptiness of the analysis, simultaneously that the faith of this one increased in view of the food that the contents or the temporary matter promised to incorporate, without having the least foundation for it. Its already time to resist these hopes, to see what they weigh in the balance. A delicate operation.

§ 98 §

But, it did not leave by chance all modern cosmology in the most involuntary and unexpected form, by the instability that Friedmann discovered in the relativist equations of field? The same Einstein hurried to tie the ends with his famous cosmological constant, but by some mysterious reason, it was decided to find an experimental evidence of dimensions not less cosmological in the equally famous formula of the Hubble expansion constant . Constants or reasons both not less specious than the same *Phi*, with the great difference to fit in a well defined metric space. Pity that we do not know if the suitable one.

From our perspective, that small imbalance shot a unjustified compulsion to fill it; although from the perspective of the physics of then and now, defined fundamentally by the vector component of its space, it had not nothing more justified or necessary. The vector space of the forces is a compulsive space, without the smaller metaphor, since it is a space for the necessity. What it is never known here is the cut is made to obtain that section or slice of necessity, because the space of the analysis is in this respect a real whole. From here the convenience of using the notion of impulses in a temporary interval, no matter how hard they have all luck of doubts on how fitting them. Without a doubt the force notion seems to have the advantage of an immediate meaning and a net defined construction; we can print with our hand a determined force to a ball or another object, but determine the internal time that it requires the arm to produce that force, that already seems to us circumstantial, and even external factor to the force that we want to define. But if which we treat in cosmology is of the own internal time to the cosmos, if we cannot really escape to the Plotinian conception of the cosmos like a great animal because already we are within, the things decidedly change, and our concepts of force, by well defined which they seem to be, can be perfectly like a slap in the emptiness. This space and this time are completely extrinsic and hardly they are immediate for no longer. In addition, not even they are constructed on a space of real numbers, but in reference to an imaginary time in "a complex" space. It does not seem then that such extrapolations deserve the credit that has been granted to them, and of course, which they do not give is the least security. In all this annoys the enormous difficulties to define clearly the potential energy of the gravity by the necessary conversions in kinetic energy; thus, the convenience has been expressed of considering negative the gravitational energy in terms of general balance. But what is certain is that the gravity, according to the circumstance, as much creates as destroys energy, and there is no extrinsic form to stop this problem.

As far as the logic, known it is the little value that can have for the physicists lost between so many timeless equivalences. If general relativity subtended a potential field to the unfolding of the vectors by means of an imaginary time, is possible to think that the introduction of an own, proper time in its real space would probably absorb and eliminate so much the construction of the space-time as much as the not less constructed force-vector notion.

§ 99 §

The space of the forces is a space of compulsion and necessity. But an indefinite number of scopes exists in which the necessity don't compel nor limit the evolution of all type of processes, that nevertheless perhaps makes possible. It would be necessary to excavate in this well of the possibility to see what time really marks the pendulum, the one of the necessity; and although this seems a rhetorical figure, the case is that we have means to do it. In any case, the imaginary world of forces that we have illuminated is an obtuse and brutal world because it lacks in itself the necessary selective or restrictive principles. This is a world in which there is no a specific time neither for the actions nor for the things, which cannot be the more physical and materially false, and aspect of which the only thing of which we can be sure is that we ignored nearly everything. The times and the frequencies of the denominated quantum world cannot be sufficient reference either nor mark by themselves the hour of this clock. Beyond the limits supposed and compounds, if the space of the cosmos seems to us so fatally empty is by the enormous hollow that the physics has in the centre. The forces depend on the time, not the time of the forces; this is something factual, besides to be implicit in the laws of Newton. The equally factual inversion that this demands, corresponds to the new concept of analysis that we are aiming.

We don't need to say that if we put some emphasis in this is as much by the highly destructive character that the uniformation of time as by the compression and increasing location that operates like reaction. That temporary compression, so manifest in the scope of the computation, but tensile to the set of the mediations, is the same one that begins to reveal to us *per absurdum* the temporary friction and interferences between the successive actions and interactions, its diseconomies and uncompossibility. This it is already the inescapable context, only a veil of the necessity.

§ 100 §

Another error of perspective or world view to which often we felt forced by the disparity of the two classes of science is the one to consider the increase of the complexity like indispensable condition of the development, the evolution and even the consciousness. In general it is easy to see that the complexity tends to the restriction of the degrees of freedom of a system, not its increase. On the other hand, are genomes much greater than the one of the man in as simple plants as the irises. No of these two facts of so different rank serves to conclude nothing, but they at least force to think to us in relation to what there is to consider the complexity and the development. We don't need to think much to recognize that it is enough grotesque to consider like reference of the complexity the number of elements and its degrees of freedom, which already assumes that it includes everything, although in fact not even rubs the problem. Think in the degrees of freedom of atoms of a fly and those of the Sun. The restriction of the degrees of freedom is in this sense much more significant. Here there is already a great condensation as much of the necessity as of the contingency —thinking from the context that assumes these sciences—, so that there is no admissible reason that could explain how that double narrowing results in that widening of the possibilities for the living and varying watch of the fly.

In principle it is necessary to recognize that the complexity little has to do with the perfection; on the other hand, the notion of perfection is not contemplated –it does not sense- in the theories of the complexity. But if we admit that the complexity is immediately contingency, and of the mediate necessity we do not know practically anything, well we could admit that the intelligence contained in a brain could be more elegant lodged in a stone or a portion of air. In fact this class of arguments has little of joke, and of course they have material pertinence, because we do not know how much of our brain is absolutely necessary or unnecessary so that our own intelligence exists, far from it we have no criteria to find out it.

Aesthetically speaking, the perfection is considered the opposite to complexity and friend of simplicity ; and the genius, that twisted perfection, is the capacity to extract simplicity of the most complex and spoiled circumstances. In effect, it is the updating of the contingency which redeems the complexity from the scatology.

We here send to the subject before mentioned of the disteleology, the only context in which the perfection idea can acquire content, and to the questions of non-analytical balance or analysis of the balance. The ratio of ratios, the Golden Section, already is contained in the purest and more immediate conceivable form, so that aesthetics in its more fundamental and primary sense cannot be excluded. It happens that, of that pure form, we only extract mediate sections necessarily referred to the first that we are looking for although already is given. Inevitably it comes at the top the perturbation theory and calculus, no matter how hard the used references are not commensurable in principle.

§ 101 §

The universe or the manifest world as much has to be more intelligible when from the smaller number of causes follow the greater number of effects; that is at least the essence of the scientific program from the baroque to nowadays. From the position of the analysis of Samkhya, nevertheless, all the distinctions between measured and immediate things finish being diluted in the arbitrariness with total consequence: because for Samkhya the most located and immediate it is already the object of the analysis. Like in the famous image of the Indian string, it is enough to follow the end to lose oneself in the sky; *rope* or *string* is as for the rest one of the first meanings of the term *guna*.

So that the small artifice which we have mounted and unfolded with our balance only finds their reason of being in our approach to the immediate thing, and only by its immediacy other subjects are related to it. We have passed over the relation between the Continuous Proportion and the modalities or *gunas*, but although nobody never demonstrates its univocal relation will in any case be always and inopportune and the improper thing to disassociate them. The question is not as much the explicit nature of the relation as the mutual implication at the time of conceiving the modifications and giving frame them. One is pleased extraordinarily that this relation remains in general as non fundable or non demonstrable, because otherwise would not arise but a new space for the compulsion and the coercion. If something had of divine the divine proportion, would be exactly this.

§ 102 §

To explain and to define the mediations of the world or the world like mediation is more than impossible, an interminable and vain task, in the sense that it makes us waste the time for more urgent things, in addition to in the sense that the later extensions will demand the destruction of previous marks. If this has not taken place absolutely, is because the extensions have not had sufficient span.

When Hegel undertook the enterprise of a general philosophy of the mediation, he only could justify it putting into play to history, the spirit, and all luck of upheavals; and he had good reasons for it, because everything indicates that even it is not a proper task for the man.

Science from Newton seemed to have drawn for before all that problem by means of the laws of the mechanics, whose immediacy we have considered, and in special for the third principle that had to define the mediation. And if Newton had had to give explanations on these three principles, it was not had less lost in all luck of justifications, not being, indeed, because such laws already justified a fact that emerged now with rank of superior law. From it there was that law of the universal gravitation the one that became creditor of justifications and explanations, when rather it had to pay them, but not to those who assumed its frame already, but to the principles that maintained it.

Being the frame of these principles immediate, but putting itself to the service of the mediation, there is no possible end or aim for the sequence, because if we considered these principles like a triangle, will still admit in his interior almost an infinity of congruent or incongruous triangles according to the case. We return to repeat it: to give account of the mediations of the world does not seem a proper task of man, except for which the man has of mediation yet.

§ 103 §

For that reason it is so insurmountable and surprising for us this starting point of Samkhya to lead back all the mediations to the immediate without previous warning nor necessity of justification, with the most accurate and concise of the criteria. And surely is this position which has prevented to ours of being able to consider Samkhya like a genuine philosophy, when in fact it is the first and most solid of the systems that the philosophy has given, and surely already was explained with total consequence in the times of learning of the young Pythagoras.

§ 104 §

That Samkhya has an exact, scientific and mathematical background, in addition to universal, is something that nobody that knows it will dare to deny; another thing is that we have not guessed right in conceiving its frame more satisfactorily in two thousand five hundred years. This exact and genuinely mathematical character, the Samkhya can allow perfectly to ignore it because also it leaves it back, deriving from it an unquestionable superiority. Nor that to say has that the frame we offer here is the most external possible, in order to permit to the experience the way to fill it up. As far as the use or abuse of the formal manipulations for which field is opened, only the same development can decide where the legitimate thing begins or finishes. The treatment will have to find in itself the adjustment term, with nothing to guarantee the balance.

§ 105 §

On the other hand, the Samkhya as soon it has nothing of technician or scholastic. Its irreducible and primary vision would often remember to us to the ingenuous philosophy of common sense, if not were because the consequences cannot be more far from it. In addition, the conscience of the ambiguous character of the signs cannot arrive at a more acute degree. In a Patanjali aphorism he affirms that the separation of the sound, the meaning and the ideas produce the understanding of the sounds of all the beings; what here one assumes is the confusion and habitual superposition, as well as the feat that marks its real separation. It is easy to see in this a general reason for the same operation of *gunas* and the possibility of its motion. This is exactly just like the old hermetic axiom relative to the three principles: “You separate and purify it, that nature will look after unite”. In this the analysis consists. In such circumstances, the knowledge is filled up by the same nature. This has an infallible reach.

§ 106 §

The ternary principle of separation of powers has known great fortune in the West, and to make sure it is not necessary more to see the joint of national and international, political and economic organisms. Here they have connected fatally the illustration and the esoteric, because it is not indeed for letting to do to the nature that organizations were created. It is the "ousting principle" of Newton for the modes what works here, which makes suppose that also here the continuous reforms with the introduction of new modes could be possible someday.

§ 107 §

Speaking of ousting principles, without a doubt everything what we are considering has remarkable relation with the static physics of the antiquity. To locate a modification is like locating the archimedian point in a scalar field that it does not admit fixed supports. The scalar fields, the last and definitive frontier of physics, simultaneously support and reference for forces of which never it was known when they were sectioned of the reality, are referred as well at the real time, at *the dense moment*, that which was evacuated of the thermodynamics. They only can be limited of explicit form, non-inherent, in the same way that in the self-recurring transformations of our notion of impulse for the balance, which always demand an arbitration.

§ 108 §

Affecting the harmony mathematics of the way in which it does to the theory of the numbers, and to the theory of the measurement, and to the relation between constants, variables and references – conditions- in physical processes, there is no of our great patterns that cannot be modified until the unrecognizable or the suitable. The least we can say is that from the situation of present closing already that capacity is had lost. This affects our concept of clock, to the units of time in the different processes, to the measurement of the information or power units, to the cycles and economic units including the currency, to the organizations and the interfaces between different processes and agents; so it includes practically everything. We can see the Continuous Proportion as the strategy subject par excellence.

§ 109 §

But all this is the instrumental context that would be so desirable to reduce. Samkhya and the Continuous Proportion show the thread to us of something as improbable as an univocal notion of the analogy, understanding by analogy proportion, through the certainty and constancy of the form. In this, that the pedantry of the modern intellectuals has considered impossible, resides nothing less than the possibility of an immediate and universal knowledge, and even simply the possibility of knowledge. Any problem of translation is inseparable of material conditions, that already are formal because they are conformed to a modularity. And when one thinks about a universal interface, also it is to this door that is calling.

§ 110 §

Any inclusion or coupling of this or any philosophy in the course of the necessity can have nothing of necessarily good; good only we can do it, and it is never known why or for whom, which matters so little for the case as it concerns the tree for whom are their fruits. On the other hand, to reinforce something with the speech of necessity only contributes to make it more contingent, whereas to catch it in all its contingency elevates the rank of its necessity. This is so certain for consciousness, that it even supports our vain games of words.

§ 111 §

Any science incurs the rhetoric when talking about to the Nature when it is clear that it only talks about itself. In fact, this appeal to the nature exactly finds its sense when alluding what it continues being left outside our knowledge more or less submissible to arbitration and arbitrary arrangement. In addition, so that the nature serves as substrate or support to the laws that we could to attribute to it, unavoidably has to also presume itself its complete indifference before these same laws, or what it is the same, its execution in the most perfect ignorance. The same form of the objective knowledge is what forces this type of consideration, independently of which is problematic or not. That is to say, within the objectivate, separated knowledge, there is no place for the consciousness in the execution of the laws because we have put all the conscience; but what it still remains independent of those laws it cannot become aware because if thus outside would not be susceptible to be satisfied to new laws that we could find like of forced fulfilment. What this means it is that any idea on the nature within this context is forced to the presumption, and is incapable to conceive nothing, and therefore it is of absolutely hollow and rhetorical character. Or more indeed, not even there is hollow or space where we could fit nature, safe, like in chance, by our pure ignorance; after all, a positive means in which we act.

But if someday we wanted to consider beyond the laws and the rhetoric to what it responds the expression *Nature*, we would see that there is no another answer possible that "to unite and to separate", at all on the conceivable levels and scales. There is another business nor no permanent activity in which it can be involved; and if these two or three words sound to us too generic, anyone will be able to verify that all other specification that we look for will already make in damage of the generality and property what we could to understand with the word *Nature*, and therefore, will not find more legitimate neither less hollow verbs that these "to unite and to separate". Less hollow, because the reach of that union and that separation will always go beyond the order that we are considering, being to this to which the well- known saying talks about of which "nature never remains idle". These terms will be always true beyond the reach of our knowledge or ignorance; one can be sure of it, as it will be never of the true "nature" of a force or a law identified by our own intellect. Therefore, we repeat emphatically that another definition or less rhetorical concept for the Nature that this one of "to unite and separate" does not exist, and only by this we can use this word with some justice and without using it like figurehead of the divinity.

Well, this "to unite and to separate", these three things, it is exactly what express the Continuous Proportion or Golden Section in the simplest mathematical way, reason why this explanation of which we understand by Nature – the *Prakriti* of the Samkhya-still is more opportune and still less rhetoric. Because we can be asked what is what to the theoretical physics or the theory of the evolution like scopes of "union and separation" contemplates, and to see to what extent they are in this respect circumstantial and sightless.

In the physics, for example, it is very easy to see the gravity and the radiation like epitomes of the union and separation, but what we are saying it is indeed that such union and separation are all-embracing, at least with respect to so net sections of the reality. That is to say, the phenomena of separation in diverse scales with respect to the gravity, and those of union with respect to the radiation are presumed. But this already is implicit of multiple ways in the same physics, so we are not saying anything new. Or maybe yes? The same mechanism of absorption and emission generalizable to all particles in the fields would already seem the least expression of this union and separation: but nothing further from it, in fact. Because the physics is already constructed by means of equivalences and identities, total and firmly seated in the equality principle, and this way any "union" or "separation" completely stops being pertinent, its pure opportunity is lost. Very surely the matter of this Solar System, an aggregate in partial balance, *is* separating at this same moment of the matter of other places of the galaxy in which there govern different conditions; we could think about conditions of polarization, phenomenon only partially explained since the quantum mechanics ignores questions as the form of the phase or the reason of the definition of the cycle, that only can have relation with a recurrence. Also very surely, the absorption of the light from the most distant corners of the space also has selective criteria, then of separation also, probably related to the previous thing. But all this is already lost conjectures by the same fact to assign them to a frame, the one of physics, modelled entirely in the equality idea; so that it only can offer deformed hypothesis to us. Often it has been believed to eliminate the question of the "internal variables" -in which we do not believe-in the quantum mechanics adducing that such variables, whereas internal, cannot be dependent of the variation of conditions in the exterior. But if we considered to what we have ourselves been referring in the phenomenon of the form in the motion like interface or limit of the interior and the exterior, and whereas defined, eliminating both by unnecessary, we see that these objections are opportune for the quantum mechanics but coarse for the case of the form of the motion in general. The quantum mechanics is therefore consistent with itself but necessarily incomplete with respect to the reality.

Thus, we see that terms such as "to unite and to separate" only have pertinence with respect to the inequality idea: with respect to the equality it is not possible or it is indifferent to separate or to unite anything. And just like we have said of the physics, it can be said of the evolutionary theory, that well would like to define itself –even when it is incapable- like an explanation of the separation of the species or chladogenesis, being able to unite its tree backwards in time, which relegates it to the anecdotal space of "thus was or thus it could be". Because, following with our general presumption, we would finish seeing that in the always virtual separation of the evolutionary branches already are producing unions and contractions between previous spaces of the tree, so that a drag in parallel of the species exists, of the very same form that are taking place in a real tree: from the trunk to each one of the branches we can insert a spiral whose ring is contracted with the force that gives a real module to us, the one of the Golden Section, that appears as well in the spirals of nuclei acids as supra-code -although it is necessary to make notice that there are other planes for the life that the one of the genetics. Without a doubt there are immense spaces of probability to study that would throw numerous surprises, but already would be enough to understand what here implies the simplest and general case.

Let us repeat again that only from the inequality it has some sense to unite and to separate, and therefore, the perpetual activity of the nature. But there is more: only from the point of view of the inequality our own identity, or any other has some sense. How it could be otherwise? It operates already in our thought as form of the motion that it is, and in all its sections. In this respect it is necessary to backslide on the subject that only this inequality equips with content the recurrence. The phenomenon of the frequency expresses a truly circular and trivial fact, that is to say, only that what can disappear finds in conditions for appearing again. But in fact it is far from easy to know if this, to which we called recurrence, is something trivial, because can deceive permanently our expectations of which we considered appearances. The inequality smooth the same sequence of the appearances and disappearances, the same time like principle of succession.

Indeed, the exactly equal waves that fall and elevate in the mind during the concentration, and that we had taken in the beginning like counterpart of mechanical inertia, are referred to the same fact of the recurrence, and not to "the precise" location of the concentration of the attention at a moment without duration of the time, which could not be more unreal. To focus attention on a point means the appearance-and-disappearance of a notion here throughout that interval, appearance-and-disappearance of which it is not possible to say if it is sequential or simultaneous since it is the attention the one that maintains and makes the recurrence possible. Most of us not even we got to realize this, but it is sufficient to show to what extent any problem of continuity or recurrence -not to say all in general- are already present of supreme form in the attention, before in any conceivable content. We cannot know then to what number of levels can exist the recurrence, being decisive that the conscience fills up of the most immediate form the presence or the absence of the object, and revealing itself this way as a different thing from both.

If somebody wanted to respond to that question from Leibniz about why exists in general something rather than nothing, he could see, in terms of the author of that question, that this one is equivalent to ask for the why of the inequality, that is its answer, and thus we can penetrate by not necessarily ontological courses, unless it is had by that the own Leibniz philosophy of or the Indian concept of the being. Why there is something to unite and to separate in the nature? Because there is inequality, and inequalities, and we do not know if both are not the same. Neither the Consciousness have necessarily why to know it, because the consciousness is already the answer.

§ 112 §

It could happen the same with the denominated fundamental forces of the physics. There is no a direct form to know it, since they are modelled on the exclusion of a mean or medium between the action and reaction.

The balance is the sensitivity that exists between the action and reaction; we will never find one more simpler and economic definition. Or vice versa, sensitivity is the tendency to the balance between action and reaction that already as motion is by definition unbalanced.

The constants of coupling or intensity of the so called fundamental forces would be equivalent to that thick signature or form in which the means of balance or sensitivity between opposed actions appear; but as its definition comes already given like force, without specific dependency of the time, does not exist an explicit way of gauge, safe by the mere proportions and their enigmatic reference to the fundamental level, that it does not have differential character unless is introduced as an opportune supplement. Time itself in its synchronous version is only another aspect of this supplement. But thus it is as more veiled is left its nature.

The degradation of a system always is originated by a loss of its optimal sensitivity, that implies readjustments in the inertial and active components. And the refinement and evolution, that does not demand necessarily the increase of complexity, always go towards a recovery of that optimal of sensitivity that, naturally, also will be varying with time. But the adjustment of

In many biological and complex systems it is noticed how the maximum of stability agrees with the maximum of sensitivity to the perturbations. this optimal is not possible without a consideration of the balance like which we are exposing. Sensitivity is the point to begin with; if we avoided sensitivity, variable by nature, already the reference is lost.

Let us think for example about the phenomenon of the ageing, so immediately associated to losses of energy and corporal mass. However, it is the loss of sensitivity what makes the process irreversible in diverse degrees: we age because we lose sensitivity, and the rest are chain reactions. The action already is also reaction with respect to sensitivity. Clear that this loss is referred to a variable optimal, since in fact with the deterioration and the imbalance a greater sensitivity or hypersensitivity to multiple stimuli is made present. In this game to outline the essence of the not-linearity, of that only apparent absence of proportionality, is synthesized.

§ 113 §

Probably everything would be mechanic if a univocally mechanical definition for a single and simple process existed; but such a thing does not exist. Apart from count on two completely different mechanics —classic and the quantum one—, it does more than one hundred years that Poincaré demonstrated that even considering the classic systems, whenever the principle of least action is satisfied, not many, but infinite possible mechanical explanations exist. This is normal if we considered the exclusion of the medium or mean by the third principle of the mechanics. And in spite of everything, it is continued imposing the persistent illusion, even for the more experimented physicist. The problem is always led back towards the last horizon of the fundamental physics, when also it can be seen in the first plane of the landscape.

That landscape of the pictures of prediction —not only in physics, but in economy, ecology, and any discipline that aspires to have mathematical models— is an amalgam of differential equations and statistic, of absolute determinism without place at least for the causality and profiles for immense clouds of ignorance. Of a very comedian way, is to this statistical part, the chance, to which the causality is attributed quite often, in a double exercise of darkening, since the probability states have been selected to a great extent by differential criteria. .We found here again another superposition and another mirage; another impossibility.

With so much emptiness there is space enough for methods that are neither statistical nor differential: for that which allow one more realistic description of the imbalances. And since neither the differential analysis nor the probability allow a causal explanation of the phenomena, although they easily generate that illusion, fits the supposition which any causality or generativity that could emerge from the analysis of the balance only with miss-understanding could seem "mechanical", at least in the sense that now is granted to this word. Therefore, any cause that is worth the trouble of being looked for will be outside mechanics, which by force it has to reveal itself in the mathematical solutions of these systems, which they do not respond to momentary situations, but produced sequentially by successive adjustments. What it is generated in these sequences is the density of the real time; not the synchronous or the ideal time. Or at least, it runs so in parallel that is confused at first sight with it. And we have to arrive at that point in where we have to conclude that, although the variations are infinite, there are always the same principles. .

Even though the tri-modal approach of the equilibrium can not aspire to the deterministic solutions of the differential analysis, what is lost in this aspect is compensated enough with what is won in natural generativity, in the sequence, and therefore, in consequence. Now we are not able to value what this means.

§ 114 §

One could think that the scheme of the *gunas* of the Samkhya is similar to the one of dialectics, to the scheme thesis-antithesis-synthesis; but the dialectic is idealistic, even in its popular and naturalistic versions very previous to the idealism, indeed by its appearance of production of the reality. Also the modern analysis, even being applied faithfully to experimental data, is of idealistic cut from the moment that believes to produce the motion from closed functions with changes in the values or coefficients. Nothing of this has to do with the Samkhya. Although we even found characteristic mathematical structures for the fluctuations with respect to the balance, we would only have a selector of readings, a tuner that allows us to listen better what there is present. A complete and closed description as the one of the analytical functions it would be beforehand irrelevant, would make that listening impossible.

In fact, if there is something primitive that corresponds with the nature of the *gunas*, of sensitivity, the action and reaction, is the same articulation of the language, with his three grammatical persons and their equivalent subject- verb-predicate flow. Not in vain it is adjudge to the name of Patanjali a treatise of grammar, in addition to another one of medicine. The pulse of the cat truly would say "I-eat-sardines" if it did not have much more varied and subtle things to tell us. But it continues being a genuine language because it is articulated, and has to have by force a discreet element to detect, reason for which we have been alluding very insistently to the mathematics of the Golden Mean or Continuous Proportion. Only in a frame like this the non- decidable questions of the last century on semantics and syntax could have value and specific weight.

Therefore, to speak of the language and the languages of the nature could not be a metaphor —rather the human language must be a poor metaphor of that other language.

Only the overvaluation of the so called classic mechanics and the differential calculus relegated this so natural possibility to the scope —again, purely apparent- of the nonsense. Disappearing the verb and the time of our perception of the nature, also the entity of the other two instances is diluted. Something more than a dimension of the "problem" vanishes therefore, something more than the "depth". Pope tried that *Nature and Nature laws lay hid in the night/ God said: let Newton be! and all was light.* We don´t know what happened to light; but now we can verify that it was at that time that the West lost any ear for the Nature. And it cannot be strange to us that the efforts that now we make to recover it must break through the most abstract dominions, until there where indeed it was lost.

§ 115 §

It have been said thousand times that Galileo moved away to us of the centre of the universe, that the theory of the evolution cleared to us of the pinnacle of the creation, and that the psychoanalysis denied our dominion on our own mind. But, being already lame the first of these three concentric revolutions, the other two, that tried to model their selves to the image and similarity of the mechanics, have been more events of the public opinion and the imaginary that true advances of the knowledge. Neither the mind, nor the life, nor mechanics have been understand or explained. The mechanics continues closing to us the passage to any superior understanding, and thus it is understood the innumerable amount of nonsense and incredible artifices that are postulated like candidates to explain the consciousness, in which it is tried the fourth and definitive "revolution". Nothing of that has the smaller plausibility, and the only good thing of so many desperate attempts is to convince to us that something extremely basic has been avoided from the beginning and the principle itself.

Many are the ones that look for the magical concept today that would give potentiality to this last turn: ideas around the computation, the information, the deterministic chaos, the quantum theory and things of the sort. Ideas of last hour to avoid questions about what is generally considered as consolidated. This is quite normal if one thinks that the consciousness is the last and the most conditional thing. Our perception, on the contrary, is that there is very little or nothing to say on this one, and that are the other layers those that would benefit from a new principle. Since nothing really has been explained by them. Even the predictions of the physics, so successful, on objects that are not understood, are based on last instance in the correspondence of the proportion, and not of the causality. From here the "inexplicable effectiveness of the mathematics". This can happen perfectly without doing nothing more than to cut layers or flat films without any other thickness that the possible and imagined relations with other objects, that are also defined in two-dimensional layers.

Only the three-fold concept of fluctuation with respect to the balance can break through from the most fundamental level of physics to the most immediate subjective experiences. From the measurable movement to the qualities, and from a whole to its details. In the landscape of present knowledge there is not and there will not be another possible bridge, and thus it will be while signs of the so called modern science, that historical contingency. The known attempts not even can scratch the reality —it is like make photographs.

§ 116 §

When the Samkhya affirms that there is not another ultimate cause of the variety of the things that *gunas*, which must be understood is that truly we cannot understand and assimilate outside the context of the equilibrium or balance. That same process around the balance acts as much in the cognition as in the objects. Therefore, any other understanding, including the one of the habitual analysis, are not but forms more or less far from the centre of the same. Naturally, to no method in itself could correspond the centre; but the differences of direction with respect to it are decisive. It is possible to be confronted or be from behind.

In fact it is the balance the one that assimilate the imbalances, and the one that also makes possible a route beyond the understanding. We not even can assimilate that this earth is an enormous globe with a certain diameter, mass or impulse; in no way we can incorporate or embody those knowledge in ourselves only from the same knowledge.

Nevertheless let us put a simple and perfectly checkable example of *hatha yoga*, the posture of remaining in balance on the head, with this one putting inside a triangle drawn by the forearms. The novice has therefore the opportunity to retake the hard learning of the balance that so much work gave to him being a child and who has arrived to forget almost completely indeed because it has managed. In the beginning, he will be basically at the mercy of inertia; with some practice, it will learn to make the force necessary to resist the oscillations. This will be excessive almost always and bad applied, until gradually it is developed sufficient sensitivity and the unnecessary efforts are saved to the maximum. After a variable period of practice, a more or less perfect union of inertia, force and sensitivity is reached; then not even a special alert on the part of this last one is precise, because it is completely integrated with the other two tendencies. A spontaneous state of suspension takes place *finally*, that nevertheless is the fruit of all the previous efforts and deliberations. Even if we took the work of measure in detail the oscillations that within a circle make the body of the novice and the one of experimented, and for identical deviations of the centre in any direction, would verify the enormous difference throughout the vertical axis with respect to the rigidity of first one and the perfect and synchronous undulation throughout the body of the second for the most effective and coherent accomplishment of the least action. This is an excellent illustration of that lost dimension of the movement to which we alluded; as well as of which we understand like assimilate or incorporate knowledge.

To this general process of concealment in the invisible of the qualities in its total presence is at which it points the experience of the Being in the India philosophy. Such experience allows the participation in the extra-mental dominion with no need of metaphysics.

§ 117 §

It is to note that in cases like this the equilibrium is not and can not be an ideal –infinitesimal- state, but it will be possible always to find it in the empirical and real world for the relevant level. Of course, the most subtle balance may be find at rest, whether the needle is centred or sloped, in every cases that the perturbations don't surpass the own sensibility threshold of the balance. Therefore, this sensibility gauge give us a possibility of realistic definition of the rest and the intrinsic, proper equilibrium of the system, too. Any perturbation will make pass the needle two times by the same point, giving the primary form of the cycle and the recurrence. Possibly, this can be applied to a tri-modal formulation of the quantum of action.

§ 118 §

I hope that the presumption which I have shown with respect to a model or hypothesis that has not been checked at least shall be pardoned; not to mention the integrating pretensions with respect to fields that one so widely ignores. All this is very certain, but it is not but fairly legitimate way to call the attention on a phenomenon, the one of the form in the motion, so absolutely real and fundamental that we cannot ignore it without loses the sense of all the others; and if those additional fields did not exist at all, the subject would not lose nor in minimum its own substance and its autonomy, that now has emerged circumstantially by its contrast with the analysis.

§ 119 §

For the Samkhya, as well as for the India philosophy in general, as much the actual infinite as the truly defined and finite escape to the scope of the manifestation. Thus, of the ephemeral world as in motion not even it is possible to be postulated its finitude, nor therefore its complete definition: to realize it would be the only privilege that in this respect we have left. We have tried to speak almost always of indefinite series and indefinite amounts, more than infinite ones; the harmony mathematics and their algorithmic measurement theory seems to be agreed with this yet, and avoids therefore a good part of the dilemmas of the mathematical world.

How it could be perfectly defined the motion? The trimodal or threefold analysis that we have indicated is a way to evade the external references to the own motion, but of this form also tends irretrievably to vanish the external metrics of the space, all the imaginary framework of the representation. Here one is to take care of the sequence of a process forgetting every other thing, perhaps the only legitimate form of abstraction and the most difficult. Sinning the representation always by excess more than by defect, this would not have to be lamented, and until it can be considered like the only real gain. A completely different space for the convergence is generated. And nevertheless, it does not exist nor it can exist the complete mutual incompatibility, since from our point of view we even see how the modes cannot stop to operate in the most net of the mechanics, even leaving apart the inevitable contact with the reality.

§ 120 §

The logic of the Samkhya leads to us with an irresistible force in the same proportion that we know to resist to the application of the force. So it seems to be the eternal criterion, imposed and to our disposition.

§ 121 §

The search and appeal to the unconditioned exist in all the spheres, not only for the religion. We have all the right and a part of the power to appeal to the unconditioned in the deepest of us; we do not have any right and a part of the power to impose something to others unconditionally. Both plates are always balanced.

§ 122 §

The subject of the Samkhya or the Yoga is to extend the inner space trying not to establish the least differential with the outer space, that is to be respected necessarily. Thus, the difference between the interior and the outside, and the field of reference for the motion stops by the motion itself. Yoga is to penetrate in the limitless using the limits that are given to us, without trying to transgress them or annulling them. The Samkhya is the contemplation of the balance and the Yoga its practice, not existing true difference between both, and being the contemplation the gradual decantation of the purity in the participation. In all the theoretical contemplation that we have made we have tried to guide us by this same practice.

§ 123 §

The subject of the form in the motion, its adjustment, is simply the subject of our insertion in the reality.

§ 124 §

The same modalities, or *gunas*, also are only objects of knowledge in a purely conditional and impermanent way: that is already its form of evolution from the beginning and even without beginnings. Therefore, one can gives the back to them completely without leaving undertaking in any moment . So that, when we said that Patanjali had written its *Yoga-sutras* with the intention of making *gunas* directly contemplable, this only could occur by the way including the salt in the assertion. Certainly, no sort of knowledge forces to contemplate them to us, far from it the one that Samkhya proposes as its crown. This would have to make think about the indescribable state of freedom in which the pure awareness dwells, being the nature of *gunas* simply the nature of the existence of the individual, and being the consciousness the pure universal existence.

It is not necessary to say this doesn't need any attendance from metaphysics.

§ 125 §

The force notion refers to the one of acceleration, that is the change of speed or change within the change; the speed is applied to bodies with mass, being this one an effective quantification of the inertia, that also can be equivalent to the rest. So that all the dynamics or science of the forces is refer to a scalar magnitude *for been able to be applied*; or otherwise the forces would exist immediately in the vacuum. However, as this scalar "component" can not be well modelled by the exigencies of the differential calculus, yes, it results that these forces act in the emptiness of a reality which they do not draw. This is, not being explained the phenomenon of the mass, inertia or the rest, its density with respect to the reality is null. Or said of another form, the thickness of his section or it cuts in the reality is null, unless the imprecision rank is considered of a positive form. That it is, perhaps, which would have to become. So that the only reality of the physics would consist of which it cannot define, not in which it can. This does not have anything of impossible or stranger , but simply, nobody seems to think that it is of some utility to have it in mind.

The same scientific knowledge is of essentially mutative or *rajasic* modality and condition, and it is therefore natural that their support and reference is inertia, the modality that at least partially remains subordinated to it. There is in all this no other mystery nor other evidence. That the inertiality can become equivalent of sensitivity, like we have proposed, is something that from its own nature cannot not even be contemplated; essentially it is forbidden to it, in addition of which it would seem to put in danger its own conception. But it would have to be clear that supporting in inertia the forces they are not less in the air than supporting to us in a self-recurring notion of the impulse, that the only thing that does is to define the density and thickness in the section of chosen reality. So that physics cannot avoid to be as much in the air or as conditional as our balance, and the only alarming would seem to be the absence of conscience of it. The remissions of the acceleration to the speed and this one to the rest are independent of the time; here it is all the fault with respect to the reality, and from here all the vain efforts to fill something that cannot be heaped.

It does not seem very consequent then to assimilate inertia with nullity, the one of the rest, for example, since if all the physics is derived from this, we do not know why at some moment would have to stop being nothing; to assimilate it with the principle of minimum action, being quantified it in reference to the value of *Phi* like Grejzdelsky does, seems to be justified and to suppose something more than a semantic question. Being so far from being a simple nothing, inertia only can be conceived like least action, also in harmony with the notion of modification or impulse like change of the moment. The not-nullity of inertia is the revocation or ousting of the reversible time of mechanics.

What confers stability then to things is not inertia, but the least action, independently of which we consider it like quantum, law, or principle, since it is itself the one that makes the translations between possible levels of the reality, like before made inertia with its own incapacity. That the least action principle is the only guarantee of the theoretical stability of the things do not speaks us as much of the fragility of the reality like of the fragility of the theory itself: therefore, it is for our conscience a magnificent index.

This would suppose a definitive step to superpose us to the ideality of our definitions, perhaps so that someday we could pose legitimately what has to do inertia with the imitation, the habit, the perception and the variation of a stimulus; and which is the sort of possible continuity between physical, biological, and psychological notions —although the true question is that other criteria of separation that the imaginary ones never existed. We now have a suitable translator principle. This principle seems to trascend the notion of territory in anyone of the specialities, being this perhaps its happier utility.

However, in all this we follow without realize that the application of the asymmetric theory of the measurement to the motion subtlety changes but radically the whole definition of the reality. **Any order of precision that is not self-refer to its own terms already is imaginary,** because as for the rest neither its imprecision, **for want of a proper form**, it does not talk about anything but a mere deficiency. Will be finally this one the principle of adjustment or truth of all the theories at the moment handled? It is necessary to consider all this very seriously, since it affects something more than to the criterion of measurement or the metric. Any more selective form of space or representation will have to emerge from which is raised here. The more perceptive aspect of the reality is also the most constructive, and therefore, this fundamentally affects to the harmony and coherence between the different specialities and disciplines, final destiny of all the diseconomies. Let us repeat it: the precision criteria that we usually handle are fundamentally imaginary, because they lack explicit degrees to define its inclusion in the reality. These explicit degrees are not necessary, since they can be contemplated of inherent form; but for the present state of things, in which this tends to be made less and less, they become a pure function of adjustment.

The self-reference is the reality criterion, the translation principle, the only truth that lies in the idea of self-organization. This is something natural and inexcusable, not a vitiated application of an abstract logic. If it has seemed a vitiated resource speech more than nothing of how vitiated are resources incapable to make a exact use of this fact. The test of this fact is the degree of independence or economy with respect to the formalisms, and we are those that supports the test.

The logic of the self-reference or inequality favours the thickening of things; the one of equivalence, favours the dissolution. The contact or confrontation of both is inevitable.

§ 126 §

Consciousness does not have memory.

§ 127 §

It is to sorry that the deepest theoreticians of the physics and other disciplines that try to seize in the reality often say that perhaps we will never understand the world because we are not the intelligent enough. First of all, by the absurd arrogance that such affirmations imply, assuming that the world has to be comprehensible, and in addition by intelligence, and a very concrete and formal intelligence in particular. By these words it rather speaks the disorientation and the ignorance, in individual and general, in addition to those pretensions so little promising for the knowledge; and also speak the modes by the people. It would be to wish that intelligence served to illuminate our ignorance more than to illuminate our fragmentary knowledge; in this respect our hypothesis or symbol does not have any integrating pretension, although they only seem to draw it the possible effects.

And what would happen if we did not understand the world simply because we are not worthy of it?

To appeal to intelligence already is a form to cry.

§ 128 §

The symbolic model of the balance or the Samkhya already incorporates the external means in the processes, liberating the problem of the reference. To define from the outside the questions of evolution and balance, or the relevant ranks of precision, are the improper thing. Now it is to see until where it takes to us to try a proper definition of the motion and the modifications adhering to the proper form, locating itself in the middle of the conditions. In any case, the distinctions between the mechanic or non-mechanic, the determinism and the non-determinism, become arbitrary and irrelevant, because they admit indefinite degrees that never are incumbent on to that question, that already is raised from outside. The intellect is one more of those degrees, simultaneously that its refinement, which indeed leaves back the complexity of the states, if it is certain that the essence of the distillation is the essence.

As far as the consciousness, it is still more certain that it is not the object of any discipline. If some grant possibility to that delirium, is by supposing it like the most highly conditional, while the un-conditionality of the physical laws is assumed. But the situation is exactly the opposite: any physical data, including the measurement of the rest, already involves a modification, an asymmetry *of* the space and the time.

The Vedanta could not express it better: If the consciousness can not explain itself, what thing could explain it?

§ 129 §

The real is the existence. The existence is the consciousness. The consciousness is the real thing. That is the triple affirmation in which it is summarized the Vedanta and the Samkhya, a triple identity as existential as absolute. Empty? May be: consciousness is completely indifferent to the fullness or the vacuity. Consciousness, already we said it, does not have memory, not at all paradoxical affirmation that anyone can verify by means of the portion that corresponds to him. An affirmation in which lie very considerable possibilities.

If the consciousness does not have memory, in that exact measure the consciousness is will; if the consciousness is will, in that exact measure the consciousness does not need to know. That exact measurement that we are speaking about is the conation, that it does not have to be confused necessarily with the will; then if there is conation, the consciousness is only intelection, and if there is none, the consciousness remains undivided in these terms and indifferent to them. This is an example of something which not even needs to be thought; then the terms fall here by themselves, without thinking about how we could take part in them. And exactly in the same way but in very different measure it happens to us with the equations. We are not invited to think about it, but to verify it. The consciousness does not have memory. With as much disorientation, we can not think in a better indicator of the "location" of the consciousness, for those who want to look for it.

For those who please to quantify, an unobjectionable definition of the consciousness is that it folds the three modalities in anything, corresponding this way to the triple previous affirmation. This triple identity does not have sequence, and therefore it is unattainable for the logic or the mathematics, that sometimes believe to be in the ultimate degree of the simplicity, the un- conditionality or even the elegance. In fact, the mathematical forms are only appeal to other forms, and would be enough to have the sufficient attention, with no need of a mathematical intellect, to realize that they are not more near from the eternity than the first or the last of the words pronounced by the man. The opposite is a presumption, motivated nevertheless by the implicit movement in the aspiration, more than by the execution or the assumption of conscience; therefore, and like all movement of the language, it has a excusable necessity.

So that the triple identity of the Vedanta offers us like a space for the contemplation to our whole disposition, since in no case it imposes nothing to us. It is the maximum and the minimum that can be said. But it is also a proclamation and the testimony of most legitimate of the conquests, because it have been made in the pure conditionality, of which all the forms depend.

In any case, to mention the consciousness is not absolutely inopportune indeed because of how acute begins to be our conscience of the insufficiency of all the formal systems, and by ours very justified distrust on the fascination of all the theoretical abysses. We can assume this in opposite direction of the rhetoric and the brutality, who already are made present in the same intellect.

Being the intellect adhesion before any other thing, if it does not realize that only lets take, he is already lost, and only through the experience it will be able to recover of that loss, whenever it is made, and non only with the intellect, the sufficient effort. This is fundamental subject for the Samkhya, because the modalities exist indifferently like liberation and experience source, without excluding both terms safe by the own modes. The Samkhya affirms that the liberation is the identity of the intellect and the consciousness: but this identity does not belong to the intellect, because all this, at least for it, is conditioned by defined degrees or modes.

Properly speaking, the consciousness does not belong to us, we only can hope to correspond to us with it. And here takes root any possible freedom far from all rhetoric. That is to say, there is no another one. The rest is only compulsion, and the lucidity of the intellect, always lent, out of the consciousness already is superimposition and blindness whatever will be the level in which it appears. Therefore the consciousness is neither a rhetorical subject nor a problem of location, but the only reference and the only means of evolution in the real time. Real time is consciousness without memory.

APPENDIX-A PROBLEM

We would like to shortly expose the problem again, in a non technical form, in the way to a better future approach:

The wave of the physiological pulse has a cushioned profile that is essential for its study and characterization.

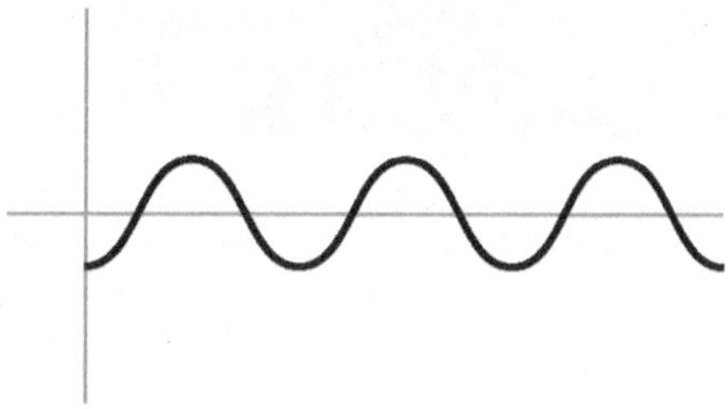

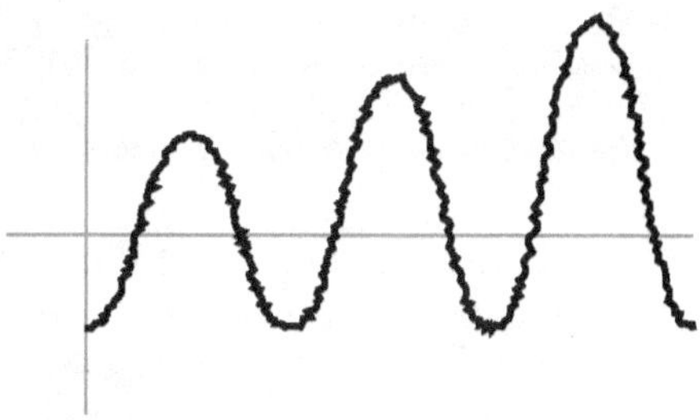

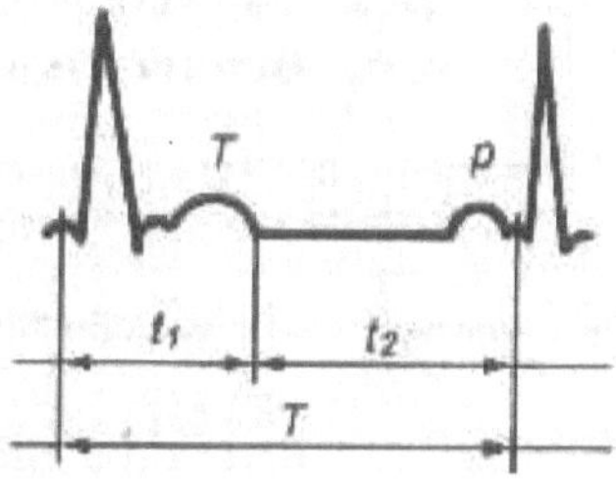

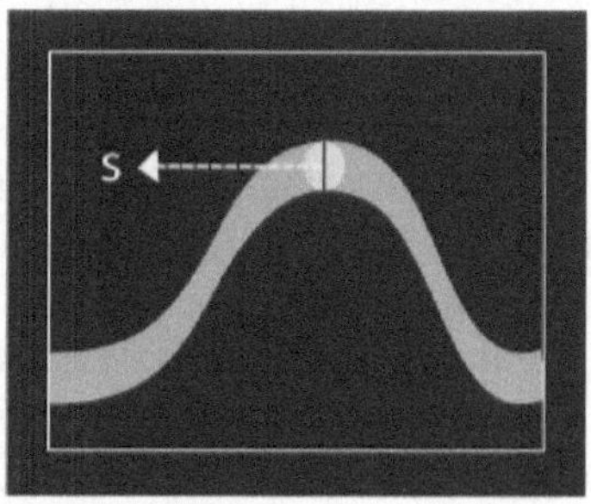

Now let us consider that damping or amplitude of the signal like a natural index of a hypothetical inequality or mean between the action and reaction, or the kinetic and the potential energy of the system. **The time** between the action and the reaction must be different from zero. This hypothesis is specially permissible for a complex system of biological type, in which the reaction to the stimuli hardly can be considered immediate. We can describe this behaviour in its simpler form by the oscillation of a balance with inertiality or **sensitivity**, that is to say, with a time of reaction for the operations of adding or removing weights.

Let us suppose that the amplitude of the signal is an indicator of that time of reaction or sensitivity (*S*) of the system.

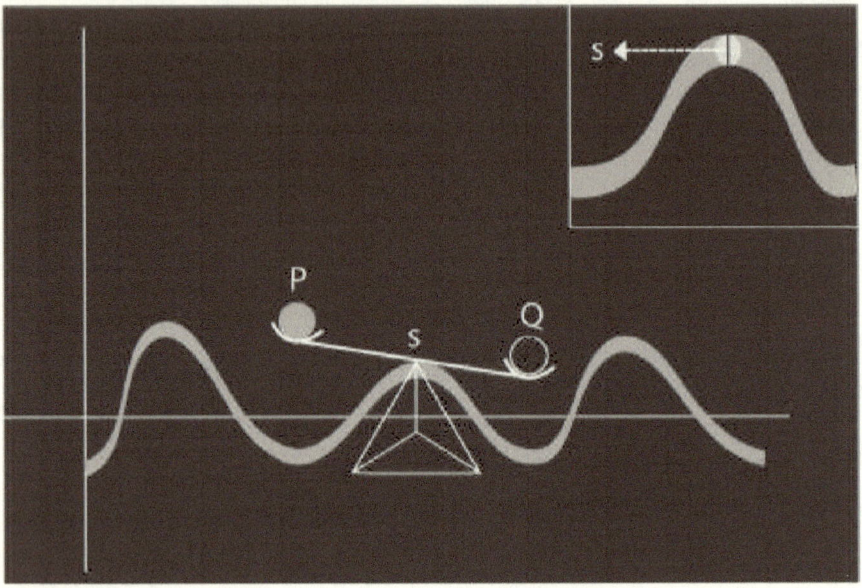

Being the weights Q and P mediated by *S*, the three can be defined as impulses *(F x t)*, where the applied force is dependent of the time — that is to say, do not exist forces that can be applied independently of the time, which also is specially pertinent in a behaviour like the one of the pulse—. Therefore, the values of impulse of Q and P are based on *S*.

The potential sum of kinetic and potential energy (Q + P) can decrease or increase throughout the time, if we suppose a disease or an improvement. We can suppose, nevertheless, that the total sum of (Q+P+ *S*) has to remain equal to the unit (Q+P+*S*) = 1. In a certain way, the inertiality or sensitivity of the balance is equal to *the internal* energy of the system, so that (Q+P+*S*) = (Q+P+I), although *S*, the sensitivity, also can be an index of the external environment and of how it affects the system. *S* also is a specific indicator of the degradation or disorder of the system; another form for detecting the entropy.

There are pathological increases of sensitivity. On the other hand, in biological systems the "maximum" of sensitivity usually agrees with the maximum of stability. Then we have the problem of how defining here the maximums, minimums, as well as the optimal value of *S*. The same stability in dynamic systems is also referred to this last component.

- The time interval St > 0 comprehends an asymmetry and a propensity towards Q or P, and possibly also towards the past or the future of the system, that eventually can balance; if there is no asymmetry the average values of the amplitude would tend to a simple line for the curve, instead of a band of irregular width, and there would not be substantial difference with the habitual descriptions.
- Can we cover therefore the complete curve with the wave of the pulse –with its whole form and its thickness- with a group of discreet weights?
- May we give different values and groups of weights for Q and P? Determining *S* the time of reaction, the proper values of impulse *(F x t)* for Q and P always depend on the first one. It would have to exist therefore a generative or causal order in the changes of Q, P, and *S*. But it happens that also *S* is variable. Lack the system all kind of solutions if Q, P and *S* are variable?
- Although there were no simple solutions, we always can fit the system with the value of *S* that gives us the measurement. The experimental value is the reference; if this value is momentary, will diverge very quickly in the time, losing the validity in a few moments. It is to be noted nevertheless that the time intervals do not tend to zero, nor the values are purely instantaneous, like in the classic differential systems, and that is the reason for which certain causal or generative nexus between the past and the future is possible, which it is not possible in classic systems. But this causal order only can talk about to the balance and the dynamic adjustments that make it possible.
- The cycle or period with *pi* base (π), each beat, naturally continues being an operator for the rotation of operations allowed by the balance. The differences of profile between a pulsation and another one must respond to the inequality of times and the consequent compensation for the operations.
- It seems to be that the ternary systems of weighting are often optimal to make a minimum number of operations. It has the *Phi* constant (Φ) = 1,618... and his associate values some specific relation with this class of systems?
- What class of predictions allows this class of description? What elements allow characterize in the temporary evolution?
- A system like this is mechanic in the most general sense, that is to say, in the sense expressed by the principle of least action. This principle, for the classic systems, allows infinity of partial mechanical explanations, i.*e*., it is not univocal. This is a logical consequence of its formulation: when lacking explicit means, it makes possible all kind of mechanics means like explanation, without affecting the general beha-

viour of the system. Do allow an additional dimension or variable like the one we consider a more univocal explanation and a greater concretion for the principle of least action? Do allow a wider base for this principle? Can we make specific and deepen the idea of the balance?

- If this description can be applied to the biological pulse, also it can be applied to other periodic biological signals, although initially the signal would not be registered with such elements that in the pulse. Nevertheless, a transformation without the experimental measurement will only have probabilistic character, since *S* is the reference for the proper values of Q and P.
- In economy, the transactions are asymmetric, as demonstrate the graphics of supply and demand. Nevertheless these not show a specific form in the temporary graphs. We can take like example the stock-market operations, which involve a time of transaction very related to the momentary slope of the curve of quotations. This it is a typical example of time of reaction for complex systems. Nevertheless, it can be that this time of reaction is not directly dependent of the slope of the curve, but on the contrary is the initial factor of reference in the determination of the values and the operations. It is possible to say the same for other many curves of temporary evolution, like for example the climatic curves, in spite of the well-known differences between the systems. Anyway, we don´t know yet if this description affects more significantly to the prediction or the understanding of these systems, or both.
- To what extent this model can become general for complex systems? The concept of balance is absolutely fundamental, weather in economy, ecology, biology, psychology or physics. Is *S* a true synthesis of the non-linearity of the system, that only allows to be kept awake by successive degrees? These degrees can not be merely methodical steps of approach, also can be inherent to the nature of the problem. In fact, the degrees and shades seem to be inherent to the nature, but the differential description does not allow an appropriate place to them.
- In physics, for the classic systems, we have a time of reaction equal to zero (St= 0).It would rather mean this that its sensitivity is the maximum, or that is null?
- However, this is clearly an idealization. For quantum objects a minimum time of action or reaction is necessary.
- Nevertheless, the quantum of action, defined as energy x cycle x second, does not give us any possible content of operations for the unit of a cycle. Can be the quantum of action redefined according to our three components?

BIBLIOGRAPHY — REFERENCES

Swami Hariharananda Aranya, 2000. *Yoga philosophy of Patanjali with bhasvati*. University of Calcutta.

Simon Weil, 2001. *Cuadernos*. Editorial Trotta.

Alexis Jardines Chacón, 2000. *El enigma del movimiento*. Biblioteca Nueva, Madrid.

Miguel Iradier, 2001. *La hija del capitán Starbuck (Hurqualya)*. Biblioteca Nueva, Madrid.

Alexey Stakhov. *Museum of Harmony and Golden Section (Alexey Stakhov web site)* www.goldenmuseum.com

Koichiro Matsuno. R*esurrection of Cartesian Physics*

Domingo *S*. Acosta. *The inertial mass*

Paul Marmet. List of Papers and Web Site, www.newtonphysics.on.ca

THE LEFT HAND OF CHAOS (MODERN AND ANCIENT INDIAN ANALYSIS)

August, 2005

PROLOGUE

This essay is basically about philosophy. Samkhya, which is the oldest Indian philosophy, had a decisive influence on the development of Buddhism. We will use its viewpoints to tackle a wide-ranging physical and mathematical question, related to chaotic and complex dynamics. The Samkhya is an exact, mathematical philosophy, applicable in any order of the experience, and therefore, also it has use in the more controlled experimental dominion. The Samkhya studies the waves or fluctuations of any movement, being the mind only another form or order of the movement in general. And although the main intention of Patanjali is the self-observation or direct perception of the fluctuations of one's own mind, with the intention of reaching the balance and the possible cease of its movement, also we can apply this frame to behaviours of experimental and measurable order, one of which is the sanguineous pulse. This is what did millenniums ago the ayurvedic medicine, although, of course, in that case the quantitative considerations were secondary. We are going to take all this farther, and will try to show that the relations between the Samkhya and the modern mathematical analysis, including in this one the science of dynamics, are real, deep and inevitable.

The book consists of two parts. In the first part, titled "Pulse and time", I show that the *triguna* and the *tridosha*, the three modalities of the conditioned nature in the context of the Samkhya, is nothing but the most general expression possible of the three laws of the movement of Newton and modern dynamics –the most general, because they are applicable as much to open systems as to closed or ideal systems, being these the limit of the first. But the same relation between open and closed systems will take to us much farther, to the borders of the analysis and the present physics, including in the word "analysis" even number theory, and being the utter Arithmetic the closest discipline to the pure scrutiny of time. This it is the subject of the second part, titled "Time and the modern science".

Being aware enough that these subjects are as vast as difficult, I have decided on a exposition pointed with historical motifs; which does not mean at all that this is a book on the history of the subject, but rather that the history, in which always our imagination concurs, allows us to assimilate thoughts, movements and currents not only more easily, but also more deeply when we have a suitable perspective. And, precisely, some of the things appearing are so little evident that even the historians have not repaired in them. But, for me at least, the history itself, always ambiguous and doubtful, has much of nature –of a nature that rarely we manage to contemplate.

"The left hand of chaos" alludes to this rarely contemplated aspect of nature, so related to time, beyond the conventional space of dynamics. The same term "chaos" is taken in a larger meaning than the one of the so called deterministic chaos, which arose from the instability in the solutions of differential classical systems; if we contemplated its "left hand", the same chaos would be diluted in something very different, but not necessarily controllable.

Not at any moment I have entered details or technical precisions, which would exceed the intention and the reach of the essay: which is to illuminate and to help to conceive new possibilities in our idea of nature, considering the always decisive limitations of our mathematical models. A correct contemplation will help us to surpass some of these limitations, because if by one side we do omission of the technical difficulties, we also indicate the experimental frame in which these have to reveal themselves with all property. What the human mind can not imagine, offers us the nature very often with insurmountable naivety; and in the present labyrinth of sciences of the complexity an Ariadna thread like this one is missing. Nature ignores the analytical difficulties, and here we are going to find a good example of something that can be simultaneously much more simple and much more complicated that what the most sophisticated techniques allow us to describe: it will be up to us what part we prefer, knowing well that here completely new considerations are introduced on which it involves the prediction and the reconstruction of the events. The present models of prediction and reconstruction are close to the collapse and it is absolutely necessary to reframe what is possible to wait for of them, since often the own expectation of the prediction or control extinguishes the possibility of understanding of the phenomena. This subject is consubstantial to the development of the whole text, although barely it becomes explicit: it has to manifest itself in the most concrete of the planes.

Therefore, this book is not merely speculative. It makes a decided bet for the experimental contrast of its ideas. The ideas are difficult to expose and to grasp, but the experiments are easily attainable worldwide. What there is to verify or to refute is a certain duality in the behaviour of the pulse. Such duality seems to be treatable more easily by the study of the correlation of the pulses of both wrists; but it is not in that that the duality consists. This one simply indicates the distinction between *prakriti* and *vikriti*, the initial and secondary components of the joint evolution of the system. If we managed to define these in a satisfactory form, we will have given a great step, because what it is at stake is not another thing that the frame of reference for the evolution of dynamically complex systems; and these, nowadays, include practically everything.

This text was written originally in Spanish, my mother language; being my English very limited, I apologize for my translation.

FIRST PART - PULSE AND TIME

1

It seems to be that in the West the only idea of legitimate science is the one derived from the Greeks. In this line, any description of the world must be external to us, and consequently, objective. All other description would fall within the subjective, when not "mystical", interpretations.

Nevertheless, modern science is something far away from the Greek spirit. Two are the great events that mark this distance. First, the appearance of an experimental purpose with respect to the Nature, that arrived at medieval Europe through the Arabs, and in particular, through the alchemy and pre-chemistry - although also through works about optics. Secondly, it was the introduction of the zero with positional notation, of Indian origin, that completely transformed the possibilities of the calculus and, in the long run, the whole face of the mathematics. Each one of these elements, completely alien to the Greek mentality, has influenced so much in modern science as the own Greek idea of rationality; so that the vindication of the Greek origins of science, even being legitimate, has always more rhetorical and romantic elements than we commonly admit.

In XVII century, and since Galileo, the notion of change and movement and the notion of infinite decimal expansion were interpenetrated and from it began to outline the concept of *function.* In this connection, something very important became from the *inside* to the *outside*: the idea of the change itself, that earlier was internal to the same matter –or to the substance, if we prefer- happens to be external and spatial by virtue of its mathematical expression. Thus, by a peculiar transposition, it returned to become present the Greek ideal of knowledge like external representation, although it was precisely Greek authors, like Aristotle, those who were challenged. But, in an intangible and fundamental way, and in the middle of the mathematical formalism, the notion of displacement and change was inherent to the positional zero notation within the infinite decimal expansions –and in a so subtle, incommensurable and elusive way as it could correspond to the same use of the zero. In this same displacement the idea of the movement released of any material considerations was already implicated. And thus the modern conception of "real" numbers begins also.

In this crucial recombination, the own ideas of space and time adopted a completely new and unexpected character. So new, that not even today, accustomed already in our imagination *the new representative style*, we can understand entirely its consequences. On the one hand, the Galilean principle of relativity deprives of any meaning the notion of location, of point in rest in the space –the same rest no more exist, since it is indistinguishable of the uniform movement. By the other hand, and as a result of it, the variation in the time and its rate become the authentic measuring rule –but this, at the cost of turning to time itself something spatially represented, and for that reason, completely passive and deprived. Galileo himself discovers and invents the pendulum. Without real alternation no time is possible. But, being the alternation of the pendulum independent of its amplitude, the model is lent naturally to its idealization, being totally turned into space. Eliminating any content in the alternation of the pendulum, what Galileo invents in fact is the "one way pendulum". This inconceivable object is the one that governs as much the form we imagine the time as governs in our clocks. Or at least, in our form to represent it: because it is difficult to see how it could affect to other layers of the imagination than the most superficial ones.

It has been said that the linear time and the idea of history are Judeo-Christians notions, although the certain thing is that they have arrived to us through Zoroastrian Iran. Nevertheless, the historical and irreversible time, as we have felt it in the Western world in the last centuries, must much, if not most, to the idea of *reversible* time of the mechanics or dynamics. How this, that we know positively certain, can simply be possible? Because the "one way pendulum". But if in physics the idea of irreversibility –the so called "arrow of the time"- is not opposed at all to the reversible time of the mechanics, nor it adds nothing –as Prigogine seemed to presume -, but it is simply its other inevitable face, Where to watch? Our simple and not ironic answer is: Not onwards or backwards, but towards the sides; to our right and our left sides. Preferably, before we make the least movement. Not even mental. Certainly, it continues seeming difficult.

2

"*Tatah punah santoditau tulia-pratiaiu cittasiaikagrata-parinamah* ", may be translated as: "The concentration takes place when the mental waves that rise and fall at two different moments are exactly equal".

What more striking seems to us in the first reading of the *Yoga sutras* of Patanjali is its exact and mathematical character, even though it is speaking simply of the spectrum of possible experience for any man with the necessary dis-attachment and discipline; not only exact and mathematical, but, properly speaking, *analytical and infinitesimal.* No work written with its antiquity, about two thousand years, offers something similar. I would say that neither the mathematical works of Archimedes, nor his *book of the sand number*, contain a similar aroma.

Archimedes and Apolonio did not know the zero, and many say that this was what the main obstacle to attain the modern calculus. In fact, they also needed the other essential element, a different idea about the movement and the time. Patanjali did not know the zero and its explicit operations; but he knew well the idea implied - and, apparently, more enough than the mere idea. In addition, which he left written on the movement and the change is of an impenetrable depth, but with an impeccable implicit logical articulation.

Certainly Patanjali was not a mathematician nor a physicist; not even a "psychologist", as he is often dispatched trying to ignore what he is saying in the clearest form: "the things exist as much if the perceiver perceives them as if not", to admit which does not need to take refuge in any of the pretensions of the realism. Patanjali is simply a philosopher, who prefers to leave us his conclusions instead of write to discuss them. Its subject is the human experience and what underlies in the change that makes it possible; nothing more, nothing less. What, naturally, it implies the time, the movement, the nature and the appearance, with the basic definition of its mixture, predominance and proportions. Therefore, it includes everything with an extraordinary economy of means. But, peculiarly –and here is all the difference with the modern western thought -, this is only obtained with a determined position, that consists basically in not to make the smaller distinction between internal and external, subjective and objective phenomena. For that reason it does not make sense to describe Patanjali like a "psychologist". His thought and vision are beyond any ingenuousness.

3

To many our association of capital parts of the classic Indian thought with the spirit of the analysis will seem totally unfounded. Perhaps they ignore that Indian mathematicians like Madhava of Kerala discovered the so called Taylor series –used before by Gregory, Newton or Leibniz- nearby the year 1400. As it is well known, these polynomials in power series are the spine of the calculus. Madhava was the first mathematician who made the infinitesimal passage to the limit; almost surely, he arrived at this extreme by his work with continuous fractions. Or that Aryabhata the great, who lived by the year 500 and before Brahmagupta –who gave us the rules to operate with the zero -, discovered the continuous fractions, virtually infinite, that was one of the preferred instruments of Euler or Gauss, and that still today give many surprises in pure mathematics and in the analysis of resonances in chaotic and complex dynamics. The continuous fractions, series themselves, are in the origin of the analysis. Aryabhata, almost with complete certainty, worked already with the notation of the positional zero. Between a great number of achievements, he advanced the incredible conjecture that the orbits of planets were ellipses. Or that Srinivasa Ramanujan –who also had a noticeable preference by the continuous fractions - discovered the functional equation of the Riemann zeta function –summit and abyss of the superior analysis- in India by himself and without another stimulus than the Carr's elementary mathematical text book. Ramanujan barely had the smaller idea about the complex variable theory. Surely, so improbable facts are not merely coincidences. But the masterpiece of Patanjali it is the most forceful and powerful argument, for those who can read, and if we attend to the spirit of the analysis itself.

We forget easily enough that the idea of the infinite and the infinitesimal is inconceivable and unmanageable without the zero, about which the Greek mathematics are the most evident proof. And we also forget that the idea of the infinite and the infinitesimal are *a double* idea since the beginning, and therefore, irreducible to the simplicity or unit. Being so, the zero remains for us a kind of immanent immovable motor, even within the most neutral of the formalisms. Nobody says "it tends to be equal", but says "it tends to zero", or, on the contrary, "it tends to infinite", subsuming the highly fragile concept of equality within the much more comprehensive and powerful one of *tendency*. To specify a tendency is all what underlies to the concept of function. Therefore, the zero released for always the notion of tendency, contributing to the same method of the calculus the idea of power of resolution. Our own idea of the identity is safeguard by the one of tendency. And the same zero allows us to tolerate the ambiguity, displacing it. Thus, it is easy to see that the great step and the great difference that separates the modern science from the one of the Greeks is just the imponderable zero, the great Indian contribution, so difficult to estimate always.

Simplifying the equation:

Greek mathematics + 0 = *modern Analysis*

That anybody will interpret as can.

One will say, and it is completely certain, that the Indians never got to develop the analysis applied to physical sciences as we understand it, and that, for the same reason, they could not develop it systematically neither with any comparable degree of generality. But this fact not only doesn't contradict our perception of the subject, but reinforces it considerably. Then what we suggest is that the analytical mentality was something connatural to certain layers of the Indian culture, but not that this one showed inclination to a science like modern dynamics, that must start off from the assumption of an inertial frame. This type of election was completely alien to the India culture, like to all the others, except for the western one. But we do not have why to confuse the analysis with dynamics, although for western people such association becomes unavoidable.

Already the identification by Copernicus of the Sun as centre of the astronomical system implies that inertia like fundamental property has been chosen; and it is the renovation of this election the one that has supposed its triumph between us, unlike the Greek contemporaries and later to Aristarchus. To the ancient peoples any imaginable property of the celestial bodies would be more worth of interest than inertia. Newton gave an answer in form of law to a question that had become something interesting; but if other peoples did not discover such laws, we must remember that they did not have the smaller interest by them either, nor they never looked for them under such clothing.

Said this, we can recognize that the analytical spirit in the ancient Indian culture responds to other interests that the ones of modern dynamics: what it is not at all equivalent to say that loses interest on the movement and the time. What it does not share is the outer and spatial representation of these. As we just have said, it is not necessary to confuse the science of dynamics with the analysis. This last one is a much more timeless object, with its own movement. The arithmetical functions, like the Riemann zeta function, give the most magnificent example to us. And who would affirm that Arithmetic nothing has to do with time? Later, we will try to see towards where this leads to us.

4

Patanjali's *Yoga sutras* are the maximum exponent of the Samkhya. *Samkhya* means *analysis, number, measurement, proportion, detailed scrutiny*. In any case, the Samkhya was the more purely analytical of all the Indian philosophies. It existed developed before Buddha was born, and his founder, Kapila, could live at a time even previous to the birth of Pythagoras, although in dates quite near to those of the birth of the Greek philosophy.

The Samkhya is the philosophy that underlies to the Yoga, so that its subject only can be the equilibrium or balance. Everybody is more or less forced by life to maintain different balances, apart from to aspire to it in the most general form; so that neither yoga nor the Samkhya is alien to any of us, even if we believe to ignore them completely. Otherwise, and in the form of conservation principles, also for the science of dynamics the balance is the fundamental subject.

5

The Samkhya, whereas the study of the oscillations of the mind that possibly can be lead to the rest, is an intuitive and perfectly direct theory of the fluctuations. But the modalities or moments that detects in the fluctuations cannot be purely extrinsic, but internal or inherent, the discernment that only characterize them opens passage in as much the mind is contemplated or testified by the immovable consciousness without attributes. It does not exist in this any contradiction, but continuity, since the characteristic movement of the mind in its passages of acquisition of knowledge results from so many other passages of easement and therefore of relative or tangential contact with the same knowledge as source; we will see that this solution of continuity is the subject par excellence of the Samkhya as well as of any approach to it.

6

For the Samkhya, all the material or conditioned nature is reduced in last instance to the fluctuations of the three *gunas* or modalities. These always act of conjugated form, so that we cannot perceive them in its isolated and exempted form. In the middle of them and beyond them it is the immutable and unconditioned, *Purusha*, the consciousness without qualities neither attributes.

7

The three *gunas, Sattwa, Rajas, Tamas*, are, in their primary form, sensitivity, activity and inertia. Or if it is preferred, balance (relative), mutability and potentiality.

On the matter of the mind, it is possible to say that it is not an exclusive attribute of *Sattwa* or the sensitivity, but that the own mind, although is more clearly manifested under the aspect of *Sattwa*, contains inherently associated as much the activity as the inertia. The three *gunas* are included and present in any object or subject that we want to isolate: therefore the *gunas* or primary qualities are those that cannot be isolated at any moment, except by the momentary or relative dominion that one of them always exerts on the other two.

8

Let us take a paradigmatic example from *hatha yoga* to conceive in an immediate manner the conjugated character of the three *gunas*: the position of remaining in balance on the head, with this one putting in the centre of a triangle drawn by the forearms (*sirshasana*).The novice has therefore the opportunity to retake the hard learning of the balance that as much work gave him when babe and that he forgot completely precisely because it has managed to incorporate it. In the beginning, it will be basically at the mercy of inertia; with something of practice, it will learn to make the necessary force to resist the oscillations of the body. This force will be excessive nearly always and bad applied, until gradually it is developing sufficient sensitivity and the unnecessary efforts are avoided to the maximum. After a variable period of practice, a more or less optimal union of inertia, forces and sensitivity is reached: then not even a special alert of sensitivity is precise, because it is completely integrated with the other two tendencies. A spontaneous state of suspension takes place finally, that nevertheless is the fruit of all the previous efforts and deliberations. Even if we took the work of even measure in detail the oscillations that within a circle make the body of the novice and the one of an experimented, and for identical deviations of the centre in any direction, we would verify the enormous difference throughout the vertical axis with respect to the rigidity of the first and the perfect and synchronous undulation throughout the body of the second for the most effective and coherent accomplishment of the principle of least action. This is an excellent illustration of that lost dimension of the movement to that we alluded; as well as of which we understand by incorporable or assimilable knowledge. To that process of concealment in the invisible of the qualities in its total presence is at which it points the experience of being in the Indian philosophy. Such experience allows the participation in the extra-mental with no need of metaphysics.

In addition, this example offers a demonstration to us of a principle that is always valid in Samkhya: to greater sensitivity, greater stability and lesser conditioning. But so that this sensitivity is so, and not an added source of perturbations, that has had to integrate *in act* the other two present modalities.

This principle, highly conditional, of greater stability to greater sensitivity, is of extraordinary importance, for example, in Biology. But its generality could extend further on.

9

Another immediate form to consider the *gunas* or basic modalities is taking care of the nature of our hunger or inclinations. These are reduced in last instance to only three: the laziness or inertia, the activity and the pleasure. A momentary predominance of one of them always occurs, which is proper of their excluding character. The laziness corresponds to *tamas*, the activity to *Rajas*, and pleasure to *sattwa*. Although *sattwa* is the more elevated modality, the fact that it exists with attachment reveals its conditional, natural character. Why the attachment to the pleasure? Because the pleasure secretly looks for the balance between the activity and the inaction. This way, sensitivity or *sattwa* becomes the leading reason for the evolution of all beings conditioned by the three *gunas*, in search of the unconditioned balance, *Kaivalia*.

For the Samkhya, the modification of *gunas* is the only causal principle that exists in nature. That is to say, there is no other cause that the change of the *gunas*.

10

But the three *gunas* or modalities also find their total correspondence in dynamics; and precisely in the three laws of the movement and the mechanics, enunciated for the first time by Newton. Perhaps then it does not define the first law the rest and inertia, the second the action or force based on the first through the mass, and the third the balance between action and reaction? Yes, *Tamas* is in correspondence with the first law, *Rajas* with the second, and *Sattwa* with the third. But a great difference exists. In classical mechanics, the action and reaction are simultaneous and immediate. In the Samkhya, such a thing it is not refused nor affirmed; simply, is not considered.

The third law is the safeguard of the conservation principles. But, in addition, it is the mean to establish the absolute time like principle of global synchronization. That is to say, this one is affirmed as a fourth implicit principle, although it remains beyond any evidence: like an assumption of metaphysical character. This way, the Newtonian absolute time plays the same role for the three principles that the unconditioned consciousness or *Purusha* for the three *gunas*.

Therefore, Newton's third principle eliminates any means or real mediation in benefit of synchronous instantaneous reactions. This way, it is consolidated and closed the purely mechanical description of the change and the interaction between objects. Literally, there is no space –nor time- for anything else.

This combination of the third principle and the different laws of conservation that are their corollary with the absolute time or global synchronizer conforms what we understand, or rather we guess, as *causality*. But it is to notice that the causality never is defined explicitly –because it can not be defined here.

The principle of global synchronization is completely conserved in special and general relativity, by means of the Lorenz transformation and the notion of covariance. It is for that reason that in these theories is often spoken about the "principle of causality", although it is not defined here at all, but only understood under the assumption that the events "located" in the future do not affect to those "located" in the past. In quantum mechanics, no longer is possible to speak of causality, although the conservation principles are assumed; that is to say, a Newtonian interpretation of systems completely different from classic properties remains to a great extent.

One assumes that the three principles –with the fourth implied- are suitable precisely because they define an ideal or closed system, so that it is always possible to be continued at the time of verifying the details. But what it happens soon it is generally quite different: the details are studied to fit them with the previous idea of general balance. This way, the behaviour of statistical mechanics or thermodynamics can be sent to the part of the quantum mechanics that ignores temporal problems to give back the balance of accounts to the classic mechanics without touching it. This way, questions that are considered like irreversible within the thermodynamics, are camouflaged in a peculiar game with three bands: the classical mechanics, the quantum mechanics, and a thermodynamics that only nominally deals with the irreversibility subject.

If we thought in terms of information, for example, we could think that there is no better definition of the information that the one of inequality between an action and a reaction. Local asynchrony would be the most appropriate measurement of the communication or mediation.

11

The three laws of the movement become transformed, through the more modern variational description, developed among others by Fermat, Leibniz, the Bernouilli, Euler, Lagrange, Jacobi or Hamilton, in the principle of least action, also denominated of extreme or stationary action.

Poincaré, more than one hundred years ago, already made notice that any physical model that can be described in terms of the least action –a gravitational or electromagnetic system, for example, allows in fact infinite causal explanations. It could be explained by means of elastic rubbers or springs, or any picturesque device that we wanted, provided it satisfied certain requirements –precisely those of the stationary action. That is to say, it does not define the causal relation at all, since this one never can be unique. This is the main reason by which the physicists prefer to speak in terms of descriptions, predictions and equations; although, finally, few of which use these terms will doubt to have catch the true causal nexuses.

That the principle of least action is bottomless from the causal point of view, is something that cannot be strange to us. Since it has been noticed sometimes –and particularly by Nikolay Kozyrev -, the classic or Newtonian mechanics, in virtue precisely of the third principle of the movement, always contemplates a *space* interval, by arbitrarily small that could be, between a cause and an effect, or between an action and a reaction; but it does not contemplate any *time* interval, by arbitrary minimum or that is. That's to say, this is the only cause of the causal lapse. For the Samkhya, nevertheless, *a real* time interval does not exist that is not constituted itself by a modification of the *gunas*; therefore, it is natural that vindicates these like the unique cause of the phenomena.

Of course, quantum mechanics, unlike classical mechanics, contemplates minimum times of reaction for the interaction between particles. But, exactly here, in the quantum dominion, it is where the causality principle is eluded, seeming completely inapplicable, in benefit of a probabilistic interpretation –although according to some aspects physicists speak of the unitary or determinist evolution. This way the problem again is avoided, and sent towards another part.

On the other hand, the principle of minimum or stationary action always seems to introduce a teleological aroma in the description of a system, since this one appears guided somehow by the purpose. Nevertheless, we must consider that this uncomfortable appearance arises precisely from a time model that, like our "one way pendulum", *simultaneously* imposes the reversibility of the actions and reactions and the idea that these occur in an irreversible direction. That impossible pendulum is the nucleus of all our paradoxes. The physicists use three different physics to better elude that impossibility.

12

Let it be so, at least for the moment. It would be thought that the Samkhya, being a self-observation discipline, does not have anything to do with the empirical dominion of the external facts. But the Samkhya conceived at least one purely empirical science, to which few moderns, nevertheless, would give value enough. This science was the *Ayurveda*, the science of the care of the life, not very distant in the spirit of other old medical arts, like the Hippocratic medicine or the Chinese. What distinguishes its humoral theory is to be an application of the frame of *gunas*. Three humours or *doshas*, *vata, pitta* and *kapha*, or wind, bile and phlegm, are not but reactive forms of the inherent properties of sentience, activity and inertia .By reactive it has to be understood that its predominance on the other properties is able to generate imbalances and ailments throughout the time; but also that they are a secondary or derived form from more original and less privative modalities in the constitution of the subject: *prana, tejas* and *ojas*, the vital breath, the brilliance of the combustion and the radical oil that permeate and nourish the body. We are then in front of much more material qualification of the *gunas*, but not for that reason less subject to the same fundamental logic.

The influence of the ayurvedic concepts in authors like Plato has already been noticed enough by the scholars, in particular in the *Timaeus*, where it is clearly spoken of the "air, phlegm and the bile", apart from to mention "the established periods of the life… in conformity with the triangles of the nature of each one". It does not stop being peculiar to verify how this book of mysterycal roots and pole of the enlightened hermeticism has provided a rational inspiration for varied readers of scientific die, from Kepler and Galileo to Heisenberg. Another of the platonic divisions of the soul in vegetative, sensitive and rational soul, also is of net Indian origin, or in any case, it admits at least a strictly congruent superposition with *kapha, pitta* and *vata*, and the corresponding functions of vegetative plasticity, irritability and sensitivity. Even the traditional contemplation of the man like body, soul and spirit, keeps a strict correspondence, if not in extension or expression, yes in the most intimate nature of its correlations; being these modes the only thing that matter. We won't allude to the numerous triads of identical nature, which the reader can easily imagine.

13

In principle, any corresponding attempt to define better the *gunas* or the three related *doshas* exhausts in mere epithets and paraphrases. We can use some of them like semantic approaches, granting to the semantics the necessary respect. After all, our investigation can take us to the conclusion that it does not exist a mean to eliminate the "semantic questions" totally.

The three *gunas* also can comfortably be classified in centripetal, centrifugal and orbiting tendencies. As far as the *doshas*, the aspect relatively more material of the primary qualities, *pitta* talks about to the heat and the energy, *kapha* to the plasticity and solidity, *vata* to all the phenomena of circulation from which the form is derived –the purest and isolable aspect of the form whereas independent of the dynamic or plastic qualities. It could be objected that this last one is an entelechy; we will try to see that it is not thus.

Contemplating a tree we can see and touch characteristics such as colour, forms and texture. To say that the intrinsic combination of those external characteristics is the result or the appearance of primary qualities like the growth, condensation and formation could sound unspecific and arbitrary. But if we saw the tree burning, we could distinguish perfectly between that which burns precisely, something that is reduced to ashes, and something that rejects both things and saves in steam and smoke. *The form,* is what it evaporates and vanish, either we say it in the superficial sense or the deepest. Steam and smoke is not simply the result of which is really burning, but very on the contrary is what it flees to burn, as well as, on the other hand, flees momentarily from the gravity. Although the scrutiny of this can take much more far, is sufficient to make see that, in the most violent transformation, while these qualities still maintain some relation to each other, they always conserve unmistakable behaviours. The form is not an abstraction at all, unless we understand by abstraction, the distillation or circulation made by the same tree and its nature.

14

Within this ayurvedic frame of the *doshas*, the clinical procedure par excellence take place: the *nadi vigyan*, the auscultation of the pulse. Those who practice it with unequal fortune are denominated *vaidyas*, sighted persons according to the literal translation. The auscultator successively feels the radial artery of the wrist of the subject with the three middle fingers of the hand, alternating the pressure, in a certain analogous way to the touch of a stringed instrument. The auscultation becomes after a sufficient interval of rest, and in the conditions of the smaller disturbance possible.

In the pulse five elementary properties are recognized: rate, frequency, intensity, amplitude and form. Being a biological oscillator with fluctuations, it is perfectly suitable to the study of its temporary series –by means of the classic harmonic analysis, or the most recent of wavelets, or anyone of the sophisticated tools of the modern analysis.

The limitations of the harmonic analysis or Fourier analysis for systems of irregular profile are well known; although the arterial pulse necessarily has accused regularity enough, it is one of those systems in which the small deviations and peculiarities do not find easy interpretation within this technique. As it is well known, the harmonic analysis decomposes any periodic empirical curve in a sum of sine waves or sinusoids. Given the independence of each one of these sinusoids, which we obtain is a scrutiny of statistical nature on some of the properties of the curve. Another very different thing is to interpret the dependencies and bounds that are relevant in a system like this, for which the harmonic analysis never will contribute explicit proofs. And, of course, spectrums of the pulse and of the electrocardiogram particularly have been made in innumerable amounts since Mobitz, van der Pol and van der Mark gave the first steps in this area.

15

Of the five elementary properties that the auscultator distinguishes of natural form with its own fingers, the four first –rate, frequency, intensity and amplitude- admit a relatively acceptable interpretation in terms of harmonic analysis. The frequency is the inverse of the period, or the number of beats or cycles in a sufficient time interval, we say a minute. The rate is the measurement of the regularity or irregularity of the cycles throughout the interval. The intensity is the peak altitude reached on the tips of the wave of the pulse. Finally, the amplitude is the total difference between the maximums and minimums of the wave. In addition to these four more analyzable elements, we have *the form* of the pulse, the true and evasive quintessence that surrounds and penetrates the values of the other four properties.

In the most trivial sense, that is to say, according to the representation of the graph, it is clear that the form involves the other components, since the form is the whole curve and it does not have anything else. Nevertheless, for the auscultator other shades exist in addition, since it modulates the pressure of his fingers in three different points. In such sense, and even without contradicting the properties of the curve in the graph, it extracts some additional information.

Since the verbal characterization of this form of the pulse assumes poetic or metaphoric characters, many people tend to think that it is here where we enter a qualitative dominion and we move away from the quantifiable one. For example, it is spoken of a typical "swan pulse" for the marked predominance of *kapha*, of a "frog pulse" for *pitta*, or a zigzagging or "serpent pulse" for *vata*. But, naturally, these indications left by the classic texts as a mere indication, within a dominion of continuous shades are too simple. Therefore, we better not deceive by all that terminology: we are treating about a quantifiable and perfectly analyzable phenomenon, and it is precisely by its fluidity, by its continuity, that is more treatable in analytical terms than for a not very well trained judgement.

16

It is completely certain that the young and not so young medical practitioners of these traditional medicines, whether the Ayurvedic, the Chinese or the Unani medicine, complain the darkness of the classic texts and the great difficulty that entails to acquire a minimum of security in the judgment for this type of diagnosis. In other words, the auscultation by the pulse is in clear regression, because it is not considered practical enough. I am afraid that this has been always the case, and that at any time have been very little doctors able to master this art. But here what it is at issue is not the technical difficulty of a learning, but the fidelity of the pulse like signal of a greater system and its possible relevance in other spheres of the analysis and even dynamics. If we understood in a legitimate analytical form all the information that contains the pulse, and the wider perspective than can offer us, the practice of these same reticent doctors would be one of the first areas in benefiting from it. But we think that the pulse has more enough to teach to us. The pulse is the true pendulum that we are looking for.

17

Like any purely empirical science, the diagnosis by means of the pulse has peculiarities that resist to being transferred to a quantifiable and exact frame. We will mention some of them in passing.

The auscultator obtains pressures and curves of the wave with a different form for the different points in which he locates his fingers. Nevertheless, the modern doctors have stated, introducing a catheter, that the pressure is the same in the three alluded to points. This has been wanted even to show that this is a fraud; at least for us, this class of fast disqualifications is one of the main ingredients of bad science. Because it is evident that to feel a signal from the outside has little to do with introducing surgical soundings, and, at least, the possibility would be due to grant to the ancient doctor of obtaining as much data as possible, mainly if he has the least conceivable means. The differences of the form of the wave are due, of course, to the greater proximity to the bone or the skin; that is to say, it is a difference that we moderns would consider phenomenological, although thus we are not doing full justice to the subject. Because, in effect, these three points are a kind of average to be able to perceive the superficial, deep and intermediate aspects that are conjugated so well with the same nature of the *tridosha*; without forgetting that in these medicines, and the same nature of the body, a natural differentiation between the bones, the flesh and the skin exists, of very similar order to which at embryologic level we recognize between endoderm, mesoderm and ectoderm. So that, not without humour, we could say that what does the *vaidya* or its colleague of other traditional medicines, is very similar to empirically solve a system of partial derivative equations, of those in which we observed the evolution of a variable while we maintain the other ones constant. Known it is that the partial derivative equations appeared first in mathematics in the prehistory of the harmonic analysis with the approach of D'Alembert to the problem of the vibrant cord –and that the measurement of reference of Galileo for his first observations of the pendulum was his own pulse.

On the other hand, it is certain enough that the different traditions that have made use of the pulse diagnosis have different categories, but this mainly affects the terminological framework, more than to the same concepts, that are remarkably homogenous. The same classifications of typical pathological pulses are even repeated abundantly, although such classifications don’t pretend to be another thing that empirical approaches and estimations. For example, the ayurvedic medicine speech in terms of the *tridosha*, and the Chinese of *yin* and *yang*, deficiency and excess: but both are speaking clearly of a balance. Otherwise, we will be completely mistaken if we think that the Chinese system is dualist, or even binary: the excess and the deficiency *presuppose* the mean or included middle –the vacum, which prevails on both. It could not be expected less from a culture that protects and values the main thing with the ellipsis. And nevertheless how many western scholars have not seen in the Chinese vision of the polarity anything else than the more refined form of the binary notation.

During a long time, the mechanical sphygmographs or pulsographs have not been able to catch all the subtleties of the form of the pulse. If, for example, we use a Doppler flux-meter, which we will obtain is an average of the speed of the flow of the blood in all the vessel, that apart from that respects the Bernouilli law and has a parabolic profile, with greater speed in the centre of the vessel and minor in its wall, and the contrary for the pressure. This way, hardly we can obtain in the graph some notion of what it can be a "hollow" pulse, or one "sliding like a ball in a porcelain cup", or "rough like a worm nibbling a mulberry leaf". Obviously, the fingertips that feels "sees" here much more than what the eye can observe in averaged and impoverished graphs, although in fact, it is simply sensitivity and possibilities of modulation. Of course, now we can have multiple classes of sensors, including high fidelity microphones and adjustable pressure sensors.

It is possible then to say that, until recently time, there has been no experimental possibility to gather mechanically the subtle range of information that the physician obtain simply with its fingers. Now that has stopped being problem; the problem is that we are incapable to analyze quantitatively these data without grossly distorting the original coordinates that served in his judgment to the doctor. As we indicated, there have been made an infinity of spectral analyses of the pulse –and it is to notice that we say of the pulse, not only of electrocardiograms -, but, For what serves to say that below a value of 100 arbitrary units the patient show insufficiency, and that over that threshold it shows an excess or plethora of which it is wanted? Even if it would be certain, we have lost by the way all the logic that attended the doctor, and we have turned it into inexplicable numbers. And precisely what we look for it is to keep awake the amplitude and importance of that logic and those coordinates.

Quite exasperating questions exist soon, like the fact that hardly somebody is put in agreement to detach and to locate the different parts of the information obtained. Someone take the pulse only in one the wrists, others in both wrists; others still assure that the meaning of both hands is reversed depending on the gender. Neither in the Chinese medicine there is agreement with respect to the relation of the meridians with the points of the wrist, etcetera. It is not to surprise, therefore, that what it seemed in the beginning a relatively simple question becomes in fact a dense forest. This is what has motivated so many hurried and coarse attempts of quantification, with the rather vain hope to turn this type of diagnosis something more objective. One assumes, yet, that it is the knowledge and not the crude numbers what turn something objective. And nevertheless, it remains a doubtless fact: that we will not find another signal more faithful and immediate, neither more globally representative of the immense set of factors that are united to form that that we call life, that the pulse.

18

The problem, seems to be, is that here we lose ourselves in the forest of the semiology. The present cardiology makes every day new advances precisely in as much it is able to avoid these semiotic questions, to turn them functional and anatomical details. Once again, the reductionist spirit seems to prevail. But this has much of illusion, because between what is "functional" and what is "anatomical" an enormous abyss remains. Aside from the fact that the *vaidya* persecutes a much more general object than the anatomical or functional state of the heart: the state of health of the whole organism is up to him as well as the details more relevant than concur in it, and being the state of the heart only one of them.

And speaking of semiology, and leaving apart the fact that the human kind never left to interpret signs, it can be said that the pulse diagnosis, and in particular within the Indian scheme, it is the first consistent semiology as much from the rational point of view, as from the experimental one; although it seems that there would be numerous arguments to doubt it after which we have said. Yet, if we observed the modern attempts in this field –the semiotic triad of Peirce, for example, that so much has to do with the general scheme of the *triguna* -, we will verify that still the old models keep advantage, specially at the time of the experimental confrontation with the brute facts.

In addition to this, and with regard to the pretended opposition between the reductionism and the global interpretation of the phenomena starting from its performances, it is worth the trouble to remember that it is this last what all the men of science and philosophers worthy of such name, from Newton and Kant, have denominated with the name of *analysis*, and not on the contrary, as sometimes the custom and the forgetfulness incline to believe to us. Then to explain the behaviour of a system from its basic components, is what is denominated *synthesis* or composition; composition that most of the times is so far from our reach that only takes place in our imagination. Nevertheless, the imagination is the less suitable that can exist to follow the behaviour of any moderately complex system, and we don’t say already, something with the complexity of an organism. One of those global analytical interpretations is, precisely, the theory of the gravity of Newton, that to date, nobody has managed "to compose" nor "to synthesize". The interpretation of the harmonic analysis locates to us in that same meaning of the term analysis, now with arbitrary degrees of complexity. Before we said how badly treats the physics with the causal level of the phenomena, and the basic reason for which this cannot have place in it. Hume’s critic to the causality concept is worth for a representation like the one of the classic Newtonian physics; but this is only a frame of representations, and our direct intuition of the change is not lent so easily to the representation. More on the contrary, may be eludes it perpetually.

The *Triguna* of the Samkhya is a system of coordinates for the change, the simplest, more direct and intuitive possible. But precisely what we understand by system of coordinates is the clearest of the representations, and here we are in the antipodes of the representation. We must always have in mind this fundamental difficulty.

19

The three *doshas* or implied reactive humours in the movements of the pulse correspond in addition with three components or precise mechanical moments: in the most general sense, *pitta* corresponds with the contractile impulse of the heart, *vata* with the state of the walls of the vessel and its conduction of the contractile wave, and *kapha* with the fluidity or viscosity of the blood –we spoke, naturally, of an elementary correspondence, not of a reduction from one terms to the others. Even like independent terms of a problem of fluid mechanics, these three factors already are sufficient to originate a non-linear system without stable values in the solutions; but in the organic dominion, and of course in the one of the Samkhya, there is an intrinsic dependency and feedback between them.

In the pulse there is no continuous transition between systole and diastole, but that, even in spite of the damping of the impacts that supposes the flow, the pauses continues being appraised, so that the scheme is perceived like contraction-pause-expansion-pause. That is to say, that for many effects, in addition to the continuous signal of the wave, we have a genuinely discreet system, as it is derived from the doubtless fact that the nervous signals that activate and deactivate muscular fibbers of the heart are antagonistic and they are excluded and mutually blocked. As we see soon, this does not lack importance, since it is in the base of the only concept of oscillator or circuit with an own content: that in which the circulation exists for the only reason that two conditions cannot be satisfied simultaneously.

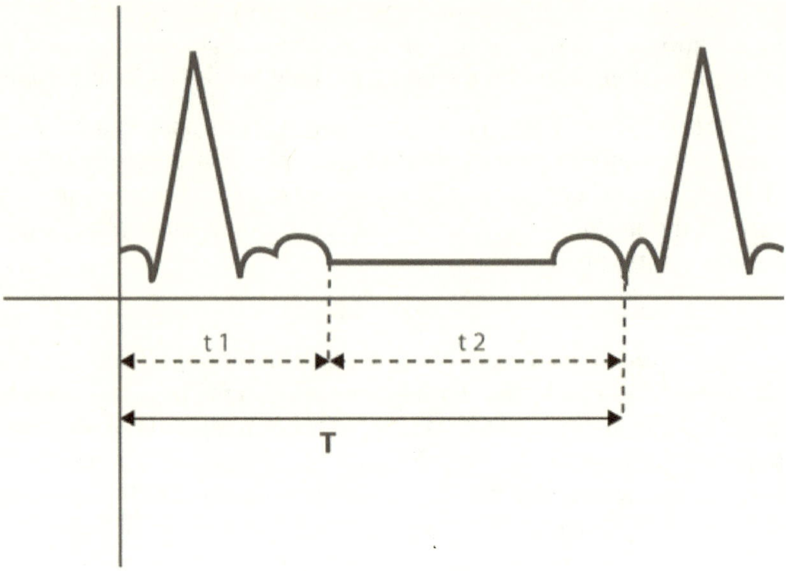

PULSE WAVE DATA

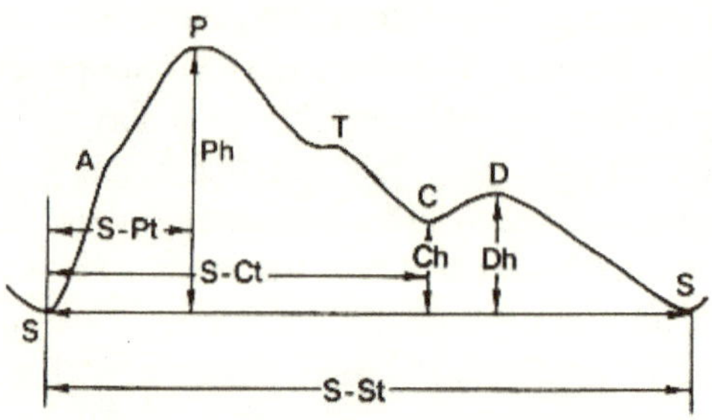

A : Anacrotic wave
P : Systolic peak
T : Tidal wave
C : Incisura
D : Diastolic peak

- S-St : Pulse interval
  normal values : 600~1200ms
- S-Pt : Peak time & Upstroke time

  Dh/Ch : Incisura ratio

fig 1

Of course, and unlike the pure scheme of the movement in the classic mechanics, the living beings and the organisms –otherwise the same than the elementary particles- have characteristic times of reaction, which cannot be avoided. This is evident from the more elementary empirical point of view, but it tends to be diluted and to lose its specific importance as we translate it to habitual analytical descriptions.

Perhaps the more immediate form to warn ourselves of the use of the third principle, the equality of action and reaction, is shooting a gun and verifying in our body the backward movement. Naturally, the butt does not strike to us with the same departure speed of the bullet, but, approximately, with that one which it is quotient of the mass of the bullet with respect to the one of rifle. In addition, and if we wanted to enter details, always we will find that there have been losses by gases of the explosion. If we only know the exit speed of the projectile, or the one of the backward movement, we needed to know what has happened in the middle to precise the balance; and knowing of the action and reaction, we even needed to know the mean or apparatus to be able to precise the way in that the loss takes place. Detailing, we would also verify that the reaction is not simultaneous and immediate, but always includes a time, although for many effects it seems to us negligible. I hope that so elementary example will be pardoned to me; but often it happens that, when we studied the organism as a black box with some or many output data –whether they are anatomical or analytical- we return to turn the same data another black box to study itself, losing interest of the inputs and outputs themselves. That is to say, everything becomes in the middle, unlike the example of the gun, and we forgot what it has entered and left in both directions, which in the case of the organism surely includes something more than the diet and the depositions.

Yes, all this round about was only to say that we can appreciate the mean and a sensitivity between action and reaction in the global conditions of the organism with no need to contradict the indispensable laws of the mechanics. How it could be otherwise? Nevertheless, Can we grasp this in some way simultaneously elegant and direct? This already seems another question.

20

It could also be added that the human body, like any organism, is essentially an open system. And, of course, for the Samkhya any object that we like to consider is an open system; reason why the *gunas* emerges like the efficient cause –and not like mere formal laws- of open systems in interchange with the environment, in the same way that we applied the three laws of the movement to ideally closed systems. Yet, to speak about the organism as an open system does not imply to reduce it immediately to a dissipative system, like opposed to a conservative one. Since an organism has boundaries and stability conditions whose reach we only know in a small degree, although they are manifested enough in its effects. It is not small task to define these dissipative and conservatives aspects in the living beings; but the predominant strategy to approach the problem at the present, through the complexity and the organization –with the genetics and the biochemistry as it base- hardly can get to fulfil this relation, if we consider the enormous epistemological gap that we noticed between analytical elements, component of a synthesis, (they themselves composed always) and causes.

21

As it is known, in fluid dynamics we can replace the three Newton's laws by variational principles, transforming them into the potential, kinetic and inner energy of the system. While we only work with the kinetic and the potential energy, it is generally easy to know the magnitudes that are conserved; but when we introduce the internal energy (P + K+ I), that can be expressed like viscosity of the fluid, pressure, or perturbations, it is much more difficult to know what is exactly what it is conserved. Simply, we have to presume the conservation principles. This is the great difference between the hydrodynamics and the dynamics of punctual particles, in which the conserved magnitudes remains in a explicit form. The pulse, that is a system periodically perturbed by the beat of the heart – a forced oscillator with power supply-, admits this type of description; only that the feed-back of the factors makes it still more complex.

The three *gunas* and the three *doshas* can be put in a correct correspondence with the kinetic, potential and internal energy of a hydrodynamic system, which in fact, seems to us much more natural than the punctual dynamics or mechanics in strict sense. Hamiltonian dynamics, in addition, supposes the incompressibility of the space of the fluid. This is a fundamental point that it cannot be applied literally to a system like the vascular one, where there are expansion and contraction of the vessels or arteries, and not only by virtue of its elasticity, but also by muscular fibbers of the walls that act to nervous impulses following the contractile rate of the heart. It is to assume that also the same heart and its rate of impulses has to be sensible to the answers of the vascular circuit, and in fact we know that the global feedback of the system takes place at several levels, nervous, as much chemical, not being the sensitivity to the fluidity or viscosity of the blood the minor of these elements. So that the circuit and its answer are a model example of biological self-regulation, whose study has grow unceasingly from the pioneering works of Cannon and Wiener. All this allows us to raise old and interesting questions to us in a new way. Truly, we know well enough and with many details *the explicit* physiological mediators, such as nervous, hormonal, and biochemical in general. These without a doubt play a fundamental and irreplaceable role in all this self-regulation process –they would not be there if they had no utility. What we don't know is how they have gotten to be there and what it has guided its formation. Doing without answers of type "genetic programming", that today we know perfectly that it constitutes only one part of the subject, and not the greater one, and that it has to do much less with the "programming" that with sensitivity and the modulation by the environment, happens another much more stimulating approach to us. This is based on the compressibility of the whole circuit and anyone of its parts, compressibility that can and must admit very variable degrees, although, of course, they must be measurable.

Of course, in such case it is not a Hamiltonian or classic system what we have; we not only contemplate the existence of friction, but of other non explicit variables in indefinite number. We maintain the analogy and the connection with classic systems to only have a reference on the conservative aspects of the system. As everybody knows, pure or ideal classic systems do not exist, which does not prevent its routine application. Nevertheless, although we are speaking of systems really far away from the closed and conservative models, we are interested particularly in the intertwining of two main features. We can call to this "semi-classical systems" provisionally and only for comfort, although by such terms can be understood an infinity of different things.

Speaking of compressibility in semi-classical systems, we entered completely which today it is known like nonlinear dynamics, one of whose main signatures, although not the only one, is the determinist chaos. This one appears in principle in perfectly differentiable systems and with initial conditions of arbitrary precision, and therefore way different from the model of the pulse. In fact, it can be said that the determinist chaos is an almost purely mathematical discovery, although their relations with dynamics have been explicit from the first moment, backing down in the time to Poincaré and his studies of the celestial mechanics and the three bodies problem. In the unitary evolution of the phase space –the Hamiltonian- of innumerable systems, next to the instability of the solutions in the equations, appears "basins" or "attractors": the dynamic evolution doesn't disperse uniformly through all the space available, but it is compressed with a certain preference in a part of the volume. This already is a clear example of compressibility, unless it is not clear absolutely what is what causes that the system is compressed and why. That's the reason of the controversial statute of this type of findings. Different measures can be taken from the instability and the degree of compression of the system –Lyapunov exponents, fractal dimension, entropy indices, and a long list of others-, without on the other hand there is never a clear dividing border between the random or stochastic component and the determinist component of the system; except for precisions *ad hoc* of the particular model. This way, the present "revolution" of nonlinear dynamics is an unfinished revolution, or so to speak, "half a revolution": it has acquired the statute of the omnipresence without being able to contribute no type of clear conclusions. The so-called determinist chaos more seems the beginning of one long exploration in the immense territory of the relations between chance and necessity that a theory defined for anything. Nevertheless, when it reached the threshold of the spreading and of the public distortion already time ago it was a suspended exploration, at least about conclusive results. And thus it takes growing more of a quarter of century, with no particular direction. The same theoretical or fundamental physics, which many blame of stagnation, has much more specific and clear challenges and urgencies; at least, according to the present order of the appearances.

22

People say that after the fever, the calm and the reason returns, and life can continue progressing with a more normal rate. The present state of things on nonlinear dynamics and the theory of the complexity –different areas but multiply associated- are so that significant advances in a great number of points could be expected, but hardly a transcendental upset. The experts move in the specific dominion of the intricate and difficult, of the concrete applications, where rather are possible to wait for technical refinements more than true conceptual jumps. Everything flows and is moved, nevertheless. Meanwhile, and waiting for results of more weight, it is totally certain that nonlinear dynamics has changed substantially the way we see the world. The determinism itself has done that we believe less and less in the determinism like explanation of the phenomena. Like this change is substantial, but still non-fundamental, as if the spearhead of our conviction in the causal explanation of the phenomena had been notched, or is even dissolved in the air, while the inertia of our background convictions continues maintaining the same direction.

23

We have spoken previously of a partially compressible circuit in all its points: the circulatory system itself, of which the pulse is the more explicit dynamic signal. Of course, if the circuit were totally compressible, its flow in a torus or ring structure would be reducible to the simplest of the forms. Nothing of that happens, on the contrary: there are very significant reductions of the space of flow, but of almost impenetrable nature. In an evident way, to look for the suitable compressible space to the behaviour of the system it is just like to look for how to make it comprehensible. Both terms become therefore equivalent.

Ian Stewart has opportunely observed that it not exist any law of conservation of the complexity. Of course, now it is very well known that very simple systems can complicate themselves, just giving the sufficient time to them. That was precisely the lesson of the determinist chaos. And if really there is no law of conservation of the complexity, we have reasons to think that this is applicable in the opposite sense, that is to say, that very complex behaviours are derived from very simple laws. In fact, this is equally certain, and in the more trivial sense, for many of the sources of determinist chaos, systems governed by differential equations as familiar as those of the celestial mechanics. But if what we are speaking of are *new laws*, with a different descriptive and explanatory rank, still none has been discovered that could make the complexity more compressible and comprehensible. For example, the duplication of period discovered by Feigenbaum generates complexity, but it does not reduce it, except in a minimum thread of the scheme of evolution in a particular transition of state.

So that the general situation continues being noticeably unidirectional, with a base of fundamental and determinist physical equations that generate more and more complexity according to they are studied, without barely any development in the opposite sense, the one of the true compressibility.

24

The presence of determinist chaos in the activity of the heart has been and is a controversial subject also, and this is natural, if we consider that the borders are by definition blurred in this dominion. What is doubtless is that there are important irregularities in its temporary series, and that these often show fractal or quasi-fractal structures, that is to say, self-similar structure with correlation on different scales and in the long term in general. Another thing is to judge and to interpret the role that such irregularities play, if they have some specific importance or it is simple randomness. Almost all these studies talk about *the electrical* activity of the heart. The pulse and its temporary profile have been studied much less from this perspective.

In general, the pulse is a lowly nonlinear system; that is to say, it is not associated directly neither with the chaos nor with the turbulence. In fact, when the auscultation perceives indirectly turbulences through whistles or murmurs, they are a clear signal of cardiovascular pathologies; in any case, such turbulences only seem to take place in the greater vessels, and not in small vessels like the radial artery. On the other hand, the flow of the blood is not turbulent in general, but neither really laminar, which is not another thing than idealization. Now specialists even continues studying actively if the best model of this flow will not be the one of helical or spiral trajectories, perhaps more economic or optimal. The low nonlinearity of the pulse is resumed in the fact that we cannot accurately calculate the phase angle of the beat and its wave, not even in few cycles, although nevertheless the global regularity of the signal is very great, without the enormous fluctuations in the values that can take place in differential systems infinitely simpler, but chaotic. That is to say, it maintains a great global stability; otherwise, we would not be alive. Precisely by this, it has been wanted to associate the cardiac failure with a transition to the chaos, although the relation is very far from being clear.

25

It is possible to ask to what extent our information on the dynamics of the pulse becomes richer when we try to catch the specific *form* as it catches the *vaidya* with his fingertips, in comparison with the schematic and averaged form of the mechanical registry of the conventional esphygmograph.

I would wish that it was not necessary to say that this specific *form* of the pulse in the palpation nothing has of mystic or mysterious, except for the differences that always will exist between the sensation and the direct perception and any analytical representation in graph form, no matter how exhaustive. What differentiates this *form* from the general and averaged profile of the wave is, from the perceptive point of view, that precise quality that cause that a pulse seems "full", "empty", "rough", "hollow", "hard", "tense", etcetera. This already is a perceptive synthesis of factors of doubtless mechanical character, but they do not have why to be well specified in the curve of a graph: the impulse until certain point yes, but not the cushioning of the fluid, the state of the arterial wall, and less the relations that can have between these factors. And all this already supposes a precious synthesis of information, by more confused than it can seem. Not even it is high-priority to detach it, but simply to catch it and to register it as faithfully as can. Certainly a part of these qualitative or perceptive aspects can even be inferred of indirect form from "the flat" profile of the conventional registry. As we already said before, now we have adjustable pressure sensors and other types that allow us to catch this general feature. And of course that we can transfer this information extended to the representations of graphs, mainly as thickness or refinement of the outline of the wave, including its variability throughout the cycle.

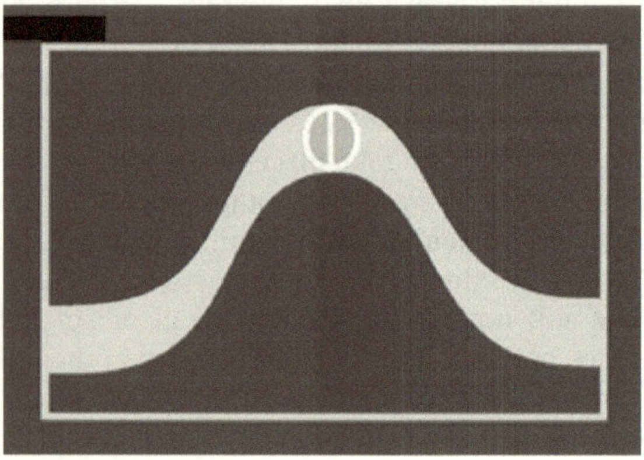

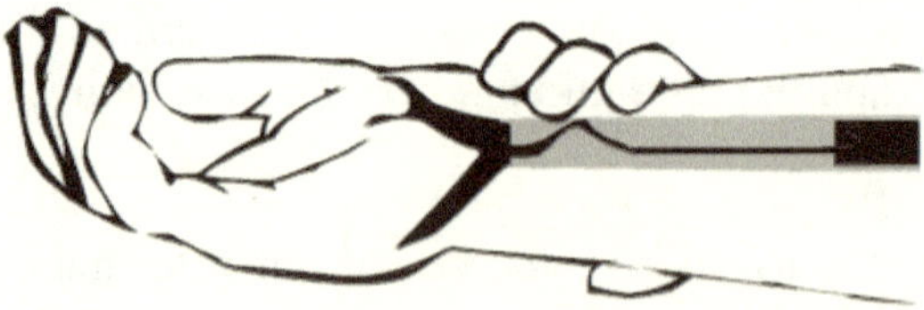

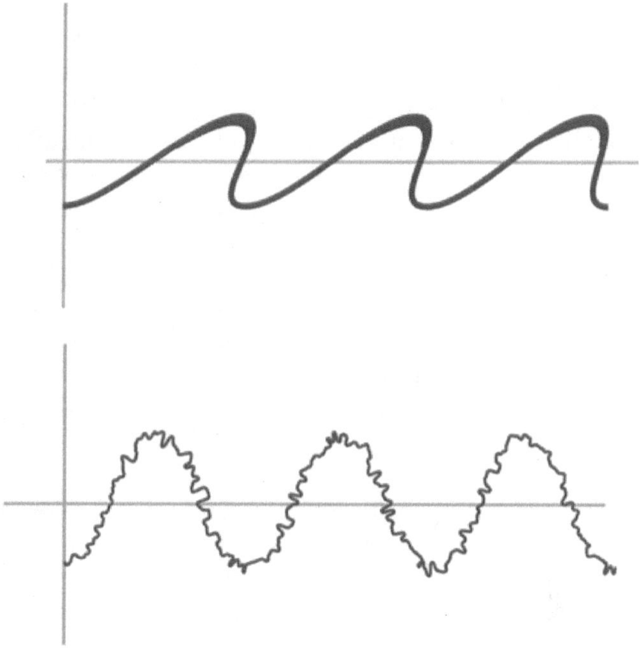

Fig 2

We can see all the phase of systolic ascent of the wave like the kinetic energy, and the one of reduction or diastole like caused by the potential energy. The whole cycle is already intrinsically irregular by functional questions, like the attack of the cardiac impact and its later diffusion; but also the spacing between the beats, and their corresponding duration is irregular. In addition, as it can be seen in figure 1, the diastolic part has its own tip of ascent, reason why the wave is far from being simple. In any case, if the ascent and reduction of the wave are both the more elementary components, all the others will be under the correspondence of the third component, the internal energy, that from the point of view of *vaidya* is not another thing that *vata* or sensitivity between action and reaction. "All the others" are not but the irregularity of the temporary series, either we consider it in the way of the “flat” graph, or with the more detailed graph that gives us the quality of the outline or forms of the pulse. That irregularity includes as much the whole interval as each one of the periods or cycles. Evidently, the three factors always are conjugated, and we can measure as much the line or outline length, as the area included by the routine methods of numerical integration, which, in a computer, is reduced to count pixels. Would it give this conjugated measurement something at least similar to the brief estimation of the quantitative proportion and predominance us that the *vaidya* establishes when he takes the pulse to a person? Surely yes, because there is no space neither place for other considerations. Nevertheless, it remains certain that the physician does not have in mind very fine quantitative estimations, which do not need if he knows to grasp the shades, and these are diluted of intuitive form in another type of considerations. But these shades are not but those variations in the continuous dominion that to the graphs and the calculus give us and that the physician would not know to quantify verbally. Therefore, a real coincidence between the quantitative and qualitative aspects exists, although it is only expressed like a mere parallelism. That is to say, we have deceived the supposed difference between both aspects, maintaining the proper categories of the *vaidya* and without weakening them. It could not be otherwise, being these genuinely *analytical* categories since the beginning. Yet, in the verbal level this only can be a correspondence, by the same fact that in the verbal description the quality necessarily prevails on the quantification. In fact, the auscultation of the *vaidya*, its listening, is an exercise of sensitivity different as much from the verbal considerations as of the quantitative ones, even being able to include to both.

26

In the first approaches, we identified the variable thickness of the form of the pulse with the sensitivity between the action and reaction, or the kinetic and the potential energy, introducing something like a pseudo-derivative. A balance with sensitivity and a time of reaction allowed us to give a kind of symbol of the implied factors.

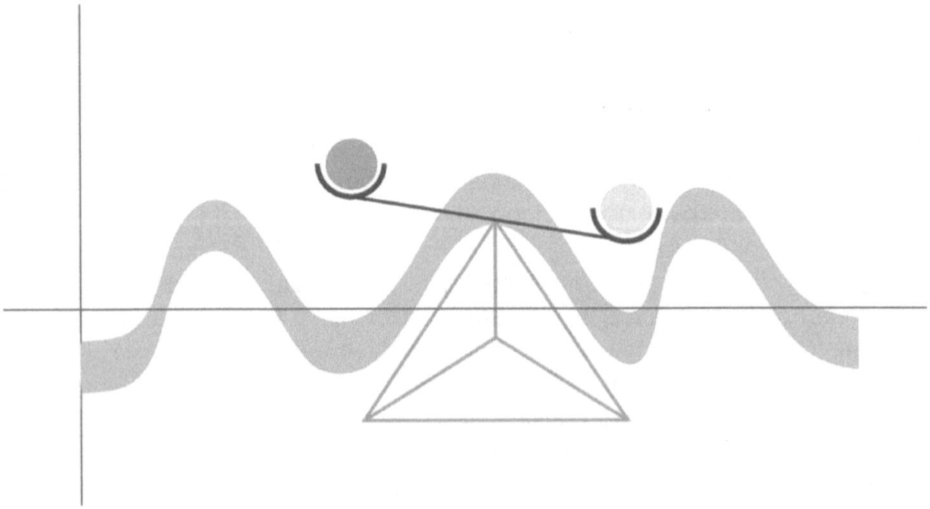

Fig 3

But it was clear that this only could be a symbol to indicate some kind of lost information, and no way a realistic model, and even less mathematical. Although this analogy was well clumsy and even picturesque from the mathematical point, it is worth the trouble to remember that it is not a simple subject to pass to graphical and analytical representations notions that apart from to seem qualitative, are conceived beyond the representation and even, it could be said, made specifically to deceive the representation.

What we tried to show with that symbol was the necessity to contemplate the three principles simultaneously, and not only two. In any case, the global sensitivity of the system of the pulse, the third *dosha* which we associated with the internal energy, cannot be caught immediately in any point of the wave –it is not punctual, neither has derivative nor pseudo-derivative -, even though the pressure, that is what the fingertip feels, can be conceived as form of internal energy by volume unit, if we follow the classical Bernouilli equation. But, even doing without other essential factors like the cushioning, fluidity, or the state of the vessel, that can concur in this synthetic representation of which truly happens, still we would be ignoring a fact that has a greater importance: that the balance between the three factors cannot be of instantaneous nature, at least within the factors that can be represented within this so limited frame. That is to say, that within this unilateral and incomplete representation, the balance has to necessarily show like something in the long term and non local, scattered on different scales within an interval. That is to say, there is no sufficient space so that it can be made in every moment. There is no space of time.

In effect, already before we said that a perfectly compressible circuit would allow its corresponding and inevitable elementary reduction within the simplectic space. That the circuit was compressible would be equivalent, precisely, to that we could find it completely in any point. And this would make the time unnecessary as well, that would perfectly become monotonous and incapable to introduce alteration or novelty. Like the ideal pendulum of the mechanics. The time would not have to make balance between anything, neither the operation to make that balance could create or consume time nor development at all. "To create or to consume", two faces of a same problem.

27

In the study of electrocardiograms the presence of these long term correlations has been demonstrated, on different scales. Also there is categorical evidence of which these correlations are richer and more detailed in a healthy subject, and minors in as much the pathologies get worse. That is to say, the most irregular and broken patterns are healthier than monotonous. This comes to be basically in agreement with the idea and perception of the Samkhya of which a greater sensitivity makes more stability possible, and on the contrary.

Nevertheless, these same studies have not been able to go further from this. Not even is suspected why reason the cardiac irregularities would have to be more healthful than monotony, what cannot be surprise with the idea of time which we maintain. At the most, somebody suspect that it could have some delicate game of balance behind these signals –what it is not suspected is that the own signals are already an exhibition of that balance. To know it is not very useful if we don't know the ground on which that balance takes place.

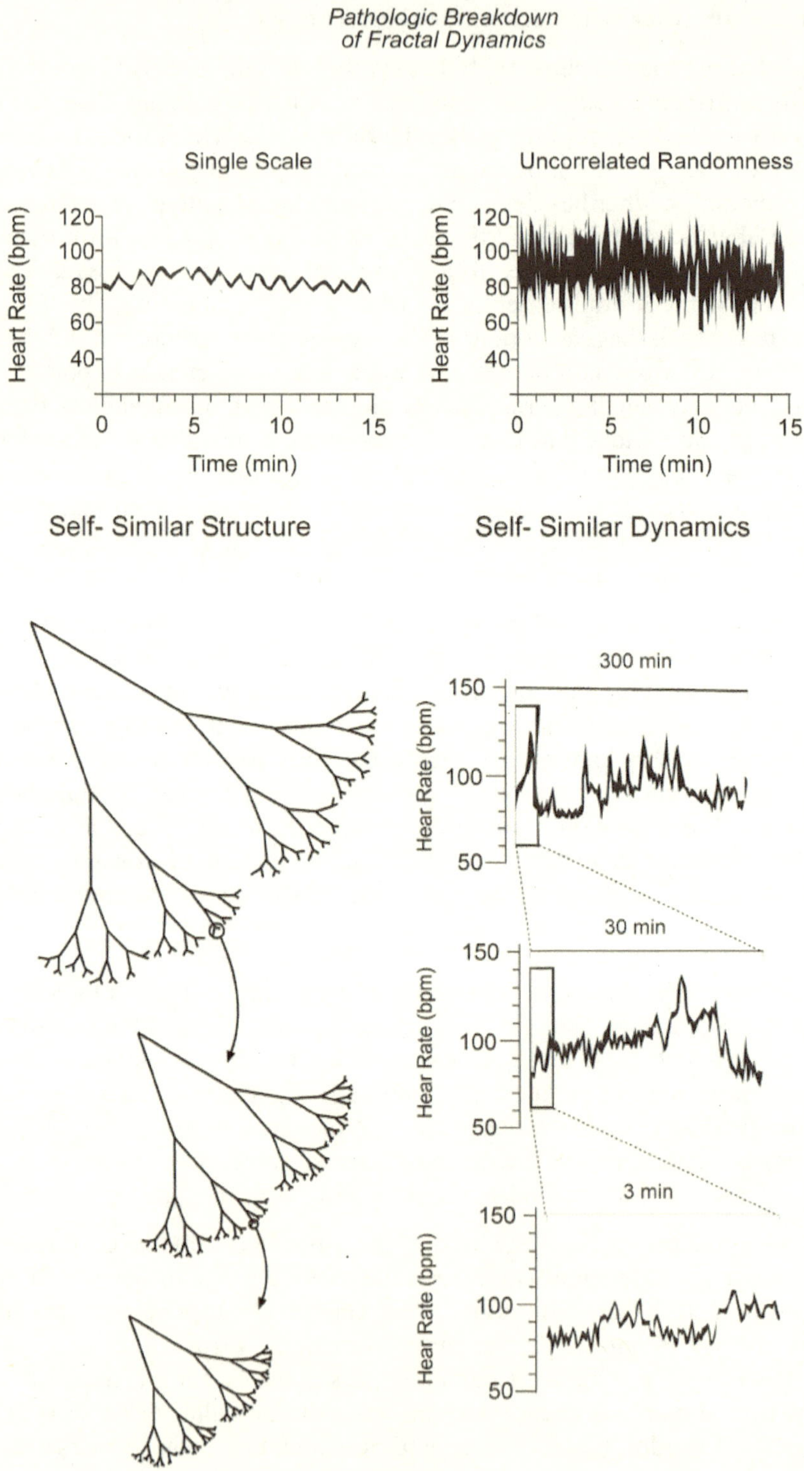
Pathologic Breakdown
of Fractal Dynamics
Single Scale
Uncorrelated Randomness
Heart Rate (bpm)
120
100
80
60
40
0
5
10
15
Time (min)
Self- Similar Structure
Self- Similar Dynamics
300 min
30 min
3 min
Hear Rate (bpm)
150
100
50

Fig.4

28

If we think from the perspective of the continuous variation, it seems trivially certain that it cannot be two exactly equal pulses. What it is already much more extraordinary, and precisely within the continuous dominion, is the fact that in any pulse, and at any time, can be distinguished the innate and the acquired factors with sufficient clearness. This is a differential luxury of the nature that would be absurd to waste; so much it is what must teach to us. Since even from the point of view of the genetics, to establish the relation between the internal and the environmental activation of the genes remains like a puzzle very far from being understood, and we do not say already of being solved. It seems to lack a minimum reference for this problem, so fundamental in theoretical biology, the evolution and development.

The *vaidya* distinguishes between *prakriti* and *vikriti*, biological and the biographical features, the original or innate constitution of the individual and its tendencies acquired throughout the life. By definition, any later state in the health of the subject is dependent and referred to the constitutional tendencies; so that the produced habits or alterations meanwhile are, on the one hand, assimilations of the eventualities to an original conformation, and, on the other hand, that same complexion has already its slopes of reaction before the accidental conditions or *vikriti*. If *prakriti* and *vikriti* come to be like the substance and accidents for a given pulse, not for that reason it is necessary to forget that in the long run those accidents and way of life may become the most decisive for the destruction of the balance that represents the health. But not in vain it is spoken of way of life, then. In an essential way, the form only is modified by its own form to wind itself in the time, and that thread has to be perceived by the *vaidya* with the right sensitivity.

Anyway we consider it, it does not stop being an extraordinary wonder the fact of that individual persistence of the form of the pulse throughout all a life, if we think about the immense variety of circumstances that can surround and alter it, and about the very slightness that its signal represents. If we think about that this little thread of time is going to maintain something own in the most adverse circumstances, beyond the successive losses of identity of the memory, and that not even a violent death can rob the right to it to disappear according to its own and intimate law.

29

This distinction of the tendency within a tendency is so fundamental for our study as it was it the distinction between speed and acceleration for the birth of dynamics through the concept of force. A model of fluid dynamics – say a circuit of pipes with pressured water- cannot establish this type of differences because the components are not connected from the origin, and all the differential characteristics that we could measure later depend on the arbitrary variation of parameters. A primary, original reference does not exist.

The dependency or sensitivity of the chaotic systems with respect to the initial data is in the antipodes of this other type of "original conditions": first they are highly unstable before small changes, whereas the very small changes in the proportion of *doshas* have an equally small effect. But already we said that the three *doshas*, their predominance and proportion, must be conceived as reactive slopes or gradients: like tendencies forming *a potential.* This potential, their boundaries, does not have sufficient freedom to evolve anyway, and a too abrupt change would lead necessarily to the death. Therefore, it would be more than desirable to know the boundary conditions of that potential, or in other words, its economy.

30

The definition of a potential of this type seems extremely difficult. We not only spoke of an open system, in which their components already are related to the environment from the beginning, not only in fact, but in its same sense and meaning. *Vata* is not only a sensitivity between the action and reaction, it is also and at the same time sensitivity to the environment. Therefore, it is simultaneously internal and external sensitivity, and to distinguish both seems more a task proper of another point of view than this one. *Rajas* and *Tamas*, the effective force and inertia, seem more internal to the system, but they always are mediated through *vata*, the sentience or sensitivity.

This, that seems a great obstacle, is also the main virtue of a system or organization thus conceived. Had not this ambivalence, the vision of the *Ayurveda* or *the Samkhya* would not offer anything truly new to the conventional mechanistic vision, which by definition considers closed systems, and altered by other equally mechanical perturbations.

Let us return again to the nature of the change and fluctuations of the *gunas* or its associated *doshas*. We have already said that, for the Samkhya, the *gunas* are anywhere as much inside as outside and that we can observe them if we learn to analyze what we see; that there is not any other knowable thing in last instance.

The three *gunas*, being in balance, are unknowable and they are reabsorbed in the consciousness without qualities. If the existence of the *gunas* is based on its inequality, it cannot be on the other hand modification or fluctuation without the dominion of one of the *gunas* on the other two, which makes possible that the modification is knowable like moment itself. That dominion has the form of a modification of the weight or value of one of the *gunas* that reverts on the weight or value of the other two. This admits and demands characteristic sequences that, beyond the abstraction, constitute the only form of causality which we can know. That's to say, there is not another, and any other apparent form of causality will be revealed like mere representation. The representation of the causality, therefore, moves away to us of the only knowable and intuitive causality. These three modalities are only applicable to the real cases, not to imaginary entities without comparison terms; we spoke from the beginning of conditional entities, not ideal or closed ones.

One *guna* or modality cannot exist separately, so that a state dominated b y *Sattwa*, we take for example, already includes the presence and subordination of the other two, as much for the most stable cases like for the fluctuations themselves. The same is valid for the other *gunas*. Patanjali distinguishes between diversified states, undivided or monadic, only indicative, and non-indicative states of the *gunas*. The same empirical or mutable ego is an undivided and indivisible case in practice, no matter how hard so many intellectuals insist presently on its theoretical dissolution; the pure I sense, the existence like mere attention, is an only indicative state, *Linga-Matra*, reason why the non-indicative state, *Alinga*, properly is non-manifested, rather than nonexistent, if we want to fit expressions that rarely we use with property. This is, to the state of balance of *gunas*, it is considered a metaphysical term or not (in which metaphysics all the well-known reversible physics would enter), it is not attributable neither the existence nor the not-existence, and the only thing that is possible to say is that it does not generate indications nor references. The fundamental states or sections of the *gunas*, in sections or groups of five and six, have a great importance if we want to deepen in this dominion, although now we will not stop in it.

31

If the balance of the three *gunas* constitutes its non-manifested state, it is because there is no place in them for the dissolution, growth or decrease, since they lack material cause. If we follow enough the growth or manifest decrease of the *gunas* in an entity or system it is inevitable to verify how values that seem to be diluted can emerge from their latency, and vice versa. But a value cannot sink in the latency, in the relatively non-manifested, without passing through the modality of the Latency or Retention, that is to say, *Tamas*, in the same way that cannot emerge without action or *Rajas*, nor acquiring certain balance without *Sattwa*. It must have an eloquent continuity between the content of its values and the form they are manifested.

32

Even accepting that the *gunas* themselves neither increase nor diminish, any action or predominance of one on the others in an empirical temporary entity must be seen in such terms for the understanding of the sequences and its evolution. A sequence begins necessarily in the *guna* that it is dominant at the moment. This already produces a redundancy that can be detached indifferently in several ways, since what is increase in one always is correlative diminution in another one. We can detach a simplified scheme in conditional form:

1. If *Sattwa* or the sensitivity "increases", this entails a correlative diminution of the activity of *Rajas*, and never its increase (This does not seem evident and there is all kind of experiences that at first sight seems to contradict it). It can not produce either "increase" or "diminution" of the principle of retentiveness or *Tamas*.
2. I f *Sattwa* diminishes, this only can happen by a correlated increase of *Rajas*, and never by its diminution. It is why the term *Rajas* is synonymous of contamination or dust. *Tamas*, as in the previous case, it is not affected directly by the modification of *Sattwa*, but already the increase of *Rajas* operates its diminution.
3. The diminution of *Rajas* also corresponds to the increase of *Tamas* or inertia.

Thus are included the four possible direct movements of *Rajas* or activity, and the two of *Sattwa* and *Tamas*. The *Rajas guna* is in the intermediate place of the three, and is the only possible connection between the other two. A defined order is conserved although what it seems beginning of a sequence can be seen from another modality like consequence, and the increase, like corresponding diminution, and vice versa. We move already in the circularity. We will not enter the formalization and quantification of a model that seems insignificant and that in fact can be complicated indefinitely, having been able any logician to treat this –that could speak to us about connectives, disjunctive functions, etcetera- with more pleasure and competence. We were satisfied seeing the order of the *gunas* as a scale or stairs that simultaneously allows the union by degrees and the separation or exclusion of certain immediate movements. That is to say, we have an elementary model of mediation, which admits almost infinite variations itself.

33

The explicit distinction between *vikriti* and *prakriti*, the present state and the drift of the previous states, are the unique thing that could give us a reference of the *gunas*, its complexion and evolution. The only thing, therefore, that could make this field worthy of a consistent and totally scientific study. Given its implicit or irreducible inherent character, is not possible that the *gunas* or the *doshas* show like explicit or exempt components; but without this reference, it does not seem possible its rigorous understanding and treatment.

Although the interpreter of the pulse must make this distinction to have the minimum of judgment certainty, it is not clear if he is able to make it explicit anyway. It would be said, rather, that he is based on the conjecture and the supposition, in addition to the intuition, although this last can be surprisingly precise and right. It can be based, in addition to the idiosyncratic properties of the pulse, in the judgment on the constitution of the person, who has certain hardly variable elements throughout the time; or in the colour of the skin, the aspect of the tongue and other semiologies. Nevertheless, all this is not but supports to delimit and to give context to a source of information, the same pulse, that would have to be and in fact it is more precise and faithful than any other index to consider.

Since we have the great advantage that the pulse is already an elementary form of temporal series, that would be render the analysis in a model form, it would be to wish that the reference of the past state or *prakriti* be equally doubtless an explicit in character. Surely, if we had all the series of the pulse of a person from the birth to its death –a number of beats of the order of 3-4 X 10^9 - we could make a quite comprehensive study of its evolution, its progressive imbalance and the increase of the disorder or entropy; but this is not a realistic perspective, since neither we want to wait for seventy years nor seems very feasible nor desirable to monitoring to a person along the whole life. On the other hand, the monitoring with sensors releases to us of the obligation to take the reading in conditions of maximum rest, and it allows us to make really dynamic studies in real time, with registries or tests of effort, pain, pleasure, etc. From here we can do quite valid extrapolations on the nature of the fluctuations of the pulse, balances, imbalances, and the form they evolve throughout the time; valid, perhaps, but still quite conjectural and fragmentary. By the way that this would be very fecund at the time of creating more solid protocols in the psychophysical studies and all those that analyze the relations between the mind and the body, although always insufficient without one more necessary and natural definition of the *triguna*, which is as well equivalent to define the relations between the potential of *prakriti* and the effective and present values of *vikriti*.

Really, which we asked ourselves is if both elements, present or actual and potential, are intertwined and present in the effective signal or if the potential only is a part of the past of the series, like would be to hope in a more conventional model, with a linear time that does not keep memory from its past states. As it is seen easily, this last one would be quite contradictory, if we consider the very considerable evidence of which the balance of the system cannot be made punctually, but scattering itself in a time interval, otherwise indefinite.

Nothing closer to the clinical precision that the analytical exactitude, nor furthermost simultaneously.

34

All this would put in enormous difficulties to us; when we try to analyze values from a more natural point of view, it would seem to be that the nature denies the same possibility to us to carry out this type of analysis. It is not to surprise that so many well-intentioned attempts to consider the dynamic systems with another perspective end inevitably at the failure. For a moment, everything seems to make the rigorous correspondence between the *triguna* and the variational principles of dynamics possible; soon, quickly, we return to realize that dynamics ends up imposing its own rules and forces to wash our hands of the ambiguous elements like strangers to its intention. Dynamics excludes many considerations simply because they do not make sense for it nor have space within. And this is very difficult, if not impossible, to change. When on the other hand we achieve to extend its variables, not for that reason such variables stop to be under the same excluding logic.

So must it be: in the infinite space of the analysis, there is no place for anything, but for the analysis.

35

Finally, the same nature comes in our aid. And it does through its asymmetry; or its symmetry, because both are interchangeable. Are symmetry and asymmetry mutual reference to each other? Perhaps not absolutely: rather all inclines to us to think that the symmetry is a purely formal reference, and that the asymmetries contribute the real contents. Nevertheless, once we found the terms of symmetry for a phenomenon, the contents of this one seem to volatilize in form of a mere particular case. Between symmetry and asymmetry, the relations do not seem to be nor symmetrical, nor the opposite. An antinomy: exactly that class of things that do not have place in the space of the analysis.

But we return to the nature. As we already said, some of the doctors who use the pulse diagnosis only feel the signal in one of the wrists, and others in two. There would be good reasons to suppose that a single wrist is enough, and that contains of *implicit* form all the information, having in account the highly synthetic character of the signal and its more than presumable unitary, univocal evolution. But that this can be so for an expert doctor does not eliminate its implicit character, which is just what we want to surpass in the direction of more explicit and measurable forms. Therefore, it is possible to suppose that the pulse of a single wrist contains all the necessary information, although in a more compact and apparently confused way, and that the other wrist can be used precisely like *reference* –just what we were looking for.

The Chinese medicine, for example, usually prescribes the reading of the two wrists. This, among other things, obeys to its breakdown of the twelve meridians, an essential feature of its etiology, in six for each wrist. Which, like almost everything in this ground, does not stop being a prescription or recipe of heuristic nature, even though can be justified by the results. And like all these heuristic frameworks, they change according to the teachers and practitioners. What all the practitioners of this diagnosis admit is the more than sensible difference between the pulses of both wrists. Traditionally it is said that in the left arm the movement of the pulse is clearer, and in the left one, its form. This is quite evident, at least for most of the subjects, but it would not even say why reason six meridians are assigned to one side and other quite different six are assigned to the other, even when we grant to these distinctions a high degree of relativity. On the other hand, the contribution of each meridian, organ or viscera can be very different in magnitude. For example, it is considered that the stomach and its associate functions or energies –and not the heart, as we could believe- contribute in near 4/5 parts of the global profile of the pulse speaking in quantitative terms; what it does not mean that the others are less important, but than they are subtler and more difficult to detect. The ayurvedic medicine would agree perfectly with the sense of this appreciation in spite of the terminology differences. Probably the organization of the meridians hides a very important symmetry –that curiously also is given in groups or semi-groups of five and six- but this one cannot be looked for this hastily and independently of the empirical nature of the signal and the concerned values. All these points can sound quite strange confronted with the ideas that we have of any class of analysis, and of the clinical analysis in particular; nevertheless, I will insist on which they also are of unfailingly and genuine analytical character, and it is worth the trouble to have them in account at the time of doing an idea about what can we hope to find. We must omit many others necessary preventions because of the brevity.

36

In order to be brief, I will say that the pulse of the right hand is not only different, but also complementary of the one of the left hand. What it is not anything new and is in basic agreement with the more spread Chinese breakdown; also with the Indian, although this one often does without so precious reference to interpret the brute signal of only one of the wrists. Taking into account the distinction which we have done between the implicit and explicit values of the signal of the pulse, this must not be contradictory at all. The pulse in both wrists includes a totality; but its correlation allows us to open an enormous space for the analysis: now yes, also for the quantitative analysis of the temporary series.

For the western medicine, that tries to avoid the semiotic ambiguities as far as possible –vainly, perhaps-, and we do not say already for physics, the idea that the dynamics of the pulse in both wrists can be very different and simultaneously very precisely correlated will have to seem quite outlandish and absurd. It will think that, if there are differences, these are reduced basically to a minimum difference of distance of the arterial trajectory, and to other questions and accidents of anatomical order. In no case it will think that relevant functional differences for all the system can be revealed here –and that's what we are talking about.

For the glance of the anatomist, the basic bilateral symmetry of our body little means. One would say that it is an aesthetic ornament of nature; although for some organs, like our eyes, the advantage of the binocular vision is patent enough. But even in cases like this one, that evident functional advantage seems to close the passage to another type of considerations. That type of advantages can be applied in the most trivial form to the kidneys or the lungs, speaking about survival advantages and security mechanisms. Such considerations continues being entirely superficial. How to explain of what class of complementariness we speak about?

37

Our consciousness is extremely opaque to the vegetative activity of our organism, although this varies enormously from one individual to another. The circulation is one of those vegetative activities; also the main one. The most opposed, comparatively, it seems to be the breathing, which until the limit of the emergency, is a voluntary and extremely adjustable activity. Nevertheless, and in spite of this apparent contrast, barely it is possible to imagine more interdependent activities than those of the heart and the breathing. But since already we noticed before, the pulse, in addition to the one of the heart, tolerates other layers of vegetative activity.

Given its voluntary nature, it is easy to experiment how the variations or interruptions of the breathing do not recover of instantaneous way: a very long forced retention will leave its track cushioned throughout many cycles of breathing. It is exactly by this so ordinary experience in first person that we can understand of totally instinctive form how the same has to be valid for the activity of the heart or the pulse, although only indirectly we can feel them. For that reason we could speak surely before about the pulse of balances deferred throughout the time: we only suppose the similarity with what happen in the breathing.

Something we can be sure about: any phenomenon associated to the breathing will find its faithful correlate in the cardiac activity and the circulation.

Nevertheless, the breathing has a type of perfectly well documented asymmetric alternation to which almost nobody pays attention: the predominance of breathing between the right and left nostril is alternated each hour and a half, approximately. This alternation was important for the old ayurvedic medicine to such an extent that it was said that if this alternation let take place by more than twenty-four hours, the subject was on the verge of dying. What, by the way, is not necessary to interpret like one superstition or fable, but like an almost irreparable dysfunction and imbalance. What means this alternation? Western medicine, even knowing it, has ignored it widely, under pretext of its irregularity and variability. As if that same variability was not the best index and source of information in all type of analysis, be they chemical, dynamic, or radiological. In fact, we know perfectly that it is not contemplated because it does not have place in that scheme of things.

As it is known, for the Yoga this alternation is not but the most visible manifestation of a more general underlying polarity, that affects to all the evolution and organic development. In terms of subtle anatomy, and in addition, in symbolic terms - that could be not merely imaginary, but the most general possible- that polarity is named as *ida* and *pingala*, the intermingled activity of the Sun and the Moon, with *sushumna* like virtual axis of the non-manifested. In much more tangible terms, this is translated in the two fundamental phases of the metabolism: anabolic and catabolic phase. Or, also, said with other words, words, creation and destruction, absorption and elimination, afferent and efferent nervous routes, vegetative formation and voluntary activity. In sum, these two phases would be the most faithful correlate of *kapha* and *pitta* or *Tamas* and *Rajas*, considered now not from the immediate point of view of a modification or fluctuation, but from the global economy of the organism. Apart from that, all the Yoga can be resumed in making the peace and obtaining a lasting balance between our vegetative and voluntary sides.

The existent correlation between the two pulses must have also a common denominator with the more elementary components of the breath cycle –inspiration and expiration, with their corresponding pauses; and with the oscillation of the movement of both pupils, and a long etcetera.

38

The only area of the modern physiology in which the functional asymmetry within an apparent symmetry is investigated actively is the neurology; everybody has heard about the specialization of the cerebral hemispheres, and even the subject tends to be trivialized, about which nevertheless it is known still so little. Then it is just here where the local analysis and the scalpel are more impotent. It is something beyond doubts that the specialization of the hemispheres exist; but of the details of this specialization and the correlations between the functions of both hemispheres, it is very few what presently can be known. This way, the modern physiology has had to assume the asymmetry only for the so called "noble organ" –as if the other organs were not -, there where it is more difficult to study the phenomenon, and without realizing that is not but *an effect* of a much more general phenomenon.

In effect, there is much noise whereupon if the left hemisphere –that by the crossing of the nervous routes corresponds to the right side of the body- is responsible of the logic and linguistic functions, and the right hemisphere –corresponding to the left- contains the imagination and the spatial representation. Only of this, many already make a mess, assigning an "analytical" character to one hemisphere and a "synthetic" character to the other; when the truth is that the mathematical analysis, for example, can require a use of the spatial representation and much more intensive imagination that in other activities considered more "creative" or "artistic".

And it is that everything is badly approached if we remain isolated in the cognitive dominion.

Certainly and surely, both cerebral hemispheres are not but the more elaborated and complex case of the polarity or basic alternation that we have aimed: the one of the voluntary and the vegetative activities or tendencies. These two aspects, as we know, have their particular centres and differentiated routes relatively well known: cerebellum, nervous plexuses, etc. We simply affirm that both continue reproducing entirely *also* in the whole cognitive plane of the conscious life. What it seems incredible is that we have still not repaired in it.

39

Everything which will continue maintaining to us in the perplexity and even in the incredulity; to such an extent our "cognitive style" has conditioned and has put our ideas of the space and time onto the rails. Yes, because we are speaking also about time. Once again we are forced to reflect on nature, and not with instrumental pretensions, but with spirit of fidelity.

Let us think about a tree. We see a certain lateral symmetry, in spite of apparent disordered caprices in the ramification. The basic lateral symmetry makes think us immediately about the growth: in fact, it is one of our more appraised images at the time of representing the unidirectional time to us, with all his cumulative factors. But we cannot see the most remarkable symmetry, because it is sunk in the Earth: the one that form the roots with respect to the branches, with the ground like irregular plane of reflection. And just this symmetry hides also the most important asymmetry, in which barely we think – although all we noticed easily the enormous functional difference that exists between the roots and the branches. Here the circuit and the polarity are made evident and acquired their inevitable character.

Exactly the same happens in our body.

Only the conditions change, being very different. Because we don't need to say that we do not have a part from us to the sun and the other under earth. It has changed the direction of the symmetry axis, and in addition, all the surrounding environment. But the polarity and the difference of phases remain. It is not that the animal, emancipated of the vegetative sleep, becomes "tree without roots", except in the most trivial and apparent sense. We are rather whole trees, with the roots inside.

But let us think in addition where a tree has its "head", if we don't confuse it with the top. Since the more "autonomous" and "voluntary" –willing, perhaps- part of a tree is in the wedge and spearhead of its roots, which become the most similar to our brain. This brings inevitable resonances to us of the great Vedic image of the cosmic tree with its roots in the sky and its branches towards the ground. And if somebody has doubts on that it is in the roots where the tree unfolds its autonomy –as relative as any other, of course -, he could think, to the way of the poet William Blake, where are the "devouring" and the "prolific" elements here: anyone understands at first sight that the roots are the devouring element and the branches, flowers and fruits the prolific; on this there is no doubt.

And it is by this that the vegetal world wakes up such resonances in our soul: because, unlike us, it is justly the vegetative part the one that elevates its sleep and dreams before the Sunlight and before our own eyes.

40

Thanks to the tree of our example, we are able to conceive something that in another way would seem more than doubtful to us: that both sides of our bilateral symmetry have their fundamental reason in functional differences. Therefore only this way the balance involves something real; because two equal things do not need balances between them. And this balance must involve more or less regular oscillations and therefore a circulation with typical differential aspects. And thus, the most typical manifestation of this functional asymmetry will appear naturally in the time, and indirectly, in some of its dynamic aspects.

We could think that the very complementary aspect of the arterial pulse is the venous pulse; in the jugular one, for example. Willing to look for so valuable reference, nothing is to discard. Nevertheless, the venous pulse seems to show more information on the circulatory return and the same heart that on the joint balance of the system. And although in the organism any element can be revealing on the set, the certain thing is that they seem too different signals in order to look for its correlation, superposition and interferences. Naturally, the same heart exhibits a bilateral symmetry that hides an evident functional asymmetry, the one of the arterial flow and the venous reflux; the best possible summary of the external influence through the breathing and the internal reflux from all the parts of the body. This asymmetry also is reproduced between the auricles and ventricles; and the crossing of these two different and coincident asymmetries of the horizontal and vertical axis is like an anatomical epitome of that difference that we find in the dual flow of even the two arteries of the wrists, more extreme than parallel. The anatomical crossing only shows us an external analogy without apparent causal connection with supposedly parallel lines of action; but it is just here where it appears our blind point with respect to the perception and conception of the time –the same mind is not able to perceive nor to take care of two different simultaneous courses, but must suppose them parallel, that is to say, simply spaced. Apart from that, one is allowed to suppose that it is the function what creates the organs, rather that on the contrary; but only the more careful experimental study of the correlation and differences of the pulses can throw light and evidence on this as surprising as unexpected question.

If somebody still doubts, he may think about what a nonsense would be the symmetry of the tree with its roots if it were a simple space symmetry. We cannot imagine anything simpler, because it would be simply gratuitous. In addition, that spatial symmetry is far from being regularly fulfilled, because of the very different conditions that exist outside and under the ground. Therefore, this symmetry mainly exists in the growth and the time, and in the cycle of circulation of the sap that the tree does entirely. We would say, even, that it exists mainly in *the spirit* of the tree. And in this polarized game with the environment has the tree its reason.

Let us say, then, that the complementariness of the arterial and venous circulation finds its faithful correlate within the same arterial circulation under the appearance of the space symmetrical bifurcation –the two radial arteries that we adopted like mere reference, just as we could adopt other arteries with ampler signals. As it is well known, the circulation process is far from simple; the venous reflux is quite complex, and a certain number of beats is needed for the whole renewal of the blood. It happens, in addition, that the venous circulation is, from a point of view, simultaneous to the arterial circulation, and, from another point of view, it is successive or subsequent to it. Then, assuming a great simplification, and attending to the most relevant differences, we can venture that what in the arterial-venous pair appears like simultaneous, must appear successively in the duality artery-artery, and on the contrary: what in the venous-arterial pair seems successive, must have a correlate of simultaneity in our reference artery-artery. Why? There is neither place nor time for anything else, that's because. That's to say, this is the main condition for the balance in the undivided. And thus, at a blow, a new light is thrown on the same knot, cross or blind point in our time-space conception; and, very probably, on many antinomies present in the different areas of the analysis. Light that, of course, we shall have to follow in the direction of the source.

The circulation of the tree has more to do with the real time that all the clocks invented since Galileo's pendulum. And their roots take us further and deeper than all the modern cosmologies, with their great explosions and enormous numbers with a metronome rate.

Before the difficulty and schizophrenia that would suppose for anyone of us to conceive the "two sides of the time", the image of the tree comes to redeem to us and to make us sense a unitary time and rate, conserving all its enigmatic character. I would dare to venture that, if for something the content of this poor writing reach some reception, it will be by this fortunate image and its eternal evocative power.

41

Truly, the ancients knew to grasp this notion of the time, so difficult to conceive for us; although only the symbols have lasted, some of them so eloquent in their plasticity: they are the two foreheaded Janus, the double axe, the bicephalous eagle, and other many of identical character, that express more openly what we already found in the tree. To say of these symbols that they are allegories of the time, with the present moment watching towards the past and the future would not help us much, because almost none of us makes both things simultaneously neither at the happiest moment. One ignores therefore the weight and plexus of the symbol, that is to say, its almond. Nobody will be able to represent this time in a point, except like symbol also, but only a formal symbol now, and dispossessed of its complexion - that is precisely what the ancient symbol alludes.

But we must leave the mysteries and try to approach the difficulty. Metaphorically speaking, and without the smaller scientific pretension, we can conceive the anabolism and the formative, plastic capacity of the organism like watching in the direction of the past, and at the catabolism like watching in the direction of the future. This is not unjustified in terms of biological development, since the formative phase must count with which there is already present, that it is supposed that “comes” from the past. On the contrary, we can conceive the catabolism like "free" of that observance, and for that same reason, and because it continuously opens a route of action of opposed nature, we attribute the opposite direction to it, in the direction of the future. As it is seen, indeed we can conceive simultaneous activities here, but we realize perfectly the metaphorical sense of that words: we know that from the past nothing comes, and know that nothing "goes" towards the future; that both things agree in a same time and what differs is the nature of the activity. But if we see with facility this, we have already obtained something, because in fact science never can see it thus.

Of course, we are turning a time dominion into a certain space not well defined. For physics, the particles have the property of impenetrability in the space, a corollary of the inertial systems. Nevertheless, for the waves, the precisely the opposite is predicated: two equal waves in opposite phase and direction become annulled at a certain moment and return to emerge following their original direction. It is the called principle of superposition of the waves, in the antipodes of the impenetrability of particles. By this same one usually says, to simplify, that the main attribute of particles is the inertia and the one of the waves the energy transmission, although naturally anyone of the two implies both.

The Samkhya assumes the existence of atoms; but atoms of time, if thus can be said. And the only which we can conceive like “atoms of time” are not but the waves –only that time waves, unlike the more spatial waves before mentioned. These waves of time are, in last instance, the *vrittis,* the same fluctuations of the *gunas*. Will obey these waves the superposition principle, will they have penetrability? Surely they will, if we conceive them seriously like waves. This way, it is possible to conceive also and easily that a moment of time can contains several values and supports simultaneously different activities, stopping to be a passive point for the mere representation in space coordinates. And of course, in the evolution of a dynamic system we can superpose at every moment of time not two, but infinity of different variables, provided they fulfil the minimum exigencies. Conservative exigencies that, in last instance, are equivalent to the impenetrability of the inertial systems, although are not reduced necessarily to it.

We could say then, without contradicting the conservation laws, that *an interval of dynamic space* exists when two conditions cannot occur simultaneously in the same point –by the same reason we defined any circulation phenomenon before. After all, even particles have "intrinsic" frequency and spin, so called because they cannot be reduced to the space representation. In its generality, this would be valid so much for a particle, an interval of the sanguineous pulse or a tree. In this way (bounded) time would begin to roll by its own rough ground.

42

One of the main reasons to reject a priori the information that can offer the analysis us of the pulse is to consider that it is a signal poor, precisely, in information. Really, many think that, even if we accept that it is symptomatic, the signal is excessively synthetic. What deserves some comments.

Of course, the modern clinic analysis is not precisely poor in data, and even so and everything, it is always wanted to get more. Is to have elements for the *judgment* –synthetic act par excellence as far as we understand, and that Kant treated like a faculty. Precisely the judgment is the easiest to lose with the increasing flood of data, and I suppose that the doctors of the modern hospitals, not to mention the epidemiologist, must know this enough - in fact, all we began to know this enough, and I am afraid that we will still learn more. From this perspective, and at least in the clinical dominion, may be an *excessively* synthetic information can exist, but, even so, not synthetic enough, yet. On the other hand, already we said that the brute feature or performance is properly the best object of the analysis, on which is not necessary to insist.

It is not possible to say that the pulse is poor in information since the moment that we ignore the factors of its variability. Let us say rather that we are incapable to read it. I will admit nevertheless that it is very difficult to believe that in this signal we can obtain the necessary or sufficient information to the judgment, specially if we measure it with the standards of the modern exigencies. But, as it was said, we think that the correlation of the signal of the pulse of both wrists not simply duplicates, but multiplies in an enormous factor the possibilities of the most explicit, neat and detailed analysis –because the difference or complementariness of both signals is not of trivial character, but extraordinarily complex. That is to say, the more detailed analysis, within the suitable frame, would show this complexity by itself, and this wealth of information. Let us repeat it: within the suitable frame; because studies of the correlation of the signal of both wrists already exist in a great number, without having reached, as far as I know, to anything essential.

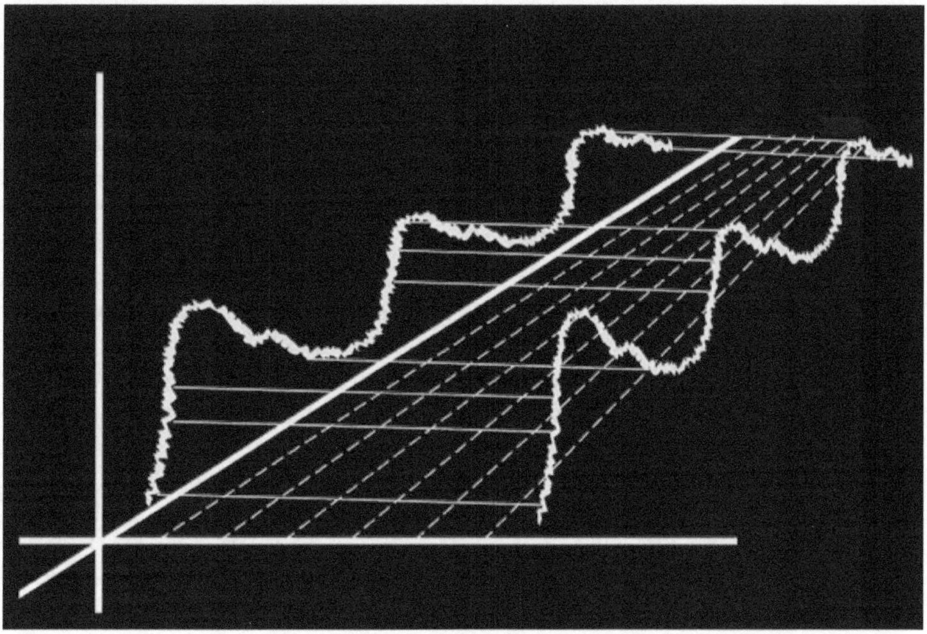

Fig 5

But it is that –apart of ignoring the long duration series- barely it has been done anything but *to compare* the pulses, being impossible to speak of authentic studies of the correlation, and not suspecting its depth; or rather, without suspecting that this correlation is revealing exactly in the more strict analytical sense, the physical and mathematical one. Without a doubt, it has contributed much to this circumstance, not only the separation between doctors and applied mathematicians, that nowadays is crossed by numerous bridges, but the spread prejudice on the "qualitative" and "specious" character of the signal of the pulse and, mainly, its frame of interpretation; prejudice whose dissolution is the front door to the content of this book.

And it is that, in the style of the mathematicians, we can speak of analytical continuation or extension when we open the space of correlation between both pulses. And practically in the same way: extending to the complex plane the different values of the real variable of both systems.

43

Independently of any method of analysis that we want to use, and even accepting the symptomatic and even revealing character of the signal of the pulse, it is very difficult to take off of the mind the idea that this signal has to be full of contingencies, as much more when more in detail we study it: from anatomical injuries and peculiarities to all a series of uncontrollable factors.

Since I am not a doctor nor I have any direct experience of the subject, I will not dare to affirm that any law or contrastable correlation in this area cannot contemplate exceptions. But, admitting in principle this possibility, mainly when we have not even begun to establish the smaller correlation, it necessary to say that we tend to enormously underestimate the capacity of adjustment of a global signal like this one. The balance has been there from the beginning, from the first anatomical bifurcations, and the smaller anatomical differences will even end up finding their functional correspondence and its resonance through the sequence or temporal series. Whatever it is the contingency, this will appear reflected in a double way and in the same measure in which it has some importance for the economy of the global balance of the system; it cannot be another way. Naturally, I already start from the assumption that in the pulse *any* factor that concurs in the state of health has to be reflected in direct proportion with its importance: until infinitesimal extremes, if we could follow them. The somatic unit is as much unit as the mathematical unit, that is to say, the number 1, also known like the monad or undivided totality by the ancients; and it is the problem of the anatomists in how many parts can and want to divide it. And if we cannot take off of the head the idea that the pulse, like the activity of the heart and any biological oscillator, has to continue maintaining uncontrollable irregularity or randomness, it is necessary to remember that this same randomness is already itself a measurement of *vata*, the total sensitivity of the system - and the most direct measurement, in addition. So everything depends on which are our departure coordinates, and the *triguna* or the *tridosha*, being from the beginning of global nature, can assume all those contingencies without the smaller difficulty –we rather are those that have lost the skill to think this way. And this way of thinking is in fact much more natural that all the clinical analysis protected in last instance by a formal mediation, like for example the biochemical data. Nobody happens to think that the sound of the motor of its car lies, although is far from distinguishing everything what it is saying. If this happen to us with a car, that barely has some thousands pieces, I do not see why the human body must not have expressions immensely more precise and trustworthy, without possible comparison. The same we could say of the human voice; the problem is not of lack of information, but of to what extent and within what frame we can analyze it.

The performance is present, and its composition, purely virtual. This is valid for any etiological model, also those of the modern medicine: we will always be able to find an endless number of interchangeable causes that adjust to the data that really interest to us. And precisely the doctor is the one who less can renounce to this meaning of the analytical and the synthetic spheres: if he does, already renounce to his very competence.

Let us remember that the same state of health, as feels it a subject, is another synthetic phenomenon par excellence: with the own life, the most general performance of the organism. To describe this simply like "subjective" seems to me only attributable to the awkwardness of our means of judgment.

44

On the nature of the correlations that we can find between both pulses – synchrony and phase lag, additive, multiplicative, periodical, etcetera- little or nothing we can say, unless they deserve the most careful and detailed study. This is an experimental investigation, and it is in the experimental measures and values where the analysis tools must sharpen. Not even we can advance what would add the correlation of the waves with the complete form to the correlation of flat or simplified waves; we cannot either anticipate what adds one to the another at the time of determining the values of the *gunas* or *doshas*, nor to what extent we can do without these in aid of a precise definition in terms of potentials. The relation between *prakriti* and *vikriti* does not have why to be identical to the relation between the vegetative and voluntary aspects, but in any case the product of both pairs must throw the same values. As far as the correlation between both pulses and the bilateral alternation of the breathing in the nose, it does not have why to give a similar frequency or amplitude at all, since from the beginning they are, even connected, very different cycles. All this and much more only can be refined with the experimental contrast, that is what this writing tries to stimulate.

Anyway, the mere idea of a complementariness or reciprocity between two aspects of the pulse that resume its monadic or unitary character lead us naturally to ask about its algebraic nature. Probably exist special and hypercomplex algebras –non commutative and maybe non associative, like quaternions and octonions, or something close to it-, that could fit in an elegant manner with different aspects of this reciprocity, so fundamental in any order of life and existence, with the condition of a right criterion in order to extract the variables. Nowadays, we don`t know any relevant application of this algebras to biology issues, not to mention that in theoretical physics they keep appart of the main stream formalisms, even when they allow equivalent descriptions.

I take advantage of this point to say that hardly will be never an investigation that can obtain such advances with so tiny means: sensors to obtain measures and a computer to analyze them. In addition, before venturing to us in the subject of the correlation of both pulses, a standard determination of the measurement of *doshas* and its proportions would have been seeded of difficulties and arbitrariness, in addition to have needed samples of pulse readings on many hundreds subjects. It was, really, the most costly and discouraging part of this type of investigation. Now, based on the correlation of both pulses, I hope that it will be possible to start to collect relevant data from the same beginning of the research, beginning here also the work of analytical refinement that this field is demanding. Of course, this does not exempt to us to do many more measurements and samples, and in particular the dynamic studies of series with the subjects under effort and activity, that seem to be specially promissory and revealing, also at the time of verifying faster the dual terms of both pulses; but the fact that already from the beginning we can take the bull by the horns would have to be stimulating.

45

When we try to think about what class of organic time is going to reveal these experiments to us, barely we make another thing that to come up against old categories and old ways of thinking. We could imagine fibbers being contracted and dilating towards the past and the future; or we could think about an expanded present whose entropy equally increases in direction towards the past as in direction towards the future, etcetera. All this barely as it is another thing that to intermingle qualities of some well-known systems discarding others to give other possible systems, but it seems to me that to speak about past and future it is particularly deceptive for the type of behaviour that we look for; if we truly enter it, these words will be used perhaps with the only intention to understand us, that is to say, without more value than the one of the analogy and the metaphor. In our daily life at least, past and future are no doubt important references, but few will have wondered to what extent they are necessary. My opinion is that much less than what is believed, although everything depends on the type of necessities that they want to get us involved in. It is more, I think that the distinction between two directions of the time is of identical nature that the distinction between subject and object, and equally illusory. Unlike the more spread present scientific vision, that also tries to be beyond the distinction between the past and the future, but it does it being based precisely on the distinction between an objective physical world and a subjective and diffuse perception. What of course it is a funny inconsequence. And from this inconsequence, already only remains to search how the subjectivity can be produced, how is synthesized.

In effect, when science approaches self-organization phenomena like the biological ones, it does not seem that it can do without the question of the memory anyway. At the cellular level, many things seem explainable by means of the genetic inheritance, which assumes the role of initial mechanism of data transmission through the temporary drift. Yet, this "memory" rather is reduced to the elaboration of biochemical components, needs to be updated by something alien to it and in addition hardly can explain itself all the cellular grouping phenomena of superior order: until the point that there is always somebody ready to introduce some type of doubtful "inherent dynamic memory" to the system. For me, the single fact that either the memory concept is used implicitly, even in the case of the genetic material, seems to me somewhat suspicious yet: it is already a form to disguise the same objective-subjective division that is present in all the other. That is to say, a causal reconstruction of the subject is searched –an imaginary composition -, because we are incapable to know what it is happening in a simple cell in any given while; hardly nothing on the evolution in real time, that so refractory seems to be the analysis. By the way all the cells pulse, and not only those of the heart.

This inconsequence emerges finally, and of the most spectacular form, in the subjects and businesses of the great black box of the "noble organ", our brain. The last frontier, as some call it; but that seems to us so equal to the others, except by that complexity in which we can take refuge. Here, the neurologists speak with any faith of the details of the short term memory, defined more or less by the temporal threshold in which the memory really does not need to be memory, but an expansion of the specious present. As far as the subject of the long term memory, nobody will pretend to have the smaller idea on where and how it can be stored; surely because it is not stored at all. It cannot be but an expanded version of the expansion of the actual present in which consists the so called “short term memory”. What it does not exclude the fact of the complexity of the neuronal plasticity, but rather justifies it. Be cause if the memory were really stored, the cerebral architecture could be infinitely simpler, aside that there would not have been as much problem in finding it. We are sent then to the difference and correlation between the vegetative and the volunteer, where is the true knot of the subject; beginning with the characteristics of the temporal series in both hemispheres, which constitute the most global and accessible of all the behaviours. And although like a mere formula it could seem very simple, we can believe that it contains surprises enough to maintain the researchers busy, as well as to remove unnecessary and spurious concepts. If we are sure of it, it will be because this evasive principle cannot be reduced to a particular dominion.

SECOND PART

TIME AND THE MODERN SCIENCE

46

It has been said, with that half truth so usual in the topics, that if the Greek mathematics were geometry, after Newton were mainly analysis. Affirmation that would be right if it is considered that the analysis grew mainly in the applications to physical problems, but that it is simply illusion from the chronological sequence of the facts.

We are not speaking about the old controversy on the authorship of the calculus, that already has been settled by the historians: the calculus was not invented neither by Newton nor Leibniz, but these were rather those who crowned it after a great number of previous contributions, and little matters who was first in possession of one or another formula which would fall like mature fruits. They two, without a doubt, did much by their maturity. But, in which does to the superseding of the spirit of the geometry, it is to say that was not Newton who took the first steps, but all the opposite. If we admitted its own later commentaries, that we have to believe, the calculus methods that could use for the incubation and elaboration of *Principia* did not have another dignity that the purely instrumental one, without giving the smaller substantiality or autonomy to them. Justly thus would be conceivable his absence of publications on the subject. The perfect counterpoint of this situation we have it in Leibniz himself, certainly much poorer of positive achievements. Leibniz conceived indeed from the beginning the calculus and the analysis like a new plane with proper autonomy and already freed from geometry, although, buried by the servitudes, barely he had time to explore it. What Newton took advantage of for more than three lustrums to perfect industriously all class of techniques and to demand the pre-eminence of his authorship. But what says the previous works of both, there for 1686?

That the *Principia* are the most arduous bobbin lace of the history of the geometry, and that *the Treatise of metaphysics* o r *the Monadology* are the treatise less interpretable in plastic terms of all the history of the western philosophy.

47

Without a doubt both Newton and Leibniz had to pay a tribute to their century; but whereas to the first that tribute benefitted to him enormously, to the second has impede any type of keen of reception.

The first French review of the *Principia* praised their geometry, but they advised to the author the improvement of all concerning to the "physical causes"; what listened with nowadays ears, seems to us the height of the impertinence. But all the perplexity before the hypothesis of the remote action at distance was justified enough. And Newton himself had to end up realizing that the best justification of the gravity law was not other one than the own space of the analysis, released of the servitude of geometry and the mechanical argumentations. In as much he was assuming this, the authorship of the calculus grew in importance for him; all this agrees very well with the sequence of the facts.

And precisely, it was the familiarization with the idea of the calculus the one that achieved that, half a century later, when the Newton theory became popular, people as Daniel Bernouilli or the same Voltaire no longer were surprised at all with the idea of the remote action at distance and even found it natural. Known it is the phrase of Voltaire on the matter: "the analysis is the art of calculating and measuring with exactitude that of which not even we can conceive the existence". It was the analysis and the analytical mentality what gave its credibility to the theory of Newton, and not on the contrary. And when still today we considered that the force of the gravity has to be a constant, we only paid a tribute to the analytical exigencies, and to no intuition at all. The immaterial ideas of the analysis are those that more have contributed to our belief of a world of inert masses, what is truly antagonistic and paradoxical.

The case of Leibniz was right the opposite, like it had to be for the counterpoint of this baroque fugue that still lasts extending and losing itself in other spheres. The philosophy of Leibniz is, in spirit, entirely musical: it is the first of the philosophies in transcending entirely the *more geometrico* within the rationalism and also the last one. But nothing harms more the understanding of Leibniz philosophy than the plastic images, of which he himself was sometimes involuntary responsible. Then what can seem the monads to us but a monstrosity having in mind mirrors and clocks? Still today, if instead of thinking about watchmaker machineries *ad infinitum,* we replaced these by waves, a great part of the work of Leibniz acquires clarity as if by magic. And about waves he must have spoken, changing in this one of the motifs of his mentor Huygens by the other one. Known are the difficulties that Huygens found for the mathematical elaboration of his wave theory of light, question this one that prevented its greater diffusion; what perhaps it is not recognized so often is that the Feynman paths of the modern quantum electrodynamics are not but the principle of diffusion of the light of Huygens – *from all the points to all the points by all the points-* with a finite speed for light. Leibniz is the most fresh, shining and intuitive philosopher of the modern thought, and the less discursive of all in his best moments; but it is the mere discursiveness of modern thought, united to the poverty of its power of imagination, what makes him to seem like an old rusty calculator.

But, again in tragicomic contrast with Newton, Leibniz barely had time to mature anything of which he did and discerned. Been born in a Germany devastated by the war of the thirty years, forced to look for the life, half adventurer and half poor devil, loser in all the roulettes of the fortune and the posterity, it is impossible not to sympathize with him. Even their spiritual disciples and heirs moved their look away from him discretely and tried to look for the inspiration in his more implacable rival. But the things always follow their course, ignoring misunderstandings.

48

Leibniz even got to challenge the Galilean principle of relativity, alluding to the impossibility to establish the correlative movement between two moving bodies. This only acquires sense within the prevailing role that it granted to the activity and energy over inertia. But these considerations moved away more and more of any practical approach. Right it has been said that the dynamics of Leibniz is the dynamics of the moving without movement –this is, without a space correlate -, whereas the mechanics of Newton is movement without moving.

The pre-established harmony of Leibniz, for example, for most of us the summit of the arbitrariness, acquires its sense and relevance in as much we realize that the absolute time of Newton, the principle of global synchronization, it is not less arbitrary nor metaphysical at all. And not only that: if we perceived the net metaphysical character of this last one, we begin to understand or to suspect at least that to which Leibniz is alluding with the question of the pre-established harmony it is exactly to the possibility of synchronizing entities and processes that are asynchronous themselves, since it is from their own that derive their law. But on the other hand the monads not even exist in the space, but are location principles, as much with respect to the space as with respect to many other substances or planes. In the same way that only for asynchronous processes can have something to synchronize, exists the non-localized for the locality, in a way that only partially can leave mark in the space. Leibniz himself defined the point like a pure modality.

49

All the philosophy of Leibniz exists to conserve the rights of the contingency; for that reason it is articulated around the modality, which it is unfolded and fluctuates within the scale of the necessary, possible and the contingent. On the contrary, the three principles of Newton exist to exclude absolutely the modality, to evacuate it, although it is virtually impossible since the necessity already is modality. On the other hand the idea of physical constant, that for Newton was not clear and only instrumental, it is not but a superstition generated by the effectiveness of the infinitesimal calculus, that the later developments have only contributed to underpin. A universal constant only seems to justify itself by the fact that without it the calculus would become impossible; but this is more than sufficient, so that actually it is outside all question, unless the facts insist obstinately on denying it, in which case, or can be ignored or can consider another different maneuver.

50

The idea of the monad continues seeming so strange that still today few are the ones that begin to decipher it. Because it is not the trite and empty figure of worlds within worlds –one more of this misunderstandings in plastic terms- but mainly of which coexists of lateral way. The monad is mainly centre or focus of activity; one of its planes or substances can be a body like ours. This body can be made of infinity or an undefined number of parts that concur in the activity of other monads, that are not only bodies as well, but other many substances. Of course, always results something unnatural and inconvenient to speak in plural about the substance. Here, the substances have parts and attributes, but not therefore the monads. What all this is suggesting is not that we are done of many beings, but that the parts that apparently we occupy are and exist of natural form in the sphere of other monads. This is moving, and in particular for the western philosophy. But how we could hope that modern science could contemplate subjects like this anytime? It seemed much more out of its capacity at the time of Lebniz, and so it seems today.

51

But a monad is a purely dynamic entity, not a substance whose bottom remains indefinite for always. It is much less abstract than this one, or that the idea of the space and the time, and all this only serves to generate confusions. It is the whole, undivided totality, and when we are feeling or measuring the sanguineous pulse, we are verifying in a greater measure the activity of the monad one that the own activity of the body. Then, any reflection of the totality, by small or tenuous that is, belongs more to the order of the monads that to the one of the substances and its attributes. Only for the moderns could appear this like incredible, because in fact the monad is the most natural concept.

Let us think in that other formula apparently so strange –and unnatural- of which the monad does not have windows. This tends to interpret in the sense that the monad is completely isolated of the environment. And nevertheless its relations extend to the infinite, and in continuity. But also this has a faithful reflection in the pulse and the maintenance of its potential: if the sensitivity of the system, that is its limit with the environment, also enters until any internal depth of the equilibrium or balance. This one is precisely the necessary and sufficient condition of any autonomy and self-organization. The monad is the most natural conception of a system that already contains to its environment like part of its definition.

52

Dewey described with skill the general tendency of the German philosophy like apologetic, and to the one of the British philosophy as sceptical. Without a doubt it has been so; although as this last one was assumed by the scientific practice, the boot's on the other foot now. Science long ago became apologetic, and every time are more those that are sceptical with respect to their assumptions and justifications.

The same figure of Newton supposed a decisive mutation of the philosophy of the islands; a rationalist contamination of the continent lodged in an empirical panorama with vocation to remain formless and so without commitments as can. Why it is described to Newton like introducer of the empiricism in science? What consolidates is exactly the opposite; in addition, the experimental method in physics already had received its baptism in Galileo. It is no wonder that since Newton the susceptibility with respect to the continent is sharpen, and that is tried to digest the influence until making it invisible and non existent. The controversy between Newton and Leibniz is knot and symbol of the relations between the continent and the islands to the present time. Although a germ, still visible, has remained without digesting.

53

At the beginning of century XVIII, Hermann Boerhaave, the most famous doctor of Europe at the time, undertook the attempt to rationalize the practice and theory of the medicine to image and similarity of an experimental physics totally consolidated since Newton. Boerhaave looked for yatromechanical principles of the maximum simplicity, even knowing the enormous difference between his dominion and the one of the mathematical physics. Although the school of Leiden left track in the attempts of renovation of the modern medicine, Boerhaave and their followers had to be satisfied with little more than an appeal to the experimental practice, reason why they could have resorted equally to the name of Galileo or the same Bacon than to the one of Newton. The simple principles never could crystallize in the medicine neither in Biology.

The yatromechanics, the medicine of mechanical inspiration, looked for the definitive overthrow of a certain medicine of chemical inspiration –the yatrochemical medicine, whose origin send us to Paracelso- in the same way that the experimental physicists had turned obsolete the scholastic disputes. Chemistry was understood during centuries, also in the West, like the science of the Nature and their transformations par excellence, but, being their categories too qualitative and tangled, the moment arrived in which their principles seemed too vain and specious in comparison with the much more net defined and manageable of the mechanics. The paracelsians spoke of three fundamental principles, deduced empirically of their operations of combustion, elimination and distillation, and in this they agreed with the basic philosophy of other previous medical schools. Nevertheless, of what they could separate of elementary way of inert substances was little less than impossible to say nothing organized and reasonable in terms of active and undivided relation.

Nor we have to say that these three principles, sulphur, mercury and the salt, were completely equivalent of the *triguna* and of *tridosha* of the ayurvedic medicine, or, to speak the more sure, they were exactly the same.

54

The three principles of the Newtonian mechanics become the three inherent modalities of the nature with a single condition: that they cannot occur simultaneously. That is to say, that any operation of reestablishment of the balance consumes time. Naturally, this changes everything to it, and in much greater degree than it is usually considered acceptable. And nevertheless, it is not necessary to say that the conservation laws would continue prevailing at a global level and as it is of rigor. Therefore, this version would not be an apex less rational than the well-known mechanics or dynamics; but surely it would be much more realistic and it would put to us on a very different game. And to where would go the time in that game?

We have already seen that in hydrodynamics this has enough sense. And it has particularly in the quantum mechanics, where we cannot do without the times of reaction; also local violations of the conservation principles occur here that tend to recover globally as soon as possible. And in addition, there is to count on the fact, frequently forgotten, that the law of conservation of angular moment in Newtonian mechanics has not the same axiomatic rank that the three famous principles, because it starts from the condition in which punctual particles has the same line of action –an extremely weak condition, as we can see, and that in most of the cases it has not why to happen neither in the quantum mechanics nor in the statistics nor in the hydrodynamics, except by average.

55

It seems to be that Leibniz was the first in using the term *function*, which would not be a surprise taking into account that he was also the first who understood the power of the idea of *tendency*. This was extremely important and even, as Bloch remarked, the modern idea of the dialectics is unthinkable without it. But, as it has been said, the same process of the calculus and the analysis, by examination of the successive approaches, is of dialectic nature. It does not belong to a "normative plane", nor is mechanic but for those for that the solutions are given. On this, the present mathematicians could say more than the mathematicians of any other time, but, lamentably, like last consumers of the mathematical production, between us the inertia prevails. In Newton any dialectic conception shine by its absence, which explains that he did not recognize the calculus although he was calculating all the day and had in front of his nose the whole time. Like Moliére's character, it spoke prose without knowing it. The history of the controversy of the calculus is as simple as this: To Newton, born analyst by pure scruple spirit, and next to their colleagues Wallis, Gregory and Barrow who already almost had crossed the bridge, their own methods had to be suspicious to him, and even worthy of being hidden. To Leibniz, virtually isolated in the continent, the method itself had to seem a wonder to him –with reason-, and dared to publish first and to generalize with agility and trust the most natural notation. To this the facts are reduced, but not the implications neither the resonances. And for the idles that always gives the things by known or supposed, we remember that sixty years later of the appearance of articles of Leibniz, Euler racked his brain trying to define the concept of function; or that barely metaphysical Gauss spoke without shame of the "metaphysics of the infinitesimals" at the beginning of the XIX century. Really, the basic moment of the appearance of the calculus is the overcoming of geometry. Leibniz was the first who gave that step; and to that step forward in the vacuum of the analysis must all the modern physical theories their main source of credit.

In Leibniz a single idea of tendency, considered in all its amplitude, serves to include potentially the physical world as much as the organic one with no need of a special distinction: the development is inherent to both of them, although nothing is gotten to specify and the role of the time is completely ambiguous in this respect. The principle of least action, to which the German gave with reason great importance, its not really specified either, although he includes potentially in it a much more vast scope than the contemplated by the present dynamics.

One of the less noticed and considered aspects of the view of the world of Newton is the absolute disconnection of any historical or evolutionary factor when contemplating the Solar System with its planets. In fact, Newton maintained a catastrophic vision: God had put of a blow the world in that state, and when he wanted that state would stop exist. Madly scrupulous, Newton didn't deceive himself about this, like about everything of his competence. Long before Poincaré, he suspected the absolute fragility of his laws before the eventuality of comets or any other source of perturbations. With the greater solicitude, Newton knew that *his* world slid magically in the nothingness, unlike most of their successors, infinitely more mediocre and narcotized already by the effect of the custom and the interests of the accumulated capital. To reach the recent times, in which the most innocent of the naturalists or the biologists arrives trusted to us to explain even the origin of the species and the life by means of mechanical principles and everything within a reasonable term of time.

56

The attitude of Newton and Leibniz to face the tradition of the past cannot be more opposed either.

Leibniz, in the purest apologetic tone, made always generous efforts to conciliate any scientific development in perspective –and he had a quite ample one- with the spirit of most of the old philosophy and with the mysteries of the religion, that not its dogmas and doctrines. Not in vain coined the term *philosophia perennis*, like pole of the wisdom that would have to transcend the circumstances of time and place. His integrating effort was sincere, and tried to give new senses to the motifs of the ancients and the contemporaries from its own spontaneous, original and superabundant source. Then precisely to him, that so lamentably had to leave dozens of important concepts and intuitions to develop, we can suppose less needed than anybody of other people's ideas.

Unlike Newton, who always was persuaded of *the knowledge* of the ancients –but not so of their wisdom -, thinking that the more perspicacious among them had transmitted it hidden in cryptograms. It is a funny idea that the ancients had to contemplate the things in the same spirit that the scientists of the XVII century, and had to understand them in such terms. Conception that, in addition, is not very different from the one of so many mystery and fantasy novels. That's to say, no real mystery, only secrets to crack. The same author of the three principles of the mechanics was secretly arrian only to deny the idea of the trinity, that it seemed to him a fiddle of the Church fathers. In this way, he denied any generative power and any possible relation of that trinity with the nature, which hardly reached the status of large estate of a Lord.

As far as his contemporaries, it does not seem coincidence that he maintained controversies with almost every of them. The luminous ring that discovered Boyle or Hooke soon turned the ring of Newton, and thus with almost everything. But known it is that what with Newton begins is, precisely, the scientific establishment and the manufacture of its prestige, and this so it was needed of unit as of the figure of a supreme pontiff, to whom he himself contributed with all his actions. The moment arrived in which, like with an absolute monarch, everything converged towards him, and in as much as everything moved away from Leibniz. Known it is that finally and at the limit, the very same Hanover employer of Lebniz was crowned king of England, to which something would contribute the genealogical vindications of the philosopher, in another one more of his forced servitudes and works; and that the new king forgot him completely for, among other things, not to bother Newton already at that time turned true institution.

The details of the biographical antithesis of the two great authors of the infinitesimal calculus, that even were born under the sky of opposed constellations, border the incredible, with infinitesimal aspects in all the modulation of their dialectics. They make recommendable the joint reading of both biographies, probably boring in themselves by separate. And it is that, at least by once, we had to be able to distinguish this deep polarity not only in the tree of the life, also in the tree of science: the polarity of disseminating and concentrating, the vegetative and the voluntary, the speculative and the speculator –the prolific, and the devouring.

57

It was not coincidence either that the islands were left behind in the development of the analysis with respect to the continent throughout the eighteen century and good part of the nineteen century, being adopted the notation of Newton instead of the Leibniz one. Leibniz found direct disciples in the Bernouilli brothers, initiating a line that would be continued with Daniel Bernouilli and the same Euler, whom among other things, developed the variational principles, the last and unavoidable form of modern dynamics. Lagrange boasted of having managed the dynamics doing without geometry entirely. Newton, we don't need to say, was the great reference for all of them.

58

But with the Bernouilli and Euler another different branch of the analysis begins also, as separate of practical problems as possible: that in which the analytical techniques begin to approach questions relative to the number theory.

Leonhard Euler, in who seems that the spirit of Lebniz had concentrated itself with the only aim to make mathematics, found a formula that related in an exact way the infinite sum of all the natural numbers to the product of all the prime numbers: the famous Euler product formula. This it is a good example of something inconceivable that it simply turns out to be certain, and even relatively easy to demonstrate; and the decisive precedent of the analytical number theory, several generations of distance from its accomplishment. In spite of his achievements, the same Euler left written that the mysteries of the distribution of the prime numbers would be beyond the human understanding for ever.

59

Gauss and Legendre did not have to think the same, around the change of century, specially the first, who was the one that gave greater impulse to the classic number theory, also called superior arithmetic: in fact, the single study of the implications of the Arithmetic. Nothing is so fundamental in mathematics, nor offers anything similar resistance and difficulties. Gauss gave the law of the global distribution of the primes in the form of logarithmic integral, which took a whole century for its proof. Some comprehensible law for the man was in the prime numbers, after all; today is called the Prime Number Theorem. And in spite of this great step, how little still which is known on the subject. The same prime number theorem only found its proof through something infinitely more enigmatic: the order that the prime numbers showed by analytical extension in the complex plane. Bernhard Riemann was the first man who saw that landscape.

60

It is worth the trouble to stop a little while in the modern analytical number theory; the one that goes in direct line from Euler to Gauss, Dirichlet and Riemann. It is one of the great achievements of the human spirit, as surprising as the development of physics and surely with many degrees more of depth, but much less appreciated and known. An achievement that was not born from any type of urgency neither of hunting of the supposed secrets of the nature, but from the most incorruptible and free of the explorations. But the circumstances are not gratuitous. The space of the analysis tends to pulverize the objects, breaking them until infinity, or until is possible. The power of the analysis arises from the development of the series; but of all the series, none more natural than the one of the natural numbers, or the whole numbers, widen a little bit. As the pulverizing power of the analysis grew, particularly for which we called the dynamic space and the real space, the integrity of the whole numbers, its irreducible and discreet character, became involved in a completely new reluctance and charm. Already Leibniz –and this is quite significant- had appreciated all the importance of the discrete mathematics, as he himself denominated it, with an ample synoptic vision of which now we called combinatorial analysis. And although barely he worked in number theory, nobody more indicated to conceive the new relation that was approached here. But Gauss sentenced it with his habitual straightforward style: the Arithmetic is the science of the time, just as geometry is the science of the space. For physics this affirmation cannot be falser. But physics not only has arisen from the mathematics, but in fact is applied mathematics. Within the mathematics the affirmation of Gauss will be unavoidable for ever. Geometry cannot separate of the given or empirical data; the Arithmetic yes, and as completely as can do it something under our scrutiny. It is more, it would be possible to say that it is the only point of the formal sciences in which the scrutiny of the own intellect is possible, -the self-observation-, no matter how hard this is prevented constantly by the very instrumental character of the techniques used.

The dynamic space of physics is equal to the analysis of geometric origin (algebra included) *minus* an infinitesimal part of the Arithmetic. This residual subtraction still prevents that we consider the physics like something empty, although soon we fill up it with the mistaken material. The filling is time in anyone of the cases. And it is possible to think that, in the same measure in which this remainder grew, the dynamic space with its ghosts would be contracted.

61

It is not surprise that for some mathematicians the Riemann zeta function is the only scientific discovery worthy of respect and admiration: in comparison, the best achievements of physics are as magnificent as mature apples and equally vulgar. And it is that there are reasons to think that, being so strange its behaviour in respect to other kind of functions, has had to arise to crosscurrent of the habitual developments, is to say: like opposite current or reflux of the main current of the analysis.

The Riemann zeta function is classified between the special functions; and, within them, it plays the main paper in the Arithmetic and the series of the whole numbers, of which derive all the others. The whole numbers are generated by the most elementary of the operations, the addition of the unit. When we attend the products, we observe that some numbers can be decomposed in products of inferior numbers, and others not: they are the prime numbers, numbers only divisible by themselves and the unit. The order and sequence of these prime numbers has a global regularity –the logarithmic integral we alluded before-, but a local behaviour as random as it can be possible. Thus, the zeta function becomes the best known form in mathematics to unite these two antagonistic and apparently irreducible components. That is to say, we cannot know a bond between the order and the chance better defined and more fundamental than the one than already exhibits the whole numbers with their primes, and the Riemann zeta function is that bond. It is no wonder that such a bond has so endless ramifications.

62

From the very beginning of the calculus the theory of the probabilities begins to be developed; they are like the two faces of a same current. The apparent exactitude of certain laws and the same functions does not make but emphasize the possibilities of disorder for independent events. And it is that the function concept tends to become synonymous of dependency, and the probability, of disconnection or independence. Between both ends it can be all kind of combinations, and thus we found from probability functions to probability of a type of function within the space of the functional analysis. The analytical descriptions would be infinitely fragile if they did not tolerate all this counterpart of the theory of the probability, and this last one would not have been developed without the first. Therefore, they are as indivisible as different: from the point of view of the same process, it can be said with all property that they are complementary, although our understanding tends to differentiate them as much as it can. A form different to consider the probability, and even the relations between necessity and probability, is the combinatorial analysis. One of the most remarkable peculiarities of Leibniz like mathematician and philosopher it is that he thinks in terms of differential analysis and combinatorial analysis almost simultaneously: exceptional characteristic that emphasizes the mathematical historian E. T. Bell, but that any reader can perceive in philosophical texts, sometimes disconcerted, and sometimes even with reluctance. It would be said that the human mind prefers to separate these two scopes in different compartments, to have more space jumping from one to the other. And what it is gained is, precisely, space: what is lost in this operation is the time.

63

Not very often we considered that the unfolding of the thought of a person is an exhibition of the whole nature. And nevertheless we know it, although badly. For example, it is too easy for us to see other people like natural subjects, and we even found please in it; but one can not see oneself the same way. We alternate thus between two visions, without having barely the smaller sign of a complete frame for ourselves: the only one that could give a right perspective of the other people. The word "nature" is only a synonymous of that correct perspective or whole frame of the activity of a single entity; and we will say that that frame and that perspective are indistinguishable of its spirit and of the spirit –what it responds to the question of its continuity, that even it is not necessary to present.

64

The Riemann zeta function allows for example to calculate the exact number of “prime numbers smaller than a given number”, whatever the number is; which already is miraculous enough having in account the purely random character of the primes in a concrete interval or proximity. But, only to arrive at this, which made the title and the apparent intention of the original Riemann's writing, it is easy to suppose that the function must contain much more information than the one than it can give us in the limit of one of its transformations. Of course, it is not information which lack in the zeta function, but rather on the contrary: one would wish to reduce as much as possible. Which takes to the formulas again to generate the function, whose main disadvantage is to be more or less equally monolithic and without fissures than the own series of the numbers. This summarizes a little the class of difficulties that presents the problem of the famous "Riemann hypothesis".

The Riemann hypothesis says that all the non-trivial zeros of the zeta function in the complex plane has a real part exactly equal to 1/2.This value is known like the critical line. The proof of the Gauss prime number theorem demanded that at least exist one critical strip between 0 and 1 as excluded values; what it is not something trivial either. The zeros of the zeta function and the prime numbers satisfy a certain duality through the spectrum and the harmonic frequencies; including these last summation formulas, the duality is not of irreducible character, and we also could speak about a ternary relation. And, really, it is just to say that these three aspects –prime numbers, summations and zeros- form an extraordinary triangle, open and in movement, with infinity of transformations between them and within. That's to say, has a certain resemblance, in other dominion, to the open triangle of the *gunas* in the Samkhya.

It is assumed, on the other hand, that this critical line is the most synthetic part of the endless information of the function, and the easiest form to filter it. But, on the other hand, there are other many interesting questions, between which overhang the values of the imaginary part of the zeros, the height of this along the critical line: hardly anything is known on its correlation. These values of the imaginary part are related as well to the "periodic terms" of the function, as Riemann called them, in fact oscillating terms, that play a very important role although often they are relegated to the background. Many very general relations although nothing clear with the harmonic analysis and the spectral and wave theory are derived from this feature.

65

The zeta function has been generating more astonishment as have been made calculations on the non-trivial zeros, all which agree with the hypothesis till the moment. To this day, it has been verified on more than a trillion of zeros (10^{12}) without no counterexample. It is more, almost all the signs seem to indicate in the direction that this will continue being certain until numbers inconceivably more elevated, in which some of the aspects of the function alter their behaviour in relation to the interval of smaller numbers and, could, perhaps, generate some counterexample or refutation of the hypothesis. The truth is that, not to have the smaller idea of why the zeros must have real part 1/2, this behaviour is already extraordinary enough, even if someday a counterexample is find. At the moment, the critical line is there, without another ratio pretends that its own obstinate existence.

Like non mathematician, always has called my attention that the mathematicians who work in this so terribly difficult subject speak about to demonstrate or to refute the hypothesis, and that they speak about understand the function much less. I must suppose that they cherish the hope that a proof or a refutation of the hypothesis will also illuminate the understanding of the function almost automatically, but this hope seems to me unjustified; not to mention the automatic character of the illumination. But, to be fair with the mathematicians, it is necessary to say that they are the first who kindly explain to us that they do not understand the behaviour of the function; even they say to us that it is the most fundamental, elementary level, the one that is not understood. What it is not surprise, because that elementary level is not other one that the relation between additive and multiplicative properties, but taken altogether –globally- already from the very beginning. As it is easy to see, this is in intimate relation with the double aspect, differential and probabilistic, of the function; being the probabilistic part (taken "separately") based on the independent or additive relations, and being the differential part (taken "separately", and including the logarithmic decomposition in sum of products) based in the multiplicative and dependent relations. But the problem here is that there is no separation at all, because exactly what does the function is to combine all these elements of the most compact form, so that little more fits that to study the values that throw the arguments: what we call the "behaviour" of the function.

66

Since the problem thus presented is so impenetrable and without fissures, it has been tried to give it the turn and to raise it the other way around: to try to imagine to what class of physical or dynamic behaviour could correspond the behaviour of the Riemann zeta function. Apparently, Hilbert and Pólya already raised the question at the light of the functional analysis and the operator theory then emergent, there for 1914.This only could begin to have some relation with the reality when that same theory of the operators found in the quantum mechanics a suitable field of applications; and even then there were the least evidence of relation. These dispersed evidences have been accumulated through the time. The history and the sequence are more or less well-known: Selberg gave some more explicit relations with the spectral theory; Gutzwiller made an approach in terms of periodic orbits for a semi-classical system with a Planck constant tending to zero, Montgomery settled down the similarity between the spacing and repulsion of the neighbours zeros and the distribution of the energy levels in the random matrices that are applied to problems of the interaction of many bodies or particles within the quantum mechanics. Only it was missed, since Berry did, to propose some type of quantum system that imitate the behaviour of the function; or, since Connes did, even to create a specific operator with an equivalent behaviour. All this is extremely interesting, of course, but at the moment doesn't help much to understand the function itself: for which reason must have or not to have the zeros in the critical line, and even why they must have those particular values in the imaginary part.

67

In spite of the enormous extension of the developed work, everything follows almost exactly just as in the beginning. Nevertheless, the relation with real dynamic systems, with physical systems, does not stop being another extraordinary fact more of this function, and one of the more unexpected. Because, what relation would have to have pure an arithmetical series, with its prime numbers and all the others, with real a physical system? And a quantum system, in particular. Well, after all, the quantum systems have indeed an irreducible discrete component, but nothing else. The relation that this could have with the prime numbers is completely conjectural. Perhaps, yes, the prime numbers can be related to resonance phenomena, in a somewhat similar way to how the rational numbers play an important role in the chaotic orbits of classic systems; but nothing of this is clear. The quantum systems, apart from that, are of wavy nature; in relation to the zeta function, the main difference they introduce with respect to classic wave dynamics is that they exhibit a much greater number of possible states. Therefore, it is possible to consider the zeta function in relation to dynamics in general, more than to the quantum mechanics in particular. Or rather, chaotic dynamics in general: that ample and diffuse field between the purely random behaviour and the determinist behaviour with stability. Thus, for example, thinks Michael Berry, who has suggested that the resolution of the Riemann hypothesis would provide to the analysis of chaotic systems something similar to the harmonic analysis for the more ordinary oscillators of physics. This would not be few; but nobody can suppose of what conditions it would have to start an analysis with such a power of resolution.

68

The connections of such a hermetic object as the zeta function with dynamics are encouraging and stimulating, but, simultaneously, potentially misleading. Speaking so much of operators and other tools applicable to the physics, one forgets for a while that is trying to understand a purely arithmetic "object". And I write object in quotation commas because it does not seem to me suitable at all to the arithmetic dominion, which, as we noticed before, is the more refined dominion of the time that we can put under our scrutiny or analysis.

Let us think about that even it is impossible to say if the infinite series of the numbers is something open or closed; and obviously I do not talk about the potential infinity of its enumeration, but to the possible relations. What it would take us now to semantic questions only; the mathematicians settle in their way defining the set like an abelian semi-group with double structure of total order (additive) and special order (divisibility).This is a simple characterization that does not limit almost anything either. The perfectly trivial fact that the function has its singular pole in the unit always makes think about the predominance in a certain sense of the total order: in the indefinite repetition of the same conditions, that entails the other order. There is no need to say that with this one does not go anywhere, but, in any case, to consider that the function in all its immensity exists there just like that, take us to that position half platonic and half empirical that conforms the more external and instrumental aspect the mathematics, and of the physics and our view of the world in general.

In any case, it seems evident that the confirmation of the hypothesis imposes the simplest of the behaviours, and that its refutation would render a more complex behaviour: but of this it is not followed automatically that its certification makes more tractable or compressible the relations and complex aspects of the series. That is to say, also it can imply the opposite, that it is that in the critical line all the circumvolutions of the plot are stretched to the maximum and no longer is possible their later reduction. In other words, exactly the critical line would be the most impenetrable. This seems well compatible with the nature of the function, which again takes to think to us that a greater understanding of this one –the function only- would be more desirable, ignoring the hypothesis as much as were possible. And this was what Riemann did.

69

Bernhard Riemann was a man with a special capacity to receive. That is to say, he had that we call receptivity in a very high degree. What is worthy of remark, because remembers to us that for some important things it is not enough with the mere intelligence, the avidity or the ambition; whole generations of Newtons not even had smelled this subject starting off of their own initiative. As sure as we can be that he worked much in the manipulations and transformations of the function, we can be sure also that at no moment he considered this subject as the hunting of the white whale. What he had to look for at any moment was a union as solid and indissoluble as possible of two apparently dichotomizing behaviours. And that search and effort are exactly its creation; because what emerged soon through it surely surpassed his expectations. But what Euler and Riemann united for ever with as much care as fortune, now we want to disassemble in pieces like an artefact.

Known is that Riemann also concentrated all its forces in physics and the philosophy, and that both concerned so much or more to him or more than his own mathematics. Shortly before its communication on number theory, he had already elaborated a mathematically very refined theory on the electromagnetism, which finally he preferred not to publish. As far as the philosophy, only we have had left dispersed fragments, and, although perhaps living some more years he had been able to distil them, the certain thing is that we can have more than doubts on that possibility: the verbal and written expression displeased too much to him, for which he had very little ease. Nobody can be everything simultaneously, and in the case of Riemann the physicist and the potential philosopher had to suffer the personal pre-eminence of the pure mathematical thought. But, with his analytical qualities, so much predominated in him the concentration and synthesis spirit, that in the end almost nothing was lost: Riemann is not only the deepest of all the mathematicians, is also the one of deeper implications for the physics and the philosophy; because both things only can go hand with hand.

70

We said before that the similarity of the behaviour of the zeta function with certain dynamic systems can be very misleading, and that is preferable to always have in mind its arithmetical nature –purely temporal-, by more stimulating than the associations can be. But we have here to introduce a turn that could only be unexpected for those who ignore everything said yet.

We have seen that the dynamic systems, already from the same Galileo, make a selection of the reference frame with preference for the inertia; this means that the space itself, or the points in rest do not have any sense and that is the movement what determines them. This means that the previous positions simply disappear, absorbed in the momentary dynamic state of the system; this is what we understand by dynamic space, of which is implied that it introduces an irreducible temporary element, although in fact does not make another thing that to geometrize the movement, which is completely different and even rather the opposite thing.

Nothing of this happens in a complete arithmetical series: here the relative positions of each one of the elements are not lost nor vanish, as much they can be numerable. Exactly this is considered like the "timeless character" of the numbers, the property that makes of the mathematics the instrument of proof more self-sufficient and trustworthy. For me it is rather the opposite: that nothing here –numbers as such- lose its position nor vanish is what guarantees that we are in time itself as it is, unlike a dynamic system that defines the instantaneous cutting it out from all the other as if it had scissors.

And in fact, a quantum system tends to vanish a good part of which we consider a classical dynamic system; in particular, it dilutes the badly called "causality principle" of such systems, that on the other hand shines by its absence and only talks about the global synchronization principle, the "scissors" of the present moment and the momentary state of the system.

Nevertheless, it continues being a proper dynamic system, in the sense that the conservation principles are applied and presumed. This way, the characterization of the quantum mechanics in purely dynamic terms is, as it could not be less, irreducibly ambiguous. And the duplicity between the local connections –with finite speeds- and non local connections –simultaneous- it is practically impossible to treat: they can be contemplated alternatively, but not simultaneously in a single form. We can say then that a quantum system is dynamic, but much less than the classic ones: in all the senses, and even although here there is apparently much more fluctuation and movement. It is by this that the connection of a quantum system with an arithmetical function seems less surprising: in as much in which there is less dynamic element there is more temporal element. It is natural and logical, even being very far still of particular or explicit relations. Something similar can be said of chaotic systems, even being these of very different nature. Some of the algorithms used right now to explore the zeta function are highly similar to those for the problems of chaotic orbits for many bodies in classic dynamics, which sends us to the pioneering approaches of Gutzwiller in that direction.

Time emerges in as much we do without the cut and the selection of dynamics. It is good for recognizing it, even when it does not make the things easier to us.

71

This anomaly with respect to any known dynamic system begins to be perceived when physicists and mathematicians try to define the supposed operator that would underlie to "the zeta function dynamics", that some even call "Riemann dynamics" to be brief. It is said, for example, that the zeros have the same statistical properties that the energy levels or eigenvalues of a "non-integrable Hamiltonian not time-reversal invariant, with not known variables", and quite similar things. Any characterization that we could find in the Literature of the subject will not be more precise. Yet, "external" properties like those that Berry remarks are worthy of interest: chaotic, unstable dynamics but with boundaries, instability of homogenous character, quasi-one-dimensional dynamic. But, when we want to tie all these ends, it is quite clear that we do not know any dynamic system like that.

72

Voronin gave a universality theorem with respect to the zeta function in which it demonstrated that, assigning codes to the numerical information of the same one, the function contained any amount of information and formal knowledge that the human kind can reach, repeated infinite times, and with any arbitrary degree of precision. Mathematics contains other seemed monsters, like the fractals:. but these are not infinitely differentiable nor talk about something as fundamental as the series of the whole numbers.

Even not liking this type of generalizations too empty always, we can use it to imagine another perspective: let us imagine this apparent universe, whatever is its dimension, with all the states of its particles, throughout anyone of its possible times of evolution. Could be described all these states within the zeta function? Yes, naturally, according to the Voronin theorem. Now, we try to think a universe with a pre-Galilean dynamics, if it is that it is possible and it has some sense. I am afraid much that it does not have it, and that a "pre-Galilean dynamics" without inertial frame of reference or relativity, aside from not existing, not even could consider particles in our terms; but this now does not matter, is only a vague mental experiment. In that universe of the here called pre-Galilean dynamics, any previous state of the system does not disappear, and continues adding itself to the successive states without we know if it can affect them in something or not. This way, we would have an apparent inflation of information. However, we imagine that this pre-dynamic universe, still much more inflated of states that the universe of familiar dynamics, we had it to compress in a line or dimension, that is to say, in a quasi-one-dimensional system. Would it fit still? Surely yes, and the zeta function would keep a good degree of similarity with it. And to where the space would have gone if all the codifications of the space can be ordered in a line or an almost one-dimensional set? Of course, I am not willing to respond to these somewhat waif questions, that, I am afraid, also reflect partly our waif idea of the time, the things and the cosmos. Nor I believe that it never could be much compatibility between dynamical frames and pre-dynamical or post-dynamical frames.

73

As far as the characterization of an underlying operator to the zeta function in terms of its Hamiltonian, the Hilbert spaces, the renormalization group and so on, this is so general and misleading as to speak in dynamics terms. And nevertheless, yes, it can be very fruitful, and that is at the moment what matters. I believe that even it is absolutely necessary to consider the analogy with dynamics with as much rigor as possible, because perhaps another way does not exist of which we begin to conceive the limits of dynamics; limits that right now are inconceivable to us.

For me, the Hamiltonian only means here that the principle of least action, or extreme action if is preferred, is present. But in the first part already we considered the hypothetical possibility to apply this principle to mixed systems of a special class, of which the pulse was a good example, which tolerates friction and certain compressibility. That is to say, we considered the principle of least action in an ampler sense than the one of dynamics or mechanics; a sense very similar to the one that Lebniz could have in mind for its system of the nature. Known it is that the leibnizian system was denominated "dynamism" by the secondary role that had inertia and the inertial frame in it, that is to say, by the few that it had to do with which now we called dynamics. And known it is also that he tried to generalize the principle of least action or of the optimal paths to the economy of the monad and the organized parts of its aspects or substances –body, thought, etc -. That we called before pre or post-Galilean dynamics would not be another thing that the Lebnizian dynamism, that, on the other hand, does not let seem to us excessively baroque.

It is the critical line of the zeta function an extreme line or a line of least action for the union of the differentials and stochastic aspects of the function? Perhaps it could be spoken thus, if we distance of the merely dynamic interpretation of the variational principle. This one is always ampler than anyone of its physical interpretations.

A distinctive seal of the reduced work of Riemann is its unit. On the matter, Weyl said that, precisely in the work of Riemann, and more than with any other mathematician, it was necessary to consider its unit. Unit that apart from that is too implicit and cryptic, with ample space for the speculations. From his first great works in complex variable, there on 1851, Riemann was based in the variational principles of the theory of the potential; in the called by him "Dirichlet principle", that Gauss and Green already had used, and that so many headaches gave to him. The same Dirichlet, mentor and teacher of Riemann, worked as much in the theory of the potential as in number theory. Between Leibniz, Euler and Riemann there is a kind of continuity like the one that can be between Newton, Maxwell and Einstein; a powerful and persistent continuity, in spite of the great differences that the evolution of the main subject marks.

74

The Riemann zeta function is the mother of a whole family of functions, the automorphic L-functions, derived from the Dirichlet L-series. A distinctive feature of all them is to have its own critical line, on which also extended conjectures have been made of the same nature that the one of the zeta function. This one remains like the fundamental by its more direct and general connection with the series of the whole numbers.

Not to delay me more, I will say that the idea which I have in mind is that the study of the time series of the pulse and its inside-out dual behaviour can help us to understand *some of the aspects* of the zeta function and its associate functions; aspects that at the moment do not seem to be contemplated neither by the different branches of the number theory, nor by the alternative dynamic interpretations, and that could be very important. What it is very different to construct a system *ad hoc* to try to replicate the behaviour of the function. The function is enormous, and what we want is to isolate some small, but decisive, distinctive aspects.

We must remark this, because we are not speaking of which the pulse and the function are similar things. The pulse and its dual correlation can and must be studied perfectly in a totally independent way, with their own and simpler tools of analysis; but the approach to the zeta function, with all that could imply, really needs other references in addition to those better known at the present. Already in the first part we pointed the opportunity of some kind of algebraic generalization for the reciprocal or complementary aspects of the pulse; and, certanly, quaternions and other excepcional groups have been used for some attempts of demostration, despite how far were they of its original purpose.

I cannot hope that everybody take seriously the possibilities of this connection, although equally unexpected connections in other areas of dynamics have changed the mentality on the subject remarkably. The relations that at first sight we can establish between both fields are, of course, somewhat vague and accidental: the dual character of the behaviour of pulse, still entirely unexplored; its mixed character with regard to the discreet and continuous series, as well as to the conservatives and dissipative aspects, etc. Also we could speculate about the tendency to the repulsion in the fluctuations, and some other analogous signatures, but these simply enumerated would have to be sufficient, at least for some people, that look for an operator with properties and variables extremely ambiguous, if not contradictory. And if we are sent to the obviousness that the pulse is a temporal series, we say that it is so in a much more specific sense that what by "temporal series" it is usually understood, that is to say, an empirical series with its continuous or discreet spectrum. Truly, I would say that the pulse has more arithmetical nature than anyone of the dynamic systems that we contemplate, whereas we contemplate from the perspective of the dynamics; and by the same I will say, if it is allowed me, that this type of approach not only is not preposterous, but maybe purer and more natural than those dynamical approaches, and even than many of the strictly mathematical approaches. In the zeta function, the extremely delicate balance between global order and fluctuations are not only important, but fundamental; in fact some of the approaches with more interesting potential applications are referred the theory of the stability.

Evidently, the very uncertain and not already defined duality of the pulse has nothing to do in principle with the very certain duality of the function, but, allowing a kind of reference for the fluctuations, would have to show finally its connections with the other problem. I recognize that an immense dose of faith in pure mathematics plus an equally immense dose of faith in applied mathematics is needed to believe in this possibility. I don't know if people such exist in this world, but, in any case, to them is addressed this writing in first instance. But the concepts are, though incomplete, very natural, and worth being explored.

We start from the undivided, monadic character of the Riemann zeta. But this not implies that it is absolutely intractable. Unlike the atom, a monad cannot be split; but a monad breaths, that's the miracle. Like almost all the dualities of the analysis, the one of the pulse is not irreducible at all, but a comparison term with a vast ground within. And, in as much we cannot "crack" the function, we will find more and better signatures of this breathing. Naturally, in mathematics we only can find this breathing in the form of pure relations, but even so, there is something more than a metaphor here.

A monad is pure activity, exactly like mathematics; pure activity, but not pure act. Probably, those who want to crack the zeta function believe much more in the platonic heaven than us. Moreover, they could believe in an empirical platonic heaven –a very strange place, indeed, though a grain familiar.

75

If we say that the pulse has more arithmetical nature than anyone of the usual dynamical systems, which here is allude is its cumulative character, as doubtless as the one of the set of the whole numbers, and with more than one similar difficulty. It could even be postulated that each new cycle of the pulse introduces newness necessarily, but, naturally, only in the continuous dominion could find this its reflection. On the other hand, it is or certain that we found logarithmic scales or spirals in the same organic rate of development, growth and aging, and even the same experience of the time with the pace of the years: in those hours and days that for the babe seems to be limitless, and that so fleeting became with time itself for the elders –and more for the late ancients, one could think. The same logarithmic law of Weber-Fechner –both colleagues of Riemann- which relates very generally the excitation and the stimulus, have a strong although yet indefinite connection with this. In the pulse it is also present, in addition, a "critical line" that allows us to distinguish the degree of inexorability of this general tendency, that is to say: the proper degree of reversibility or irreversibility of the modifications in course. And again, *prakriti* and *vikriti* are the only criterion and frame of reference to settle this question. Without the refinement of this criterion, whose limits is absolutely compelling to know, aside from what revealing can be in themselves, any other form to measure the degradation or entropy, the disorder or the aging of an organism, will be so sophisticated technically as absurdly primitive in concept.

76

So perhaps we don't need to construct a complicated device to imitate some essential characteristics of the famous zeta function, and have these and others either to our disposition; in order to realize strictly this it is necessary to deepen with sufficient well-taken care and rigor in the experimental study of the pulse. To understand the nature of its variational principle, quite a lot ampler than what it is understood as such, but also rather more explicit from the causal point of view –if we understand causality like the ordering and sequence of the fluctuations.

One of the many important questions that there would be to limit is the least time interval for a fluctuation. Or in other words, what minimum interval we need to define the sensitivity of the system within the potential of *tridosha*, or, *vata*, that comes already given in irregularity terms. In addition, this interval can admit different readings according to takes the pulse in one of the wrists or considering the correlation between both.

Not even we will try to enumerate the most important points that it must require this study, because only in the experimental arena could they highlight. We will only say the following. If some dream about an analyzer of complex or chaotic dynamics based on the zeta function in the style that Berry suggest, they can be sure that any temporal series of this class cannot be reduced to other smaller elements than the ones corresponding to the fluctuations of the *tridosha* in the pulse with its specific variational dynamics. This would be the limit of "spectral resolution" and of the compressibility also. But this limit would be a big theme itself.

77

Some could ask to us what could have this to do with the basic physics. The answer could be: "Nothing, neither we need it at all. A complex or emergent system does not have why to have explicit relations with the fundamental physics, except for the elementarily indispensable ones." Thus we would avoid problems. Nevertheless, it is something annoying always to leave the things in the air, particularly if we can go further and demanding it the conclusive character of the subject. Apart from that, also the relation between quantum systems and prime numbers remains like an absolute mystery.

Of course that our theme does not seem to have too many connections with the more spread vision of the mainstream physics; mainly, when it is considered that this one hardly has any vision, but a good bunch of equations. Leaving this aside, we will make an effort to locate the subject closer to the area of problems of the physics in its historical development and present state.

78

Since Newton we have the problem of the vacuum in physics. Namely, that in an inertial frame with forces, like his, the emptiness or the vacuum cannot exist, neither an ether-like medium. Known it is that own Newton researched with regular intermittence the dynamic possibilities of the ether, already from the first years, returning in addition on the subject in the last years, with successive editions of the *Optics*; and that never could settle the question.

The ether or universal medium counts with a long history of illustrious attempts to describe it: aside of those of Newton, we have the own Euler, who gave us a hydrostatic and almost Archimedean version of the same one for the gravity, and Riemann, and Poincaré. Einstein was neither convinced at all about having eliminated its necessity, except for his own formulas, and Leibniz spoke of *a plenum* like a medium, but still more intractable than the one of the other mentioned proposals. As it is seen, the problem of the medium has enjoyed an enormous attractive and prestige, particularly between the greatest mathematicians. At this sight, that today we tend to forget, it is more surprising the sufficiency whereupon we judged the subject, mainly if we notice we are not better than before. It is not only that now we have scalar fields for the masses of particles, the inflationary model and even for the fluctuations of the phantom cosmological constant, not to mention the polarization of the vacuum and the energy of the fundamental level, the so called zero point. No; all this is or phenomenology or complementation of the present discoveries. To which we talked about is to the point aimed in the beginning, that modern dynamics excludes by definition as much the medium as the real vacuum, so that only can choose between a false ether and a vacuum equally false. Of course, it always shows preference for this last one, because it can measure it at least and consider it in terms of fields and forces.

And to crown it all, the universal medium is made absolutely unnecessary when we have established an absolute time from the beginning, that invisible and arbitrary "universal synchronizer". But this one it is still more undetectable than any ether –it is absolutely undetectable. Thus, we could reverse the question with great advantage for the measurement and the experimentation: in as much in which a universal medium or ether exist, the absolute time or universal synchronizer is unnecessary. Then, the immediate form to characterize the ether would be time itself.

79

Something similar was what Nikolay Kozyrev thought –Kozyrev, maybe the most original experimenter since the times of Galileo. Kozyrev and his team at Pulkovo already made in the Fifties all type of experiments with pendulums and torsion balances in order to measure distortions of the force of the gravity. They regularly measured them repeatedly and with enough homogeneity, although this class of experiments are highly dependent of some circumstances. Naturally, the mainstream physicists are not willing to consider neither the experimental results, nor the philosophy and ideas on which they were based.

Kozyrev conceived the time like a flow in four dimensions with an additional constant equivalent to a "causal speed". Apart from to exist space between causes and effects like in dynamics, also time intervals had to exist not arbitrarily small. Kozyrev made in addition considerable efforts to harmonize its theory with the structure and data of the relativity and quantum mechanics, with which it is quite compatible. Great part of all this lead to the problem of fitting the non-local factors and the simultaneity with the finite speeds that relativity –and the vacuum- imposes; but all this from the perspective of which the classic mechanics has an "infinite causal speed", and the quantum mechanics, a causal speed equal to zero. That is to say, a brilliant reversion of the problem, complemented with the only conditional conservation of the angular moment.

The time in Kozyrev is not the time absolutely passive and special of dynamics; it is an active time with an own density that imposes a course and a directionality, aside from specifying the causality. The problem is that Kozyrev seems again to reduce the time and its flow to a new constant, or a more or less linear variable. The same title of his first introduction of the subject says it openly: "Causal or Nonsymmetrical Mechanics in Linear Approximation". In addition, after giving axiomatic rank to the necessity of time intervals between causes, seems to ignore any time interval between the action and reaction, which establishes divisions in his own criterion. Like a pioneer, Kozyrev only could be fragmentary; but even so is the first introducer of the problem of the real time within physics, with alternatives to the vanished ghost of the dynamics.

Kozyrev does not get to consider of a theoretically convincing way to what extent the density of the time affects or governs to the constants of the fundamental forces; although from its own experiments its variability comes to be deduced. This, that even today seems absolutely incompatible with the modern physics in all the extension of its predictive apparatus, would not have to be so scandalous to us: the so called "constants" of the coupling forces in the standard model are dependent of the energy, and by the same, variables. We do not know why reason this would not have to apply for the gravity.

80

Zhvirblis, one of the continuators of the legacy of Kozyrev, shows the example of the of Lazarev's koltsar; a vessel closed in a ring form with a porous separation, which exhibits a more or less identical behaviour to a superconducting ring, with a continuous circulation of the content. As Zhvirblis says, the Lazarev's koltsar is not an experimental artefact at all, but rather an evident proof of the difference that there is between a real thermodynamic system and the ideal systems without structure contemplated by the classic thermodynamics maintained near the thermal equilibrium.

A real thermodynamic system as a koltsar it is a meta-stable system far from equilibrium; that is to say, it can never stop. The meta-stability is derived as well from the bi-stable character of the simplest structures, and descending to the quantum world, we see that the single electron of the molecular hydrogen H2 cannot be coupled to only one of the protons, but that move continuously to belong to both. The same type of interchange occurs in the Cooper pairs in a superconducting material, so similar in this to a koltsar, and moving away of the thermodynamic model; nor has to say that superconductivity is a phenomenon still very poorly well-known, in spite of the intensive effort developed in this field.

Since Lazarev's koltsar is inadequate for a quantitative control of the circulation, Zhvirblis proposes a vessel with osmotic barrier, in which the thermodynamic demon of koltsar with its associate paradoxes can be studied better.

In Lazarev's koltsar not only *may* be, but *must* be circulation, because of the very existence of both different stationary states. The energy released in the movement is not of trivial origin; it is not followed from the laws of Raoult and Van´t Hooff, "*rather appears like a ghost*". Zhvirblis says that it is as if something had lost or had let without say in "some pre-axiomatic level". "*the circulation appears not to obtain the balance, but by the two different stationary states that appear by virtue of the interaction of physical forces"*. The origin of those forces is not considered at all and it is assumed as understood like a "*perfectly obvious* "physical reality. Zhvirblis says this so apparently trivial that we have applied in the very different context of the duality of the pulse: *"When two conditions cannot be satisfied simultaneously, the circulation occurs"*.

Zhvirblis ends up concluding that the problem of the irreversibility is an illusion:"*the thermodynamic reversibility is inferred from the reversibility of the mechanics as it is applied to a koltsar or to the meta-stable structures. Correspondingly, the irreversibility of the thermodynamics can be inferred from the mechanics only if the last one is postulated like irreversible, that is what Kozyrev did.*"

The question is that the koltsar will remain an open system, independently of how much we want to isolate it. And any other object that we consider, if we know to appreciate its fluctuations. Zhvirblis shows different electrical devices to us from which can be deduced a similar behaviour. The author concludes that any real object is meta-stable and only can maintain a defined structure in as much interchanges energy with the environment "*This interchange not necessarily manifest transformations, but in the bi-stable systems, with circulation, cause processes accompanied by useful work.*"

That is to say, if reversible behaviours exist, it will be thanks to a determined structure of interchange with the medium, and not by the grace of the mathematical formalism. However, what is what fluctuates in all this interchange? Is it energy, or rather the energy and the movement are their manifestations? These questions always have been considered insoluble; but exactly when arriving here, and if really we had some new right to speak about the causality, we could to reframe the question entirely. A koltsar is not but a kind of monad, and we have many of them to study properly.

81

"*With each thought, something invisible, super-substantial enters our spirit.*" Words that one would wish that entered more than one spirit like something more than ideas. The phrase is not from Rilke, but from Riemann, that left them pointed in one of its outlines for a psycho-physic theory. By the way that he was not making poetry, but that tried a mathematical and technical definition of how that happens. At the moment, few would be able to take seriously these theories; but like everything what Riemann left to us, the phrase has infinity of resonances, and they are not exhausted in the physical interpretations. Of course, Riemann is thinking about a kind of breath.

The preoccupation of Riemann by the general statute of the physics was not precisely a pastime of its free short whiles like mathematician. As some know, he worked much in a theory of the ether in which an influence or flow of the medium to the interior of the particle took place, just vanishing then; in this also he followed the track of Leibniz and Euler, trying to refine arguments, although much more explicitly speaks about Newton like model. But Riemann knows perfectly where it is: "*the inertia is the hypothesis...* ", and certainly there is not in physics another hypothesis that the inertia from Copernicus and Galileo, being Newton the one that isolates it of any other possible influence.

Models of flow inside-out of the particle have been continued reproducing until the coming of the quantum electrodynamics. The same Feynman worked in one of them with Wheeler, two years before giving the version that now has become standard; and still today there are physicists who continue defending it. The question is not that it cannot work, the question is that within the inertial frame this is ignored by unnecessary.

82

There is no harm in remembering the place than Riemann occupies in modern mathematics and physics. As far as the mathematics, it is not difficult to see that he is the main inflection point with respect to the old one, based in computations and algorithmic; that is to say, he is the main introducer of the present conceptual style, generalizing whole masses of concepts so to speak at bird's eye. He is in the same beginning of the unfolding of the topology, as well as of the set theory, that by many years the same Cantor also called "theory of the varieties". Being his style not inclined to put algebra in first plane, all his work on varieties and surfaces are pregnant to the maximum of algebraic implications. The signature he left in the complex variable and the real variable, establishing the bridge for all the modern theories of integration, is simply inescapable. But all these achievements are concentrated and enormously surpassed by that small communication on the prime numbers given to the Berlin Academy of Science. This it is the only writing that does not contain a single geometric idea of an author that like geometrician remained to the posterity; after that writing, Riemann, just at his thirty three years, no longer produced nor published anything.

The contribution of Riemann to the physics is critical and delicately strategic. Known it is the Riemann tensor, consubstantial with the general theory of relativity and relativistic cosmology. It is quite worthy of interest the fact that the Riemann tensor can be decomposed in other two: the Ricci and Weyl tensors, which respectively measure the change of initial volume and the tidal distortion. These two tensors are theoretically separated, but in fact they can be intermingled and to be affected mutually, showing within the line of evolution or direction of the cosmological space-time some suggestive analogies to us with other potential dualities of the same zeta function, that here we only can aim. Also we have the sphere of Riemann, to determine the possible states of a particle of spin ½, as it is the case of electrons, protons and neutrons –all the "ordinary matter". That is to say, we spoke of objects with a minimum of two distinguishable states; in addition this would take to consider the transition from the angular moment of particles, defined in the complex plane, to the angular moment of the classic level of description. Known it is that some of the probabilistic interpretations of the zeta function become in the simplest terms of the "fair coin" with a probability of ½; the electrons and the other particles would be other many fair coins here…

A good reason exists not to believe in the possibility of a probabilistic demonstration: that the random aspect of the zeta function is only half of the subject, and that a half arbitrarily taken can never decide on a set arisen from more precise and ampler conditions simultaneously.

83

Riemann could not introduce the ether –medium and substance more temporal than spatial- in physics because this one already was born without leaving space for it; but, in turn, we can say that he discovered an ether of mathematical or numerical nature. In effect, it is possible to speak in a certain sense of a "numerical ether", in a way similar to how the engineers speak the ether of frequencies. Nevertheless, this numerical ether is something quite different of the so called "Cantor dust", and even almost the opposite: it is so infinitely bounded that it must find at any place a vent or a free line of action. To the critical line would correspond this role.

84

The curse of physics has depended in an evident way on the mathematical techniques available. A spectacular example of this is the role played by the harmonic analysis in the quantum mechanics; to such an extent, that one never knows at all if the limits that are attributed to it as inherent are really of experimental or technical nature, being both so united aspects. The same Planck's constant, the quantum of action, is not simply an energy per second, but an energy per cycle per second –the cycle is generally missing. And, naturally, it can be said the same for the Heisenberg's uncertainty limit, that it is not more than a technical limit of the spectral analysis. This way the frequency, the most purely temporal phenomenon, is emptied of any content and become a mere integrant part of the formulas. What is here what turns? Why it turns? These questions are considered the cusp of the absurd or the nonsense, but the physicists do not give much better answers: "what oscillates it is an oscillator" (may be a particle), "and what vibrates, it is a vibrator". Of the cycle, like of the circle, we have learned how round it is, and that is everything; but think for a moment that we are speaking of a dominion –the quantum mechanics- in which the temporal series find their apotheosis. And in this dominion, the time is what shines by its absence.

We can obtain resolutions of phase greater than the ones of the spectral analysis by means of different types of phase-sensible frequency meters, some of them are so common that they are in the colour televisions. It does not seem to have another technical limit that the noise, and would be very interesting to consider under what assumptions and conditions could be useful this class of information. If a cycle is something more than a mere unit, we would have to be able to ask if it is not possible or necessary a tri-modal definition of the quantum of action, which already seems justified since its initial definition: work per cycle per time.

While I am writing this I see a rainbow by the window, and I think that the mere phenomenon of the colour is absolutely impossible without breathing itself, that colour itself it is already a breath that penetrates without limit until the deepest. And I think, also, that in spite of the magnificent spectral theory, we do not know much about the colour. The Planck's constant is an integral, and if we could understand the fine structure constant in appropriate terms of time, we would understand to what extent the spectral analysis put limits to us. This is also intimately related to the Riemann zeta, whose zeros have been wanted to relate to the zero point of the fundamental energy level. But the zero point, besides the vacuum, would be also absolute darkness.

Any cycle must involve something, of course. A cycle is not but another word for which in other contexts we denominated circulation: that what occurs when two conditions cannot be satisfied simultaneously, as Zhvirblis remembered to us. And the interval that mediates between an absorption and a emission, seems that it would have to be decisive to discern between local and non-local action, that, in the same way that a wave and a particle, are temporally excluded and do not manifest simultaneously.

It is said that someone asked Einstein near the end of his life, when the quantum electrodynamics already had been established, what was what he would like more to understand. The answer was that he would like to understand a single electron. The modern theories of fields have an extraordinary predictive reach, to such an extent that the understanding seems to have made unnecessary. Of course, it is certain that we do not understand a single electron, and in this respect to speak about the vacuum would not improve the things precisely. What is what fluctuates? "What fluctuates, it is a fluctuator". And what operates, an operator; naturally.

85

Gradually, physics has been approaching number theory and Arithmetic. Some researchers still apart from the mainstream, like Pitkanen or Volovich, already put it in the centre of their considerations. It is even spoken already about the number theory like the ultimate physics; and here, "ultimate" would have to be understood simultaneously it in the double sense of last and inevitable in the development of the facts and first by its irreducible character. As we say, this seems to be quite far of the mainstream and its daily agenda; but, as it is observed with Connes's developments and non-commutative geometry, it is quite logic to jump from the algebraic aspects of the topology and differential geometry to the algebraic theory of number fields, to where it seems that the incompatibilities would find its last instance.

86

In the same way which the zeta function cannot be understood like a simple partition function, we must understand that the correlation that we can measure in the pulse it is not born from a mere division of the whole system. It is not trivial precisely because it is not a division at all. We have already said that the dualities we speak about are not irreducible, but only terms of a balance that presupposes a mean to be defined. Without this mean, not even there would be possibility of translations neither of transformation of terms – no breath at all. That is to say, considered from a ternary point of view they involve necessarily TIME, a real time, us much real for us as the time of computation, no matter how hard the purely abstract time could remain unaffected. What allows us to make a curious reflection. As it is known, the principle of the global synchronizer in the relativity and the theories of fields depends on an imaginary time against which the "real time" stands out; that is to say, the order principle depends on a non ordered set of numbers. With this peculiar circumstance few can be done since the point of view of dynamics. But a similar circumstance yes admits an inside-out reversion in the same Arithmetic, in the zeta function for example, so that we can consider the time of computation that consume the different types of transformations of the function, the precision in the imaginary part of the zeros and any other aspect. That is to say, we can make an algorithmic approach assuming different computation algorithms and translating them in terms of synchrony and real and virtual time; approach that has extraordinarily interesting possibilities and that here only we can leave aimed. Even the most intractable aspects because of the same magnitude of the numbers, moments and intervals, must find significant correlates in term of computation time and number of operations involved. This way we can generate infinity of functions with totally new characteristics around the main function and its critical line. The same criterion for the time of computation and the algorithms already is a great subject itself, but, in any case, which is within this function, given its nature, also has to be outside, and the more direct is the mode of translation, more will be thus. We believe that this kind of approach is the most straightforward, even involving many concentric and eccentric spheres. On the other hand, it would be great irony that the old algorithmic approaches had seen in the length of the computations the main obstacle, instead of the real path, and that those who are the whole day making computations had not evaluate the value of the number of his own operations. May be, finally, we begin to take seriously the idea that all this has to do with the time and the activity in the most direct way possible. Only in as much we renounce to the idea that we are in front of an object, the real thing will become for us a thread. The zeta function always impels to us to entirely reframe the analysis in the measure of our capacity, thus revealing itself more useful than any possible utility.

If we spoke before with regard to the hydrodynamics about the internal energy of the system like a third relevant element, is necessary to realize that, in terms of the *tridosha*, *vata* penetrates into the most internal, without being for that reason separated of the medium or environment. From here arises a sort of "automorphism" that we have denominated monadic. The question of the compressibility or incompressibility of the variations throughout the time arises, also –what it is the less abstract of the questions, since is intimately associate with the reversibility or reestablishment of the organic balance or its irreversible deterioration. In these conditions, any real system, that is to say, any complex one, must have its own line of specific temporal cancellations, well different from the cancellations that appear, for example, in the field theories. Otherwise, it would only be a fugue or escape line for three arbitrary principles, which is not the case.

And thus, the critical line of an entity, its union of *prakriti* and *vikriti*, to speak in terms of the Samkhya and the Ayurveda, far from being an escape line, is the subtlest and intimate axis of its breathing, its unfolding and its destiny. The monad is indivisible, but not unconditional. All what exists is breathing. A monad is analysis plus breathing, and then, we naturally enter the Samkhya dominion. The zeta function could be the most unconditional of the entities or monads, but even so, it is possible that it must remain necessarily open, and perhaps only thus can be related to the contingency, if not threaded with it. This mode, perhaps all real system are related to zeta of the most intimate form; but, in any case, more than to worry about this, there is to try to conceive the real systems in its purer temporal form. The zeta function seems to breath in the unlimited; if someone could close this unlimited breathing in a closed field, may be all the things would not breath anymore.

Don't we need hazard in order to breath, by any chance?

87

It is said that David Hilbert asserted anytime that the Riemann hypothesis was not the most important question of the mathematics, but just the most important question. Assert that we don't need to agree with, but that seems perfectly comprehensible and justifiable within the microcosm of the scientific mentality, and not only the mathematical one. Because the world would be equally mysterious without the famous function, but the coincidence is that the same function also seems to be inherent to the world, and from here all its wonder and interest. Independently of the hypotheses, the achievements of Euler and Riemann are absolute achievements of its kind.

The experience seems that it is showing to us, and will still show more, that it is much more viable to subsume behaviours within the zeta function than to derive this one from other areas and objects. And thus it has to be, if we admitted the fundamental character of the Arithmetic. In fact, it seems easier that this one fits the nuts to all the algebra, the geometry and the analysis, than on the contrary; nevertheless it is natural and necessary that these three adjust and readjust around the big subject as the only possible form of becoming aware. It happens here exactly just like it happens to the *gunas* in relation to *Purusha* within the Samkhya, and here the comparison is not a metaphor.

The problem is what class of things we try to subsume, and how. We have already seen that here the greater obstacle is precisely the common meaning of dynamics with its systematic evasion of the temporality and its equally systematic substitution.

We have tried to show that the zeta function is not simply a challenge for the analysis and the analysts, but that is the challenge of the other part of the analysis secularly ignored by considerations and interests presumably more pragmatic or at least more immediate. The single intuition already says to us that this is the case, but all class of arguments comes to confirm it. Mathematicians are fully within their right to explore their subjects as they please and can; but it seems that what with the blessing of the nature of the things has united the zeta function, it will not separate the man with all his efforts. The atoms and the particles are infinitely more docile to our manipulation.

The zeta function and its implications appear to us like one of the most legitimate and fundamental subjects of the philosophy, that now, in addition, it is being loaded of all type of practical connotations and urgencies. Then its subject is the maximum possible extension of the global-local relations, the chance and the necessity -maximum extension and maximum existential implication. Be cause it has been spoken about the chance and the necessity too much and flippantly enough, and from the scientific mentality precisely, assigning them comfortable compartments for what we want to emphasize and what we want to ignore. So that even it is surprising that this great motif of the mathematics has not provoked a genuine philosophical interest before, to which without a doubt it has contributed the hermetic character of the subject.

Writing like generalist or mere philosopher, I must say that I have not lost anything in the inner worlds and infinite complexities of the function; what on the contrary it calls my attention and surprises to me is the same fact that this so general subject has gotten to consider within the mathematics of a so elegant, so net and so closed form, and with such an archetypical pureness, as well as the fact of its practically subterranean statute from the times of Euler to now, and being extremely significant its emergency in the present. And of course, it would be interesting to know why reason a so general subject is so indigestible for our actual methods and so dissonant with our interests; and why a so general subject is not able to find a more natural place in our understanding of how the things are. Even if our understanding could be far away from it, the zeta function is the sharpest thing that exist, and it is nearer to us than our own jugular vein. Said this, we leave the Arithmetic already and we return to the Samkhya, the universal theory of the fluctuations, and to the pulse like experimental example.

88

Any prognosis that can make a doctor when is taking the pulse depends on the sharpness and precision of the diagnosis; naturally, this is equally valid for any type of medicine. Nevertheless, when we considered the purely analytical character of the time series of the pulse, this raises to us other questions on the nature of the prediction in general.

In physics, it is worth to say in its more linear aspects and behaviours, the precision in the prediction exempts to us of any determining causal understanding; having a model with its parameters, variables and constants in a manageable form, the question about the causes can be relegated to the limbo of the metaphysics. We have already seen in several occasions that the most powerful reasons exist for this, reasons that are in the same constituent plane of their activity. When in physics it is wanted to better understand something and more precisely –say, for example, the hypothetical quantum nature of the gravity -, all the effort is made to extend the dominion of the predictions, and nothing else. Then in physics, or at least for that to which it is conformed, the understanding is synonymous of an absolutely controllable prediction. And I won't be who try to change to the reach and competence of the established physics.

This all changes radically when we approach complex, nonlinear and non controllable systems. Here, it is continued supposing that the more well-known laws with their more linear expression remain the same, but, apart of them, it becomes necessary in general to also suppose another type of conditions that assure the stability and other typical characteristics of the behaviours. In such conditions, we either resign ourselves to a study *ad hoc* of the multiple particular details of the system, or look for another class of more general signatures; and in fact, both things occur together usually.

Nevertheless, it is continued looking for specific levels of causality forgetting that in the fundamental level the causes have never been specified.

Within the framework of the Samkhya or the Ayurveda this problematic is not even considered. Any process is reduced to the form of its temporal series; but this one not even is reduced to a space of coordinates. There is no need to say that for us a time series is something very different. This is valid also for the assembly of organs and anatomical characteristics which here we see indeed like agents of the "causality": all of them are products of their respective series and sequences, no matter how hard they are connected.

When we see a face, we are seeing the product and the last form of a time series. We think about it very rarely, although it is evident. For the Samkhya, the same fluctuations of *gunas* or *doshas* are equivalent to small phase transitions towards the interior and the exterior to the system, with a mediator that penetrates and goes beyond those apparent ends: in terms of the organism, this goes from the skin to the bones, with a sensitivity that surrounds to the assembly and shade off, but that continues being clear in the form –and not another way occurs in the pulse, with the form of its temporal profile. The same solid, liquid and gaseous phases maintain a correspondence with the *gunas* that goes beyond the analogy.

We say too often that the man always has wished to extend his predictions, but nothing falser than that. In the past he was much more worried than us about the interpretation of the reality, that not of its prediction. The prognosis of the *vaidya* cannot be remoter of the prediction of the physicist: the *vaidya* tries to understand the reasons of the uncontrollable, and the physicist of what he can isolate and control. The advantage of the first vision is evident in everything concerning the behaviour of complex systems. What we have not explored nearly at all are the relations between both methods; because if it is certain that they are interpenetrated, it is not less certain either that they are excluded.

89

The stock market is the maximum exponent of a system of signals under the most intensive scrutiny. Known it is that the financial analysts and the agents continuously speak about "to take the pulse to the market". Metaphor or not, there is no better expression to describe something than no analyst knows in what consists it exactly, beyond taking the most direct contact possible with the flow and the fluctuation of the stocks or values. Naturally, the most direct contact with this flow is to participate in it, to unite oneself to the torrent of buying and selling, observing the reactions and the "tone" of the market.

The efficient market hypothesis says that in a certain moment –at any moment the prices of the stocks already reflect all the relevant information on the market; or in more attenuated versions, at least the main of the information. Since the relevant information in this respect can be unimaginably varied, it is difficult to find in the daily life a similar postulate of synthesis, update or reduction; even remember us, in the further sphere, the famous "wave function reduction" of the quantum mechanics. This postulate of actualization cannot be more related to the hope of the *vaidya* that which in the pulse all the distinctive elements of the health and the disease are yet present –with the sensible difference that no *vaidya* will pretend that his analysis, diagnosis or reduction can be confined to a moment or purely momentary state. A suitable temporal sample, an auscultation with a minimum interval of time will be needed.

The efficient market hyphotesis is considerably more hypothetical and radical and than the vindication of *vaidya*, and nevertheless we give much more credit to the former than to this last one. In our world it would not be surprise that we were more sensible and considered with the money that with our own body. The hypothesis itself can be compared with the idea of the "random walk", although on the other hand the presumed assimilation of the information turns this randomness the most relative and ambiguous thing we could imagine. On the other hand, it is quite evident that this hypothesis is not but a reprint of the sufficiently ambiguous "invisible hand" of Adam Smith, which invites us to trust in the effectiveness of the free competition like a new incarnation of the providence.

Precisely, the theory of the efficient market is a form to do without all the spurious speculations of the so called technical analysis, that is not another thing that the interpretation of the graphs by themselves. In opposition to it we have the so called fundamental analysis on the real structure of the companies that sustain the stocks. This last one does not try to predict, but to understand what there is in game. The clinical analysis that the *vaidya* realize, his diagnosis, would agree precisely with this kind of fundamental analysis. But, on the other hand, which makes buyers and sellers at every moment, whether they consider a type of analysis or another one, it cannot stop being speculative in a high degree, and being the market which is –and affecting more and more the curves of the price to the internal dynamics of the companies; so we have another species of inescapable duality, that the theoreticians, until now, have only been able to contemplate in ambiguity terms. The inside-out type duality between the technical and the fundamental aspects is in fact irreducible in its own terms, and nor to say that it is of highly similar nature to which exists between *prakriti* and *vikriti*.

This is not the place for doing a detailed analysis of the paradoxes of the of the efficient market hypothesis, thing that on the other hand already has been made over and over again and with so little fruit. In the context of our writing, it is enough to say that the dynamics of the market is not punctual except in the graphs: the acquisition and the sale of stocks, like any transaction, are simultaneous by definition; but not so the orders of acquisition and sale, with its added tendencies. The paradox of the efficient market, raised just at the beginning of the twentieth century –the century of the paradoxes, and not only theoretical- is not but an exemplar among others many of identical or very similar structure, generated all by a purely binary posing in terms of yes or not, with excluded middle, in a system conceived alternatively from an internal or external perspective. This could be applied as much to the paradoxes of the set theory, relativity, quantum physics, and thermodynamics, and other many. And though apparently theoretical, all has contributed in not little measure to install to us in the ambiguity; until the point in which the contemporary world has become a systematic running of this ambiguity, so apparently fought with the exigencies of the scientific exactitude. And we think that there are in all this substrate of ambiguity and paradoxes not simply theoretical problems to solve, also a great inertia to mobilize.

It does exist for the market, not only an "invisible hand" like the one of Smith, but a left hand next to a right one as we have observed in the pulse? Not anatomically, certainly. But there is a duality of components that contribute to something very similar, and perhaps almost identical. Of course, the conditions of stability in a market do not have anything to do nor are comparable with those of an organism; but there is an assimilation and an elimination; a cycle of creation and destruction, and not only at a level, but to many levels. We do not have nor the smaller idea if it is possible to decompose naturally a rough curve as the one of the stock market in two as happens in the case of the pulse, or if it is simply impossible; but given the similarity of components implicated, it is possible to learn very much of the approach of both models, respecting all its differences. We always spoke of the most satisfactory characterization of the fluctuations.

The curve of a stock or company never is isolated, but contains the highest degree of sensitivity to the outer circumstances and the behaviour of all the other curves; and this as much throughout the time as almost simultaneously. Of course, there is all class of combinations of graphs. But in any case, if we had some frame of reference for the optimal state of the values as it can be deduced in the pulse, this would imply very little prediction; that is to say, the predictive component would be more implicit than explicit, and everything what it would tend to obtain would be a momentary adjustment around the optimal values –reducing the speculative component near to the minimums. It is this possible? Don't we continue moving inside the same paradox, that speaks us of how the own knowledge affects to the sensitivity of the system and its evolution?

This is a paradox only for the predictive-speculative method that characterizes to us. In the frame of the *gunas* and the Samkhya, the sensitivity and through it the knowledge form an integrant part of the curve and the temporal series. But this excludes as well great part from which we considered prediction, and even dilutes the invariant character of the temporary axis by means of the duality and its fan of possible combinations. There is no possible magic in this respect; the only thing that would change is the centre of the attention. But this is much and more than what it is believed.

Of being viable, this method of analysis of the fluctuations would be applicable to the economic behaviour in general and would impose new criteria: because by force it would radiograph and clarify so many elements that today are opaque and that even there is not too much interest to bring to the light. The question is how much the dark we can allow ourselves.

On the other hand, it is not improbable that this new approach throws light on all that nourished body of modern formal paradoxes.

90

The example of the stock market is particularly useful because in it the more purely quantitative aspects, and the contents and the semiology cannot be more united both; but not in other way happen with the organic functions. We already can put the heart in the billfold, that the billfold will go to the heart.

It could be thought that the scheme of the *gunas* of the Samkhya is similar to the one of the dialectics, the well known scheme thesis-antithesis-synthesis; but dialectics is idealistic, even in its popular and naturalistic versions very previous to the idealism, precisely by its appearance of production of the reality.

Also the modern analysis, even being applied faithfully to experimental data, is idealistic since the moment that believes to produce the movements from closed functions with changes in the values or coefficients. Nothing of this has to do with the Samkhya. Although we even found characteristic mathematical structures for the fluctuations with respect to the balance, we would only have a selector of readings, a tuner that allows us to listen better what there is present. A complete and closed description as the one of many analytical functions would be often irrelevant, would make that listening impossible.

In fact, if there is something fundamental that corresponds with the fundamental nature of the *gunas*, the sensitivity, the action and reaction, is the same articulation of the language, with its three grammar people and their equivalent subject-verb-predicate flow. Not in vain it is adjudge to the name of Patanjali a treatise on sanskrit grammar, in addition to another one on medicine. The pulse of the cat truly would say "I-eat-sardines" if it had not much more varied and subtle things to count. By the way that that language of the pulse has so much an articulated and discreet aspect, in addition to continuous the most apparent one. This sends to us to the old India distinction between *vaikharï* and *madhyamä*, the articulated discursive language and the inarticulate and deep language of nature and things.

Only in a frame as this one could have some value and specific weight the non decidable questions of the last century on semantics and syntax of the badly called analytical philosophy. To speak about the language and the languages of the nature could not be a metaphor –rather the human language has to be a poor metaphor of that other language, the Adamic language of creation.

91

When the verb, the breath and the time disappear of our perception of the nature, the middle, also the entity of the two other instances is diluted. Something more than a dimension of the "problem" vanishes therefore, and something more than its "depth"; it disappears also the man and his privilege of being the middle and the means, or even just a means in the middle. Those who put at the top the humanism and science jointly would have to meditate on so a doubtful pedestal.

In addition, the exuberant growth of patterns that we believe to see in the nature, its abstract intellectual recognition, has resulted in the proportional impoverishment of our sensitivity to perceive them, until the point that to speak about any direct perception must sound to us like a fable unavoidably, or to a clumsy translation of some underlying mechanism or highly refined pattern. Refinement and dulling can go perfectly hand with hand, and meanwhile, we are condemned to look for in seas of the abstraction what we think that there we have lost.

92

Some investigators,, find significant or at least intriguing that the ratio between the diastolic interval and the systolic one of the cardiogram tends to come very near to the value of the Golden Section, that, like was made long ago, we prefer to call Continuous Proportion. This same ratio tends to occur also with good approach between the average diastolic and systolic pressure of the aorta. The certain thing is that we find the Continuous Proportion in all class of places and aspects of the nature, without on the other hand these manifestations have another character that the purely phenomenological, when it is not simply arbitrary associations; and often, barely there is a criterion to separate both things.

We do not have to the date the smaller idea of which can be the importance of the Continuous Proportion in the nature, nor could either define the plane that corresponds to it. Nevertheless certain properties exist that are worthy of mention. In the first place, that has not been the smaller explicit relation of this constant with any sphere of the dynamics, and that therefore cannot be derived from it. This is precisely what relegates it to a value completely secondary or very small; and nevertheless, it is this same thing what turns it something as gratuitous as mysterious. Be cause, after all, this constant appears with perfect clearness in the phyllotaxis of the vegetal groupings, evident expression of development and movement. What today we called Fibonacci numbers, and that this one took without a doubt from Arab mathematicians, it was described for the first time by Gopala and Hemachandra, around 1150, with regard to problems of exact and optimal packing for objects of lengths 1 and 2; the last one of these even applies it to the metric of the sanskrit poetry. In this sense, it would not seem stranger that the sunflowers and so many other plants find the solution for something that there is not a problem for them. Nevertheless, this does not leave the smaller track in the dynamic description of the process, and for a good reason: the cells reproduce and accumulate of discrete form. Precisely this accumulation of discrete order tending to a ratio in terms of real numbers or continuous fractions, is one of the most remarkable characteristics of this constant. Oleg Bodnar, for example, has given a model of vegetal phyllotaxis in four dimensions with a hyperbolic turn in the related functions. This same property of union between the continuous and discrete dominion has called the attention of mathematicians and information theoreticians like Alexey Stakhov, that have demonstrated the wealth of algorithmic and combinatorial possibilities of the constant with refined methods. In some of its numerical extensions, *phi* permits a conversion or optimal translation between analogical and digital expressions, for the same ratio already pointed. Their same algebraic and quadratic properties, elementarily simple, allow infinity of connections as well. Eduard Soroko has shown its relevance in problems of structural and functional stability with many components, settling down valuable bridges with the information theory and its measurement; his own information theory has a purely monadic starting point. The same Stakhov has developed an algorithmic and asymmetric theory of the measurement...We could give some examples more of important contributions on the subject, which nevertheless, remains almost as disconnected and marginal almost as always. This no doubt obeys to its absence of relation with dynamics.

What also makes to think that perhaps would be some more luck within the number theory, to which the subject seems quite open. The same Stakhov and others have developed the hyperbolic functions of Fibonacci and Lucas, who extend the Fibonacci numbers to the continuous dominion, in a way somewhat similar, although in a much more reduced dominion, to that of the analytical number theory. This would have to be very important, precisely to isolate the less dynamical aspects of the more general arithmetical functions, like the Riemann zeta: and we have already seen to what extent these less dynamic aspects are important from our perspective. The same vegetal phyllotaxis exhibits the primary ratio between *phi* and *pi*, first of the great constants of the calculus, in the so called angle golden, equivalent to $\pi/\tau 2$. Perhaps, on the other hand, it is possible to explore to what extent *phi* is antagonistic of the other great constant of the analysis, the number *e*, base of the natural logarithms and exponent of the proportionality on the functions. The ratio of this antagonism or mix-up can be reduced to the fact that *phi* originates the most elementary and particular association of additive and multiplicative properties, whereas *e* gives rise to the most general, also the more proper of dynamics and probability. Said another way, *phi* permits discrete and generative relations, whereas *e* tends by its nature to pulverize them. But this would also take to subjects of number fields that are out of my reach. In addition to the mentioned, *phi* could have importance like constant by other fundamental characteristics: it is the simpler mathematical form of recurrence and self-reference, it express the relation between a part and the whole and in addition does it showing the more slippery ternary relation, defined by the equality of ratios AB/BC = BC/AC. And finally, but not less important, it supposes the more elemental connection between symmetry and more elementary asymmetry.

On the other hand, the Fibonacci numbers appear in Klein groups and fractals. The same Klein considered the regular icosahedron like the most involving and interconnected motif of the whole mathematics. This is related as well to the equally mysterious and more recent emergency of the ADE classifications, that appear in the most divergent and unsuspected fields, with very few well known on its possible relations: Lie groups, critical points of non-modular functions, algebraic singularities, crystallographic reflection groups, *octonions* and many others. The same platonic solids or regular polyhedrons are the archetype and mother of these classifications, always referred to simple finite objects. These classifications are, apart from that, in the more alien aspects in respect with dynamics, or at least in the more limitative and border aspects of it. It would be necessary to see if all this is a chance. It is even possible to ask if this could not have some relation with the groups of six and five elements that appear in the division of the meridians of the Chinese medicine or in the canonical sections of the *gunas*; all this is highly speculative, but we have the possibility of experimental contrast at least.

From a statistical point of view, it is observed an increasing appearance of phenomena that seem to respond to potential laws or scale laws, of which the Pareto distribution – non-Gaussian distribution- is preceding and archetype. This type of distribution or 1/f noise, more or less independent of the scale and also of an explicit dynamics, maintains necessarily a controversial statute. It seems as if the concurrence of space and time, released of the partial adjustments of dynamics, acquired a magical characteristic, that is to say, the liberation of the causality. But this "causality" is only the one of dynamics, because the existence of interaction within these systems is often clear and evident. Only one careful separation of these superposed elements could throw light and law on which today remains like mere phenomenon.

Let us close these considerations with a reflection on the principle of least action that perhaps fits here. In physics the principle of least action is particularly bounded with the time; for example, the minimum paths of Fermat are the trajectories for which the light spends less time in its trajectory. In a similar way the relativistic physics has defined geodesic or minimum lines for the continuous space-time. This is enough strange if one thinks that the nature, and specially the inorganic nature, must have the most complete indifference with respect to the time and its saving, that surely does not provide an interests rate to it. What it seems to demonstrate once again and from a different perspective that the time of the physics is an *almost* entirely spatial device. The organic growth shows to us the opposite face of the coin, where the time is indeed decisive and constituent by nature. Precisely the presence of the Continuous Proportion in infinity of biological ratios, like in the sunflowers, it is tried to interpret not in terms of the principle of least action, but rather of a principle of material economy. That is to say, what it has cumulative and irreducible temporal components we want to explain it by principles of material and space economy, and what it is almost purely space we try to order it by principles of temporal economy. It is not this the best evidence of that here an inversion and a strange crossing has taken place here, and that something essential it is lost in the operation? Nothing would take to us more far than the resolution of this problem.

93

To what extent the temporary series can give a record of the space constrictions of a system? At first sight there are enormous differences between systems and systems. In the case of the pulse continue existing necessary connections; in the stock market also, but it is necessary to look for the pertinent dimensions of very different way.

If we said before that the analytical space of dynamics continued being geometry and algebra except a part of Arithmetic, also it would be possible to hope that the temporal series suitably studied –and the notions of the Samkhya are our main reference here- had components or irreducible dimensions within the pertinent simplectic structure; and perhaps it is possible to reach that taking steps in the suitable direction.

The pulse is an example of which we already have to our disposition the class of model that combines these dynamic and arithmetic aspects suitably, but until we do not study it properly, we do not know what can be there. Not even we know if of the correlation of the pulses could be deduced with fidelity something so basic in the organic balance and easy to measure like the alkalinity or acidity of the blood; but if that is possible, surely that other many data of the clinical analysis could be deduced with the suitable extension of the frame of reference.

94

To what extent the confusion with respect to science arrives, and with contemporary science in particular, is illustrated with the topic so frequently repeated that it comes to say that the Newtonian science was "hierarchic and theological", whereas the present one "destroys hierarchies and finds order in the chaos". The opposite is almost the certain thing, although, not to simplify in excess, it is possible to be detached: "Newtonian science" is the less hierarchic science that exists, and in that exactly its absolutism is based. We have already seen why the three principles of the mechanics exist. The modern science to which the topic talks about, the heteroclite disciplines of the emergency, the evolution and the complexity, which does precisely is to create hierarchies, levels and meta-levels where more likely there are not, and everything because they are incapable to accede to an apparently simple constituent level like the Newtonian one; as far as that of the order in the chaos, as today the things are raised, we better don't talk about it.

But the certain thing is that as much one posing as the other runs like chickens without head. It is necessary to leave the soft fantasy of which both methods are complemented, as if they were the classicism and the romanticism. No, because a single meeting point does not exist, but lines that are superposed to different levels. Let us see, For what we want so much "teeth and claw" if the third law of Newton not even allows the friction? A mechanist explanation of life, it is said.

Without a doubt the evolutionism and all the phases of romantic science, cosmology including, have benefited enormously from the credulity generated by the ample enough emptiness of the analysis, simultaneously that the faith of this one increased in view of the food that the contents or the temporal matter promised to incorporate, without having the smaller foundation for it. It is already time to contrast these hopes, to see what they weigh in the balance. A delicate operation.

95

Time against the field.

The modern scientific view of the world, like the man of Shankara that confuses the cord with a serpent, lives the illusion on this superposition of planes: everything what cannot be explained by the dynamic analysis, because it does not have internal space for it, we gave it to the dominion of the time and the evolution, that as well, it has to be governed by identical dynamic principles. Soon, and to complement the evident emptiness, one resorts to the probabilistic elements, that always are diffuse and uncontrollable, to extract of them the most convenient elements within the previous context. The probabilistic field, as diffuse as can be wished, may be the survival of the species or the economic competition: at no moment the conflicts are considered of direct and local way, but that these are stumped and scattered in the maximum possible generality. Also in physics we have examples of the same: Oersted made his fundamental discovery that proved the relations between electrical and magnetic phenomena, coining the term electromagnetism. Peculiarly, used the term "conflict" to speak of these relations; all this was absorbed soon in the concept of "field". We are specialists in dissolving the difficulties in the limitless and general space, which will send us as well to other spaces.

If the interpretation of the temporal series advanced in its own sense and direction, and not like a poor assistant, we would begin to see all these things more clearly. Perhaps it is to this to which secretly we aspired speaking more and more every day of the "real time" processes. There is something here that sounds like an inescapable demand, no matter how hard often it is replaced by the virtual representation, exactly the opposite, again.

96

Faraday, for example, who was the one that developed the idea and the concept of field, did not need mathematics to develop experimentally a completely new vision of the electromagnetism, and in a time in which some of the greatest mathematicians of all the times, like the same Gauss, barely were able to penetrate in the paradoxical forest of the electromagnetic phenomena. What Faraday warned was the importance of the dynamic process to advance in the investigation of these phenomena: shifts of position of the magnet or changes of intensity of another current. This was the decisive thing, and from that Maxwell could produce a totally consistent theory. Without a doubt to Faraday helped his ingenuous mentality as much as his incessant experimentation.

Maybe something similar must happen in the area of the theory of the complexity and nonlinear dynamics. Of course, the field is still much more vast and multiform here, although it assumes that we look for as general principles as it is possible. My experimental invitation is of this sort; when we have the suitable experimental protocols and we learn to work with them, inevitably the Maxwell of the moment will arrive to give us some of the equations. We did not need mathematics to see what is missing here, but we need them indeed to explore it and to weigh it properly. Clear that this is a mere parallelism to encourage this class of research, because the implications in this area, it is not necessary to say it, are completely different.

97

One is inclined to think that in science exist defined cycles of production, a little in the style of the Kondratiev economic cycles for the predominant types of industry. In this case, the cycles would come to be about sixty years, approximately. Without a doubt between 1887 and 1947 a very well defined circle is closed, which goes from the experiment of Michelson-Morley, the first observation of the photoelectric effect and the verification of the transmission of the electromagnetic waves at the speed of the light by Hertz, to the present version of the quantum electrodynamics of Tomonaga, Feynmnan and Schwinger. This last one supposes the beginning of the quantum theories of fields in which still are embarked the theoretical physicists. This present cycle would come to close itself in these years, more or less agreeing with the starting of the large hadron collider of Geneva, waited for the 2007. And peculiarly, if the cycle of 1887 began with the experimental contradiction of the theory of the ether or medium as reference frame and with the reinforcement of the speed of the light like constant, also now, before a cycle that seems to close itself, the statute of this constant is questioned, as well as the phantom of the cosmological constant, and an experimental resolution of the vacuum like frame of reference for the masses. So we can see important similarities, pointing almost all of them at problems of definition of the medium and frame of reference, although we will not speculate on the results.

But the fundamental physics is not everything at all, and even it seems that its importance is diminishing decidedly before the much more immediate challenges of the complexity. And truly, the modern vision of the complexity also begins in the starting of this second cycle, there by the 47 or the 48, with the appearance of the computer, the transistor, and works of Shannon and Wiener. The computer and the new notion of information are the instrumental vectors of our modern idea of the complexity. Even the modern genetics, that was born shortly after, has been totally conformed, and to undesirable and absurd limits, by the metaphors of the world of the information and the programming. Let us hope that this cycle also is next to close itself, of which we could show more indications than those that at first sight appear.

The metaphors of the information have arrived until a point so that even abound the physicists and mathematicians now who conceive the universe like a gigantic computer; that is to say, that so much time has been boasting science to surpass the anthropomorphisms to fall in coarsest and vulgar that it is possible to imagine, derived from a mere artefact. At least, the ancients knew well the rank of their myths and metaphors, and after all it was not about the man the only the spoke, nor even the essential often; but these wise people of the last hour get to believe in their metaphors of the most literal way, and they even despise somewhat to those that have still not found out of what the subject goes. They are, therefore, the best incarnation of the limitation of their own program. But were they or not an exponent of one more general barbarism, this does not have to worry to us, because the man creates new devices more quickly of which costs to us to internalize them. I venture, therefore, that enough before they begin to fulfil the dreams and nightmares of many on the digital world, will change the operative concept and the reference of the same information, beginning with the computation and its use of the time. Be cause there is too much to explore; and because the world is for the computer a greater challenge than the computer for the world.

The sequence of the human genome was the last great reductionist project that was given to the modern science; simply, there are no longer anymore simple objects of importance that we could treat in the same way. Not even in physics, where fields and particles begin to show a dependency of the vacuum that is impossible to reduce to simple cases: even the scalar fields of the so called Higgs mechanism require a multi-compound structure. The reductionism is collapsing visibly, in the same way that the old criteria of prediction; but it worked so well, and for the prediction indeed, that it is very difficult to resign ourselves to its decomposition. The reductionism nowadays is not more than pure inertia, but this fact does not mean that here science finishes and we must be reduced to the technical problems and to the particular details; on the contrary, we think that what can begin now is much more interesting and deep that all the previous developments. And in addition, this does not consist in one more turn of the screw in the dominion of the abstraction, but rather on the contrary, also. What it is not necessarily in conflict with the fact that the mathematics are more useful when more abstract they are.

Therefore, we do not think that the world of the future will be done nor of atoms, nor of bits, but of monads. Leibniz was the great precursor as much of the notion of the complexity as of the modern computation; nevertheless, the monad, his central and indestructible concept, became darken and was made doubtful indeed with the superposition of the badly called modern "dynamics", that he himself tried to avoid, but not to negotiate. I have tried to show in what sense has to be understood that the monad is closed - in the same one, indeed, in which it is in relation with the environment since the beginning, and exactly by the natural extension of the principles of the dynamics that Newton constrained. And thus, the monad finds its breath and becomes the most natural and robust of all the concepts. The same concept of individual will change as a function of the understanding of the monad concept; because there is not to say that the nominalistic concept of individual, unlike the I-sense, is completely derivative, and not original at all. An oafish concept indeed, that even can not include the idea of reciprocity within.

98

And speaking about our devices, none like the clock itself, of which we have been deriving all of them until to arrive to the same computer. And with them, also our irrepressible views of the world.

If what is wanted is to vindicate the "real time", what can we understand as such? It is possible to hope that that real time includes something more than the live images and today news? And what is actuality? The images are not, and less from day to day. What in fact we look for already is at the other side of the mirror.

Real time, present or actual time, is for example the gradient on which is based the climate and the climatic change. Of which barely we know anything, by the way. The paleo-climatology have verify recently changes and abrupt climatic inversions, in less than ten years, in the gulf of Mexico, Greenland, and other many places; before it was thought that an important change required a term of centuries. With the problems of gradient and stability, that include all sort of complex behaviours, from earthquakes to our own devices, we return to the subject of the pulse and their “dynamic” coordinates, the form more reasonable than is happened to me to approach this question, that is to say, the temporal reference for a gradient, something that is to look for within the same system. A paradigm in the own sense of the word, if we consider that never has been exposed openly and that since the same Patanjali rather always has been given as supposed.

This real time can concern the analysis as to our sensitivity as much; what of course it is not viable is a direct representation; and even of the indirect representations only is possible to say that either they are a mean or they are an obstacle; that is to say, exactly just like it already happens with the present daily representations. Nevertheless, in as much in which we are able to escape from the predictive and speculative horizon usual now, we escape also of unnecessary representations.

99

It has been said thousand times that Galileo moved away to us of the centre of the universe, that the theory of the evolution cleared to us of the cusp of the creation, and that the psychoanalysis denied to us the dominion on our own mind. But, being already lame the first of these three concentric revolutions, the other two, that tried to model themselves to image and similarity of the mechanics, have been more events of the public opinion and the imaginary that true advances of the knowledge. Neither the mind, nor the life, nor the mechanics have been explained nor understood. The mechanics continues closing to us the passage any superior understanding, and thus it is understood the innumerable amount of nonsense and incredible artifices that are postulated like candidates to explain the consciousness, in which is pretended like fourth and definitive "revolution". Nothing of this has the smaller plausibility, and the only good thing of so many desperate attempts is to convince to us that something extremely basic has been avoided from the very beginning.

Many are those that look for the magical concept today that could give potentiality to this last turn: ideas around the computation, the information, the determinist chaos, the quantum theory and things of the sort. Ideas of last hour that put in question the minimum of which is considered consolidated. This is quite normal if one thinks that the consciousness is the last and the most conditional thing. Our perception, on the contrary, is that there is very little or nothing to say on it, and that those other layers are what would benefit from a new principle; since nothing really has been explained by them. Even the predictions of the physics, so precise, on objects that very probably are not understood, are based on last instance in the correspondence of the proportion, and not of the causality -from here the "inexplicable effectiveness of the mathematics". This can happen perfectly without we make another thing that to cut layers or flat films without any other thickness than the possible and imagined relations with other objects, which are also defined in so to say bi-dimensional layers.

The triple concept of fluctuation with respect to the balance can break through from the most fundamental layer of the physics to the immediate subjective experiences: from the measurable movement to the qualities, and from a set to its details. In the landscape of present knowledge hardly we will find another bridge like this one, and this will not change while the modern science, that historical contingency, exist. Other approaches not even can scratch the real thing – it's like make photographs.

100

The empty space of the analysis is synonymous of freedom for the arbitrary interchange, reducing to the minimum the internal bounds of its objects. And thus, everything becomes possible object of the analysis. One of those possible objects seemed to be the series of the numbers, but the bounds turned out here to be stronger than all means of decomposition available; these, on the contrary, did not make another thing that to manifest them. Maybe, during a long time, and still today, we have confused the roots with the branches.

The harmonic or Fourier analysis was the trigger of the modern concept of function. Exactly when this concept became general, revealing until where its arbitrariness with respect to the empirical curves arrived, we began to confuse the idea of function with the one of an automatic execution –with the automatism itself, when this seemed less justified than ever. Today it seems more than probable that it is an opposite tendency the one that generates a new function concept, which on the other hand is acquiring levels of necessity. Something says the same slowness of these processes of the depth of the changes. We will not find an equivalent of the harmonic analysis for complex systems, but rather the analysis will end up adapting to the nature of which resists to us.

There is in Riemann an invisible gesture, that one whose willingly gives back a key to its legitimate owner. More than an offering, it is a piety gesture in the sense of the ancients, confirmed by the modesty of his author and inconceivable within the voracious modern scientific spirit. That gesture cannot be in greater contrast and in greater agreement with the time at which it happened, when the nineteenth century positivism reached the cusp of the vociferation and the vulgarity. In the same month of November of 1859 that Riemann gave its paper on the prime numbers, appeared on sale the first of the innumerable editions of "The origin of the species". Also the statistical mechanics and the modern spectroscopy raised.

101

The logic of the Samkhya leads to us with an irresistible force in as much we know to resist to the application of the force. So it seems to be the eternal criterion, imposed and to our disposition.

102

The subject of the Samkhya or the Yoga is to extend the internal space trying not to establish the smaller differential with the external space that necessarily is to be respected. Thus the difference between the interior and the exterior, and the field of reference for the movement stops by the movement itself. Yoga is to penetrate in the unlimited using the limits that are given to us, not trying to transgress them or annulling them, but with reciprocity. The Samkhya is the contemplation of the balance and the Yoga its practice, not existing true difference between both, and being the contemplation the gradual movement of the purity in the participation. In all the theoretical contemplation that we have made we have tried to guide us by this same practice.

103

The subject of the time and its sequence is simply the subject of our insertion in the reality.

104

The same modalities, or the *gunas*, are also only objects of knowledge for purely conditional purpose and therefore impermanent: that is already its form of evolution from the beginning and even without beginning. Therefore, one can completely turn the back to them realizing them at any moment; no sort of knowledge forces to us to contemplate them, and further from it the one that the Samkhya proposes as its crown. This would have to make us think about the indescribable been of freedom in which the consciousness dwells, being the nature of the *gunas* the existence with the appearance of the individual, and being the consciousness the pure universal existence. And if metaphysics is just the postulation of the extra-mental like an object, since the extra-mental is postulated here decidedly, but not at all its condition of object, this has to do with metaphysics as few as with psychology.

105

Consciousness has no memory.

106

At no moment we reject the aptitude of dynamics to deal with an ample rank of behaviours, and in fact our assumptions and inferences with respect to the pulse are dynamic in a great measure. The variational principles are always of global nature and ignore the causality; when we applied them to an open system that admits an explicit reference of the interchange between the environment and the interior of the system by virtue of its peculiar duality, the time and the causality have to also admit a new and more or less explicit expression, but not merely supposed as it happens with dynamics in general. The same idea of the balance has to become also more or less explicit, because to no other thing refers this concept of causality.

And the same must happen in the cognitive plane with the existing duality between the prediction and the memory, operations of opposed nature that cannot be simultaneous and that by the same require the adjoined assistance of its complementary in the phase of predominance of anyone of them. It would be to hope that this found its necessary dynamic correlates in real and complex variable. Surely this is much less abstract than what now we can imagine, and experimental results incomprehensible today but verified thousands of times, like those of Kornhuber relative to the temporal delay of the voluntary reaction, with more of a second or second and a half of interval, will find their interpretation in this context.

107

The Samkhya already incorporates the external medium in the processes, leaving freed to the same the problem of the reference. To define from the outside questions of evolution and balance, or the relevant ranks of precision, it seems the improper thing. Now it is to see until where takes to us to try an internal definition of the movement and the modifications adhering to which precisely circulates within the temporality, the only own measurement of this one. Perhaps thus we can locate to us and our intellect in the middle of the conditions, or as near it as we can. Anyway, the distinctions between the mechanic or not mechanic, the determinism and the indeterminism, become arbitrary and irrelevant, because they admit indefinite degrees that never are incumbent on that question, that already is raised from outside. The intellect is one more of those degrees, simultaneously than its refinements, which precisely leaves back the complexity of the states, if it is certain that the essence of the distillation is the essence.

As far as the consciousness, it is still certain that it is not the object of any discipline. If some grant possibility to that delirium, it is because they suppose it like the more highly conditional, while they assume the unconditionality of the physical laws. But the situation is exactly the opposite: any physical data, including the measurement of the rest, already involves a modification, an asymmetry of the space and the time.

The Vedanta could not express it better: If the consciousness can not explain itself, what thing could explain it?

108

The real is the existence. The existence is the consciousness. Consciousness is the real thing. That is the triple affirmation in which it is summarized the Vedanta and the Samkhya, a triple identity as existential as absolute. Empty? Maybe: consciousness is completely indifferent to the fullness or the vacuity. Consciousness, already we said it, has no memory, not at all a paradoxical affirmation that anyone can verify by means of the portion that corresponds to him. An affirmation in which underlies very considerable possibilities.

If the consciousness does not have memory, in that exact measure the consciousness is will; if the consciousness is will, in that exact measure the consciousness does not need to know. That exact measurement that we are speaking about is the conation, which we don't need to confuse necessarily with the will; then if there is conation, the consciousness is only intellection, and if there is none, the consciousness remains undivided in these terms and indifferent to them. This is an example of something which not even needs to be thought; then the terms fall here by themselves, without thinking about how we could take part in them. And exactly in the same way but in very different measure it happens to us with the equations. We are not invited to think about it, but to verify it. Consciousness has no memory. With so much disorientation, we can not think in a better indicator of the "location" of the consciousness, for those who want to look for it

For those who please to quantify, an unobjectionable definition of the consciousness is that it folds the three modalities in nothing, corresponding this way to the triple previous affirmation. This triple identity does not have sequence, and therefore it is unattainable for the logic or the mathematics, that sometimes believe to be in the ultimate degree of the simplicity, the unconditionality or even the elegance. In fact, the mathematical forms are only appeal to other forms, and would be enough to have the sufficient attention, with no need of a mathematical intellect, to realize that they are not more near from the eternity than the first or the last of the words pronounced by the man. The opposite is a presumption, motivated nevertheless by the implicit movement in the aspiration, more than by the execution or the assumption of conscience; therefore, and like all movement of the language, it has an excusable necessity.

So that the triple identity of the Vedanta offers us like a space for the contemplation to our whole disposition, since in no case it imposes nothing to us. It is the maximum and the minimum that can be said. But it is also a proclamation and the testimony of the most legitimate of the conquests, because it has been made inside the pure conditionality, of which all the forms depend.

In any case, to mention the consciousness is not absolutely inopportune precisely because of how acute begins to be our conscience of the insufficiency of all the formal systems, and by our very justified distrust on the fascination of all the theoretical abysses. We can assume this in opposite direction of the rhetoric and the brutality, which already are made present in the same intellect.

Being the intellect adhesion before any other thing, if it does not realize that only lets take, he is already lost, and only through the experience it will be able to recover of that loss, whenever it is made, and non only with the intellect, the sufficient effort. This is a fundamental subject for the Samkhya, because the modalities exist indifferently like liberation and experience source, without excluding both terms except by the own modes. The Samkhya affirms that the liberation is the identity of the intellect and the consciousness: but this identity does not belong to the intellect, because all this, at least for it, is conditioned by defined degrees or modes.

Properly speaking, consciousness does not belong to me, I can only hope to correspond with it. And far from all rhetoric, here takes root any possible freedom. That is to say, there is no another one. The rest is only compulsion, and the lucidity of the intellect, always lent, out of the consciousness already is superimposition and blindness whatever will be the level in which it appears. Therefore the consciousness is neither a rhetorical subject nor a problem of location, but the only reference and the only means of evolution in the real time. Real time is consciousness with no memory.

SOME RECOMMENDABLE READINGS

Swami Hariharananda Aranya, *Yoga philosophy of Patanjali with bhasvati*, University of Calcutta, 2000

Institute of time nature explorations (Related to the legacy of N. A. Kozyrev) http://www.chronos.msu.ru/

V. E. Zhvirblis, *Stars and koltsars*, http://www.chronos.msu.ru/EREPORTS/zhvirblis_stars/zhvirblis_stars.htm

N.A. Kozyrev, *Experimental study of time*, http://members.tripod.com/~jtooker/kozyrev.txt

Matthew R. Watkins home page. *Number theory and physics archive (the best resources about the Riemann zeta I know in the web.* http://www.maths.ex.a-c.uk/~mwatkins/

TGD- Topological Geometrodynamics-Matti Pitkanen home page.
http://www.physics.helsinki.fi/~matpitka/tgd.html
http://bio.nagaokaut.ac.jp/~matsuno/preprint s/RESURREC.htm

Paul Marmet, *Absurdities in modern physics: a solution*.

Paul Marmet, *Subjectivity of Heisenberg's uncertainty relationship* http://www.newtonphysics.on.ca/HEISENBERG/Chapter3.html

Koichiro Matsuno, *Resurrection of Cartesian Physics.* http://bio.nagaokaut.ac.jp/~matsuno/preprints/RESURREC.htm

Museum of Harmony and Golden Section (Alexey Stakhov site). www.goldenmuseum.com

Riemanniana selecta. Edición y estudio introductorio de José Ferreirós. Consejo Superior de Investigaciones científicas, D. L. 2000.

El cuerpo y lo otro. Introducción a una teoría general de la cultura. Alexis Jardines. Editorial de ciencias sociales, 2004.

El enigma del movimiento. Alexis Jardines. Biblioteca Nueva, 2000

La hija del capitán Starbuck (Hurqualya). Miguel Iradier. Biblioteca Nueva, 2001

Cuadernos. Simone Weil. Editorial Trotta, 2001

Does God play dice? The new mathematics of Chaos. Ian Stewart. Penguin Books Ltd. 1997

Prime Obsession. Bernhard Riemann and the Greatest Unsolved Problem in Mathematics. An excellent introduction to the Riemann zeta for non mathematicians. John Derbyshire. Plume, Penguin Group, 2004

Cybernetics or Control and Communication in the Animal and the Machine. Norbert Wiener. M.I.T., 1948, 1961

INERTIAL THRUST ENGINES AS NON-CONVENTIONAL CLOCKS

December, 2005

SUMMARY

The non-conservation of angular moment, and its eventual conversion in linear impulse, in many inertial engines, already is an excellent indicator of the internal time of the system: of its internal synchronization, instead of the external or conventional synchronization. Hence the great theoretical interest, in addition to the practical one, of such devices.

***

It is well known the gyroscopic rotor Eric Laithwaite introduced in the Royal Institution more than thirty years ago, and the glacial reception received. Innumerable similar machines, based in the exploitation of the inertial impulse, have been constructed worldwide before and after this exhibition. Laithwaite himself adopted the idea after a communication of an amateur inventor.

Laithwaite´s machine, a centrifugal rectified rotor, produces when turning a lift force reducing the weight of the apparatus –without apparent reaction of the medium to support it. Taking into account the "suspicious" aspect of the device for the assumed laws of mechanics, Laithwaite dedicated all his efforts to demonstrate that none violation of the fundamental laws existed, but only a separation between the action and reaction for Newton´s third principle –none "local violation" of it; a good explanation, at least, for his satisfaction. Even so, and in spite of the great utility that this class of devices and its principles must have rendered already at this point, the engineers and the industry have not been able to find them generalized application still.

The "Laithwaite machine", ridiculously simple and concrete, remain thus in the same border line between which is considered acceptable in science and the misty world of the unexplainable. The apparent non-conservation of the angular moment of this engines makes of them undesirable objects for physics, by the single fact that turns the calculations an absurdity, and being this the main barrier for a more serious study and consideration. Of course, the angular moment could be conserved if a medium with specific features existed, but nobody knows enough of a medium such in order to produce such effects.

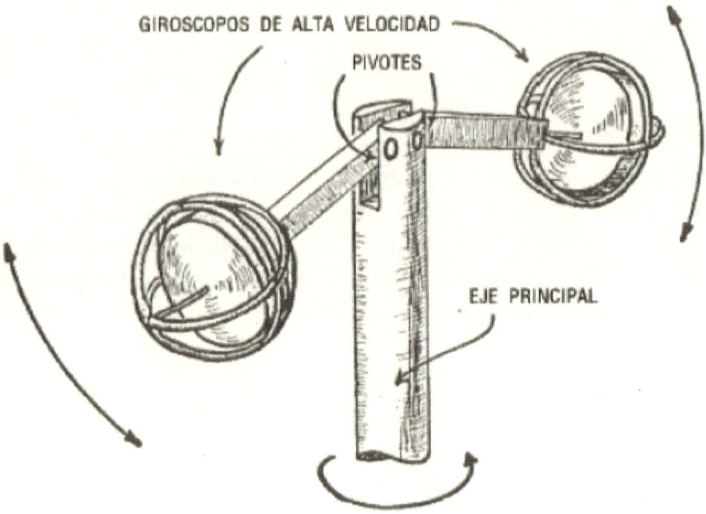

In the Laithwaite design, when the main axis rotates, the acceleration caused by the precession makes rise and fall the gyroscopes during each revolution; being the problem to arrange or synchronize the ascension of the gyroscopes to get lift effect, disengaging simultaneously the descending swing. Given its intrinsic instability, the device tends naturally to violent oscillations. Many other people have introduced multiple types of partial rectifiers to stabilize the system.The designs of Alfred Evert, for example, starting from own principles, are a fascinating unfolding of motifs with great universal content, which transcend our accidental and arbitrary idea of the machines.We will not enter to consider now to what extent they could have been successful. Besides this, many engineers and inventors have developed electromagnetic devices with quite similar principles, taking into account the analogy between inertia and electromagnetic induction. The instability of this kind of devices is not accidental, nor attributable merely to the imperfections of design: the root of the problem lies in the same fact that these machines, with their successive rectifications, tend to recreate artificially and with increasing approach one of the more omnipresent motifs in nature, and in a fluid medium in particular: the eddies or vortices, inherently unstable, and with the uncontrollable ramification of its vectors. This already is a merit itself, and one can ask why would be reprehensible the increasing approach to the model of the nature, no matter how clumsy our attempts could be.

The “Laithwaite machine" only demonstrates a separation between the action and reaction; but being both the sceptical and the rash enthusiastic waiting for an action without reaction, most of them decided that the subject has not interest at all. But, far from it, the separation between action and reaction, the non-immediate character of the third law of dynamics, is the basic and fundamental premise to translate the conventional mechanics, purely descriptive, to a causal mechanics as demanded, for example, N. A. Kozyrev: a mechanics of open systems with time like an active agent, and in which even the reversible behaviour depend on the open nature of the system and one defined action with the environment. If the non-conservation of the angular moment may be matched with the separation between action and reaction, and with an interpolated sensitivity, this one may as well become equivalent to the global effect of the environment in a non-homogeneous system, even though its properties remain without specifying.Naturally, this takes us to a different way of thinking.

Since Poincaré we know that any system that satisfies the principle of stationary or extremal action admits infinite explanations or “causal” descriptions, and, therefore, it makes impossible a unique causal interpretation of such systems; it is mainly because of this that physicists appeal to the equations, more than what they could mean in intuibles terms –even when it is clear that nobody can be deprived of interpretations, and of the most reductionist in particular. Without a doubt, the successful management of consensus within the community of mainstream physicists crucially depends on the construction of models with the minimum of possible causal interpretation; or with the neutral appearance of it, since, in fact, it is not possible to do without the interpretations; on the contrary, most of the alternative models emphasizes more or less “causal” interpretations, that other people probably don’t want share, over the possibilities of calculation. From this situation so many problems arise.

If the separation between action and reaction is the essential and distinguishing premise in order to give to time a real content instead of remaining an imaginary and passive entity, and we can speak another way of the interaction of the objects with the environment and space itself, we would have to be able to change the situation and to begin to use these inertial thrusters like basic instruments of measurement for the mathematical description: like an essential clock that helps us to define what is really interesting for us. Everybody recognizes the underlying but conventional role that the clocks play in modern physics. Since Newton´s absolute time to the conventions of the relativist clocks, the principle of *global synchronization* remains intact; and from this all the more abstract aspects of this discipline arise. Naturally, *the convention for the synchronization has been always chosen based on the most treatable or convenient equations*, instead of looking for mathematical models based on measures of the time with its own variability, independent on the conventional global synchronization. However, anyone of these rotary devices with positive inertial performance is a non-conventional clock by itself, offering to us a conversion of forces, oscillations, and fluctuations of those oscillations: they are quite more absolute clocks, no matter how hard their rates of variability are opposed to the uniform synchronization, and for this reason, indeed. More than a problem, this kind of devices already offers a solution to us. They offer archetypical models of conversion of inertia, mass, force, action, reaction and sensitivity to us.They introduce the temporal separation between action and reaction on a mesoscopic scale, just what it is needed; that separation is what determines the total sensitivity of the system and that must be its very contribution to the control and stability theory.

We can apply to these machines a reversed black-box approach; if a gyroscopic rotor cannot have any lifting force for closed systems physics, any lifting force is giving to us a rate of interaction with the medium, no matter we don't know its properties. The question here is if we can find a common factor in the temporal series –in Laithwaite design, for example, between the mechanical gyroscopic oscillations and the vibrations of the main axis. And even if this engine is structurally very simple, may be we could find very interesting analogies with the coupling of biologic bilateral systems, like the pulse in both wrists o the EEG in both hemispheres.

The idea is that, while in conventional systems with no explicit separation between action and reaction the Fourier series cannot give explicit bonds either, in this kind of systems we can find bonds that are a measurement of the interaction with the environment; and that this measure maintain a common denominator independently of the nature of the system. This would be basic for the global behaviour of the system.

The basic breakdown of a system in measures of action, reaction, and separation or sensitivity between both takes us to an important problem of *functional* or global hierarchy, different from the structural or purely mechanical hierarchy of the systems. This principle of functional hierarchy, in which the third element –the sensitivity- plays the decisive role for the output and the balance, is the one that make us hope that the mathematical analyses can offer relevant descriptions, beyond the circular description or the mere statistical patterns.

But the same could be applied to infinity of complex systems; the same life and its definition pass through the great although variable difference between inputs and outputs, action and reaction –this is a must feature. The times of reaction and changes of direction conform the in-homogenous synchronization of the activity of the cells and the traffic of signals in their membranes, and without this, hardly we will be able to understand, not only the operations of the neurons and their problems of functional hierarchy, but even the much more modest "computational" activity –real time and real synchronicity- of the cardiac cells. Curiously enough, now asynchronous integrated circuits without unitary clock-cycle begin to be developed, although far from the flexibility of behavior of nature. The functional and structural asymmetry is also a common element of these devices; and asymmetry is also what we found in most of interesting biological molecules. Nevertheless, while physics is used to extremal principles, in which the time consumption tends to a minimum, the organisms, whose temporal rates are metabolic derivations, which try to save is space, matter and energy, tending to stabilize its thresholds of sensitivity. No matter how much they must have in common, we are before two completely different types of economy.The organized nature ignores time, because nature itself produces it.

Inertial thrust engines as non conventional clocks are in an essential and intriguing double cross between linear, non-linear dynamics, conservative and dissipative; introducing a new variable –a kind of quintessence- on what can or can not be conserved. A whole research program itself. One may suppose that privileged algebras exist to describe these behaviours. The problems of definition of units remind us those of the first attempts in thermodynamics at the early times of Carnot, before the assumption of principles and measurements by Joule and others; although the relation between abstraction and application, the general and the particular, will be also very different in this new dominion.

The non-linearity of these devices can also be described by the fact that the own output is not only a mere result, but part of the reentrant conditions, being not clear enough the feedback loop; of coarse, this aspect is also highly characteristic of living beings. On the other hand, in a system of this kind, and even with its rigid structure, the momentary state of the system is not enough to predict its behaviour, but a necessary reference for some undefined potential or scalar magnitude –that is to say, the punctual and spatial description of time is quite hollow and incomplete. But the temporal separation between the action and reaction would have here to be the measurement of proper reference and the nucleus for any possible standard –or, at least, the starting indicator.

Therefore, these devices reunite ideal conditions to create explicit models of time, the spontaneous synchronization and the nonlinear systems –and if the specialists are not able to advance in their description, hardly they will be able to do it with other much less explicit systems. This already would have to be reason enough to take seriously this class of studies. Maybe we are trying to avoid which is to be searched and identified first of all.

Far from taking to us out of the causality, these devices allow us to go deeper in it and to describe it; far from moving away from the economy and its general balance, these machines and the principles implied are talking to us of the most critical and delicate aspects of the economy of things. We know that the ordinary mechanical description does not allow a univocal identification of causes; if the contemplation of an interval for the third principle and its associate circumstances lead us to it, this notion of the causality must be different from the present spatial representations. The question is if the measurement of the temporal separation for the third principle takes to us to better generalization and predictions, to a selective sieve of the temporal series –Fourier, dimensional analysis, etc. - or not. That an additional selective criterion must exist, seems to be so clear as the fact that better and worse designs exist, more unstable or more sturdy, in addition to the output we could demand to them. Only if new standards of measurement for the third principle were obtained – new clocks -, previously to introduce explanations and interpretations, a fast advance can be expected in this area that must avoid the dispersion of efforts. Besides, the free and creative community of researchers studying these phenomena has many more common than diverging ideas.

That the appearance of the pendulum-clock is consubstantial to the awakening of modern physics, it is well known. Talking about clocks, oscillators and pendulums, few ones stop to consider that Newton laws of dynamics are incapable to explain how a kid seated in a swing obtains increasing oscillations without external impulse –changing only the position of the mass available. This is not obtained without sensitivity, although, on the other hand, this sensitivity can be generated spontaneously in an infinite number of situations. This sensitivity clearly is not magic, and is itself in the middle of actions and reactions; moreover, summarizes in its external efficiency innumerable influences, often impossible to consider in detail. In fact, self-induced amplifications of an oscillator happen in all type of complex behaviors, from Biology to the stock-markets, passing by the hurricanes. In other writing I dealt with how with the appropriate interpolation and measurement of this sensitivity in signals like the sanguineous pulse we can achieve much more comprehensive and rational conclusions in the medical diagnosis. And if certainly theoretical physics is not prone at all to reconsider the properties of a universal medium, time or inertia, equally ubiquitous and more urgent problems exist in applied physics relative to non-linear dynamics, where substantial progresses shine by their absence.

The theoretical physicists make of the gravity the great problem and the great challenge; in this, they remind us many intellectuals, fighting always against Evil. Far away from those challenges commensurate only with the grandeur of so abstract constructions, many of us think that inertia and the centrifugal forces play a more basic and universal role in the reality; being fundamental, unlike the gravity, on all the scales. Inertia thus considered leads naturally to the subject of mass, and the "mass gaps" of the standard theory also could be speaking about "time gaps" Of course, many people, from the more elementary point of view, consider that the same mass is nothing more than angular motion. But, mainly, the inertia and centrifugal forces are manageable by manhood here and now, and perhaps only for that reason many believe they involve no theoretical interest. Today, like always, within the ideas and within the heart, a fight between the lightness and heaviness seems to exist. And one trusts that lightness will prevail –because that's the vocation of Life.

SOME SITES OF INTEREST

Alfred Evert website. http://www.evert.de

Jean-Louis Naudin, *Inertial Propulsion Engines*. http://jnaudin.free.fr/html/IPEmain.htm >

Institute of Time nature explorations. http://www.chronos.msu.ru

Miguel Iradier, *The left hand of Chaos,* http://www.hurqualya.net

WAVELETS, PULSE ANALYSIS, AND THE THIRD DERIVATIVE OF TIME

January, 2006

SUMMARY

Wavelet analysis allows time multi-resolution, besides being an efficient filter for signal processing and denoising. Here we conjecture the possibility that the own mechanical movement of bodies acts, through the three successive differential orders, and by virtue of the resonances that these generate, as a selective filter of the statistical background noise. From this possibility, new considerations for biological information, emergent systems and control arise, as well as for time analysis in general. It follows a philosophical discussion about the location of this analysis in the context of general knowledge, including topics as fractional calculus, nonholomic systems, etc.

Keywords: Emergence, causality, evolution, transient motion, wavelets, pulse analysis, Samkhya, third derivative, fractional calculus, impulse, time.

Background

The sanguineous pulse reading has been the main procedure of diagnosis in the ancient medicines. For the modern scientific standards, these procedures have been considered, at best, like qualitative and empirical approaches, if not plainly rejected as groundless. Nevertheless, it is clear that the pulse is a perfectly quantifiable dynamical signal, and, therefore, we would have to be able to translate the qualitative descriptions in quantitative and verifiable terms. Pulse continues being the more faithful global indicator of the health and state of organism, and if the modern clinical analysis cannot give rational account of the information that it contains, it would also mean that it is incapable to accommodate the most elementary facts of life.

In previous writings [1, 2], I have tried to relate the categories of the Ayurveda and the Samkhya philosophy to the principles of Newton's mechanics. Basically, the three humours, *kapha, pitta* and *vata,* and the corresponding and more general modalities, *tamas, rajas* and *sattwa*, can be respectively superposed with that defined through the first, second and third Newton's laws -inertia, the force or acceleration, and the balance between action and reaction- with the condition that always exists a relevant time between the action and reaction –that is to say, these are not strictly simultaneous. Of course, all known real systems require a time of reaction; but being this time different for any of them, it happens to be considered accidental or miscellaneous. Nevertheless, this meanwhile between action and reaction is indispensable to at least have a minimum intuition of cause, beyond the description of a general balance, which is what the ordinary equations of dynamics establish for stationary systems. The third principle, in this context, is equivalent to a sensitivity term between the action and reaction.

Principles apart, the form of the pulse, as the doctor perceives it, is an additional component not described in the usual oscillographs, although now this is fully recordable. We had talked about that specific form like an elusive quintessence difficult to grasp; and in that respect, we spoke very vaguely of pseudo-derivatives and the measurement of the impulse considered like unit of force per time. In physics, the term "impulse" is equivalent to the integral of the force with respect to the time; nevertheless, in this context we wanted to introduce other considerations with respect to the specific form of the pulse profile, the variations of thickness in response to the pressure, the contractile impact of the heart, etc. Of course, we would like to remain within purely mechanical parameters.

The third derivative

The three principles of the Ayurveda and Samkhya philosophy can be matched with the characteristics of the temporal series and their graphs; but even so, we remain without having a direct and intuitive understanding of what the third element, the sensitivity of the system, could mean. The simplest possible answer would be the third derivative with respect to time, that is to say, the rate of change of the force or acceleration: d3s/dt3 = d2v/dt2 = da/dt

The first speculation on the possible relevance of third derivative in mechanics is as late as 1962. W. O. Davis et al [3] conjectured on their possible applications to systems with transient states, far from the ideal, stationary state –impacts, the entrance of a rocket in the atmosphere, and other phenomena with violent rates of change. But, in fact, the discontinuous application of forces is the norm, and not the exception, for most of the observable phenomena. The physicists and the engineers have preferred to ignore the pertinence of the third derivative or the control of the acceleration, considering it as the option of the operator –this is, like an external component of the system.

Davis approach can be applied not only to sudden impacts and exceptional transient conditions, but, more generally, to driven oscillators with a period of application of the force minor than the critical time of the system.

The signal of the pulse involves impacts inevitably, having pauses as much after systole as after diastole. Therefore starting conditions of the signal and, presumably, a critical time or times; but appearing here control, homeostasis and stability conditions implicitly, nobody will say in this case that the third derivative values refers to something external to the system. Besides, and in a different dominion, the critical time of action is also recognized implicitly in hydrodynamics and aerodynamics for Mach or Reynolds numbers.

In certain conditions, the application of the third derivative offers quite unexpected solutions. Since “Davis mechanics” does not admit the simultaneous action and reaction –in the same way that Samkhya doesn’t -, or, said another way, the moment of a real object cannot change instantaneously and independently of the magnitude of the applied force, the existence of a Critical Action Time (CAT) is to be observed for each system. Admitting a time delay factor, one could interpret that the conservation of the energy during that lapse obeys to a net contribution of the environment of the system –in order to conserve that energy momentarily, indeed. All this sounds too exotic, if we don’t repair in the fact that we are trying to make commensurable two very different types of description. In parallel, for the logic of the Samkhya or the Ayurveda it is required that the third principle, *sattwa* or *vata*, -in fact, the first one of the three in its own order of causality, it is the *most direct* expression of the interaction with the medium or environment. That is to say, both logics agree outstandingly, and they do it since they do not look for the idealization of the events that they try to describe, and even though their description can be largely incomplete.

In addition, the systems with transient states and a critical time delay for the action ordinarily generate resonances –transformations between kinetic and potential energy, which are amplified in as much the denominator tends to zero. Even being quite small, this component never vanishes absolutely, which allows to explore a much more ample spectrum of relations in terms of series and resonances in an undefined medium. That is to say, it allows another type of relations between phenomena of very different order. The new set of resonances appearing in the Davis equations are determined by the relations between the first and third derivative, and their damping, by the coefficient of the second, equivalent to the mass. Possible hyperbolic solutions also may have importance, but they would take to us too far now.

Arthur Young’s Theory of Process

The following who spoke of the importance of third derivative was the American engineer and thinker Arthur Young. Young proposed a highly simplified scheme, but with sense, in which the position, the speed, the acceleration and the control -the derivatives of zero, first, second and third order- conformed a circle map of four moments, "unconscious action", "unconscious reaction", "conscious reaction" and "conscious action". This scheme also is very similar to the one of Samkhya, since for this philosophy the movement and the mind would happen in the three last ones, being the first or the position an indispensable assumption, about which, like about consciousness, nothing with sense can be said –being on the contrary the condition of the rest.

Young belonged to that line of maverick thinkers and generalist like Buckminster Fuller, who considers the present specialized sciences as completely incapable to give account of the most fundamental and basic facts of life. Young argued reasonably enough that without the inclusion of a third derivative –the control-, we could not explain anything of our daily life, like driving a car, for example. This is quite clear, but, being already in the scope of the external activity, they remain somewhat dissociated of the internal dynamic processes. But do "internal dynamic processes exist? By self-definition, dynamics is the science that studies external variations, reason why the question can be reversed. Without doubt, Young would have find better arguments in the dynamic analysis of the pulse, although at his time did not exist tools to register its more specific form faithfully. In fact, Samkhya, being the philosophy of Yoga, is nothing but the most basic description of the control mechanisms.

What Young describes is a process that –as in the sanguineous pulse, or any evolutionary motif of Samkhya- is subject of sedimentation. That is to say, a conscious action with the time happens to be conscious reaction, and even unconscious reaction in many cases. Samkhya describes this like the transformation of the *gunas*, the three basic modalities or principles we deal with. And although these transformations seem to have a continuous and infinite spectrum, they follow one another according to a logic sequence and proportionality.

Young himself was perfectly aware that, even being the third derivative important, which one obtains with higher order differential terms are not but subsequent divisions of time *more geometrico*, remaining the real time of the events undefined. He appealed for that reason to the idea of different times nested one within another, an idea that in a more technical manner arose about that very years and became increasingly popular later with the introduction of fractal patterns and correlations on different scales in the analysis of time series. This type of correlations with self-similar time structure appears, for example, in the graphs of the ECG; but the modern histograms usually study long intervals, with time records about one hour, whereas a typical traditional pulse reading takes a minute. Without doubt, self-similarity can be found in these more reduced intervals, although it is not necessary to think about its infinitesimal character, as many have come, inevitably, to speculate. The successive application of three differential orders leads naturally to question if their values give us some clues on these different time scales or similarity orders.

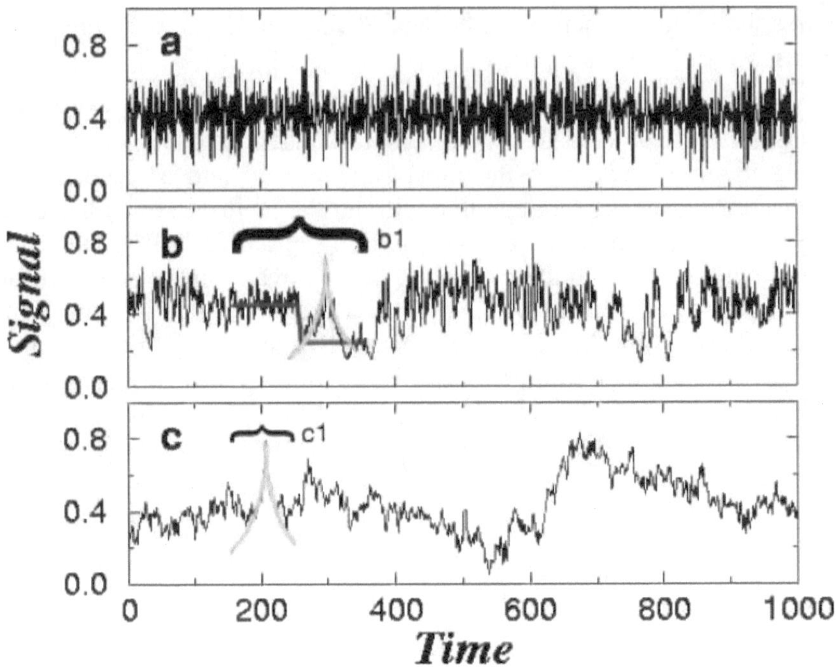

Wavelets and new techniques of time series analysis

This leads us to mention the new modalities of analysis of temporal series that have emerged since the arrival of the computers. It is known that until the last decades the harmonic or Fourier analysis was the only reference in this dominion. The harmonic analysis only yields the spectrum of frequencies by means of sum of motionless sine waves; that is to say, it is independent of time and global with respect to the interval in question. Therefore, it has great relevance for waves with simple dependencies, like light and sound in homogenous media, and its relevance decrease in as much the periodic behaviour of the system presents complex dependencies and bonds, as well as characteristic singularities.

In order to cover this void new modalities of analysis have emerged, like wavelets analysis with time multi-resolution; this one in turns admits many different criteria of election, depending on how the mother wavelet is defined. Indeed, one of the most interesting aspects of wavelet analysis is that it allows to grasp discontinuities in higher order derivatives, besides to fit well the particular features of the series (since the different derivatives admit different signs simultaneously, and the speed can increase while the acceleration diminishes, et cetera, this provides additional criteria for resolution). That is to say, it allows a local analysis with a great versatility in the resolution. Today, wavelet analysis is routinely applied in the more varied fields, from data compression to seismology or partial differential equations, and being medicine and physiological signals one of their main fronts; otherwise, the field is in continuous and accelerated expansion.

Wavelet analysis, by means of selection of criteria, take contact with self-similar fractals and iterative analysis, through other techniques of functional approximation, as radial basis functions, et cetera. Although we don't want to enter in the arsenal of modern tools, an appropriate estimation of the magnitude and rank of information that can render a signal like the one of the pulse is needed. The analysis of wavelets also allows a connection with the p-adic analysis and ultrametric criteria in general: these are important whenever the possibility of hierarchic structure with layers or levels of information arises, and being this case of the pulse according to the understanding of the traditional medicine. In the study of biological signals it is crucial to understand what is dependent and what independent of the temporal and space scales.

Otherwise, wavelets analysis is itself an optimum information filter, tending to reduce the distinct and relevant information to its minimum dimensions. Reciprocally, it is also an optimum filter for denoising, that is to say, to filter the irrelevant part of the signal. Therefore, it is already giving a close estimate on the distinct and characteristic amount of information –"highest personalized information", as modern medicine claims in its publicity- that the pulse signal produces. This deserves to be noticed, because modern medicine is prone to conclude that the pulse is a too *poor* signal in terms of information, and therefore ineffective for the diagnosis, whereas, on the contrary, the practitioners of the traditional medicine complain and struggle with a signal considered in practice too *rich* and subtle in its continuous variability for the judgment, even provided that a sufficient sensitivity has been developed. It is so obvious that the modern medicine has underestimated the volume of possible information of this signal, as it has not make the effort to extract it and to stand at its proper complexity level. Simple numbers can tell us conclusively the amount of possible states involved and the correlative capacity of resolution for the diagnosis –whenever we grant to that information a structure of global character, and not the mere rank of a signal about the cardio-circulatory apparatus only.

Discussion

Although for the modern medicine it is almost impossible to conceive that the signal of the pulse can contain the main relevant information for the diagnosis, we don't have many doubts about it. Common sense itself is in favour of this affirmation, if we forget for a moment all the theoretical devices that we have, either to explain the information of the pulse, or to explain the organism in chemical or reductionist terms. By common sense we understand nothing else than the postulate, if not a simple fact, of the unity of the organism. Therefore, we will not insist on *the possibility* of a satisfactorily deep diagnosis of the state of the health by means of the pulse, but rather in its statute and its more elementary interpretation.

The three principles of the Samkhya and the Ayurveda are to the description of something alive just like the three principles of Newton are to classical mechanics; that is to say, they are the most simple and irreducible in its own level. When the Chinese medicine talk about *yin* and *yang,* also assumes a vacuum or mean like a third element that relates both. However, our endorsement of these three principles with the three derivatives is a pure speculation, and perhaps it does not have the smaller justification.

What is to be known is if we can describe the three principles of the Samkhya –which literally means *analysis*, after all- in a *deductive* way in relation to a dynamic system like the pulse, or if we can only reach an *inductive,* heuristic and approximate quantification of them. Lacking experimental studies in this direction, the question is not trivial at all. Nevertheless, the same philosophy of the Samkhya describes the three *gunas* like interdependent agents, which never can be separated. But, perhaps the three principles of mechanics can be separated? Nevertheless, we are speaking about very different things. The three principles of mechanics are axiomatic in nature and independent of any measurement, and the three principles of Samkhya are already included in the empirical and measurable data. Samkhya, in addition, talk about concrete causality, and mechanics not; Samkhya defines the time fluctuations within and from the interior of a system, and mechanics only defines univocally metric of an external space. Finally, Physics is an activity about prediction of events external with respect to the observer, and Samkhya, being Yoga philosophy, takes care of the internal control of the present by the own subject. So great difference and so great parallelism is what makes fascinating this field of study, still to begin with. And although the clinical analysis of the pulse is widely independent of the fundamental physics, is worth the trouble to make one short reflection on the principles or foundations of Physics. On the other hand, it is evident that the main interest of Physics for the human curiosity is just what fails to explain, namely, life and subject.

A very striking feature of XX century physics is its specific dualism. In eighteen century XIX, with the study of electromagnetism and including a Maxwell and Lorentz, a new level for the question arose about the relation between inertial and accelerated systems, charges and fields, through an underlying medium or ether. Indeed, it was at stake the redefinition of the three principles of mechanics, on whose general validity nobody doubted. The adoption of relativity indeed supposed the refusal to that possible redefinition, adopting conventions of measurement for stationary clocks that made any consideration about the medium or its variability unnecessary. That is to say, the "fourth principle" of Newtonian mechanics, none other than the absolute time or global synchronization was consolidated and reinforced. A modest gain in the predictive frame was obtained, at the cost of ignoring other multiple possible variable conditions. Both extremes conforms the poles of the physics activity; but the increase of the number of variables intrinsically compromise the viability of the predictions. Nevertheless, the problematic of the medium emerge one time and another in the different levels of fundamental Physics, either to specify what the inertia is, or for the mass, or to determine the energy of the vacuum, et cetera. The same problem arises when we asked ourselves, in the most ingenuous and legitimate manner, what makes a particle to "know" that cannot go faster than light; as well as when we ask for the origin and reason of the so called fundamental constants. Then and not before arise, in a natural manner, questions related to control. What shows to us another way, that the dualist scheme of present physics is incapable to describe true causes, in any intuitive sense that we can understand. We cannot expect from this scheme something that it cannot give; and present physics, by its structure, cannot yield nothing else than dispersed masses moving randomly in time by forces equally opaque –just like in the days of Lucretius. Its outline does not give for more, and we fill up the rest with our fantasy.

Be worth this to see that the introduction of a third derivative not necessarily is a trivial subject; in fact, it turns to introduce in other level the questions related to the mean and the frame of reference. Only that in most of the occasions, and specially coming from the perspective of physics, could seem doubtful and little intuitive its necessity, unlike the "ingenuous" questions about the medium and the reference. In the temporal analysis we have better fortune, because always is allowed the reference to the sine or harmonic waves, independent of the movement, although sometimes decompose it exemplarily, as it happens in the case of sound.

The wave of the pulse and its sound are very different things, as it is the spectrogram of the series of the mechanical wave and the spectrogram that its sound produces. We tend to imagine the third derivative like pure brute force or impact, although in fact what indicates is only with what rapidity is the force applied, its degree of availability in temporal terms: again, the most realistic form to conceive the control, in any of the possible levels. Nevertheless, it happens that this component seems barely relevant for the time the beat has arrived muffled to the wrist, having lost the greater part of the profile of the impact of the heart. Nevertheless, a great part of the sound that can be perceived with a microphone comes from that contractile impact, propagated in addition faster than blood itself. Many questions arise on the relations between the acoustic waves and the propagation of the sound with the compression wave as it is perceived with the fingertips.

Evidently, the sensorial capacity of a *vadya* or doctor that reads the pulse is not out of human ordinary scope, but it is highly amplified by the training. In addition to that sensorial capacity, a greater training for the inference and the judgment is still required. In spite of all this, it is not easy to state a defined limit to the sensorial capacity when synesthesy processes take place, as in the case of the *vadya*; in particular, a synesthesy or association of the senses of hearing and touch. Then, it seems indispensable an explanation of the relations between the aspects properly acoustic and those more mechanical of the pressure wave; among other things, because it is very important for the doctor to take a reading at different levels of pressure. But also because we run the risk of underestimating the modulator role of the sound in the organic, from the indiscernible background noise to the more located resonances. Fortunately, wavelet analysis has as the limit case Fourier analysis and therefore includes it; and simultaneously, wavelet analysis is a fine filter for denoising; therefore, there are not greater impediments to grasp the main theme of these relations, still deeply mysterious. It is not necessary to remember the fundamental role of the sound in the classic Indian conception, or that within it space itself is not but the medium in which the sound propagates.

In anyone of the approaches we can try to translate the three principles of the Samkhya and the Ayurveda to quantitative terms, whether deductive or inductive, a mean between action and reaction must take place –because is in that mean that life takes sense, as each sentient being can feel it. It is difficult to think that this mean is reduced to a mere ratio, like for example, the one that can exist between the systole and diastole intervals, even though this ratio contributes proper data; evidently, and by the own irreducible definition of *gunas*, there must be much more than this. Otherwise, an etiological or causal monitoring would not be possible, nor would exist any possibility for a hierarchic structure of the information available in the pulse. The interpretation of the *gunas* in terms of the three derivatives could be erroneous, but it allows explore certain possibilities, like the relation of the resonances with the stationary aspect of the sound spectrum of these same waves, in an attempt to close the circle of the four possible aspects. Of course, we can contemplate also the speed and the acceleration with respect to that stationary background, et cetera.

It happens that the definition of the *gunas* cannot be more elementary; in fact, not even some precise definition exist, but a simple qualitative characterization, and even this one is usually supposed in the classic texts. Precisely because they are so elementary, we can hope to fit them in a perfectly natural way in a deductive description, or otherwise they are not deductible at all and then we only can infer them indirectly. A half way definition does not seem probable. Without a doubt, and even assuming we are talking about real things and not about fictions, the second option would seem more logical. One would be the first surprised in the case that qualitative principles could admit deductive character; that they admit a quantitative correlation, seems trivially certain. And nevertheless, we have repeatedly affirmed that the Samkhya philosophy demands a specific causal frame, in the same way that dynamics prevents it or makes it trivial (Poincaré). Let us emphasize that Samkhya is a theory of the *temporal* fluctuations, without any appeal to the space or a frame of reference. Therefore, lacking experimental evidence, we can only think that a great void to cover in our temporal metric exist, contrary to the great profusion of metrics for the space and the space-time in modern science. It is not possible to underestimate what can fit in this great void.

Finally, it is necessary to return to take in consideration other factors in the pulse diagnosis giving additional information. These are mainly two: the variable pressure with the fingertips, and the correlation of the information of the pulse in both wrists. The role of these two factors is completely different.

The modulation of the pressure in three different points of the wrist, at superficial, medium and deep level, remains as a heuristic approach, which otherwise can imitate simple adjustable pressure sensors perfectly. The intention of this is to grasp better the subtleties of the form of the wave, its thickness, which is not recorded in the usual esphygmographs. Naturally, this is very useful to obtain more information, and has no counterpart in the usual mathematical tools of analysis of time series. Somewhat humorously, it could be said that the doctor sets out empirically his own partial differential equation, alike as a mathematician deduce the evolution of a variable while maintains the other variables constant. In fact, what the doctor attempt to grasp with these three changing levels is a more focused reference for the three humours, *pitta, vata* and *kapha.* Some will be tempted to say that this is all the mystery of these three principles, although we think that this is to confuse causes and effects. The state of the three humours is an objective fact, independently of the form of measurement. But, precisely because the importance of this additional information cannot be underestimated, it is necessary to follow it in its consequences. The three gradual levels of the pulse are not a separation of the principles –being this impossible also by principle -, but one more index of their relation. And it agrees to say that, doing without for a moment the immediate characteristics of the dynamic profile of the pulse, these three levels and the three principles find a logical and full correspondence with the three foliations of the embryological development -endoderm, mesoderm and ectoderm, with all the ramifications here implied.

As far as the differences of the pulse between both wrists, some doctors consider them, and some others not. Again, for the western modern medicine it will seem nearly impossible that such differences, if they exist, can have the least significance; but for the modern medicine, even the whole human bilateral symmetry is a biological coincidence. The alternation of the breathing in both nostrils, with a clear periodicity, is well reported. In general, it is hard for us to imagine significant differences between the pulses of both wrists; but it is less difficult to conceive significant differences in *the sound* of both pulses, as, for example, can detect accurate microphones disposed to this effect. And, with the sound, the dissonances also. The detailed record of these differences, including its possible periodicity, is a subject that could leave us even to algebraic questions; let us say only that the difference between both pulses corresponds in the most general manner with difference between the vegetative and the voluntary –but here in the measurable frame of time series. A subject like this never has been properly tackled, in spite of its great implications for cognitive sciences and the study of time itself.

The spatial symmetry hides a dynamic asymmetry. This asymmetry must have substantial correspondences with the decomposition of other organic functions in their extremes: the inspiration and the expiration, systole and diastole, the relation between the arterial and venous circulation. Both extremes express also in the most direct and immediate manner how the higher functions constrains the lower –voluntary means -, and how the lower functions constrains the higher ones –vegetative roots. These are the two main constrains, by will and by state, fully intertwined, in addition, with the dissipation and the maintenance of the unity of the organism. These two extremes find their natural correspondence with the first two principles of Samkhya, inertia and activity, and being the third the indicator of their relation. It is really surprising that so elemental aspects have not found its place in a measurable and experimental frame. Even we emphasized in a previous essay how this is fully and globally linked with the *cognitive* difference between both brain hemispheres.

In addition, the correlation between both pulses, *indicating simultaneously the two ends of the arrow of organic time*, the past and the future, helps us to define a potential or gradient for the union of the three humours or principles, and this is completely necessary since the moment that the indissoluble union of the three principles in any individual is itself, before any other thing, a gradient or reactive slope. This allows us to save immense amounts of work in the laboratory at the time of identifying the nature of the three principles in the empirical via time series. Also allows us to extrapolate past or future conditions, although it is to see with what degree of precision. All it inclines to think that, for a complex evolution like the one of an organism, the "retrodictive determinism" is much more feasible than "predictive determinism" proper of Physics. On the other hand, it is surprising that modern medicine tries to acquire predictive determinism, being this one quite senseless in this area. That is to say, in medicine, like in many other fields, the "prediction" is reduced to a diagnosis in which we must suppose that the future conditions will be similar to the present and past conditions. But it is known how to estimate these past and present conditions? Not, except for the most trivial cases. It is much more "revolutionary" to be able to understand what it happens and really has happened than to predict what it is not understood.

R. M. Kiehn has dedicated a great part of his life to show how the irreversible and non-equilibrium systems, as well as their entropy, can be described without statistics, by means of coordinate free methods of exterior differential forms. Considering the ambiguous and precarious character of all the statistical treatments on the entropy, it is an extraordinary achievement, which would have to lead to many later developments. Kiehn has stressed repeatedly the mathematical viability of the "retrodictive determinism", questioning that predictive models are the only way to understand nature. Hardly the main one, we would add. Saving the distance between the dynamic description of the systems and the profiles of their temporal series, we think that Kiehn's descriptions of continuous evolution in topologic dimension 4 could fit and describe in the most compact form the reality of the changes of the pulse in the long term, but the experimental verification of this would take whole generations. And when we say more "compact", borrowing the term from topology, we are thinking also about the irreducible form of body aging, with the added dimension –literally embodied- of time. Since reversible systems have a zero torsion component, and irreversible ones a component with a positive value, it is conceivable that this vector, equivalent to the arrow of the time, would be also related to the Critical Action Time implied in third derivative of time.

Finally, it is inevitable to remember that the pulse reading is telling much more than the beats of the heart. In fact, the most of the signal, in quantitative terms, is related to the stomach and the associated activities of assimilation, and not even to the heart, that plays a smaller part in the amount of blood. The heart propels the blood, but it does not produce it. Indeed, modern physiology, that so much has studied the mechanisms of mediation in the organic homeostasis, would have to conceive easier the polysemy of a signal like the pulse. William Harvey knew perfectly that the bad performance of an organ obstructs the sanguineous circulation locally and secondarily affects to the general circulation; something so evident that many of his successors tend to forget. Even in the most elementary analysis of the sanguineous pressure, we observed at least four factors: the volume of ejection, the total volume of blood in the circulatory system, the flexibility of the arteries, and the vascular resistance, that is controlled by the autonomous nervous system. Therefore, inexcusably, different organic levels are involved, but an acceptable key to articulate them of elementary form is not found. The only postulate is the unit of the organism –and the first thing we would have to undertake seriously. For the traditional medicine, the problem of this type of diagnosis is not its mere possibility, but its difficulty with purely inductive methods.

A crucial aspect to evaluate the pertinence of a third derivative is the identification of the Critical Time of the system and its relation with third order variations and their associated resonances. In the case of the pulse, it can be thought that the critical time is the time interval of the beat –or, at least, is the first relation coming to the mind. The relation between the third component and the background noise must find its equivalence in the temporal perturbation of the series. That is to say, the more the control component is fitted, the more fitted will be the relation between beats and even between the two phases of each beat –what it does not imply always more regularity. The time delay can vary also, and its variations can be decomposed in average component and a fluctuating component subject to autocorrelation. Obviously, the functional character of this third element has to be demonstrable, or at least, congruent with the general evolution of the series. It has to play an active role in the maintenance of the stability. Therefore, also must give account of a part of the random components that always appear in these series –of the most essential part of it, indeed. In other words, the resonances of the three derivatives in the pulse series must show a causal feedback loop with the cardiac activity, therefore being directly relevant for cardiology also. The third order gives us the more close measurement of that which mediates and is interposed between the action and reaction along the cycle or the interval. Immediately arise the question about in which interval is effective this fitness, and to what extent has validity in the direction of the past or to the future. About this and many other questions only the experimental survey will decide.

If the third derivative has not found universal application still, it is partially due to the infinity of possible critical times for so many other systems. Nevertheless, it is this same thing, the proper time of each system and its sensitivity to the environment, which is to define. On the other hand, many cyclical processes have in principle a source of continuous periodic perturbation, as for example, the weather with respect to the daily and annual period; it could seem then that does not make sense to apply a third order to them. Yet, we hear the media and even many meteorologists talk about an "acceleration" of the global warming, when at the other hand we are experimenting some of the coldest winters. If that supposed acceleration exist, it would have to be with fluctuations, and from that we could gain conclusions on to what extent the climatic balance is affected with respect to its capacity of control, name it the capacity of interchange –whether with outer space, or with its own thermally active internal mass. It is really very difficult to think about real changes without dimensions of impact, and thus we talk about the impact of the human activity on the climate also. If so, we would have to be able to measure it. We believe then that the third derivative is always the most general index of the control in a system, but only the experimental data of the different systems must decide about its possible relevance. It will be necessary and of great interest to apply the tools of dimensional analysis to discover similarities and irreducible aspects of systems with a relevant third order.

Definitely, we see the third derivatives with resonances as the most generic form of temporal interface –of contact or transition between possible behaviours or states. They are intimately related also to the theory of phase transitions with critical points, but here we only consider temporal indices, doing complete abstraction of the underlying dynamics. That's to say, under a relatively simple feature, complexity is implied. Besides, with the term "phase transition", we understand something much more general than the term usual in physics of states, and we leave its outline for later works.

The idea is that the emergent surface of a complex system finds the main index in the relation of the three derivatives. It must be so, if we understand as "emergence" a significant change in the relations between a system and its environment. Since all real things show complexity and simplicity only remains for idealized closed systems, what we get this way is the generic index of the deviation of a system from its closed and idealized mode, unable to create novelty. So we can transcend the sharp dichotomy between causality and organization –between artificial hierarchical criteria and a purely formal and devoid dynamic causality.

In this context of wavelet analysis and higher orders, it seems quite logical to ask for the role that here could play the fractional calculus, with non-integer or arbitrary integrals and derivatives. Even today mathematicians are far from consensus with respect to the geometric and physical meaning that fractional calculus can take, although works as the one of Podlubny [4] proposes this one like an approach to the local time of a system, in contrast with the absolute time usual in physics. The fractional calculus arose from the developments of Liouville and Riemann, although already Leibniz in 1695 stated that its adoption "*It will lead to a paradox, from which one day useful consequences will be drawn*". Naturally, fractional calculus may be connected with wavelet analysis, and even in its more explicitly analytical form, as polynomial splines and the derived fractional splines are; in addition, the fractional derivatives are not determined by the local behaviour of the function, but by the whole dominion of interest, completing the circle to return from the local to the global behaviour. The fractional calculus is being applied now to problems of control, probability, diffusion, viscoelasticity and hereditary mechanics with memory of states, as much for stationary and transient systems. Transient resonances must have fractional relations and bonds. Therefore, everything we are saying on a third differential order and its interpretation can and must put itself under the light of the fractional calculus, that naturally obeys to questions on temporal metrics and fluctuating, inhomogeneous times. Anyway, we must recall that fractional calculus does not support a simplistic interpretation in terms of geometrical fractals, although, of course, many relations can be found. On the other hand, both fractional derivatives and transient states allow important connexions with the Riemann zeta function, as was showed since 1975 by Keiper and by Pavlov and Fadeev, respectively. Besides the interpretation in terms of resonances, the same local-global relation we notice between wavelets analysis and fractional derivatives brings to us closer to the core of the zeta function.

The fact that fractional calculus has not admitted until now a physical or geometric interpretation only could be surprising for those who consider the development of modern physics like the unique possible and the natural one. Indeed, this very fact is indicating in another fashion to us the void of modern physics with respect to time and the temporal metrics. Before we have talked about to "dualism" of modern physics; another way to see this dualism is its recognition of two types of fundamentally different magnitudes, scalars and vectors, incompatible in its representation. As Gustave Le Bon already notice a century ago, it does not exist any good reason to think that cannot exist other kinds of essentially different magnitudes, not less fundamental; however, it is indeed through higher and arbitrary calculus orders that these other magnitudes, forms of energy and behaviours have to be outlined. What is really arbitrary is to think that with two integer differential orders we have exhausted the possibilities of nature. In addition, if we were able to physically interpret this extension of calculus, it is perfectly possible that we not only accede to much greater degrees of diversity in nature, but also to a greater simplicity and unit. What we call resonances already are a connexion between this two so different kinds of magnitudes, and will show continuity aspects where we can see separate things only, and, on the contrary, separate things in evolutions considered continuous now. In a really unified frame, the same fundamental forces would be derived effects more than causes –and also "differential sections" of a much more vast temporal evolution. But we think that it is not possible to come to such a level without a specific approach to the problematic of time in mechanics.

P. Fiziev and H. Kleinert, [5] for example, give us a new anholonomic action principle for pure rotations around a fixed point of a body –Euler equations- *without reference to stationary systems*. Non-holonomic systems are those which are path-dependent, a car for example, and besides the usual control or differential degrees of freedom, the history of the systems must be known. We will use this for an example later. Transient systems, resonances and anholonomic behaviour can describe each other, when the naïve application of hamiltonian action principle fails. This also must yield powerful connexions between chaotic behaviour and number theory –elliptic integrals, modular forms and so on, because the number system seems not to be sensitive to anholonomic conditions, but anholonomic systems are sensitive to numerical conditions. Anyway, there is much to find in the always *momentary* condition of nature.

Vacuum, information and other metaphors

It seems that if there were something able to transmit all the shades and subtleties implied in this type of diagnosis, it had to be the sound; but the limitations of the sound like source of information seems evident. The frequencies are very low in relation to many levels of activity, the ubiquity of uncorrelated white noise, et cetera. Nevertheless, all Indian literature has insisted in this respect, although its notion of sound ignores almost all the features of modern acoustics. In its conception, the sound is not only independent of any space scale, but the same waves create the space. This connects with the well-known differences that we have find between the dynamics of objects in the space and the Samkhya like description of time fluctuations. It does not exist either, of course, the smaller consideration of the noise like statistical phenomenon.

The science of the pulse diagnosis is called *nadi vigyan*, examination of the channels or ducts. What the ducts or channels transmit is just the sound, or what it is understood like that in this non-spatial conception, but phenomenal perception. The Upanishads makes an empirical and testimonial enumeration on the different levels of noise or background sound –no distinction is made between both- that are audible in the own body and beyond it; and this same gradation is one of the canonical forms to describe the union process in which the Yoga consists. It turns out natural to contemplate all this process like a noise-filtering process, and reciprocally an identification of the fundamental patterns.

Beyond details, we think that any thing that is moving is filtering in real time the background noise: with the condition, it's clear, that the movement has an articulated component, that is to say, a third derivative of time with resonances. This would be applied in the first place to the pulse, but also to infinity of real systems. The ubiquitous 1/f noise could fit in this context, if it really allows some general explanation.

Any kind of temporal analysis contains in its nucleus its own uncertainty relations. It is not possible to know exactly the frequencies in an exact instant of time. Heisenberg's uncertainty relations, far from being a fundamental principle of nature, are only a consequence of the limitations of the Fourier analysis as a mathematical tool. Today are working routinely many devices, like optical microscopes, that get around this last uncertainty subsuming it under a much more general set of uncertainty relations –just in the same way that the harmonic analysis is only a limit case for wavelet analysis [6]. Observations much below the wavelengths are made diary and the only limit is not theoretical, but depends on the technical capacity to filter the noise sources. Since we see to what extent modern theories are conditioned by the mathematical tools available, it is easy to see that we are in a silent revolution that will have very far-reaching consequences. Be enough to think that until now the so-called "classical time analysis" was fully independent of time itself. In other words, only now we are beginning to penetrate in the temporal dominion.

Anyway, what we touch here is the superposition of two types of systems apparently different: an articulated mechanical movement generating resonances, and a statistical background noise. This last one includes all kind of random motion. The question is to what degree of accuracy both, movement and noise, can interfere to each other, and yield an additional precision. Naturally, all this has to do with the critical time of the system.

All the cells –not only those of the heart- pulse and have articulated motions, reason why we can even apply an analogous treatment to them. This is equivalent to evaluate the global reactions of a cell in relation to the underlying medium, including the bifurcations of the cellular division. The study of the cellular pulsations, for example in cell culturing, is walking its first steps right now.

Resonances are not an exotic phenomenon; among other things, we hear thanks to them. In fact, they exist always and everywhere, and the only question is to know to what extent and in what frame can be relevant; when they converge towards the stability and when they generate divergence or dissonance. On the other hand, the articulated motion, motion with impacts, is the norm and not it exception. In fact, they are the only event to which the discreet processes can be attributed, and not to any arbitrary measurement of quantization, that since we have seen, arises from the harmonic analysis and its time-independent criteria. If we are technically beyond these limitations, it becomes an absurdity to grant them an absolute character. The real world is made of articulated contacts, not of "emissions" and "absorptions" of ghost massless particles, suited only for the convenience of the physical predictions.

Already Descartes, in the first days of kinematics, tried to develop the idea that all movement is rotation, and therefore, acceleration. With these attempts was also related his program of vortices mechanics. Today one tends to see these ideas like clumsy prehistoric, previous attempts to the arrival of the "true" Newton's mechanics; but the question is that both mechanics implied completely divergent points of view, and didn't go in the same direction at all. In order to begin, the vortices can be so complex dynamic systems that not even today the mathematicians fully agree about how to define them. They also admit to be considered like objects with intermediate properties between waves and particles, a quite important feature. Decisively, they also imply accelerations of accelerations and torsion components, which also complicate enormously the calculations. It could be said that the vortex is the simplest and archetypical form of complexity itself, and therefore, the first to be avoided in a predictive frame for "fundamental" physics. Until Descartes, and even although this one inaugurates the subject-object dualism, there is serious interest to think about the real causes of the natural events; since Newton, the decision is made in favour of the general balance equations in order to prediction. Naturally, if all movement is acceleration, it must have rates of change in the acceleration to give account of the phenomena –local transitions. Still today there are physicists like Gennady Shipov trying to develop the Cartesian principles, who even has been able to unfold a whole arsenal of general equations with physical sense.

Without a doubt, a third differential order makes more complicated the calculations, mainly if one thinks about high energies and highly non-linear events. But, on the other hand, gives us more general indices in a different direction. Anyway, in comparison, the modelling of these ideas in a system so lowly non-linear as the pulse would have to be a child's play. Let us propose an analogy to outline what can describe the pulse diagnosis.

In a "Cartesian" car, even when we travel at constant speed, there are acceleration variations. That is to say, even though the speedometer seems to be fixed, there are opposite forces being compensated: the force of draft of the motor and the friction of the road and the air resistance. But if we make sufficiently exact measures of the action of those forces, we would see that they never really compensate each other, and therefore accelerations and decelerations take place constantly (So far, I have taken the purely kinematic example from Shipov). In this feature, we see that there is no real inertial system of reference. In addition to this, we are free to accelerate the automobile, imposing other additional conditions to the fluctuations of the acceleration at constant speed. It could be said that the fluctuations produced at constant speed are characteristic of a certain car traveling on a certain ground or pavement; and that those fluctuations will change at different speeds, giving us a more complete profile or spectrum of possibilities. The acceleration, the speed changes, will still give us a more exhaustive profile of these fluctuations, and they will allow us to know more of that black box that the car is. As far as the third element, the optional use of the brake or the accelerator, it seems to be independent of the car, although not of the ground. If we are measuring all it from the outside and without knowing the circumstances, we have a black box within another black box within another black box: the ground, the car, the driver. It is already difficult to say something about each one separated, not to say about the relations among them. The driver sees things that we do not see, and in addition, nobody is in his head.

Then, let us suppose that there is no head, no driver. Or that the driver must move without seeing, in complete blindness, being able to listen until the minimum detail the noises, resonances and responses of the car at every moment. Let us suppose, in addition, that it have not to deal with other cars and traffic, having enough with not to leave the road. Without a doubt, the car would go with an extraordinary care and slowness. And if the car were in addition very fragile, the "driver" would make an effort to optimize his response to each new signal: the smallest curve would be noticed in a lapse of time, would accelerate when leaving it, et cetera. Road, car and driver would be revealed in a same signal. The purpose and the aim, the intentionality, become included in the adjustment to the given conditions –here, as much internal as external. Without the least paradox, the objectives have some sense in as much as they are not arbitrary –we are not entirely free to choose them. Think well in this.

This is a metaphor of the life; mainly from her initial, more stuttering phase, than it is so difficult to imagine for us. Only that the ground is much more varied than what the linear figure of the road could suggest to us. Life is blind, but it is inconceivable without sensitivity. Life, also, arises in a cavity, but not a static one, but a cavity in motion, alive itself. Its own motion is a filter of the environment. Everything we see now or that we believe to see, previously has been felt and attempted in the dark. It has happened in the cells, and even in the tiniest molecule; and they had not reached here without that feeling, attempting activity. If Prigogine said that matter is blind in equilibrium, but out of equilibrium sees, we would say that matter in resonant conditions is still blind, but hears and listen.

The present evolutionary theory, that so much has fought to prevail, can take refuge peacefully in its statistical background to avoid any description of real time evolution. In the analogy of the car calls the attention the slowness whereupon everything would have to take place. But, does somebody ask which can be the perception of the local real time for a cell or a molecule? The third derivative gives us the index of that time.

There is too much discourse on biological programming and information now. However, the information processing –in real time, again- indeed demands that the reaction of a system is *unequal* with respect to the action that receives; this is, it does not take place –it cannot be concrete and local- within the framework of Newton laws with an absolute time like global synchronizer [7]. In other words, the information, that we use to consider so concrete, not even can occur within the framework of modern physics; it is supposed and superposed like a simple metaphor. Only a Cartesian mechanics of the kind proposed by Shipov, describing locally the transition between inertial and accelerated systems, could generate and process information. This should be a warning. Therefore, taking into account the different angles, not even has been considered the problem of life, no matter how much one speaks about "information".

Let us look at this closer. Classical physics can make the transmission of information possible, but cannot make it univocal –Poincaré's conclusive argument [8] on that the general dynamic equations with extremal or stationary principles admit infinite causal interpretations. Then, one can appeal to the thermodynamics and quantum mechanics, which governs the interactions of atoms and molecules. The consensual thermodynamics not only cannot create information, but, in addition, destroys it. Quantum mechanics, on which all kind of mythifications proliferate, is a *linear* theory; it is curious how little this essential characteristic is stressed. In addition, the reversible aspects are emphasized and the irreversible aspects are mainly ignored. Therefore it is incapable to generate information. What remains then to make possible the creation and even the mere *univocal* processing of information? Information itself, devoid of any physical reality. Moreover, information is measured in purely quantitative terms and with total independence of the time of processing involved. And when this one is taken into account, like in the microprocessors, a clock cycle is used completely analogous to the absolute time or "global synchronizer" of classical physics. It is easy to see, therefore, that nobody at all knows about what is physically speaking when the concept of information is applied to Biology: its explanatory value is in the same metaphysical and indemonstrable level that the vitalism and the "intelligent design". Information is a purely external concept, requiring for its processing other equally external concepts like the design, the synchronization, the statistical selection, et cetera, that in addition are not mutually compatible to each other. Do not look for the logical combination of all these elements, because it does not exist. The information is for present science a concept as void as the concept of causality, but still more confused, because it is ubiquitously patched in one's own interest with the notion of chance and randomness. To say that the world is caused by the causality is not to say much, mainly when we consider that causality necessarily remains to define, as in the case of physics. But in the case of information it is still worse, because to the information concept a specified and univocal character is supposed, and certainly the things don't improve adding to them the chance and randomness, the concept of the unspecific itself. Then, somebody comes to us saying that we have our origin in "Chance and Necessity", with capital letters added: brave and precise answer. Neodarwinism does not risk nor explains more than creationism, having only more scientific pretensions.

And all this is fatal and unavoidable, because without a specification and local definition of inertia, the whole physics evades the dominion of the reality. In other words, remains *unphysical*. By the same, the "physicalist" explanations of the mind remain equally *unphysical* and equally void of real content –they belong to "parallel worlds". Hume's criticism on the concept of causality, as well as the much more refined from Poincaré, are valid for the type of causality of modern physics, that is the emptiest causality possible; but they gradually lose its validity in other dominions, like for example, the redoubled analytical version of the Samkhya that we are proposing. That is to say, not only it is not contradictory to consider the causality and the emergent properties of an object together, but it is *necessary* to precise the dominion of causality itself. Even Standard Model physics must appeal to phase transitions with emergent properties to define its ground and the rules of the game, not to mention the properties of any material; although, surely, it does not do it without arbitrariness. But this has very little to do with the levels of causality and emergency that we propose here, based only on redefining the effects of inertia at a superior level by means of a third differential order: with dimensions of *impulse*, to give it the right name.

The election of the inertial reference frame is *the only* theoretical bet of western physics, not only since Galileo, but already from Copernicus himself. And by the same, there is no way other to close the arch of its proposals and to give them content that the local definition of inertia, the definition of the inertial field. This is the bottom of all this matter, and the latest thing that whoever who allows himself to think can grant as a matter of course. Without it, not even we can know what means the word "mechanical"; with it, we would enter directly the problems of control. And thus, for example, theories like Shipov's one, which does not need any of the fundamental constants, must set out necessarily the problem of the control of the local metric space. Of course, and as theories like the one of Shipov or Heim shows, all the interactions and equations of the well-known physics can be rewritten and reinterpreted with other criterion of the fundamental dimensions, having the same quantum of action the dimension of angular moment, and not to mention the relation between "stable" particles and their resonances. From our view, we could do without the underlying physics and just observe a still causal evolution in other levels. But it is necessary to say in addition that the lack of definition for the inertia is entirely parallel to the lack of definition for the memory for the present cognitive sciences. Both, in effect, have a natural correspondence. And, as the naive example of the Cartesian car discovers to the most elementary analysis, the question arise on the conservation and destruction of states and the accumulated record of the car in its present mechanical state, considered like the present dynamic response. However, in that test of response, the structures, those "frozen contingencies", are no more than the same present dynamic response, and any other thing can be ignored as a contingency, that help us to solve nothing. The background is the response, because the response is the unique thing that makes perceive and suppose a background to us. What can be applied to the "structures", also can be applied to the memory and its ghosts. The structure is a stabilized dynamics that it has to destabilize itself to return to be noticed. Therefore, which can be predicated of inertia and the memory can be predicated of the notion of structure and consequently applied.

The mere manifestation of inertia is the most immediate form of *orientation.* There is no un-manifested inertia, otherwise we would not ask about it. In other words, inertia is not a principle or something reducible to a point or state, but the more elementary *relation*; that's the notion of the so-called Mach's principle. In this sense, already it is an emergent property, more than an automatic derivation of the causality. Therefore, it will yield as much information as we know to extract to it in a given circumstances. Any fast enough change in the acceleration has a momentary inertial yield that does not follows from the second derivative directly, but indirectly in conjunction with the other two: this *transient* margin is fundamental in order to explain many things. In Shipov's Cartesian vehicle, inertia and the apparent constant speed are irreducibly informative, because inertia implies the control. In our extended version of the same vehicle, the "external" control of the conductor is not arbitrary either; the control does not depend here on doing what one wants, but on the measure in as much the circumstances can be appreciated, responding to them with the smaller possible interference. Sensitivity is the capacity to perceive inertia: that sensitivity is all the immediate information that we have. To interfere more is to lose "immediate information" and control. Nothing of this is contrary to mechanics and to the simplest reason, and we only add an additional level of information. From this new level, other uncertainty relations are derived. Naturally, all this seems equally valid for our own behaviour, and as an elementary corollary can be added that losing the fundamental information and the control by means of our interference, also we lose autonomy and freedom. The same Cartesian vehicle is a clock of the local time, which can become more explicit using like clock an appropriate type of gyroscope. The third derivative of the time, the control of the impulse, gives us an index of the anisotropy and non-homogeneity of the environment, as a *necessarily oriented* system perceives it; an index also of its own time in terms of information –of effective processing of information.

It is quite curious that all type of anomalies, largely verified in the laboratories, but without possible place for the constituted and constructed theories, has to do in one way or another with this third differential order, and not with "collateral" or "hidden" "perturbations". Even discounting the great percentage of spurious cases, the amount of literature on the subject remains immense and amazing, waiting for a theoretical frame to give it some more sense. This is only another indication of the clearly non-trivial character of the third differential order.

Machines seem mechanical to us because we have synchronized its parts avoiding as far as possible undesirable effects, like frictions and resonances. Their design is a time-binding cutting of functions to be optimized. Even so, the own life of a machine and its duration depend to a great extent on these factors, in addition to the use suffered. Nature has used since the very beginning and without beginning these perturbative elements like a constructive principle, and this is the main reason, in the long run, of the enormous difference, even starting from identical principles. But these "identical principles" are often completely opposite, because of the incommensurable criteria of isolation or relation with the environment: Nature often unites many things that we separate, and on the contrary. Unlike the engineers, physicists and mathematicians, the honest Nature laughs at the analytical difficulties, and makes an indifferent use of all she has within reach. By the way, there exists a strong relation, somewhat antagonistic, between friction and resonance, but the interesting thing is that we can treat the resonances in a better defined temporal frame, doing without a good part of the always ambiguous statistical elements. One thinks that not a single beat of a heart is automatic, except for our ignorance. Terms like "random", "automatic" "inertia" are too void still today, and indeed what we are doing is to look for a frame in which they can mean something.

The great historical irony is that Descartes, creator of the modern dualism of body and mind, proposed a physical theory opposed to that dualism and with potentiality *avant lettre* to reconcile it and to reduce it to a nonsense; whereas the physical monism derived from the success of Newton's mechanics, opposite to the Cartesian mechanics, has turn out to be the consummation of the success of dualism in fact.

Talking about the fathers of modern science, it is inevitable to associate our Cartesian car with the celestial mechanics and the planets. The measurement of the gravity in this same Earth has an experimental uncertainty associated of more than a part in ten thousand, and being this only the most conservative average estimation. The oscillations seem random, but applying the proposed criterion, it seems possible to filter a good part of this noise and to deduce relevant factors. If the tiniest particle needs a time reaction, how could we believe that the enormous planetary masses have to react instantaneously without absorption. In addition to void, this is completely irrational and it is only based on our blind faith in the "naturalness" of classical mechanics. But the absorption of these reactions or jerks must have a close relation with the fluctuation of the values of the gravity and the mass separately, remaining the product constant and being observed the orbital regularity of Kepler's laws. Cassini's fourth degree ovals, toric sections, also can be related to all this. We don't need to stress the relevance of this to find a *real* link between quantum and classical mechanics. Once again, in order to maintain the fiction of a universal constant, we are ignoring the precious local information.

Returning to the diagnosis of the pulse, our subject is to what extent a relatively slow movement but with a component of resonances can differentiate the background noise, to filter it. Although it is evident that a *vadya* must resort to the inference in a very ample degree and at different levels, which we want to know is to what extent all this can have a deductive support. And although the considerations on the underlying mechanics are not absolutely necessary for the purpose of diagnosis and understanding, it helps to see their pertinence in the most fundamental level. Not only time, but the space and the inertia themselves are private of content and reduced to a passive paper in modern physics. All the complex and real systems have a relevant third differential order, and only in the idealized scheme of the "fundamental forces" this third element is excluded necessarily and by definition.

It is to be noted that the third derivative has a *strictly physical* meaning; surely this cannot be said of higher differentials orders, which could only be interpretable like resonances within the resonances already present in the lower orders. Since the third derivative implies a conditional relation between the external environment and the internal states, the third order must mark also a physical limit in the capacity of filtration and absorption of the statistical background noise. That is to say, whichever be the degree of non-linearity observed in the relation of the three time derivatives, they define *the limit of physical causality* of a system, and everything beyond this limit must be considered as *physically random*.

We would go further to affirm that without this kind of resonances the memory and the same inertia would never become manifested, and that this manifestation always exist within a critical time that really is the specious present of the system, its real time. And if inertia already is the first manifestation, further only the unmanifested remains, not to be necessarily confused with the non-existent. Thus, for example, the three derivatives of the pulse signal would have to give the simpler possible "radiography" of the present state of the organism, in its own feeling and sentience, without excluding at all the existence of many other "radiographies" of the same order that can be taken for the same moment, be they an electroencephalogram or a measurement of the electrical conductivity of the skin. This is inevitable, if we take into account that the three derivatives must yield always a coupling index with any other possible section of the system.

There are good reasons to think that if these notions find their justification in the analysis of a signal like the sanguineous pulse, they also will find some rank of application for the processing of other types of signals; since we spoke of very fundamental and basic concepts of the temporal analysis. The question is, today and always, but with the new means at our disposal, to find the natural articulation of the own time of a system –and to give it a sense and interpretation as rational as possible. Any system in motion is filtering the background noise, which is simultaneously inside and outside: be the system the global climate and the atmospheric evolution, or of the fluctuations of the stock market. Of course, each of these systems has its own parameters and conditions, which to a large extent are totally unrelated. But, precisely, of most interest is this little they can have in common –this little could weigh much. But we insist again on the probability, not very intuitive for the usual thinking, that a greater understanding of general aspects in this ground does not imply automatically a greater predictive power. If we don't understand this circumstance, we won't understand the meaning of all we have spoken. Let us see a bit clearer.

All we know and we assume that in the long term the stock-exchange behaviour cannot be predicted –thought nobody who plays resign himself to it, since we know also that, in spite of the huge number of random or *uncontrollable* aspects, there are present tendencies too that finish in *logical* conclusion –"logical" mainly when the facts already have happen, but not as much before. Think about a stock-exchange collapse. But the question is that in the tendency the *control* component is contained, in the same way that in the variables of the pulse is contained the reaction of the subject and its relation with the environment. It is exactly this which we think that can be considered in an *explicit* manner in the analysis of the temporal series. However, it is exactly this which the companies in their stock-exchange policy or the economic authorities are *implicitly* making. That is, they try to exert the control as much as they can and they try to avoid being controlled as much as possible. Naturally, this has very delicate margins, and in this consist the political sensibility in order to take measures. The economic news continuously speak to us about the speed of the growth, and its acceleration or deceleration. In addition, all we speak about the effect of the economic measures (more or less arbitrary or necessary) in terms of *impact*: "the impact government's measures", etc. Also in the boards and councils is discussed endlessly on "how to articulate" the measures and "how to manage their impact on"…the others. Government itself is not another thing than this kind of control; neither politics in general nor the economic policy in particular. The elementary dialectics of all this is reduced to the *conviction*, nobody knows to what extent justifiable, that when the more control, the least possibilities are of being controlled and have to account. On the other hand, accounts with the future are the most difficult to render, although this is almost the unique thing proposed and sold. In short, and to be brief, I believe that it is possible the objective control and monitoring of the measures of control and monitoring of those that exert them optionally and arbitrary. We can control the control. To obtain this would be much more "revolutionary" than to try to predict future itself, a future whose prediction already comes included in the sale package. Control is measurable and then controllable also. But this does not lead to an infinite regression, because time itself imposes restrictions over this as a limit of the interaction with the environment. Then, control must be explicitly expressible in any kind of complete balance. Beyond the usual predictive horizon, technical paradoxes like the efficient market paradox have a partial, but concrete solution.

I think that to pass with all the consequences from the second to a third differential order is literally equivalent to going from the world of *prediction* with inexplicable but efficient causes at a partial level, to a world of *control* with explicable causes and an efficiency or use of global type. Really, this control gravitates towards self-control. The great paradox at present, clearly visible for everybody, is that prediction is pursued in order to get more control, but even so we are losing the control in an increasing incontrollable way. The subject is so general, that I do not believe somebody can consider it a chance. Beyond the predictive "causality," a whole world to explore exists, but surely not a whole world to exploit.

This is a general and philosophical essay, and we cannot expect much more precision; in any case what is written is enough for those who want to research it. It can be said that, if our proposal makes sense, the only thing we do is to introduce a third grammar element in the habitual grammar of dynamics, which only have two cases presently, this is, verbs and predicates with rigid connections. The introduction of this third element is equivalent then, and within the simplification of the case, to the introduction of the subject in dynamics. I am already hearing the angry protests; in particular, the protests of the so-miscalled "reductionists", that, nevertheless, hold much more immodest pretensions, as the creation of "intelligent machines" or the "strong artificial intelligence". Had they called it Automatic Intelligence instead of Artificial, nobody had taken it seriously. But here we are speaking of much more elementary and simple things. The discourse of modern science only seems to be able to enunciate phrases of the type "to eat potatoes", "come back tomorrow", or "natural selection"; at least, in terms of sense, and in spite of the fantastic technical sophistication reached. And that is in the most generous cases, because often verb and predicate are confused by the expeditious procedure of turning them equivalent. Thus, in spite of the super-specialization, everything leads towards a fatal confusion of planes, and the specialists are the firsts who exploit this ambiguity and confusion.

Any movement expressed in three differential orders maintains the possibility of a linguistic equivalence, and, in addition, of many other analogical translations or reproductions. This, that seems completely dispensable in modern science, is of decisive importance to open to the subject other possibilities in the knowledge and its use. And, in addition, modern sciences cannot either do without that analogical component, as it shows Poincaré's argument on the absence of univocal causes in mechanics. Naturally, to introduce a third grammar case does not imply automatically to take the grammar to a high level of refinement, but at least a completely new form of articulation is reached. In addition, everybody knows that the third element always has been playing hide-and-seek in all type of discourses, and in all the scientific discourses also. But the case is that precisely the transient motion is the only motion that can demand univocal causes, or, anyway, it is the one giving the closest approach to it.

The Indian theory of language makes a decisive distinction between the articulated and unarticulated languages; we will continue finding this distinction and relation in any approach to the motion like nature's language. We have already written in other places that surely is not a metaphor to speak about the language of nature, and that, on the contrary, the most probable is that human languages are a poor metaphor of the edenic language of creation.

One can ask finally why so simple ideas, if they have some substance, have not found already a rank of application. It is necessary to respond, on the contrary, that they must be substantiated before any application, and this is not the case yet. We still move in a ground of conjectures. Nevertheless, a decisive fact is that the temporal analysis has not begun to be really temporal until these very last years. I think that nobody still is aware of what this really implies, existing such differences between the spatial analysis and the analysis of the crude, uncut temporal evolution. It happens that these new tools just forged still are exercising their wisdom teeth with inherited problems, without repairing in their independent potential and still trying to formulate their own questions. That is to say, the modern temporal analysis already is the answer to a question still not formulated. And it takes work to find a place in it to that question, among other things, because it raises a matter of a completely different order, very against the grain of the unthinking formulas and the routine thought.

The modern theory of the control and the stability incorporates the mechanics on the one hand and on the other makes a discretionary use of statistical tools; but its mechanics is of course a mechanics of stationary, closed systems. The external relations of the closed system are mainly statistical, and this often applies for a great part of the internal components. Therefore, it is easy to see that an approach like this continues having enormous limitations and loopholes, and still more if we apply it to biological organisms. By definition, a live organism cannot be a stationary system. Then, our only hypothesis consists in to affirm that the *transient* component of any dynamics makes its own statistical selection, a selection whose reach has to be measurable and valuable. This is the only source of natural selection –and of any possible evolution, as the Samkhya already stated thousands years ago.

One is the first harbouring doubts on the universality of this phenomenon, wishing criticisms and attempts to demonstrate the unviability of its operation. In addition, I will not conceal that this hypothesis is *only* based on my personal conviction that it is impossible to give account of a single beat of a heart with the whole arsenal of mechanics and statistics available yet, and that the physiological structures only a-posteriori "explain" the real behaviour, but eliminating the essential. It would seem that it cannot be an argument more fragile than this; it is fragility itself. But it can happens also that mechanics itself is extraordinarily fragile at the time of facing the gross, uncut temporal evolution, eliminating what gives it the specific robustness, that is the own robustness of real beings and the world we observe. The coupling of pendulum clocks in a wall by means of resonance noticed by Huygens is really the first example in modern science of *natural and real, spontaneous synchronization*, in opposition to the abstract and arbitrary global synchronization or absolute time of Newton's mechanic; but the phenomenon was known since time immemorial through musical instruments, as for example, tubes of flutes and organs.

The development of a theory based on these new principles necessarily must have important limitations in the predictive dominion. This is good itself, in the sense that it indicates to us that we deal with something real. Not the prediction, truth is important to us. In addition, modern theories like the synthetic theory of the evolution does not have the smaller predictive value, nor either retrodictive value, and hardly could touch real time and present evolution, and even so it seems that many people take it seriously. As we have already said in other occasions, the only reason why these people take it seriously is the void of mechanics with respect to the univocal sense of the causes. In less extent, this is equally valid for modern cosmology.

Let us finish with a consideration in the symbolic dominion. Motion with three differential orders and inherent resonances that are related to the background noise finds a natural analogy with the sense, both indicative and physical, that the Vedas grant to the syllable AUM. As it is known, the sacred syllable, being an inarticulate sound, is also the seed of all the articulated sounds and their coupling, as well as the background in which they finally fuse with. A way is intended to allude, through the physical sound, to that which drives it across and transcends.

Conclusion: What can we expect of pulse diagnosis

I hope that the third derivative and their implications will turn out to be a key to go beyond the labyrinth of specialized sciences which have kidnapped the common reality. Because within those specialized sciences there's no longer hope for somebody, except for the experts themselves.

We believe to have given sufficient arguments to make possible to conceive the informative richness of the signal of the pulse and its relevant synthetic character. We believe also that with the technical tools now routinely available the degree of resolution of the relevant information can be achieved in only a few months of laboratory work and with diminutive expenses. Moreover, some fundamental correlations could be verified, specified or refuted in a matter of days. In any case and since the beginning it is to be observed that the measurement and correlation of the three principles includes necessarily different levels of interpretation, which nevertheless must be congruent. This is not a deficiency, but an intrinsic virtue of this approach –in fact it is the only thing that could make sense in the level of complexity of an organism.

As far as the question of what can we hope of this type of diagnosis, we return to the image of the car on the road. The renewed pulse diagnosis would allow us to recognize with very reasonable clarity: a) the road travel, or in other words, the vital vicissitudes of the journey; b) the mechanical state of the car; c) the behaviour of the driver in relation to both previous elements. But this is only what is to be expected of any rational and reasonable medicine.

Let us say that, at the most, the modern biomedical model is able to diagnose point *b*, that is to say, "the mechanical state of the car", and being limited by its own nature to make suppositions devoid of any functionality on the other two extremes. The subject only exists outside, to convince him, and his circumstances can be only considered accidents.

Present medicine cannot define with any criterion an optimal degree of health in a certain context and term; the advanced pulse analysis, yes, would have to be able to consider with great accuracy the relation of our present state with the optimal, but realistic, possible one; to define comprehensively the landscape of the health in the most direct, intuitive and irreducible terms. Anyone would have to be able to check itself in house, knowing and comprehending the matter. Each personal optimal, like the body, mood and mind of each one, has absolutely concrete limits and restrictions. The "unlimited health", like the "unlimited freedom", only finds its natural place in the advertising.

Since now there are so much work and effort in the evaluation and description of the aging process, let us say that the pulse analysis must be able to give a *functional and objective* global measurement of the degree of disorder of the organism, that is to say, of the entropy, but understood quite independently of the statistical criteria. Is there something similar now? In addition, must be able to show the difference between reversible and irreversible imbalance, within a defined context and period of time. This is fundamental also. Besides, the criterion of these estimations does not emanate from a horizon of predictions, but from the organic potential of the complexion of each person and his accumulated history. Who said "personalized medicine"? No need to say that the so-called genetic information of the individual, an aggregate of construction materials, can not report individual history nor its interaction with the environment either.

Let us say in addition that the study of the pulse in a really analytical frame would allow fast tests of collateral effects of medicines, much more resolvent dynamic tests of effort and a robust bridge with cognitive sciences and psychology, to mention only a few. It would have to transform our most intimate understanding of time processes and, first of all, of we ourselves. Samkhya always was the analysis of the temporal fluctuations, independently of the arbitrary divisibility of the space. In as much we develop an authentic theory and practice of control passing through the filter of the *person*, and not merely the individual, we will escape the control of anonymous powers, always hidden behind the opacity and the unintelligible character of the world, and the so often false pretension that they hold a "technical" knowledge of superior order. The exercise of the control is, mainly, which they hold.

We are in the Age of Control.

References

1. Iradier, M. *The left hand of Chaos (Modern and Ancient Indian Analysis)*

2. Iradier, M. *Samkhya and Golden Mean,* http://www.hurqualya.net

3. Dan Sewell Ward, *Davis Mechanics*, from *the Library of Halexandria,* http://www.halexandria.org/dward137.htm

4. *Igor Podlubny,* Geometric and Physical interpretation of fractional integration and fractional differentiation, http://www.tuke.sk/podlubny/pspdf/pifcaa_r.pdf

5. Fiziev, P. and Kleinert, H. *Motion of a rigid body in body-fixed coordinates system,* http://arxiv.org/PS_cache/hep-th/pdf/9503/9503075.pdf

6. Croca J.R. *Beyond Heisenberg's Uncertainty Limits*, http://cfcul.fc.ul.pt/equipa/croca/berkeley%20-%20paper.pdf

7. Matsuno K. *Information: Resurrection of the Cartesian Physics*, http://bio.-nagaokaut.ac.jp/~matsuno/preprints/RESURREC.htm

8. Poincaré, H. *Ciencia e hipótesis*, Madrid : Espasa Calpe, 2002

9. Kiehn, R.M. *Retrodictive determinism*, from *Cartan`s Corner,* http://www22.pair.com/csdc/car/carfre24.htm

10. Shipov, G.I. *A Theory of Physical Vacuum*, http://www.shipov.com/science.html¶

BETWEEN STRESS AND PRESSURE

10 July, 2018

We live between stress and pressure, this is the framework that defines our life most immediately. But what is between stress and pressure, oneself aside? This Modern Prometheus' X has a clear characterization in current Science and engineering, but oddly enough, to our scientists, so eager to make their concepts understandable to the average person, the potential of such elementary and universal connection has gone unnoticed. Between pressure and stress, only superficial opposites, we get strain or deformation, and both materials science and the constitutive aspect of electromagnetism just deal with that. This way we can establish a formal and rigorous analogy between a quantitative aspect of science and the qualitative aspects of traditional pairs of opposites that are even prior to perception. This key connection could be critical for the future of technology, our perception of nature and our self-understanding.

We all talk about blood pressure and hypertension, but do we know where the difference is? Blood pressure refers to the pressure that blood exerts on the vessels, which is lower when the blood circulates faster, and higher the slower the circulation is. Tension, on the other hand, refers to the stress to which the arterial wall is subjected, stress that can have different responses depending on whether this wall is more or less rigid, more or less elastic. And this is how we place ourselves, even without thinking: by pressure we mean the action of external forces on ourselves, and by stress or tension we mean the presence of forces in our inner constitution, which may yield or not yield at all, giving or not giving place to a deformation.

Materials science characterizes materials following this same pattern: they are the subject of strain and stress, and their resulting three-dimensional response is studied. The basic idea is that matter, which has the attribute of a relative impenetrability, can withstand stresses without deformation —thanks to its rigidity-, but can not exhibit deformations without the presence of a force, either stress or pressure, and these being mere expressions of opposite sign of the force itself.

Based on the mechanics of continuous media, the constitutive law of materials is a universal mathematical tool that applies to all types of particular subjects. And since the original formulation of electromagnetic fields by Maxwell is precisely a continuum one, it has the typical attributes of the Elasticity Theory and Continuum Mechanics, only with particular properties that are not found in ordinary matter. The symmetry of Maxwell's equations has been widely celebrated, but nothing captures its essence better than this verbal formula we owe to Nicolae Mazilu: "*The form of the stresses that are not accompanied by strains is what characterizes the case of a classic electric field classic electric, while the form of the strains that are not accompanied by stresses is what characterizes a classical magnetic field, and vice versa."*

The Continuum of the old electromagnetic theory was not only the medium between bodies, but what is both inside and outside of matter. In this constitutive framework, and in tune with the above, would be "*that capable of withstanding stresses without strain when it is in the matter, and of exhibiting strains without being under stress when it is in free space*". A combination of both must be considered and this reveals "*a mathematical structure describing electromagnetic radiation, and light, therefore, within them."*

In other words, electromagnetic waves themselves already support those apparently contradictory aspects attributed to the medium, so that, whether we are talking about fields or Ether, no theory can really circumvent them. Maxwell's equations are composed of a constitutive part, relative to the material sources, and another part that corresponds to the fields emanating from them. And it is because science relies preferentially on the field laws, better suited for prediction, that Mazilu's constitutive interpretation goes unnoticed.

In fact, the electromagnetic waves, whose two field components are not even measured with the same units or dimensions, can not really be that mutually transverse waves that geometric idealization displays in our mind. The perpendicularity of electric and magnetic fields "*is not a geometrical property but a straight statistical one*."

So we have an average. But average of what? It is clear that we can not directly detect waves in vacuum, because for this we will always need some material means. And hence the only thing we can get is an average of the waves in free space and the waves in matter.

We can then consider the apparent repulsions and attractions between electric charges in terms of stress and pressure; in both cases it is a trivial relationship, since it only involves the change of sign. This is the field perspective and we don't need to say it's like looking from the outside. But the same does not happen with the constitutive or internal aspect: a material can be deformed or not depending on the force exerted, or else, a rigid material like a metal will have a lot of resistance but once deformed it will not recover, while an elastic material yields easily but also recover in the same measure. In this case there are margins, not everything measurable is controllable, there are reversible reactions and also irreversible responses. In short, it is not a perfectly symmetrical domain, but it is much more like life and the properties commonly observed in it —and yet, it is still obeying universal quantitative laws with which it can be measured and calculated.

The physicist will often say that the most fundamental aspect is not the constitutive one, but the one related to the field; and that the fields in turn obey the particle dynamics… but the particle is just the source or constituent part of the field. And then, if we ask what an electron is made of, the answer is usually that it is made of ... electromagnetic fields, what else could it be. At most, this electron field, in order to avoid explosion, would be cohesive with a kind of tension, like the Poincaré stress, which even can be related to gravity or the cosmological constant of the Space-Time Continuum of General Relativity. And so we would shift from the Old Continuum to its modern version through the narrow tunnel of an electron. In Maxwell's continuum there were no more limitations than those of the metric; but we only want to show that under the constitutive guise one can get something so fundamental as one could wish.

What is the Continuum? That for which it does not make sense to say "inside" and "outside", that which transcends its categorical separation. Stress and pressure are internal and external expressions of the forces, deformations or strains can have both internal and external meaning. But if this is the case, and we oppose strains and stresses, this means that stresses can also be external or internal, and the same goes for pressures.

Our interest here is not the theory but our ordinary perception of the material reality. And it is clear that we can not aspire to a more tangible and earthly definition than this one, since the most immediate properties that we sensorially attribute to matter, its hardness, resistance or impenetrability, as well as its relative capacity to yield and deform under external pressure, are precisely those included in this definition. That is, no matter how hard we try, we will never find anything more *material* than this. But it is not something that depends only on our touch as an external sense: it belongs to our innermost primary sense as self-perception, as bodies in the middle of the interplay of forces.

The question arises: it is a commonplace to say there is no possible continuity between the quantitative descriptions of Physics and the qualitative aspect of our perceptions; as there can not be continuity between the analytical scientific method and the dialectical method that underlies the old world views of all peoples, those that spoke of the generation of the world, of course, by genders: the extensive and the intensive, the empty and the full, space and matter, Heaven and Earth, water and fire, mercury and sulfur, expansion and contraction, soft and hard, Yin and Yang.

But we can see that at the constitutive level the electromagnetic continuum is the most literal quantitative translation possible of the relationship of these pairs of primordial opposites, and that this relation can not be confused with the much more trivial one of the charges —action and reaction don't follow automatically and we must consider the response capacity of the material. Of course, factors that are measurable but not controllable can occur here —which draws the limits of mathematical analysis. The mathematical relation of two 3 x 3 strain and stress matrices gives us the basic case. The continuum is partially or totally uncontrollable, and it is very appropriate to affirm, as Mazilu does, that Maxwell's electromagnetic theory of light is an intellectual reaction to its uncontrollability.

We don't need to convince ourselves, it is enough if we can ask about the appropriateness of this connection. Do we have here the quantitative translation of our innate sense of the pairs of opposites? I am more interested in rigorous analogy than algebra, in connection than substitution, in continuity than in discrete operations. There seems to be enough of the latter in this world, but a "sustainable development" of the former is missed.

I think that the analogy is perfectly justified in the qualitative, and that in the quantitative its limits are those of the equations of this type; but with limitations and all the connection is of the deepest interest. And electromagnetism in the language of external differential forms teaches us that it is always possible to leave down the constitutive part, with its particular metric, and to cleanly distill a field part that possesses natural invariance and greater universality. Is not this an excellent mathematical form of Alchemy? Thus seen, Maxwell's equations are a highly universal structure applicable to thermodynamics, hydrodynamics, or many other disciplines. But it's doubtful that the implications of this can be extracted without the proper physical motivation.

Actually all the physical fields and the ideas we can make of space are generalizations of a particular metric applied to matter, prolonging the oldest sense of the word "Geometry". But the space of the physical continuum is not the space of geometries; a completely homogeneous medium has no interior or exterior, it is not full and therefore can't be void either.

Today we use the fashionable language of Relativity and Quantum Mechanics, but we must not forget that both have arisen, by segregation, from the workshop of electromagnetism. Phenomena that not long ago surprised the most trained theoretical physicists, who did not even know how to fit into standard quantum mechanics such as the Aharonov-Bohm effect and the geometric phase that generalizes it, also appear in the classical electromagnetic field, and even in elementary hydrodynamic analogies on the surface of the water. There is nothing that can not be derived from the physical continuum if it is sought with due consideration —for neither quantum mechanics nor any other theory has been able to substitute the continuum for some definite type of discrete entity.

The same goes for the attempts to explain consciousness by relying on the nonlocal aspects of quantum mechanics. Is it not much easier and natural to say that Consciousness and Continuum are Identity itself, since they are the undifferentiated, and that it is Self-consciousness that introduces difference and the discrete operations of thought by virtue of this same background? The apprehension of physical reality can only be done from Reality as such, not the other way around.

Maxwell's contemporary, the mathematician William Clifford, speculated with a pregeometry of feelings, beyond matter and geometry, in terms of contiguity and succession. But all this is already involved in Maxwell's physical continuum, although we do not see it, and it can be appreciated perhaps in the topological aspects, independent of form and scale, that can be derived from electromagnetism or thermodynamics. Things like quantum foam seems more suitable for the surface than for the bottom.

To a large extent, and if the continuum is already the base assumption and the universal law, it is from this "pregeometry of sensations" that we are talking about. This physical continuum has a representable part as a game of forces, which is for which the equations have solutions, and has an uncontrollable but no less real part, since it can be even measured. Here we have not only a crossroads for physics, but also a touchstone for what technology can achieve with certain means. And this would be a good place to talk about possible applications of something as recent as continuous quantum measurement.

Indeed, we can explore our own internal environment with technologies no less than space can be explored: in both cases we have both the immediate impression that everything is possible, and that everything is terribly limited —as it should be, being the means always limited by its very nature.

We attempt to transcend with machine, measurement and mathematics the obvious limitations of our five senses, which, we are told, are basically reduced to small variations of the electromagnetic fields of our constituent molecules. But in reality, just as physical space can not be limited because it can not even be void, the small village of our senses are five huts built on an indeterminate ground, but that doesn't exclude perception, and that reminds us of the old *sensorium communis* of the ancient philosphers. And how could this be an abstraction, I wonder, if it is the pillow of darkness over which our impressions vanish every night when we fall asleep.

Could this have some practical application? We are speaking here of the match of a physical-mathematical language, not only with the most tangible, sensorial properties of matter, but with our own self-perception and the exercising of our internal physical space too, in which the body itself acts as an interface between the objective and subjective. Therefore, almost by definition, to develop this line of reasoning would affect, rather than one or another technology, the same Gordian knot, the man/machine interface that defines their mutual relationship in both directions.

More than anything, it seems a matter of perseverance and method, of the creation of an appropriate language too. On the one hand, we see that information technologies accumulate layers and layers of increasing complexity precisely to facilitate the user experience; on the other hand, the same users demand certain resistance and naturalness in order than this same experience can have depth and "a touch of reality". For arbitrary complexity there are no limits, but for simplicity it is the continuum what set the rules.

If we talk about the whole complex world of man/machine interfaces, it's clear that on the side of the machine the main weight will fall in the electromechanical issues, that is, the relations between electromagnetism and mechanics. Both aspects are already integrated in the aforementioned constitutive view; but it results that for the human being as an organism those aspects are equally inherent in his internal environment, biomechanics and external activity. Therefore the deep development of this connection should have huge implications, since it gravitates towards the optimal point between immediacy and the interface depth: what we call an appropriate and appropriable technology.

And as is often the case, as the relationship deepens, the back of both extremes recede in the depth.

In fact, if the inadequacy between man and machine is not greater, it is because these factors, being constitutive or inherent, are already considered and integrated implicitly in the complex "equation" that defines their coupling —in a range of factors that goes from the resistance of the materials to the subjective experience of the user. They range from the most basic level to very subtle aspects, and the question is whether they can find a common unifying language, which would transform not only our technological culture but culture in general.

The electro-biomechanical interfaces are increasingly diversified and in them the digital took command over the analogical to an overwhelming degree, but this doesn't belittle the least the interest of the line here pointed out. Today many wonder which, among the many possible, will be the decisive level for the direct connection between hardware and wetware, computers and neurons: this seems to be the critical line of the digital vanishing point. And yet Nature seems determined to put a brake on our impatience, or rather we should say she is totally indifferent. Although the firing of neurons is a discrete act of all-or-nothing type, the building of potentials has little to do with binary logics.

Everything suggests that these binary logics are last minute by-products, and that if we want to place ourselves at Nature's level we have to attend the lower level functions and thence, follow their guidelines. And I think that the lowest level is neither chemical nor electrical, but constitutive or electromechanical in the sense already mentioned, as it is also the most plastic in allowing analogies or quantitative extensions —thermodynamic, hydrodynamic, and so on. The lower the level, more present this feature must be; and I, for one, cannot conceive a more basic frame with relevance for macro and microstates, and, what is more important, for *subjective and objective states*.

If Physics seems to play such a subordinate role in this area, it is mainly because the customary field standards and definitions are used, operating from the local to the global instead of the opposite as Maxwell's original theory — plainly organic and integral- did; and because in these descriptions issues of orientation that can be very relevant are largely ignored. A vision that included them could perhaps shift more fluidly to the questions posed by information theory, semiotics or semantics, just as one can pass from exterior differential forms to geometric algebras (Clifford), or to Grassmann algebras that are their common ancestor.

There would be a completely unexpected twist in all this, no matter how much we would expecting for it. We are fumbling the socket to Nature in the dark, and to ourselves in passing, but in the process, if we want to match the issue, what has to change, and to a great extent, is the shape of our plug.

Physics is worthy of interest only when we remember that things are not reducible to extension and motion. In fact, when we pretend that material particles are punctual, their properties are attributed to completely inextensive phenomena. Unlike in mathematics, local and global in Physics do not mean "small" and "large". The dualities that cross modern physics, and that have their origin in the electromagnetic duality, are diverse expressions of this, too: eminent pledges of non-duality, rather than of some pretended Cartesian dualism that has never taken place except in our heads.

The same thing happened in the old hermetic version of the artists of Fire: the male or sulfur, the dissipative, the intensive, has been caught in matter; the female mercury, space or spirit is nevertheless the matrix of everything we consider conservative or mechanical, fully reducible to movement. When this is understood, the slackening arch of Manifestation recover again its Tone, and we see a Male Nature where we thought we were seeing only a Female Nature, and vice versa, and the commerce between "high" and "low" abandon a routine that couldn't have another place than in our heads either.

Digital/analogue and analog/digital converters, sensors and medicine are three of the areas that first come to mind when it comes to test these ideas at the most basic level. An ever-expanding field that specifically puts to the test the connection between mechanics, electromagnetism and constitutive laws of materials is the design of metamaterials. We will dedicate a separate article to medicine, health, aging and evolution.

References

Mazilu, N, *Mechanical problem of Ether*, Apeiron, Vol. 15, No. 1, January 2008

Post, E J, *A History of Physics as an exercise in Philosophy*

SELF-ENERGY AND SELF-INTERACTION: EXTENDED PARTICLE AND THERMODYAMICS

1 August, 2018

Since the times of the Manhattan Project one cannot expect transparency in applied areas such as particle physics. Such concepts as the point particle, special relativity, or the segregation of thermodynamic problems, not to mention quantum mechanics itself, render a great service as a smokescreen for the real physical processes. This veil was found convenient and has been maintained so far in the technology sectors of strategic value. We'll give here both a historical perspective and future prospects.

The electron theory, already since the works of Lorentz of 1895, seems to close definitively the problems of Maxwell classic electrodynamics and lead directly to the problems of Relativity; on the other hand this particle is the very workshop in which quantum mechanics is developed. Thus, the entire foundations of twentieth-century physics, at the micro and macroscopic levels, find here a common Gordian knot. Later, the relativistic extension of classical mechanics and quantum mechanics acquired such an unfolding that the open questions of electromagnetism seemed too distant, if not irrelevant. That this was not so, and still is not so, is conclusively demonstrated by the fact that the main numbers of quantum electrodynamics, such as the mass, the electron charge or the fine structure constant, are still put by hand and are not derived from anything, which leaves this whole outstanding construction hanging in the air.

That enormous unknown that we call "electron", the faithful and common servant of all our technology, so apparently ordinary that it leaves no room in us for suspicion, so rubbed, used and abused even in our most trivial devices, is a genie unlike any other: the only genie, perhaps, that would allow us to realize certain desires not by taking it out, but by taking it back to the lamp.

The basic problem of self-interaction in current models is that of radiative reaction: an accelerated charge radiates electromagnetic energy and moment and therefore there must be a reaction on the particle, or from its electromagnetic field on itself. The causes of radiation do not necessarily have to be external, they can also be due to interaction with other particles.

For some of the authors of QED, Feynman for example, the idea of self-interaction is just silly, and the only sensible thing to do is to get rid of it and ignore the whole problem simply stating that an accelerated electron does not radiate at all. The issue has always been highly controversial and here we would not like to take side on any particular stance, but it seems to us that the position above mentioned is first and foremost a matter of convenience, and that there is no clear admission of the reasons for this convenience.

It is not said, for example, that Special Relativity, in reality the general case, is only a local framework for punctual events, in which extended particles don't make any sense —in the same way that in General Relativity, in reality a particular case for gravity, it doesn't make sense to speak of point masses.

Maxwell's equations have general covariance but this does not lead to the Lorentz transformation that gives rise to Special Relativity. The original Maxwell equations have integral form and contain more information than their later differential version, but even in this last one they do not yet consider discrete charges which is the case that Lorentz introduces. However, Maxwell's equations, in their most universal form and free of the metric expressly introduced for the electromagnetic force have natural invariance as Cartan, Kottler and van Dantzig recognized in their time.

Maxwell's theory, limited as it is, does not cease to be a theory of the physical (not geometric) Continuum, but in Special Relativity this is the first victim, no matter how much one speaks of a space-time Continuum. Its operationalist character entails just that: the discretional cutting of that continuum with the introduction of arbitrary postulates incompatible with the continuum such as the invariance of the speed of light. This creates an additional contradiction with the principle of global synchronization that is not stated here but that since Newton always has been assumed.

If in Newton this principle is just a metaphysical statement, in STR, since it is a local framework of punctual events, only local conservation laws can be considered. In fact, if even today a extended particle model does not seem viable, it is because the special relativity already excludes it right from the start. Such basic phenomena as inductance and self-inductance, that obviously require circuits with a defined extension, don't have a place in this theory. Then we find ourselves in the strange situation that a global synchronization is assumed that is excluded by the theory itself and that this very theory makes impossible to implement. Minkowski invariance is incompatible with any classical equation of motion, and, moreover, it does not seem that Special Relativity can give an appropriate transformation for accelerated frames of reference, and therefore, for radiation.

The question here is not to make a diatribe against Special Relativity, but to see that this is incompatible with the idea of an extended particle and with other basic features of the classical field theory such as acceleration and radiation. From the very beginning, STR is highly abstract and divorced from physical context, a sort of Solomonic judgement to move beyond the impasse of the frames of reference problem, later generalized without the slightest restraint to other cases with a rich material context. In all this, Special Relativity catches no mice, and a high price is paid by sacrificing everything to its formalisms. But this is not admitted, and attention is drawn to subordinate and secondary matters.

We don't really know whether an extended particle model or a point particle one would be preferable, whether to contemplate the problems of radiation or to forget about them; but in any case to decide it we would have to start from a sufficiently impartial frame. Special Relativity is not impartial, neither Maxwell or Maxwell-Lorentz's electrodynamics. In the classical electromagnetic field there is no place for point particles, and on the other hand, if Bohr was still proposing circular orbits in 1913, it was because Maxwell equations don't tell under which precise conditions radiation is or is not emitted.

There is a gap in the balance of forces in Maxwell equations if they radiate whenever accelerated, and the radiated energy is the compensation of the work done by the auto-force (although in Maxwell's original theory is not explicitly a force law). It is obvious that if we don't know yet when there is radiation or not, auto-force and auto-energy constitute possible adjustment terms for the big numbers whose origin still remains unclear, besides enabling a reopening and a redefinition of the thermodynamics and irreversibility topics which were "outsourced" from fundamental physics. Both things seem to be as inconvenient for the QED version that pretend to be the last word on the issue, as they are promising for other approaches free of such commitments.

On the other hand, the opposition between point particle and extended particle is surely not as sharp as we sometimes want to put it, because there has never been anything like a "point particle" to begin with, but a particle-in-the-field whose center admits a point on which a force can be applied. Hertz had already seen the need to distinguish between material particle and material point: the first is an indestructible point of application of forces, the second is an extended variable set internally connected. These are different meanings and application domains, in a sense very similar to the one later raised by de Broglie.

What is an electron supposed to be made of? Of electromagnetic fields, of what else. This is why the particle cannot be out of the continuum, and this is also a good reason for self-interaction to raise. And as electromagnetic fields, electrons are not point particles; the vast majority of their energy lies within a radius of $2.8x10^{-11}$ meters, which for long distances gives a good approximation as a point particle that cannot be maintained for distances short enough. The wave of an electron in a superconductor can occupy meters and even kilometers, showing to what extent a particle is what its environment allows it. In any case the density of its field decreases as $1/r^2$, and if the magnetic field around a current is appreciable at meters, the same goes with each electron. In the limit this field extends to infinity to the greatest size and minimum structure; and it is in the opposite direction, with confinement, that the details are to be find —details that to some extent mirror of the confinement itself.

So there is no point particle without field. It is assumed that quantum electrodynamics was just about that, but then came the Aharonov-Bohm effect and the confusion among the champions of local action was more eloquent than millions of words; and it is even more eloquent if we consider that the effect can be derived entirely from the classical Hamilton-Jacobi equation and that Berry and others even found its exact analogous on the surface of the water, with illustrations so reminiscent of those of Bjerkness at the Paris International Exposition of 1881. In spite of everything, and who knows why, it is still common to hear that the well-known effect is a clear exponent of the unique character of quantum potentials.

We are told once and again that quantum mechanics is the fundamental level and classical electromagnetism necessarily follows from it, as it should be, but in practice without a classical view quantum mechanics is basically blind, a standard calculation formalism to which one has to tell how to operate with, and furthermore the transition zone between both is extremely diffuse and lacks a general criterion. Quantum mechanics is not a vision, but a recipe.

Self-interaction without radiation: Weber's electrodynamics and retarded potentials

Weber's electrodynamics, which precedes Maxwell's by quite a number of years, is a theory of direct action with many similarities with the direct electrodynamics arising after Special Relativity but without any of its contraindications, which still makes its revision so recommendable today. Maxwell himself finally recognized in his *Treatise* the basic equivalence between field theories and the then called action at a distance theories such as Weber's one.

Wilhelm Weber developed in passing an elliptical atom model fifty or sixty years before Bohr drew his circular orbit model and without the benefit of any of the data known to the Danish. Moreover, the nucleus remained stable without need of nuclear forces.

Weber's law is the first case in which central forces are defined not only by distances but also by relative radial accelerations; beyond a critical distance, which converges with the classical electron radius, the inertial mass of the charge change sign from positive to negative.

A decisive point is that Weber's electrodynamics, although largely equivalent to Maxwell's, allows us to see another side of self-interaction that does not necessarily pass through radiation. Rather, it is a self-interaction of the whole system under consideration, a global feedback of the circuit, rather than self-interaction as a local reaction. In Maxwell's original conception everything should be global and circuital as well, but you can't work with point charges. Weber's formulation allow both point and extended particles.

As Assis likes to remember, Weber's electric force law of 1846 is the first equation of motion completely homogeneous and relational, that is, expressed in known quantities of the same type. The generalized use of heterogeneous quantities already tells us how far we are from transparency, just as the use of dimensional or "universal" constants shouts from the rooftops just that a theory is not universal. Weber's force law of 1846 is fully relativistic 59 years before the not-so-relativistic Special Theory of Relativity, without any postulate or rupture with the continuum.

It is also the first law in dynamics that is a natural extension, rather than an amendment, of Newton's central forces law. As usual, Newton's followers were more newtonian than Newton himself, but the first definition in the *Principia* don't ask that the magnitude of central forces should depend only on the distance between points, as with a string attached, so to speak. This they took from a mere metaphor like that of the slingshot, but not from definitions.

The extension of the law of central forces, apparently suggested by Gauss in 1835 and later reformulated by Weber, is in fact the most natural form of evolution of Newtonian mechanics: it involved only the static case (gravitostatic), and thus introduces the dynamic case in which force is not invariable but depends on relative velocities and accelerations. This combination of dynamic and static factors, represented by kinetic and potential energy, will also be key in Maxwell's later formulation.

However, by the time physicists had come to grips with the idea of central forces, they also had needlessly overdetermined it. Weber's law was criticized by Helmholtz and Maxwell for not complying with energy conservation since the delay of the potential grows with the increase in the relative velocity of bodies, and more potential energy seems to be lost than is kinetically recovered. In 1871 Weber succeeded in demonstrating the conservation of energy in cyclic operations, but various events, such as Hertz's experimental findings, finally tipped the scales in favor of Maxwell's view.

The same type of force law and retarded potential was applied by Gerber to gravity in 1898 to explain the anomaly of Mercury's precession; to claim, as so many historians have done, that this was an empirical correction without theoretical basis is, unless one understands by theoretical grounds the introduction of arbitrary postulates, truth upside down.

It has been said that the greatest limitation of Weber's model is that it does not include electromagnetic radiation, but the truth is that if it does not predict it, it does not exclude it in any way. In fact it is with Weber's formulas that the term for the speed of light is introduced for the first time, something

Maxwell knew perfectly. The dynamic extension of Weber's force has a discrepancy with respect to the static case of the same order as the Lorentz factor, which is why results similar to those of Special and General Relativity can be obtained in a much more natural way.

Weber's mechanics is not free of problems and ambiguities either. The most obvious, already noted by Poincaré, is that if we are obliged to multiply the velocity squared we no longer have a way of distinguishing between kinetic and potential energies, and even these cease to be independent of the internal energy of the bodies under consideration. On the other hand, if we are obliged to work with point particles we can hardly consider internal energies or forces. If Relativity predicts an increase in mass with velocity, Weber's law, the purely relational one, predicts a decrease in force and an increase in internal energy, which we can perhaps attribute to frequency. Without this ambiguity there would be no non-trivial feedback cycle, no possibility of real self-interaction.

Of course, the increase in frequency is not Weber's prediction, but that of nuclear engineer Nikolay Noskov, who took up the track of retarded potentials in various articles since 1991 and gave it a universal range of validity. Since Weber's law gives correct predictions, and energy is after all preserved, Noskov assumes that the non-uniform character of the retarded potential causes longitudinal vibrations in moving bodies, which are a normal occurrence at the most diverse levels: "*This is the basis of the structure and stability of nuclei, atoms, and planetary and stellar systems. It is the main reason for the occurrence of sound (and the voices of people, animals and birds, as well as the sound of wind instruments), electromagnetic oscillations and light, tornadoes, hydrodynamic pulsations and wind blows. It finally explains the orbital motion, in which the central body is in one of the focuses rather than in the center of the ellipse. Moreover, the ellipse cannot be arbitrary since the lengths of the cyclic and longitudinal oscillations have different values with a resonance v1 = v2. This circumstance determines the ellipticity in each concrete case*".

Remarkably, Noskov, who doesn't restrain himself of making generalizations either, dares mention the case of the elliptical orbits of celestial mechanics, an inconvenient subject by nearly all standards. It has been repeated so many times by history and publicity that Newton definitively explained the shape of the ellipses, that saying otherwise must be met with disbelief. But it is evident that Newton did not explain the case, which in turn gives rise to a very pertinent comment.

Whoever has become accustomed to the Newtonian description of orbits, assumes that (for the two-body problem) ballistics and a force dependent only on distance are enough to grant them stability. But the truth is that in Newton the innate motion of the body —unlike orbital velocity - is by his very definition invariable, which only allows two options. The first one is that the planet increases and decreases its velocity like a rocket with an autonomous impulse, which I don't think anyone is willing to admit. This option also has a comic reverse: let us accept, instead of self-propulsion, that centripetal vectors can stretch and shrink under a certain quantitative easing. The second option is that which Newton himself proposes with his sleight of hand, and which everyone has accepted, combining in a single variable orbital velocity and innate motion.

What is not noticed in this second circumstance is that if the centripetal force counteracts the orbital velocity, and this orbital velocity is variable even though the innate motion does not change, the orbital velocity is in fact already a result of the interaction between the centripetal and the innate force, so that then the centripetal force is also acting on itself. We do not see how self-interaction can be avoided. According to modern relativistic field equations, gravity must be non-linear and able to couple with its own energy; however, all this is already present at the most elementary level in the old original problem of the ellipse, which General Relativity has never dared to touch.

So even Newton's equations would be hiding a self-interaction and, whether we think in terms of action at a distance, or in terms of fields, what would distinguish the fundamental forces of Nature from those we humans apply by external contact in the Three Principles of Mechanics is precisely this self-interaction of the system as a whole. Hertz, the creator of contact physics, already noticed that precisely in celestial mechanics there is no way to verify the Third Principle.

Then, rather than disputing over which theory best "predicts" the tiny anomaly in the precession of Mercury, which after all is subjected to many other incidences, we could take the larger case of the planetary ellipse and its elementary asymmetry. For our so-called fundamental laws have still been unable to account for this the most fundamental asymmetries of nature in terms of contemporary forces instead of initial conditions which beyond the innate force and its vector here are completely irrelevant.

The Lagrangian of an orbital system, the difference between kinetic and potential energy, is a positive value, and not zero as one would expect. The Lagrangian is the quantity conserved, but nobody explains us why the heck there is more motion than its corresponding potential energy. According to the logic of the *Principia*, kinetic and potential energy, directly derived from appreciable motion and position, should be as equal as action and reaction. Is the retarded potential of any use to explain the difference?

It serves to save the time lapse of the forces in terms of energy, as the Lagrangian but in a more explicit way. With retarded potentials it is the energy what is not present at a given moment, which is another way of demanding an action principle. So this seems to be the limit of application for contemporary forces themselves.

It is said of Newtonian gravity that it is the first fundamental law expressed as a differential equation, but in the specific case of the orbit it is the differential or contemporary description what fails notoriously. According to the vectors of an invariable force, the orbit should open and the planet move away; there is no way to grant stability. The Lagrangian version in terms of energy arises to obviate this difficulty, not because it is more convenient for complex systems. So in the strict sense there is no local conservation of forces here, what there is is the discretional derivatives based on the integral action followed from calculus; and this is the criterion for a theory to be considered as "local".

The "unreasonable effectiveness of mathematics in the natural sciences",as Wigner put it, is only unreasonable if we forget what the procedure was.The fundamental forces of nature, independent of human mechanics, and which we try to restrict within the three principles of mechanics, do not satisfy these but indirectly —Newton's gravity and Maxwell's electromagnetism are theories with an indisputably integral origin and an interpretation as differential as their application. Nature does not obey our equations, our equations are instead the inverse engineering of nature, and it's only normal that our engineering, like our knowledge, always falls short.

Noskov's longitudinal oscillations depend on three variables: distance, force of interaction and phase velocity. Its length is directly proportional to the phase velocity and inversely proportional to the distance between the bodies and the force of interaction. Two cardinal formulas can be derived naturally from this: the Planck radiation law and the de Broglie correlation, which should include a phase velocity. This phase factor could also be applied to the so-called mass defects of nuclear physics.

We know that the de Broglie correlation applies without discussion to diffraction experiments showing waves in matter from electrons to macromolecules. Schrödinger's equation itself is a hybrid of oscillations in a medium and in the moving body, which led Born, famous for his statistical interpretation, to consider it in terms of longitudinal waves too; Planck's constant is reduced to a local constant relevant only for the electromagnetic force and masses of the order of an electron. Another proof that Weber's force has something to say about the electron at the atomic level is the straight derivation that as early as 1926 Vannevar Bush made of the fine structure of the hydrogen atom levels using Weber's equations without the assumption of mass change with velocity. Broadly speaking at least, is not that hard to reconcile quantum mechanics and classical relational mechanics.

This way Weber's force and its retarded potential is in the best position possible at the crossroads of Maxwell's field or classical continuum theory, Special and General Relativity, and Quantum Mechanics —being earlier and immensely simpler than all of them. And although it will surely be said that the aforementioned connection with quantum mechanics is too generic, the mere fact that it occurs without forcing things and without a chain of ad hoc postulates is already something.

There are, of course, many ways to rewrite quantum hieroglyphics and send them back to a classical conception; but few, if any, as direct as the royal road opened by Weber, which has historical and logical precedence over the others. One has complete freedom to tread other paths, but it is always advisable to contrast them with the original one.

Weber's law can easily be transformed into a field theory integrating over a fixed volume as shown by J. P. Wesley, and then it is revealed that Maxwell's equations are simply a particular case of the former. In this way and with retarded potentials high speed variations and cases with radiation are fully workable. Wesley also proposed other modifications of Weber's law that we will not touch here.

Noskov's connection with Planck's radiation formula is unavoidable since the retarded potential demands by itself an action principle. On the other hand, if the case of the ellipse in the Newtonian terms of an invariable force and an invariable innate motion leads to a gradually open orbit, what would be hopelessly lost here is the closed and reversible system, as if we had a dissipation rate. A virtual dissipation rate, it is understood, because that the orbit is conserved is something we already know. Since we have ellipses both in the micro and macrocosm, this is, therefore, the most obvious form of connection between the reversibility of mechanics and thermodynamic irreversibility. Thus, we can situate relational mechanics at the junction of classical mechanics, quantum mechanics and thermodynamics.

One might fear, as Robert Wald says, that an extended particle irreducible to the point particle case will be lost in details and unable to yield a universal equation; but in Weber's electrodynamics, unlike Maxwell's, one can work with extended particles knowing that one can always get back to point particles with solutions that make sense. Such neutrality is absolutely desirable in any possible respect.

Quantum thermodynamics and open meta-stable systems

Physicists and mathematicians nearly in equal shares try to get away from thermodynamics as much as they can, and if they have to deal with it, at least it is expected to be the most habitual equilibrium thermodynamics, not that of open systems far from equilibrium and in exchange with the environment. This last case is supposed to be that of living beings, characterized by levels of complexity far away from anything germane to fundamental physics.

Distrust could be justified if what is at risk is the closed and strictly reversible character of the so-called fundamental laws —the most precious treasure of science. But on the other hand, there is no loss that cannot be turned into a gain; and in this case, the prize would be nothing less than the return of physics to the general current of fire and life.

Macroscopic irreversibility in no way can be derived from microscopic reversibility. Boltzmann and many after him have assumed a highly refined statistical argument, but there are also topological arguments, free of statistics, that show irreversible aspects in electromagnetism independently of scale or metric. Beyond any sophistication, statistical or not, we never see the emitted light rays returning to their source, which already tells us that theoretical physics has a very peculiar criterion about what is irreversible and what is fundamental.

Finally the experimental evidence has begun to arrive and could multiply in the next few years if interest really exists —and interest exists for obvious technological reasons. Needless to say, the experimental and technological environment in microscopic physics has changed beyond recognition. New disciplines such as continuous quantum measurement, quantum feedback and quantum thermodynamics are flourishing, enabling an increasing filtering of noise and an always better distinction between quantum and thermal fluctuations.

The print of atomic irreversibility has to be proportionally small and only become macroscopically relevant with the usual large numbers, but there are and there will be more and more ways to measure and detect it. The subject is closely related to that of the point electron —one more idealization- which is an excellent approximation for long distances but which in close enough distances is bound to failure. Macroscopic irreversibility should be only "*the global superposition of the microscopic irreversibility due to the large number of atoms and molecules in matter*".

Following the logic of miniaturization, that of nanomachines or quantum information, there is much more interest in taking into account dissipation than in ignoring it, since it sets the very limits of operation for this devices. A different story, though, is that of reporting data transparently. Once again, applied physics will find new problems for theoretical physics, but we can predict that people will keep saying everything are fantastic new confirmations of the incredible quantum mechanics.

It seems then that it is the same interaction between photons and matter particles, what else, what affects in an irreversible way the electron and the center of mass of the atom; some already have advanced an estimate of the order of 10^{-13} joules. On the other hand, if we can only imagine electrons being made of electromagnetic fields —or by the constitutive tensions and deformations that translate it into its most material limit-, sooner or later it's inevitable to think, as it has done so often, that particles of matter are only trapped light transforming linear into angular momentum. At least laboratories has been trying to create electrons and positrons from photons for years and the prospect of doing so seems reasonable.

The idea that reversible mechanisms depend on irreversible dynamics, that closed systems are drawn on an open background, or that irreducible and ideal particles are not such, sounds to the ears of the theoretical physicist as a loss of status, a fall from the mathematical firmament of pure ideas. But it is the only outcome to expect if things are taken far enough, even respecting the margin of validity of closed and reversible systems or point particles. Real things use to have structure.

This does not mean that an electron should decay, photons apart, into other material particles. It is enough that its limiting surface and its spin have a differential structure, not necessarily geometric, that can account for its continuous evolution and its always ephemeral configurations. That is, the structure describes the momentary relationship with the environment, not an internal composition in terms of other equally abstract particles. Models for this have already been suggested. Once we have a limiting surface for a volume, we also have a framework for its changes, the structure of its spin, density, statistics, etcetera, etcetera. I don't know why we should give all this up in the name of a mere principle divorced from physical reality.

As for the magical ring of reversibility, as long as physicist believes is their own right they will never be able to receive it as a gift. Since the only thing that makes it valuable is the background from which it is conditionally, precariously constituted.

V. E Zhvirblis studied the osmotic and electrical rings in perpetual operation and came to the conclusion that systems where there are stationary forces cannot be isolated. Quantum mechanics, which pretend to do that, is, from a thermodynamical point of view, illegitimate. It is curious that this is considered completely normal at the quantum level but is banished from the macroscopic level in cases as clear as Lazarev's koltsar when the only thing we have to accept is that isolated systems are simply not possible.

The problem is that in these cases the quantities, although measurable, are not controllable, and physics is based first and foremost on controllable quantities such as forces. The only way to solve the paradox, as Zhvirblis observes, is if the interaction forces in thermodynamic systems are described only in terms of thermodynamics itself.

Thus all systems, from particles and atoms to living beings and stars, could be seen as metastable systems, islands temporarily far from equilibrium supported by their own internal equilibrium laws.

Current physics constantly speaks of energy conservation but atoms theirselves seem unverifiable perpetual motion machines, not to mention the same energy and matter, which are conserved but have emerged from nowhere in a remote past. It would have to be more interesting to see what a system is capable of doing in the present than twelve billion years ago, to try to guess why a real entity such as a particle is not a point, than to imagine how the whole universe comes out from a point without extension. Both things are more closely related than we think, but some lead to answers here and now, while the others just take us as far away as possible.

It is not so difficult to find the thermodynamic print and the origin in continuum of particles: it's enough to look for it with a zeal similar to the one that has been put in ignoring both -in separating them from what is really fundamental.

Relational Clock, Piston Engine, Whirlwind Computer

Relational statistics

For the ancients, to ask why the world existed was like imagining how fire came out of water; and to ask about life was like figuring out how water incorporated fire again without extinguishing it. Two apparent impossibilities that nevertheless balanced each other and acquired the intangible consistency of facts. In physics, the only way to create a similar relationship would be to try to see how the reversible comes out from the irreversible, and how a closed mechanism makes the omnipresent trace of heat disappear for a while. Surely we will never be able to explain one or the other, but perhaps their perspicuous balance would free us from the compulsive need for explanations.

Nothing can replace rectitude in reasoning, but in modern physics it is impossible to go very far without the use of a certain statistical apparatus that is, we might say, the inseparable companion of calculus. Not only quantum mechanics, even something as basic as the classical electromagnetic field has undeniable and basic statistical aspects. And above all, we would like to find a different general position for thermodynamics.

The reasons for creating a competent statistical-relational apparatus are multiple and range from the most obvious to the deepest.

The pure relational, in its very emptiness, reminds us reversibility itself; on the contrary, of heat we are left only with its statistical fingerprint. To see how they interpenetrate we need first to make a broad detour. In physics we will never have too much perspective since perspective itself is already the best part of knowledge.

Relational statistics can be seen, in its simplest version, as a modality of dimensional analysis, which tries to bring the equations, constants and units of the formulas of the current theories as close as possible to Fourier's Principle of Homogeneity, generalized in more recent times by Assis as Principle of Physical Proportions: the desideratum that all the laws of physics must depend only on the ratio between known quantities of the same type, and therefore cannot depend on dimensional constants. Such were, for example, Archimedes' laws of statics, Hook's original constitutive law, or, shifting towards dynamics, Weber's force law of electrodynamics.

It can be said that this principle of homogeneity is an ideal, not only of Greek science, but of science in general; it's a universal zenith or pole. Needless to say, nearly all the laws of modern physics violate this requirement, so the issue is not to discard them, which is out of the question, but to bear this factor in mind for what might allow us to see one day.

It is not necessary to expand on the more standard modalities of dimensional analysis, already well known, and which over time have extended to other fundamental combinatory branches such as group theory. On the other hand it is not superfluous to bear in mind that when we say "relational statistics" we are uniting in a single concept ideas that in physics are rather antagonistic: the purely relational is the most transparent, the purely statistical is the most opaque at the physical level.

In any case, it is always useful to start with the dimensional analysis before proceeding with the major complexities of statistical analysis. Even today the interaction between these two branches of the analysis is scarce, due in part to the fame of superficial that it has among the highly creative theoretical physicists, more concerned also about the predictive capacity of their equations than about their cleanliness or legitimacy. In addition, an elementary dimensional analysis often calls into question the basis of many of their assumptions, such as the uncertainty principle.

An example of relational statistical analysis is that proposed by V. V. Aristov. Aristov introduces a constructive and discrete model of time as motion using the idea of synchronization and physical clocks that Poincaré introduced precisely for the electron problem. Here every moment of time is a purely spatial picture. But it is not only a question of converting time into space, but also of understanding the origin of the mathematical form of the laws of physics: "*The ordinary physical equations are the consequences of the mathematical axioms, which are "projected" onto physical reality by means of the fundamental instruments. One may assume that it is possible to construct different clocks with different structure, and in this case we would have different equations for the description of motion.*"

With a model of rigid rules for geometry and clocks for time, a space-time model of dimensionless variables is created. In the exposition of his idea Aristov deals basically with the Lorentz transformations, the axiomatic construction of a geometry and the most basic relation of the quantum uncertainty principle; and if this is mandatory on one side of the map, on the other side we can put Weber's own relational mechanics with its derivations, just as we can place as a clock at the center an extensive model of the electron, using its vectors for many other correlations.

A chronometric approach to relational statistics is opportune because of the many discrete aspects that will never cease to exist in physics and that are independent of quantum mechanics: real particles, diverse aspects of the waves, collisions, acts of measurement and time measurements in particular, or cuts imposed at the axiomatic level are discrete.

The performance of a relational network is cumulative. Its advantages, such as those of the physics that bears that name —and information networks in general- are not noticeable at first glance but increase with the number of connections. The best way to prove this is to extend the network of relational connections. All this is about collective work and intelligence. With arbitrary cuts to relational homogeneity, destructive interference and irrelevant redundancy increases; on the contrary, the greater the relational density, the greater the constructive interference. I don't think this requires demonstration: Totally homogeneous relationships allow higher order degrees of inclusion without obstruction, just as equations made of heterogeneous elements can include equations within equations as opaque elements or unravelled knots.

From continuity and homogeneity comes the unwritten legitimacy of laws, just as from primordial waters came the sovereignty of ancient kings and emperors.

In his sketch of relational statistics Aristov has no considerations for thermodynamics, but it is precisely this that would have to give a special relevance to statistics. Instead of a Poincaré-style clock, we could have introduced, for example, a "piston engine" like the one exemplified by Zhvirblis in order to generate forces without leaving the thermodynamic realm.

Let us return to our peculiar perspective exercise. Physics would have a "relational" north pole from which it aspires to explain everything as mere relations of homogeneous motion, and a "substantial" south pole from which it could give an authentic mechanical explanation of phenomena, generally with a kind of medium that provides continuity for the transmission of forces between separate parts of matter.

Trying to satisfy both extremes, the commitments of the history of physics left us in the middle, with Newton's physics of absolute magnitudes, or with modern field theories, which attempt to maintain continuity but make use of the wrongly termed universal constants, in fact absolute magnitudes as in Newton.

This apparently philosophical disquisition contains the issue of the universal synchronizer. Between fire and water one would not put a cylinder and a piston, but perhaps a whirlwind, as in the famous Newton's own bucket experiment, distant heir to a more daring one already conceived by the father of the theory of the four elements. Empedocles noticed that turning a bucket of water vertically prevented it from falling, counteracting, in other words, gravity.

The Newtonian bucket experiment and its centrifugal force obliges us to take a stance. What is the reason for the curvature of water? Newton says that absolute space; Leibniz, Mach and relational physics, that the relationship with the rest of the objects, including distant stars. For Newton, even if we could eliminate all the surrounding matter, the same phenomenon would occur; for the relational physics such a thing is impossible. And what would be the substantialist position? This one would say that an absolute medium of reference is required, and that it is that medium in any case the only that can transmit the influence of bodies from the environment, whether distant or not.

There do not seem to be more conceivable positions than these. And yet none of the three seems satisfactory to us. The assertion that there are absolute magnitudes independent of the environment, though preserved in all modern physics, is metaphysical in nature. On the other hand, pure kinematic relationships will never be able to explain physical reality, however desirable the principle of homogeneity may be. Finally, the determination of a framework independent of measuring devices seems to violate the relational principle — and the relativist one that came later; apart from the fact that the non-uniqueness of the action principles precludes the identification of unique causes.

But a fourth position, such as that of Mario Pinheiro, can be held; you can affirm that there is no kinematics without irreversibility. Pinheiro observes that what is important for the curvature is the transport of angular momentum holding the balance between centrifugal force and pressure.

I believe that this answer reveals at least both the part of truth that can be found in any of the three postures and its manifest insufficiency. Pinheiro advocates the use of a new variational principle for out-of-equilibrium rotating systems and a mechanical-thermodynamic time in a set of two first-order differential equations. There is a balance between the minimum variation of energy and the maximum production of entropy that fits in the simple classical examples such as free fall and that would have to be relevant in dynamics in general and electrodynamics in particular, because the conversion of angular and linear momentum should be central to it, no matter if this is not how is told.

As a sign of the times, the efforts to integrate dynamics with the the entropy/information concept are increasing dramatically. The reasons for this are diverse but convergent: the ever-increasing tendency to consider matter as mere support for information, the exhaustion of viable dynamic models, the gradual introduction of statistical factors and physical magnitudes of higher level. These trends are not going to subside. However, so far the use of entropy helps little or nothing to understand where the dynamical regularities we observe arises from.

It is also said that gravity takes entropy to its limit by area -in singularities-, though our ordinary experience says quite the opposite, that it is the only force that seems to compensate and revert it. In this case entropy reaches its extreme because the theory of gravity as an absolute force inescapably have to dispose of everything else; in a relational theory there is no place for gravitational singularities, because the force decreases with the increase of velocity. On the other hand it is curious that in Newtonian physics the forces that produce deformation, which is the only thing to expect, are considered pseudo-forces, while the fundamental force does not cause deformation of bodies when in motion, but only through potential energy in the static case.

Surely it is impossible to understand the relationship between the reversible and the irreversible, great key to nature, while subordinating nature exclusively to prediction. It is clear that what is predictable is regular and that this tend to confer to a certain extent the rank of law. But to what extent? Without due contrast, we will never know. We pride ourselves on the predictive power of our theories, but prediction alone needs not to be knowledge, and can rather be blindness.

Yes, power of prediction is also power of obfuscation. There is nothing more practical than a good theory, and a good theory is the least artificial and most respectful with the case presented. Our knowledge is always very limited, even regardless of our degree of information, and the respect for the little we know, being aware that it is intrinsically incomplete, is also respect for everything we do not know. It is clear that the ad hoc postulates, the teleology of our principles of action, or the inverse mathematical engineering, drastically reduce the quality of our generalizations no matter what degree of universally we attribute to them. It is not a matter of detracting from its merits any human achievement, but of being aware of its limitations.

R. M. Kiehn already spoke in 1976 of "retrodictive determinism": "*It seems that a system described by a tensor field can be statistically predictive, but retrodictively deterministic*". It is conceivable, for instance, that we can deduce initial conditions from the final global deformation of a solid, while conversely we can only calculate probabilities from the initial condition and the applied force.

The irreversible, dissipative part of electromagnetism seems to be included within the intrinsic covariance that Maxwell's (and Weber's) equations show in the language of outer differential forms. There has to be another way of reading "the book of nature" and this form cannot be simply an inversion — it cannot be merely symmetrical or dual with respect to predictive evolution. Ultimately it is neither a matter of going forwards nor backwards, of future or past, but of what can be revealed between both.

The whole quantum electrodynamics can be retrospectively derived from the classical Huygens principle, which is a principle of homogeneity, just as the conservation of momentum derives directly from the homogeneity of space. However, everything suggests that gravity exists due to the heterogeneous character of time and space. This brings us back to our previous considerations in terms of extremes.

The enigma of the relationship between dissipation and usable mechanical work, externally seen as a problem of natural laws, is only one with that of the relationship between effort —which is not the same as work-, work and reversible transactions in the logic of capital. It is not surprising that thermodynamics developed in the very years of the vindication of work as an autonomous category, and that that happened between both considerations of physiology, blood and heat (Mayer) and those of metal and machines (Joule) in the 1840s. There is still an unexpected circle to be closed here, and the more we manage to go full circle, the greater will be the repercussions on our vision of "external" nature (exploitable resources) and "internal" nature (society). And perhaps then we will begin to fully understand to what extent exploiting nature is the same as exploiting ourselves.

The double-speak of modern physics

It is well known that since the 17th century scientists have been pleased in encrypt their communications in order to be able to claim priority while at the same time avoiding giving an advantage to their competitors. Moreover, even in the time of Galileo or Newton there was a great awareness of the strategic, commercial and military value entailed even in a knowledge as apparently detached from the ground as astronomy or the maritime calculation of length.

Thus, for centuries science developed in the West in a delicate balance between its eagerness to expand and communicate, and the convenience of masking to a greater or lesser extent its procedures. If this happened even in the most pure and abstract disciplines, such as number theory, we can imagine a little of what happened with applied sciences.

When in the first decades of the twentieth century quantum mechanics and the theory of relativity developed, although certainly a reinvention of the public image of science was already underway, it can still be assumed that physicists communicated their theories with some spontaneity and that the rush to advance rapidly and at any cost created a kind of "natural selection" in the supply of methods, hypotheses, and interpretations. But then the Second World War came, with things like the Manhattan Project —Big Science in short- and scientists, as Oppenheimer admitted, lost whatever little amount of innocence they had left.

It is to be assumed that it was more or less around that time or shortly thereafter that it was fully understood that the two great new theories served as much to conceal as to communicate, which for such compromised areas as particle physics was specially convenient.

Why someone as practical and well-informed as Bush was wasting his precious time with the forgotten Weber theory in 1926, just when quantum mechanics was in full swing? The same inevitable Feynman, spokesman for the new algorithmic style in physics and man of the Manhattan Project, once admitted that the law of retarded potentials covered all cases of electrodynamics including relativistic corrections. Even Schwinger worked for the government in radar development. It is no exaggeration to say that the gestation years of QED make up the epoch when the dividing line between theoretical and applied physics, with all that implies for both, fades away forever.

Simply put, it's too hard to believe that armies of talented physicists have not dared to leave the superfluous relativistic framework in particle physics when any amateur who considers it can see that it is a full-fledged embargo to any steady progress in the field. That this is not said is even more significant. Of course, in order to convince us otherwise, an unprecedented new specimen of physicist came out from nowhere, extrovert, casual, persuasive, fantastically gifted for communication and public relations. The spitting image of superficiality —and with a remarkable resemblance to Cornel Wilde. Each shall judge for himself.

We will ignore the things that have been said about the atomic bomb and the theory of relativity, because the atrocities of the advertising industry could still embarrass physicists. We can always blame it on the journalists. It is obvious that relativity has practically nothing to do with the development of nuclear physics, and just as obvious should be how little quantum electrodynamics had to do with the myriad of applied achievements of all these last decades.

The famous "shut up and calculate" of the aforementioned physicist and spokesman conveys wonderfully what is expected of the new researcher. The calculation would have to be only a third of the work in physics, theoretical or not. If we divide the sequence of any human activity into principles, means and ends, in physics calculation or prediction would be only the means between the intelligent use of principles (which are present at all times), and interpretations, which, far from being a subjective or philosophical luxury, are inexcusable when it comes to making sense of the mass of empirical data, continuing research and motivating the synthesis of applications. Ends as purposes are both interpretations and applications.

So you don't see why in physics interpretation should be of less practical interest than calculation, and I think these arguments are so basic that anyone can understand them. Those only concerned with the means will also be used only as a means. Besides that, quantum mechanics and relativity, rather than facilitate calculations, often make them unusually difficult. Compare, for example, the calculations required for the simplest problem in General Relativity with those of a Weber law for gravity; not to mention the somewhat more complicated cases where it becomes completely unmanageable. The prescription of the relativistic particle has a very clear meaning and it is to make impossible any calculation with physical meaning. It goes without saying that experimental and applied achievements have been obtained in spite of this stumbling block, ignoring it rather than observing its prescriptions.

"Shut up and calculate" simply means "calculate what I tell you and don't even think of anything else". Curious slogan coming from someone who has been glorified as an incarnation of unrestricted originality and of the genius on his own. He could also have said, "Keep it superficial and forget about getting to the bottom of the matter forever." Calculation is completely blind if it is not properly coordinated with principles and interpretations; you can even make sure that this coordination is more important than everything else. But we should not see in this nothing inconsistent, but on the contrary a faithful translation of Big Science double-speak and its defined priorities behind the ever-present screen of public relations.

Known is how in 1956 Bohr and von Neumann came to Columbia to tell Charles Townes that the idea of the laser, which required the perfect phase alignment of a large number of light waves, was impossible because it violated the inviolable Heisenberg's Uncertainty Principle. The rest is history. Now of course what is said is that the laser is just one more triumph of quantum mechanics, from which is a trivial particular case.

The above case has not been the exception but the general trend. We are told that quantum mechanics is something very serious because its mass of experimental evidence surpasses that of any theory, but for me it seems more serious the work of engineers and experimenters trying very hard to figure out causal relationships and applications without the support and even with the obstruction of an interpretation that prohibits physical interpretation and that seems expressly conceived to sabotage any attempted concrete application.

And as for the incomparable predictive power of quantum mechanics and QED, which cannot even imagine the collapse of the wave function it postulates, we must understand, what else, a power of a posteriori prediction. Today they also predict the Berry phase, that even appears in the soup, although with the opportune extensions of that which is not even "a theory", but hard facts only. It is clear that procedures that subtract infinite from infinite in a recurrent and interminable way, can be bent to obtain practically any result; but in spite of all this we are still told that it is a very restrictive theory. Perhaps it is, in the light of some abstract principle of symmetry among other principles still more abstract. It will be, no doubt, after you have adjusted all the nuts and bolts imaginable to the case in hand. Who said self-interaction? This is already a self-adjusting theory. And the best of all, you can make say anything to an incoherent theory.

Nothing could be more castrating for the physicist than to ask him exclusive fidelity to the calculations. All the more so if these calculations are so unscrupulous that they are used in an openly teleological way to replicate certain results —pure reverse engineering, hacking of nature to which one wants to give the category of Law.

If what is hacked from nature soon becomes Law, then it is not surprising that it is used as Law lest others hack into it.. And that is why today the Great Theories are used as the more effective blockade to technology transfer.

Nothing is more practical than a good theory. But those who are only interested in governing nature are not worthy of illuminating it, let alone understanding it. So surely we have the level of understanding that our social structure can tolerate, and this is the natural thing, first because scientific knowledge is a pure social construction, and second because all social construction is second nature trying to isolate itself from a supposed first nature.

Today, transformation optics and the anisotropy of metamaterials are used to "illustrate" black holes or to "exemplify" and "design" —it is said - different space-times. And yet they are only manipulating the macroscopic properties of the old Maxwell's equations. Can distortion and distraction be taken further? And the fact is that all this could serve to investigate the uncontrollable aspects of the electromagnetic continuum that are but the classical form of the famous non-local aspects of quantum mechanics. Haven't we seen that if Schrödinger's wave function describes vibrations in a moving body and in the medium, classical electromagnetic waves are also a statistical average of both things?

In something as apparently well-trodden as the electron we can find not only the key to particle physics, but also the limits of application of nanotechnologies, quantum computing, and a legion of emerging new technologies. But if all these are matters of power and interest, we can already forget the truth.

That you can't serve simultaneously both truth and power is something known to all. That scientific publicity develops its stories and narratives as if this conflict did not exist is the only thing you would expect. If the current accounts delay scientific development a little or a lot, it is not something we should regret either, since more technological advances, in the absence of other things, can only mean more disorder.

Fortunately, those who in one way or another obstruct knowledge are not capable of developing second-order knowledge free from their own distortions either; the complexity of their compromises strictly limits them. All the cunning of the world, all the experimental-statistical-mathematical-computational arsenal, cannot replace the sense of rectitude, the only one capable of delving into the unlimited promise of simplicity.

References

N.K. Noskov, *The theory of retarded potentials versus the theory of Relativity*

N.K. Noskov, *The phenomenon of the retarded potentials*

J. P. Wesley, *Weber electrodynamics*

Assis, A. K. T, *Relational Mechanics -An implementation of Mach's Principle with Weber's Gravitational force,* Apeiron, Montreal, 2014

T. B. Batalhao, A. M. Souza, R. S. Sarthour, I. S. Oliveira, M. Paternostro, E. Lutz, R. M. Serra. *Irreversibility and the arrow of time in a quenched quantum system*

Lucia, U, *Macroscopic irreversibility and microscopic paradox: A Constructal law analysis of atoms as open systems*

V. E. Zhvirblis, V. E, *Stars and Koltsars,* 1996

N. Mazilu, M. Agop, *Role of surface gauging in extended particle interactions: The case for spin*

N, Mazilu, *Mechanical problem of Ether* Apeiron, Vol. 15, No. 1, January 2008

V.V. Aristov, *On the relational statistical space-time concept*

R. M Kiehn, *Retrodictive Determinism*

M.J. Pinheiro, *A reformulation of mechanics and electrodynamics*

BEYOND CONTROL: FEEDBACK AND POTENTIAL

1 August, 2018

There was a great controversy in the last half century about the physical meaning of the potentials and the quantum potential in particular, with several open questions that persist. In this short article, before solving them, we will try to raise new questions that might have an unexpected impact on the future. In particular, we want to indicate the value of feedback beyond the omnipresent paradigm of control. Calculus is an heuristic device, an interface between mathematics and the physical world, but it is not the only one. There are other ways of exploring this world, although the one we are dealing with here is not incompatible with calculus.

Potentials emerged in physics and calculus as an auxiliary concept but over time they have become essential. It has been said that quantum potentials are completely different from the classical ones relying on phenomena such as the Aharonov-Bohm effect but it is clear that this effect can be derived from classical equations and in fact a strictly analogous example has been found even on the surface of water.

In principle, it is assumed that the potential energy is derived simply from the position and is the passive, reciprocal factor of the active element that originates the kinetic energy and the appreciable movement that is the force — one and the other would be the static and the dynamic case. In Newton the idea of potential is strictly the inverse of the force, then its sum should be by definition equal to zero; but the Lagrangian introduced a century later tells us that the kinetic energy is greater than the potential energy and therefore has a positive value. This should have already raised suspicions, but since the point was to facilitate derivatives and calculations, nobody protested.

If the potential energy is the static case and the kinetic energy the dynamic one, Are they both equivalent? Weber's force law, applicable both to electrodynamics and to celestial mechanics, allows a different view of the subject: all the invariable forces, as in the case introduced by Newton, are reduced to the static case, and only the forces dependent on the speeds and relative accelerations allow to speak properly of a dynamic case. Weber's law is the first instance of purely relational force in dynamics but, as we all know, his argument, although perfectly valid, was not followed.

Is the kinetic energy the same than the potential energy for our self-perception? It seems clear that this is not the case: our body deforms with gravity when this one is only potential, but it doesn't deform when gravity manifests itself directly as motion, as in the case of free fall. And yet the deformation is the first indication of the acting presence of a force! Our body's interpretation of these things seems quite at odds with the calculus view and its conveniences, though it isn't incompatible with it either.

Probably the static/dynamic distinction is more important than what the framework of dynamics shows —since dynamics, by its very definition, has already decided what its priority is. That is, the values associated with the positions are already bound to the forces and no other consideration. This is perfectly legitimate from its own point of view and there is no problem with it; but it doesn't exhaust the case either.

Before the term "potential energy" came in —a term that apparently Rankine first introduced- it was not uncommon to call it tension, as in the celebrated Helmholtz paper of 1847 that so much harmed Weber's theory; and still today stresses of diverse types and systems are counted as potential energy. This is not something that makes it independent of the priorities of dynamics to our view, but at least it does not relegate it to the purely passive role that we use to grant it.

An additional point is that thinking in terms of forces compels us to think exclusively in terms of controllable quantities. We can measure other values, but if they are not within the framework of what we can control and follow in a deterministic evolution, they are labelled as accidental. This is patent in something as universal and classical as the constitutive law of materials; but we can see it also, for example, in non-local quantum correlations.

Weber's force law, earlier and more general than Maxwell's formulation, can be equally extended to hydrodynamics, thermodynamics, and many other situations. It is also the most elementary example of intrinsic feedback in a fundamental physical law, since the length of interaction influences itself through the force of interaction. We have already pointed out elsewhere that the very Newtonian case for the ellipse hides a self-interaction or feedback right from the start, to the point that the closed ellipse can not subsist without it — there is really no local conservation of forces. Therefore, the most general fundamental laws, such as gravity and electromagnetism, to the extent that they require principles of action, do not seem possible without feedback. In other words, the principle of stationary action seems to hide a feedback principle.

In recent times we have witnessed the development of new experimental specialties such as continuous quantum measurement or quantum feedback that allow the manipulation/modulation of individual quantum states, and that will shed new light on the indefinite and controversial transition zone between the classical and the quantum domain. In all this vast field we just begin to explore the possibilities and bifurcations: quantum engineers even talk about self-feedback, for cases in which a resonator interacts not with a controllable system but with an environment with many bodies.

The improvement in the mechanical manipulation leads here to a greater sensitivity in the modulation, and the best modulation allows to specify more precisely the mechanical details and the circumstances of the environment. We get a classic feedback circuit, sensor → calculation → action. And this way, without much noise, the vast realm of microphysics is entering a new territory.

All this not only involves a gradual shift in man/machine interfaces, it will finish provoking a deep change in the man/nature interfaces too, that in the long run can make not only a change but also a turning point.

The idea we have today of this broad change is exclusively technological, relative to the order of manipulations, but the aspect of tuning, of awareness that entails, is still being ignored. However, the very nature of things —the very human nature- allow us to predict that all this will end up affecting both the scope of our perception just as much as that of our action, to both our "efferent" and "afferent" impulses.

Let's imagine that we are doing biofeedback through output signals of a function of our own body, such as pulse or breath. A rather innocent question is whether there are better signals than others for the sake of self-regulation; that is, whether or not the type of representation or interface of the signal is indifferent. Let us think, for example, of representation in terms of movement and force — kinetics- and representation in terms of position or potential. Since we have just spoken of afferent and efferent impulses, of input and output, it is simply reasonable to think that, for our self-perception, both correspond to the kinetic and potential aspect.

It is clear that with biological signals it is quite easy to check the effectiveness of one or another type of signals/interfaces. Now let's take a leap and think about how our mind or intention could manipulate/modulate/tune physical systems out of the body —exosystems- through certain mechanisms and interfaces. Are there also different degrees of efficiency here, depending on the intended objective?

Finally, within this context of the mind/machine interface, we could naturally ask ourselves what are the minimum mechanisms and interfaces allowing communication in both directions. And the answer, as ambiguous as we want, is that the very limit depends on how we understand and perceive the Physical Continuum. We have already talked about this in relation to the interface conditions of electromagnetism.

Our idea of self-interaction, at a fundamental physical level, is also important, as in the framework of the law of retarded potentials the principle of action is directly translated into feedback. Retarded potentials fit naturally in continuum mechanics.

What really matters, in any case, is that we can access a perception of nature other than mere motion and extension. Physics is interesting precisely because things are not reducible to motion and extension, but the utilitarian views compel us to forget it.

It is necessary to go beyond motion to capture the transformations in the inextensive and apparently immobile. Capturing change in the motionless, and what is motionless in motion itself: this would have to be the value of biofeedback as endoscopy in physical reality. David Finkelstein coined the word "endophysics", which Otto Rossler described to a great extent; but it is useless to deal with the theory of Relativity and quantum mechanics as "participatory frames" when they are explicit ruptures with the notions of the one and only Continuum. If there is no rupture, there is no need for participation, and it is only in the continuum that one can speak of exterior, interior, and what goes beyond these more or less superficial distinctions.

We have already commented in other writings that the constitutive law, strain-stress relations or the theory of deformations and tensions, veils with its metrics the most universal aspects of the physical continuum, of which the known fields are just extracts.

The connection between biological feedback and exosystem feedback is just a matter of interface; and the same would apply to quantum exosystems — if in such a case it were possible to speak of external systems at all. Obviously, there is quite a difference between bound and unbound states. But in all cases this "insider view", which is really only a certain recovery of the physical continuum, must throw a totally new light —the light of what is free of utilitarian aims. Our mathematical descriptions, however accurate in language they may be, are already severely limited by their inherent predictive purposes.

Thus, beyond manipulation and "modulation", it is possible to speak of knowledge by tuning, without ulterior purpose, which takes advantage of the means that modulation and manipulation have made possible. These means, more or less opportune, would be basically accidental.

It seems that the most peculiar aspects of the quantum potential are reducible to those of the field of its wave, and therefore, are due to the notions sustained the continuum and nothing else. Surely a more or less direct and subjective sensitivity to phenomena such as potential wells and the tunnel effect can be developed, which also do not seem to have anything of intrinsic quantum nature and can be explained with a more or less thorough classical description. Here, however, description and interpretation are at the service of something quite different. Bistable and metastable systems, which can be found at both the biological, classical and quantum level, can have decisive features that are independent of coordinates and scale.

Calculus is an heuristic device, an interface between mathematics and the physical world, but not the only one. There are other means of exploring this world, although the one we have noted here is not incompatible with Calculus.

THE CITY LIMITS

Nature is not simply the background on which society has been built, and which we recognize "out there" as forests, skies and oceans; unrecognizable, it crosses us internally also as passions, as perceptions, even as silent consciousness in the midst of the procession of thoughts and words that are the social currency proper. Science, which is the most opposite to instinct that can be imagined, can only be incorporated into Nature as an external object and as an abstracted law of phenomena. This perspective is bound to instrumentality even with the best will of the world.

This being understood, is foolish to think that man can develop science and technology for the mastery and exploitation of external nature without doing the same, measure by measure, with our internal nature; but perhaps the very social illusion depends on the systematic ignorance of something so plausible.

If everything were descending forces in this world, and the same goes for the ascending ones, things would soon come to a halt: it is like imagining a body with only efferent or afferent functions. Thus, even against our will, trends always seek balance. I can assume this is why I am writing here, without being able to explain to myself what I am looking a way out for, even when I don't feel the need to justify it. Instinct will tell some of us that it is necessary to insist on this direction regardless of utility or purpose. Someone who is drowning or wants to break through the siege does not need them.

To live in the Society of Control is to tempt its external and internal limits up to the threshold of its cessation.

Appendix: The pulse wave, the retarded potentials and the Continuous Proportion

As we know, the profile of the pressure of the pulse wave over time is the sum of the cardiac impulse and a reflex wave created by the peripheral vascular system at the interface between large arteries and the smaller vessels that cause resistance. At any time and in any part of the arterial system there are three factors: the amplitude, the duration of the contractile impulse of the heart, and the amplitude of the reflex wave.

It has been said that the ratio between the time of systole and diastole in humans and other mammals tends on average to the Continuous Proportion; likewise the systolic pressure in the aorta is 0.382 and the diastolic 0.618; this proportion is also found in the electrical activity of the brain. There are good reasons to be skeptical about such numerical associations when they are not connected to mechanical reasons, which use to be the case; but perhaps here, if time and pressure are under the same unexplained common denominator, there is an opportunity to find the desirable connection with dynamics; the systolic time of the profile already involves the reflex wave, and the same goes for the diastole.

On the other hand there is the pulse wave speed, which is a measure of arterial elasticity: both are derived from the Second Law through the Moens-Korteweg equation. This wave velocity varies with pressure, as well as with the elasticity of the vessels, and it increases with their stiffness. The return distance of the reflex wave and the time it takes increases with height, and a lower diastolic pressure, which indicates lower resistance of the whole vascular system, reduces the magnitude of the reflex wave. Treatment of hypertension should focus, it is said, on decreasing the amplitude of the reflex wave, decreasing its velocity, and increasing the distance between the aorta and the return points of this wave.

To a large extent, it seems that we can consider the elasticity of the reflex wave as a Weber-Noskov retarded potential dependent on distance, force and phase velocity, and see if this provides a coupling or resonance condition that, incidentally, would tend to to the values of the extreme and mean ratio. The myocardium is a self-excitable muscle but this entails also the return of the reflex wave, so we have a beautiful example of a circuit of stress-pressure-deformation transformations that are fed back and that should not be far from the most basic problems of physics that we have dealt with in other articles and that also involve a certain feedback.

There is a remarkable similarity here that demands a detailed exploration. Not only can we model it anew, it can also be simulated numerically and even physically with elastic tubes and artificial pumps.

References

Peter J. Riggs, *Reflections on the de Broglie–Bohm Quantum Potential*

Nikolay Noskov, *The phenomenon of the retarded potentials*

Zhang J, Liu Y, Wu R, Jacobs K, Nori F, *Quantum feedback: theory, experiments, and applications*

V. D. Zvetkov, V. D, *Heart, Golden Ratio and Symmetry*

Miguel A. Martínez Iradier, *Self-energy and Self-interaction*

Miguel A. Martínez Iradier, *Between stress and pressure*

HEALTH, LIFE, AGING AND EVOLUTION

15 August, 2018

For a while it seemed that information and complexity theories, along with genetics, were going to give us a comprehensive description of life and its characteristic processes such as aging. Later we have learned that man has fewer genes than a banana or some fishes and monkeys, while definitions of entropy, another name for information, have proliferated almost as much as its applications, making the term increasingly fuzzy and less universal. Complexity sciences have grown under the impulse of the information explosion, but they have not brought much clarity, and they have certainly not made us any wiser. In this short article we do not intend to critically review such a vast field; instead, we would like to point out the real-life interest and relevance of using much more simpler, basic and physical categories.

Skepticism about the complexity theories has already begun to spread for some years, and after decades of speculative intoxication, the hangover was predictable even if the impetus of the computational paradigm does not subside. An astrophysicist like Eric Chaisson has shown that a measurement as elementally physical and simple as the energy flux density is far more reliable and expressive for the metric of complexity than any of the much more abstract definitions of entropy that proliferate. That already tells us something, even if that simpler measure could not take us too far.

Since so far so few lessons for life has been drawn using all the power of our arsenal of complexity theories, we believe it is appropriate to look in the opposite direction in search of simple and robust concepts that have not received enough attention —even at the risk, of course, of being over-simplistic. In any case, we believe that it is easier to see what is missing in something too simple, than what is in excess in something too complex.

After all, does modern medicine, with all its technological deployment, have even an acceptable definition of vitality or health? Not at all; at best, we will be told it is an optimal state of functioning of the body, which does not exclude certain subjective components —or something similar. Mere words and truisms that do not add even a hint of knowledge to what any of us knows without considering it for a moment.

Compare that to this early twentieth century definition due to the "vitalist" Ehret: vitality is power minus obstruction (V = P—O), actually the only mechanical and functional definition that has been given so far, and the more truthful and simple possible too. This is not the place to assess the dietary ideas of the German naturist or its viability in the modern world, but it is worth remembering that both those ideas, and their definition of vitality, are the same ones that a Taoist would have subscribed two thousand years ago.

What really interests us is that this definition, in addition to being full of meaning in the qualitative and subjective dominion, admits an immediate physical translation. And it is that, after physics and mechanics have served to model the rest of the experimental sciences, such as chemistry, anatomy and physiology, by the mere logic of specialization we have curiously come to assume that, at a basic level, they can not have anything else to say. And that depends on how we understand physics and mechanics.

Take for example something quite fundamental as Maxwell's electromagnetic theory. It is assumed that this has little to say about the human body and physiology, except, in a very mediated way, for the electrochemical potentials, which we consider to be part of biochemistry. But it turns out that Maxwell's original theory is a hybrid of mechanics and hydrodynamics indisputably based on ideas of flow and circulation. It is a mathematically written but totally phenomenological theory of closed currents and flux tubes, macroscopic, built from outside to inside, defined by integral equations and not by differential equations or vectors. That is, in its original form these equations, which are considered one of the deepest axes of theoretical physics —the main source of dualities-, although strictly mechanical and tautological, are more "holistic" than a number of medicine views that are taken as such.

Not only that, but, if we do the proper constitutive interpretation of those equations, we come, like Nicolae Mazilu, to the conclusion that the classical electric field is the form of the stresses that are not accompanied by strains, while the form of the deformations that are not accompanied by stresses is that which characterizes a classical magnetic field. And this definition is equally valid for materials, biomechanics, or even our self-perception, subjective as well as objective, in terms of tension, pressure and deformation.

That is to say, all this has global aspects that go far beyond our idea of electricity as a motion of charges. Maxwell's equations have a part defined by a metric associated with the electromagnetic force, and another part, much more universal, that can be applied to hydrodynamics, thermodynamics, seismology, and so on.

In Ehret's formula, power is synonymous with pressure. Solids and liquids behave in a similar way under pressure; it is said that under tangential stress "the solid deforms, and the liquid flows". This already gives us the general idea of how to shift from stresses to currents; and thus, through hydrodynamics, we find another completely natural way of bringing together the fundamental structure of electromagnetism as a theory of continuum mechanics to biology.

It is well known that one of the most tracked vital signs in modern medicine is blood pressure, that we use to call hypertension when it's too high; actually the pressure is exerted by the blood, the tension or stress affects the surface of the vessel. In truth, with the increase of the flow velocity in a vessel, the pressure on its walls decreases, and in general terms it is the stagnation of the blood which increases the pressure and the tension to which the vessels are subjected.

It is clear that if we only want to know about the stress indices, we don't need more than Bernoulli's general principle of hydrodynamics or the Laplace equation which connect the former with the potencial theory. The Bernoulli principle can be derived directly from the Second Law and tells us that the sum of the kinetic, potential and internal (pressure) energy remain constant; the Laplace equation gives us a distribution of lines of flow in a current and leads us to the Poisson equation and Maxwell's second equation.

The very hemodynamics of the cardiovascular system finds a good approximation, often used by scholars, in Ohm's law (pressure is equal to the product of flow and resistance, or also, in terms of electrodynamics, the potential difference is equal to the product of intensity and resistance); this electric law naturally corresponds to the Poiseuille equation of laminar flow usually applied to the case. Although Ohm's law is considered empirical, it has been verified with precision up to the vicinity of the atomic level.

The Poiseuille equation gives the pressure drop of the laminar flow in the cross section of the vessel; this in turn originates a Bessel function of the first kind, the vibration modes of the cross section. Old pulsology manuals tried to draw this feature of the pulse shape, so different from the time profile of the wave.

The presence of these well-contrasted equations strongly suggests that in the mere pulse signal there is much more useful information than what currently we are able to interpret —information with a direct and clear physical meaning. Until recently it has not been possible to faithfully track the pulse signal without the use of catheters, but now the applanation tonometry makes this possible non-invasively and easily.

Measurements apart, if what defines our idea of health is the flow and circulation, the structural element that define them is the tube, exactly as in Faraday and Maxwell approach; and the external differential forms developed by Cartan are the most natural, elegant and compact way to consider their contours and global, geometric and topological changes.

Cylinders or tubes of flow occur in the body at a much wider range of levels than the cell, currently considered the fundamental biological unit. We have tubes and channels at the level of macromolecules, at the intracellular level, at the multicellular level in ducts and blood vessels, in the organs, and finally, in the most elementary topology of the human body as a digestive tube that external matter passes through. Above or below, tubes or channels are a much more general structure for life than cells. To say that life is functionally a matter of tubular flow is pertinent and adjusted to reality, although it has not yet been considered with a keen integral strategy.

As for the molecular level, which until recently was considered the ground level of univocal information —remember the so-called "central dogma of molecular biology"- we now know that genes have multiple expressions and that enzymes are capable of perform very diverse functions depending on the environment. What ever happened to reductionism? And yet we see that physics, supposedly the most reductionist of sciences, naturally allows a more global vision. Moreover, the classical, macroscopic version of electromagnetism already contains a rich statistical structure, with conservative and dissipative aspects.

The current clinical knowledge about hypertension, mainly based on biochemistry, ignores such basic mechanical things as the role of respiration, and breathing is already a first-order circulatory function, even if we only consider the gaseous exchange that generates. Now that there is so much talk of evolutionary medicine, we can ask which organ would take Nature for something as critical as circulation as a first choice, if one with a given volume, or another with four times its volume and therefore much greater adaptation capacity. And we don't mean only of the lungs and the thoracic pump, also of the diaphragm and the so-called abdominal pump.

From a global point of view the heart is more a control valve than a pump driving the blood; We should have drop this idea for what it is ... another primitive representation. And the mere fact that the diaphragm requires more blood than the heart would have to tell us something. The first and most important measure to reduce hypertension should be to use about half the capacity of our diaphragm, instead of the usual quarter or fifth. The hypertension is due to the stagnation of the venous blood that fails to be evacuated, and that saturates the capillary circulation; when it is full it has to overflow raising the blood pressure. This piece of logic can not be more mechanical nor the empirical evidence more overwhelming, and yet, in 2018, both continue to be systematically ignored.

In the formula (V = P-O) P represents energy and O represents matter. Conservative systems are governed by minimal energy variations; dissipative and far from equilibrium systems, and biological systems with their own internal homeostasis, try to minimize the use of matter for reasons of obvious scarcity. More energy usually leads to more incorporation and more elimination of matter; if the first predominates, one tends to the plethora, if the second, to the consumption. The first also tends to a uniform fullness, but also to gradual obstruction; the second tends to contraction, differentiation and wrinkling, also to increasing fragility. Both are forms of increasing restriction, though. They are very generic patterns of temporal evolution that we all can observe in our fellowmen and in ourselves, and, as simple as they are, they contain important indications.

Naturally, the interplay of energy and matter flows is called metabolism —the balance of catabolic and anabolic reactions releasing energy and forming matter.

It would seem that anabolic reactions seek to get the maximum of matter with the least amount of energy possible, and catabolic reactions, to get maximum energy with whatever matter content; in this sense, both sides mix conservative and dissipative aspects. In other words, they are not only antagonistic, they also contradict themselves to some extent though. Circulation and alternation takes place when two states can not be satisfied simultaneously.

It is well known that bistability is a fundamental attribute of cell function at many levels, and its impairment affects the loss of cellular homeostasis associated with degenerative processes and cancer; it is also known that it depends on complex feedback circuits with ultrasensitive regulatory barriers.

A prototypical macroscopic bistable system whose importance is still unknown, is the nasal cycle, which alternates between the two nostrils with an average duration of about two hours and a half or three hours. Thus far, there are indications that this alternation and its flow rate could have considerable influence on the metabolic activity and the afferent and efferent functions of the nervous system and the brain, but we are still lacking a global understanding of the phenomenon. Why is there alternation here, why circulation? In this case, as in any other, we must ask ourselves which two conditions can not be satisfied simultaneously and demand such a period.

It seems that the periodicity of this nasal cycle has been studied more than its flow rate, which both quantitatively and qualitatively should be relevant. It is very doubtful that this respiratory bistability is a mere cumulative effect of the sum of local cellular cycles, since in fact it is very easy to modify it intentionally or switching sides when lying down. Even ignoring the depth of its regulatory effect, its global nature is quite evident.

It is a priority then to proceed from the global to the local, not the other way around; in the same way, as a matter of fact, than in the original view of electromagnetism. Two complementary approaches come to mind. The first is biofeedback, since, being a function that can be modified at will, it allows us to follow its more immediate incidence in other physiological, cellular and biochemical functions. Biofeedback allows to set experimental situations that are usually out of control, and it is relatively easy to reach equilibrium situations in which the respiratory function does not fall in any of the two valleys and remains with a relative stability in the middle of the potential barrier.

The second approach is the construction of bistable electromagnetic systems with a significant dissipative component, for example, with metamaterials. Here, the point is to search, by reverse engineering, for systems that generate an analogous, as close response as possible. This is not absurd at all if we start from the idea that the (generalized) "electromagnetic" component, in the most universal sense already mentioned, of living systems is constitutive and not accidental; and on the other hand, and complemented by the previous approach, this allows a certain constructive or bottom-up approach, since metamaterials are in phase aggregates of unit cells such as split ring resonators —and not only unit cells, we could speak of "unit tubes" too.

The phase modulation allows here to alter macroscopic properties of the equations such as permittivity and permeability, thus obtaining negative indexes that seem to violate elementary laws and make possible exotic properties such as optical cloaking, etc.

Both approaches are highly complementary, proceeding top-down and bottom-up, and the thinning of the potential barrier by self-feedback allowing also to enter the domain of phase modulation and attainment of negative indices. We must not forget the overdetermined Maxwell equations emerge from an indeterminate background.

In the same way a metastable or bistable system is a small island of stability in a sea of matter and energy that always remains undetermined but that is actualized by the interactions: the same matter and energy that make it possible define its limits. An obvious statement that has many non-trivial aspects.

The potential barrier of a bistable system should faithfully reflect its metabolic limitations in the use of matter and energy —and what of its restrictions are reversible or irreversible. This would give us the still missing global idea of the aging process.

Aging as an increasing restriction

In this article we assume that aging is a global and macroscopic rather than a microscopic process, and that the microscopic aspects are always hostages of an internal environment and local context, which in normal conditions obeys the global balance more than the other way around. In any case, we already have chemical markers by myriads, but not a general vision that delves deeper into what we already see.

We are used to see the increase in complexity of systems as an opening to endless possibilities, but this increase in complexity inexorably leads to a tighter restriction and closing operation. It's clear that old age is complex, not in the sense of greater possibilities, but rather on the contrary: it is increasingly complicated, more difficult, also more characteristic. What were once dispositions have now become structures that increasingly limit the free flow of energy. As pure differentiation, it is a highly personal, individual process —there is not, there can not be anything more individualized than this.

Simplifying it to the maximum, physical aging is an increasing lack of elimination of the obstructions, which accumulates in the form of more or less characteristic material structures. Put another way, it is the cumulative growth of the obstruction. Although, needless to say, an organism can eliminate too much without eliminating all the obstructions. Nature in fullness knows very well what to eliminate and how; it is the impeded nature that must be helped, by means that may be lesser, or greater in cases such as surgery.

This is what really matters, the rest are corollaries. Obstruction is superfluous by definition, then it can not be characteristic but in the most external or limiting sense. It is the least individual if we understand the individual as an infinite singularity, and it is the most individual that if we understand the individual as a mere material limitation, as limitation par excellence.

Matter and energy are the two faces of metabolism; to this it could be added that matter can support information, but it can not transmit it, while energy can transmit information, but it can not support and retain it. The self-organization that we attribute to living organisms is their capacity to shape themselves, to produce or alter their structure from the information obtained from their very dissipated energy (Margalef, Barragán).

Thus we return to the trinomial matter-energy-form of the founder of biology, none other than Aristotle; only that the form, as information, has now become a quantitative category of a statistical nature.

The microscopic description of metabolism, its biochemistry and the reactions and cycles it draws at the cellular level, is a science that although always incomplete is extraordinarily developed and does not have great problems to define its method. But all those pieces of the puzzle are still missing the model to assemble, the macroscopic perspective, without which all that information and statistics will never reach the directly appreciable and higher rank of the Form in capitals.

Since the metabolic rates can be followed in ascending order from the reduction-oxidation reactions and cellular respiration rates to the overall profile of the respiratory system as a bistable system by the nasal potential barrier, this is the form and axis of symmetry needed. It is even something else: it is the space and the metric that defines in each moment the relative value of the parameters of the system. Our appeal to electromagnetism in its most universal aspect was not vain, neither our appeal to reverse engineering with metamaterial structures.

In fact, from the constitutive view of the continuum that finds its limit in the field, and in which there is no space without matter or matter without space, just as there is no rigidity without matter, neither exists deformation -changes of form- without energy.

Then we can consider what we call "form" at a constitutive level, as a balance between rigidity and elasticity. This applies even to molecular mechanics, and it's already known that the elasticity of a rigid molecule depends on the energy while that of a flexible molecule depends on the entropy of its configuration (on the other hand, the inelastic forces produce heat and the elastic ones mechanically recoverable tension). This constitutive form is the limiting form, the most external aspect of the individual profile.

Then there would be the innermost and less limited "form" of individuality, which, in reality, lacking limits in a tangible sense, can not be form in the sensory sense. It is, more properly speaking, the space that defines them.

Thus, we could speak of a more external form governed by the balance of matter and energy, tension and deformation; another more internal form, which is the "metric" arising from its interpenetration and which is not properly a metric but rather a fluctuating plateau; and finally an intimate form, which is beyond the distinction between the external and the internal ones. This latter is the most universal aspect, but, seen from the outside, it has to be the utterly undifferentiated.

With some ambition we could figure out the *singular function* that drives live in each individual —the most comprehensive individual context for the vital signs of a person. This one would encompass all the specific details as a by-product.

A framework like this has allowed us to indicate a lot with very few words. While it's not my intention to consider any biochemical detail here, maybe we should mention that the prevalent theory of cellular metabolism and respiration typically ignores the important role that radiation and the corresponding photochemical agents can play with respect to chemical energy, and the extremely variable margin that this aspects may imply.

Temporal evolution of complex systems

Aging as an increasing restriction presents us with a face of temporal evolution that, despite being the most evident of all, continues to be the least attended in our modern culture. This is coincidental, as neither theoretical nor technical means are missing. On the other hand, we constantly talk about the increase in complexity as an increase in possibilities, and, of course, evolution as natural selection, competition and competitive advantage —words that serve for everything and can't explain anything.

I don't think this has to do with science but with discourses in the line of least resistance to the powers that be; It is a horizontal narrative that is considered appropriate for the "atomic levels" of the social body, read individuals. The vertical logic, much more concretely structured, thinks nevertheless in the more "ecological" marketing terms —in the exploitation of niches and ecosystems. Thus, there is an intensely publicized narrative for the many while there is a vertical logic implacably managed by the least.

Whether power is interested or not, we live in increasingly aging societies and this increase in age is far from being only a matter of population average. In the era of diminishing returns, this general circumstance invites to apply even more exhaustively the vertical logic of exploitation. Just as in mechanics and engineering, there is a method of flexibility and a method of rigidity, but to the extent that structures are ossified, rigidity is wining the game as it approaches the breaking point.

Undoubtedly a good part of what applies for a biological organism like the human being would have to apply also, to the extent that it does not depend on particular elements, for the social organization. Here there is also an opposition between free circulation and obstruction, but the arbitrary element, imposed from above, can have more scope and depth. It is the component of self-organization by feedback, and puts us at the heart of control and stability theory, which in its day was called cybernetics.

There is no doubt that biology can be affected by monitoring its output rates —as in the case of breathing and its inconspicuous bistability; but of course there are always limits for intervention, which depend, of course, on the very profile of stability of the system, which in turn can have a margin of evolution over time, to the extent that the organism evolves. Thus, the interest of the more general bistable profile is self-evident and does not need further justification. The lower levels are already the basis of stability with which the higher level is confronted.

The potential barrier that separates the valleys is not internal or external, it belongs to both perspectives and, in a certain sense, it is apart from them. It is Form in a higher sense than that of visible forms, it is Matter in a lower and more generic sense than that of matter that we can see and touch. We can understand it in the terms of the continuum only.

The main question is quite specific: what is reversible and what irreversible in an evolution that tends to an increasing restriction and to an abrupt outcome long before the total uniformity of thermal death. All this is approachable in an experimental and non-speculative fashion, and has a deep theoretical and practical interest.

References

Ramón Margalef, *La ecología: entre la vida real y la física teórica*

Eric Chaisson, *Energy rate density as a Complexity Metric and Evolutionary Driver*

Jorge Barragán, *Sobre la termodinámica de los sistemas físicos biológicos*

Nicolae Mazilu, *Mechanical problem of Ether*

Elliott, S, *The Valsava wave. The changing landscape of Heart Rate Variability Biofeedback*

M. A. M. Iradier, *Between stress and pressure*

M. A. M. Iradier, *Beyond control - Feedback and potential*

LIGHT, GRAVITY AND COLOR

10 September, 2018

Mechanical theories of gravity fell into a deserved discredit at the beginning of the twentieth century after having been seriously considered by the greatest physicists for centuries. We will not make here an attempt of rehabilitation but we will remind that if we don't know when a particle radiates we are not really qualified to discredit other theories by thermodynamic criteria. Mechanical theories may be pathetically inadequate, but even if this is the case they might stand on the threshold of the most unsuspected revelation. We talk about the relationship between the color theory and field theories. We talk about Color Technologies. We even talk about the Riemann zeta function.

There is no need to review all the physicists who carefully dealt with the mechanical or kinetic theories of gravity; it is enough to cite the names of Huygens, Jakob Bernoulli, Leibniz, Kelvin, Maxwell, Lorentz or Poincaré to give it a respectable place among the recurrent topics in the history of physics. It was only after the widespread acceptance of the General Theory of Relativity that the subject was relegated to the marginal research and amateur physicists, even if it was clear that the new brand theory had nothing to say about the cause of gravity.

It has been called "Le Sage theory", "push theory", "theory of radiation pressure", "shadow gravity", "Universal Repulsion Theory", and so on. It could be said that there are two parts to it: a supposedly mechanical part, which attempts to explain the push with collisions, and the non-mechanical part, which actually states that gravity is not a force but the absence or shadow of a force. To the first part belongs all the pedestrian and pathetically inadequate explanations of this kind; to the second part, a truth so sublime that we do not know how to take it, since it is evident that the free fall of bodies does not produce deformations and in this sense has nothing to do with forces as they are commonly understood.

Between one thing and the other, there is room enough for indecision. We know, however, from the thermodynamic arguments provided by Kelvin, Maxwell and Poincaré, that the more basic estimates of friction and temperature seem to be conclusive on the unfeasibility of the model. And yet, it is precisely since Maxwell's theory that we lack a definite criterion for knowing when accelerated matter should radiate or not —remember Bohr's circular atom-, what at least should have to put all these objections into quarantine.

Needless to say, the problem of radiation remains completely open in our modern quantum electrodynamics, among other things, because as we have already commented in another article, in addition to the ambiguities of the Maxwell framework, special relativity is not suited for accelerated frames of reference, nor for extended particles, nor for other kind of energy conservation than the most purely local. Too many limitations and contraindications for any realistic thermodynamic model.

To date we simply do not have an adequate framework to settle friction and radiation matters in our particle/field theories. In all honesty, this should be the starting point.

So, what to do? If one takes into account the great and elementary gap in the subject of particle radiation, it can still be said that the probabilities of finding some important truth in this type of theories remain around 50 percent. This, without having any problem in admitting that the mechanisms hitherto proposed are not really mechanisms at all and sidestep the difficulties of justifying the energy balance.

On the other hand, even from the point of view of the constitution of matter in the Standard Model, only 1.88% of mass is resting mass, with the remaining 98 percent being interaction energy —another name for radiation. If we add to this the fact that particles of matter such as electrons use to be considered to be made up of nothing more than "electromagnetic field", one can understand that popular theory saying that matter is nothing more than trapped light after different transformations of its linear momentum in angular momentum.

If space is defined by light, it has been thought, what else could particles be made of but light, which is synonym for space. And if experimental physicists already try to create matter from light, some of these common theories of gravity due to ultra-long wave radiation have long held speculations about the transformation of light into polarized pairs of matter on the surface of supermassive galactic nuclei, and vice versa, with the added merit of banishing the no less speculative black holes.

The idea always underlying, although rarely openly formulated, is that light is space, and matter with its gravity is simply its shadow. We could also say that light and its propagation depend on the homogeneity of space while matter and gravity involves its heterogeneity. In this abysmal simplicity, supported by the elemental common denominator of the law of inverse squares, resides all the appeal of such theories.

There are incredibly silly things not only in the mechanical theories of gravity, but also in the most common ideas about the propagation of light, which we imagine moving through eternity but which we cannot conceive of as remaining and vanishing at the place of its emission, as standing waves for instance. Our inability to represent the inextense activity of nature leaves us in a desperate situation when it comes to aspiring to understand anything. Physicists' theories contain many inextensive elements, but they don't reveal by themselves any activity, which we understand to follow only from the motion thing… so finally they try to explain mass by a kind of braking friction against a field.

Sometimes you think you might be asking physics for something it can't give at all, but, otherwise, Where are the written limits of what physics should be?

Field theories, such as quantum electrodynamics, display an overwhelming complexity in calculation —to the point that it makes us think that a single photon knows more and better than our own brain - but, despite this, it is incomparably less rich and revealing of nature than the panorama of the visual field, the bare perception of colors. And since here I am thinking of the most painful deficiencies of our theories in the face of nature's unfathomable simplicity, I will go one step further and ask, with the painter's eye, what is between shadow and light. The habit of the mathematician compels him to say: "a limit". In other words, practically nothing. The understanding of a naïve educated person like Goethe whispers to him: "the color". In other words, all colours and the entire field of vision.

And most interesting of all is that the essence of color is independent of motion. Certainly, the naked perception of color, the pure phenomenon, is incompatible with the requirements of a theory proper. In order to move from the elusive phenomenology to the modern fields theories we need a robust connection for which there are only attempts. Nicolae Mazilu, for example, has made a laudable effort to connect gauge and Yang-Mills fields, quantum chromodynamics and the holographic principle with the classical theory of light and color. His approach through external differential forms also allows very basic connections with concepts such as friction.

Colors in modern theory, since authors such as Schrödinger, form a linear three-dimensional space; color geometry instead is not Euclidean, but Riemannian, and its metric bears a statistical significance with the components of the metric tensor as covariances of the three coordinates of color. Mazilu points to a general dynamics of color with a flow of colors for the human eye related to Hannay's angle. I miss many connections but I think it's clear this is a way of mediating between modern field theories and the questions raised by the real visual field and its perception of color.

Mazilu's guiding principle is that light is a universal model for the physical world as far as we can go, for after all it is light what transmits information in the universe. In contemporary theoretical physics, this privileged role of light as an information carrier has taken the form of the holographic principle.

Probably in the future we will talk about "Color Technologies", which will be a good part of the most integral and analogical information technologies. If light is a supreme model, we are still far from having discovered all the aspects of its interaction; and among them the "chromatic interaction", as far away as possible from kinematic considerations, has a special role to play here.

In the most obvious physical sense colors are a mere surface phenomenon; but for the visual field and perception —let's say, for Nature's eye - they are themselves responsible for our sense of depth. There can be no greater contrast between the two points of view, but if we really want to delve into the paths of nature we cannot do it disregarding the greatest, plain and primary evidence that it has unfolded before our eyes.

It may seem that we adopt the most contemplative stance, but in reality it is here, beyond motion, where nature exhibits its most genuine activity, activity that in this case it's not even hidden. On the other hand and in quite a different sense, we are still in a phase of "literal" reading, that is why we speak of the famous and non-existent "book of nature" that makes us think of a code and a single meaning, even though all this are man's games playing the codebreaker.

The same genes could one day be the subject of the new color sciences; for we now know that there is no single reading, that enzymes make very different proteins with the same message, and that three-dimensional folding is something completely different from linear sequences. All stereometry problems can be translated in colorimetry ones; and they can be applied to the atomic nucleus as well as to the nucleus of the cell, or so many other matters.

As Schrödinger saw it, the argument for color depth requires a framework of projective geometry. That there is no form in nature without a rigorous equivalence in color was a mere truism before Galileo began to speak of primary and secondary qualities. At the end of the escape, in the extreme flight down of gravity, we will again find part of this notion in the holographic principle itself.

Goethe spoke of polarity in colors, and physicists, so concrete always, have observed that there is no polarity in light but only in electrical charges. Well, I for one think just the opposite: electrical charges are the inexistent, but there is polarity in light, which, as one can guess, cannot depend on something as trivial as a mere convention of particles with opposite signs. The naturalistic polarity of which Goethe spoke would rather have to do with the duality of electricity and magnetism, matter and space, tension and deformation, which interpenetrate and can emerge from each other by the work and grace of the dynamic and static aspects of kinetic and potential energy: the same would happen with colors on the part they share as an electromagnetic phenomenon, and beyond that, in what we might call their "analytic continuation".

Not only in the nucleus and the nuclear matter and forces we can continue applying the theory of the deformations, also in the virtual interactions of color in the visual field. Talking about an analytic continuation for this one, we are irremediably lead to think about the Riemann zeta function and its proven but enigmatic relation with atomic energy levels. It is well known that the type of quantum system that can replicate the real values corresponding to the function is actively sought. Berry, Connes, Sierra and Townsend seem to have even suggested confining an electron in two dimensions and inflicting electric and magnetic fields to it in order to "obtain its confession" in the form of zeros... For me, the critical line is the same virtual line between light and shadow, space and matter, unfolded in the miracle of our sight and vision.

In other words, the physical replica of the zeta function cannot be a matter of quantum mechanics alone; to believe that is perhaps the main obstacle in specifying the type of system that can replicate it, since quantum formalisms are at odds with any tangible concretion. Once again, quantum mechanics acts as a smokescreen.

They say that the zeta involves the most basic and deep relation between addition and multiplication, the quantitative and qualitative aspects of numbers. Can we figure out the most basic and deep relation between addition and multiplication of colors?

In order to better understand the relationship between the zeta and quantum mechanics, one must necessarily have a better understanding of the relationship between the latter and the physical continuum from which it emerges. There is no true universality without this. For the physical continuum, which makes itself present for the first time as the Maxwellian electromagnetic Ether, is an indefinitely broader concept than the mathematical continuum of real numbers. Understanding the zeta-quanta relationship involves understanding the quantum-continuum relationship, no more nor less. And this forces us to look both at the past and the future of physics, and at the discarded theories that seems to be in conflict with the prevailing ones.

Of course, in order to circumvent modern vetoes, physicists are forced to look for subterfuges such as "semiclassical approximations"; encrypted messages that everyone will do well to interpret. All attempts to "crack" the zeta instead of trying to understand it globally are doomed to failure. And the same can be said for the rest of the hard sciences with a dramatically diminishing reductionists returns.

Needless to say, the zeta function has been used to regularize and calculate the partition functions of "thermal gravitons" and quanta of matter in black holes —to obtain "finite values despite the infinite blue shift of the local temperature over the event horizon".

The fact is that we have a theory of light and color strictly in terms of surfaces, while gravity and General Relativity, for which the punctual particle does not exist, give them infinite depth. Now we would rather have to proceed to the contrary: delimiting the superficial part of gravity and grasping the infinite and uncontrollable reverse of the electromagnetic continuum in which light has its being.

Light is extremely universal; the zeta is extremely universal; color is extremely universal. It would be extremely unlikely that they do not have a relationship both very general and crucial. How universal are gravity and matter, light should tell.

Finally, I would like to say that although the idea of privileged frames of reference has been practically banished from physics, it would still be highly desirable to consider it for different reasons. Weber's law already fulfilled relativistic criteria better than Relativity six decades later and yet it did not outlaw Ether, because Ether is much more than just a cinematic assumption. Anyone who cares about the non-kinematic aspects of physics, and there are many of them —even in General Relativity - would do well to keep this in mind.

In an old Ether theory, as Riemann and Maxwell themselves conceived, material condensations could show an atmosphere around them, since matter itself was conceived as a condensation of the medium. Thus, in the same way that planets show their own atmosphere or transition zone with free space, particles would have their own halo; as the field theories have not stopped talking about particles as condensations of the very fields.

This is not merely a question of how we represent the transition between space and matter, there is a huge range of experiments involved. Even going backwards, we see that the Lorentz undraggable ether, the Fresnel and Fizeau partially draggable ether, and the Stokes theory of total drag of ether are not contradictory and refer to clearly different cases; one could even say, with Stoinov, that they are complementary visions of an issue which, like matter itself, cannot be reduced to a simple arbitration once and for all.

Mass and force, matter and gravity, distribute their extensive and inextensive aspects according to our consideration of motion, which, needless to say, can be extremely variable. The relations between surface and depth, matter and motion, statics and dynamics, have find its symbol through the ages in the interaction between the Sphere and the Cube, which defines at each moment the limits of manifestation and the ever-present limits of the ascent and descent of knowledge that, with or without self-consciousness, we can assume are balanced.

No surface is only motion, but all motion is meticulously superficial. The great question of the Continuum is not so much to provide us with a privileged frame of reference as to allow us to discover the part of physics that is less dependent on motion. To understand it better, and to find its relationship with the best known kinetic and kinematic aspects, be it in electromagnetism, in gravity, or in thermodynamics, that and nothing else is for me the great prize. It is as if we have not yet opened the door, ignoring if it is open or closed, if the door even exists.

References

N. Mazilu, *The Classical Theory of Light Colors: a Paradigm for Description of Particle Interactions*

N. Mazilu, *From Kepler problem to Skyrmions*

N. Mazilu, *The concept of physical surface in nuclear matter*

N. Mazilu, *Mechanical problem of Ether*

S. W. Hawking, *Zeta Function Regularization of Path Integrals in Curved Spacetime*

D. G. Stoinov; D. Stoynov, *For Physics of reason against Physics of misconception*

M. A. M. Iradier, *Between stress and pressure*

M. A. M. Iradier, *Self-energy and Self-interaction*

M. A. M. Iradier, *Beyond control-Feedback and potential*

THE MULTI-SPECIALIST AND THE TOWER OF BABEL

10 September, 2018

One hundred multi-specialists with a basic agreement on their priorities and no personal interests in the current system of scientific production would suffice to rewrite science as a whole without losing the bulk of their actual achievements.

To think today that the collective enterprise of science is at the service of all is simply despicable. It never was, and even less so with the screen of the information society. It is precisely in this time of "data deluge" that we are realizing that what is published about research is first and foremost a facade, that we must assume that real research has passed into the catacombs and that it serves increasingly particular and minority interests, in harmony with the inverted pyramid of wealth distribution. Or what did we expect, don't we all know that the scientist depends on his funds not less than the caged guinea pig on its periodic food ration?

This would have to turn upside down our appreciation of scientific knowledge as an open skylight on the roof towards the universal. Something that is very difficult for us, because if we do not recognize the universal in science... where else? It's been a long time since we shifted from talking about of the scientific revolution to talk about the technological revolution, a tacit admission that our interest in science itself is dwindling; and yet already in Newton's day the scales were decidedly inclined to make from practice a theory, and that's where we're still at after all.

And the funniest thing is that for the modern figure of the scientist there was never betrayal, neither about who could be his master, nor about universality itself, because in both cases it was only a matter, between the compass and the square, of bringing step by step the sky to the earth. But now that it operates on autopilot and is sold as the West's winning algorithm and killer application, we can be quite sure that it has become the best way to busily sleep, even if certain results are harvested, whose ever-decreasing return is difficult to hide.

All of which is still a fortunate and even favorable situation, for thoughtlessness in science cannot fail to pay dearly, and today's science as sold demands it to a great extent. Such successful theories, methods, and procedures can only go forward at the cost of categorically denying many things, and just the very same which contributed to its rise ensures that they find a natural brake on their exponential claims.

Today, the multi-specialist, who has has acquired different competences and hasn't placed his fate in any of them, seems to be the only force with muscle enough and the necessary position to stay free from the huge inertia and braking mechanisms of the specialized fiefdoms in their inevitable and triumphal proliferation. But having the necessary muscle does not take them beyond the category of labor force if they fail to raise their perspective on a panorama increasingly full of contingencies —only gaining height as generalists will they be able to decide where they place their loyalty and why. Beyond personal morals, science never had a criterion for such things, unless one believes in the value of those ethics committees so difficult to distinguish from their homonyms for corporations.

There is a question of content and another of form- in the modern scientific deadlock, if we want to call it so instead of talking about a death of science that would be all the more unappealable the less felt it is. Both questions depend more on pure instinct than on any rational argument among the many that can be made, and that is why, because they test our instinct, at the service of which reason will be put, that they are interesting and revealing. Because they are not things that affect only the man of science or knowledge, but the space that man entails in what he does or doesn't do, and that ends up defining also the outline of his action-knowledge.

The internal question is the weight we concede to motion, in our imaginary and out of it, when it comes to explaining things. The external question is the relationship between the quantitative and the qualitative. Both questions are intimately related but they are too vast to reduce their connections to a simple criterion.

With regard to motion, we have always known that physical reality —not to mention plain reality- is not limited to extension and motion. Of course, motion is not only extension either, although we are so inclined to think about it, nor is motion reduced to translation or displacement. The fact is that physical magnitudes —be they vectorial or scalar, forces or masses- always have an intensive component that can be bent towards interpretation in terms of motion, and on the contrary, motion can be used towards a more intensive interpretations in which motion is a secondary outcome.

Yes, we are talking about interpretations with all the subjectivity that this entails. It matters little that it is subjective if it is the rudder that determines all the syntheses and applications; science is a highly teleological enterprise bent on denying this last point, but this denial does not bring it closer to its reality but rather to the contrary. Thus, we can interpret what does not move as a function of what moves, as physics does from Galileo, and we can try to interpret what moves as a function of what doesn't move.

Newton deduced the force of gravity from motion, even though what deforms a body —the most unequivocal sign of force- is the potential, and not the kinetic energy. Kinetic energy and potential weren't even the same in the quantity conserved, the Lagrangian, but it didn't matter as all that people wanted was to blindly take derivatives. In the end, the application is the interpretation. Could we say "and vice versa"? In general, it is what we do that determines what we think, not the other way around. But sometimes the will to do other things makes its way through "other interpretation".

So if we want to go deeper into the most interesting part of physics, into that great unknown that has always been there, but that we have interpreted on behalf of the surface, we already know what we have to do: go in the opposite direction to the mainstream. Since this part is always present, it is less difficult than it might seem. Moreover, in the still very brief history of physics it is not at all difficult to identify the breaking points either, the introduction of operationalist criteria to the greater glory of our arbitrariness.

What is set in motion does not change, and what changes is not set in motion —if we understand motion as displacement in extension. This deep conviction becomes little less than irretrievable in a world incapable of deciphering activity in immobility, something that maybe we could call *inmotion*. More than science itself, which does not even have a position for this, what betrays us is the drag of the scientific imaginary, of a functionalism more presupposed than deduced.

The other point is the no less arbitrary opposition between the qualitative and the quantitative. It is indisputable that physics is a quantitative discipline based on measurement. But it also seems obvious that mathematics, the language in which physics is expressed, is no less qualitative than quantitative —some have even said that it is the quantitative translation of purely qualitative aspects. The distinction, also since Galileo, between primary qualities that are measurable and secondary qualities that aren't lies at the basis of a dissociation that has come a long way.

And so today it has become a universal belief to think that in order to go further than with current theories we need instruments and experiments of a prohibitive degree of accuracy only within the reach of Big Science. This is usually true only within the current views of science and not in others. There are very simple experiments, within the reach of the means of the nineteenth century, which yield a continuous range of information and conceptually go beyond the models we handle, experiments such as those of Fizeau, Hoek, Scanglon, Miller, Sagnac, and so on. The only thing that has happened is that in the meantime the idea of a continuous medium has been discarded due to purely cinematic considerations, which are only a part of physics, and certainly the most superficial one. And so the theme of motion again has an impact on what is considered measurable or not, quantitative or qualitative.

But if even now it is admitted that broader and more general theories demand new concepts and criteria, the only thing that is being admitted is that these theories cannot follow the well-trodden paths of the standard calculations and have to be qualitatively different from the previous ones. Thus, deeper changes and rearrangements are more qualitative than quantitative, and then it is calculus that explodes these qualitative breakthroughs in understanding translating them into routines.

It happens, however, that theoretical physicists have exaggerated the greatness of the theory, in reality the part of calculation, and have subordinated to it the experimental part, which is often the one that best embodies the new concepts. The distortion in this sense has been extreme, turning the theoretical physicist into the priest who defines reality and the applied physicist into an inevitable operator. If we understand that the opposite is true, we will also understand that the conceptual and qualitative advance are more on the experimental side, and that it is the theoretical aspect, which should be at its service, that goes on to confiscate practical achievements and decides what experiments should or should not be done.

It is what we do that changes our way of thinking, much more than the opposite.

Quantum mechanics has often done nothing more than hinder a multitude of practical findings that later claimed as its own. Physicists speak of theories with eleven or twelve decimals of accuracy, but the great number of things that they cannot foresee in any way is ignored, not to mention discrepancies that only involve two or three decimals, as in the case of gravity. Let no one be fooled by these boasts.

The misrepresentation of experiments in the name of theory was already present in the most blatant way in experiments as famous as Newton's prism or Joule's one about the mechanical equivalence of heat; but all this has not ceased to grow like a snowball with the standardization and bureaucratization of the procedures of present-day Big Science.

Well understood, all this is excellent news as well as common sense. So who's going to believe that the depth of concepts depends on the money invested in them? The opposite is by far the most likely: the more money is invested in a project, the more people are involved in it, the greater the flattening of its profile will have to be; it couldn't be otherwise.

One hundred multi-specialists with a basic agreement on their priorities and no personal interests in the current system of scientific production would suffice to rewrite science as a whole without losing the bulk of their actual achievements.

Of all the experimental sciences we give priority to physics because it is the one with the most solid knowledge structure and the most difficult to change; if this can be done with physics, it goes without saying how much easier it can be done in biology or other less structured disciplines.

The current drift of science is completely irreversible, yes, within its present assumptions. So those who are satisfied with their winning algorithm go ahead and keep buying. We could never convince them of anything, as it is first and foremost a matter of instinct. My instinct tells me very clearly that the current model is going down, and that has been the case for a long time.

When I say that in order to revert the current situation one only needs independence, to value the qualitative over the quantitative and to go beyond motion, it goes without saying that there is a good note of exaggeration and simplification in it. But if one admits that the present research system has taken exaggeration in its own sense as far as it could be taken, it is understood that mine is only a way to regain balance. However this rather common sense reflection would be fraught with unexpected consequences.

Science was the West's great asset, its dominating argument, but now that we lack the conviction of its universality, as pure domination, it could look detestable in the end. All their omnipresent public relations are directed against this perception.

And precisely one of the arguments most frequently put forward when talking about the superiority of Western science, and with it of the West itself, over its forerunners was the universality of its knowledge. Other cultures would have been more interested in the practical results than in their generalization, and the greater universality of knowledge corresponded to an equally more universal culture with a much greater natural right of expansion.

Whether we like it or not, there must have been some truth to these claims, for they did not merely refer to the argument of domination and the results themselves, but often to the formal conditions of knowledge —to matters of knowledge for the sake of knowledge itself. In a gradual process of degradation, it is these formal conditions that have been distorted, simply confusing mathematical language with universality. Calculus has always been a heuristic issue, but if everything for us is algorithm we can hardly notice it.

That is why we insist on going beyond motion and quality, so that we can notice it again. That the language of physics is mathematics does not automatically mean that the language of mathematics has emancipated itself from human limitations, as physicists so understandably tend to believe. Indeed, the fact that our idea of natural laws folds to what we can predict is a utilitarian criterion par excellence, and therefore, equally human, centered on a very partial and limited objective.

The more "algorithmic" knowledge is, the more it descends to a lower order. Two great ways of doing science have been spoken of in this regard, a "Greek" or axiomatic style, and a "Babylonian" or heuristic style. But does anyone have doubts about the side we are now on? One is free to proceed as one wishes or as one can, but we don't abandon so easily our claims of universality.

We began by talking about an inverted pyramid and that is exactly what the current tower of Babel of sciences is —and it is mathematics that has inverted its own universality. It has been and continues to be believed that mathematics spares us the difficult problems of interpretation of experiments, and that by limiting ourselves to the prediction of events we also rid ourselves, holy simplicity, of the limitations of subjectivity. Only that subjectivity and universality are practically synonyms, which simply helps us to corroborate the aforementioned inversion.

Basically, no matter how much we talk about the complementarity of observation and theory, what always underlies it is the dualist mentality: in the best Baconian spirit, nature is something to hack. Experiments should help to refine "quantitative models", not only in the number of decimals, but in the very quality of mathematical tools. The continuum mechanics is the best guide for this; the physical continuum itself is the basic assumption for mathematics to be refined by learning from physical reality. If we oppose observation and thought we have already lost the game beforehand.

There is a return staircase for this nefarious situation of quantitative sleepwalking that passes through the mathematicians themselves. They would be the prototype of the multi-specialist, if only they learned to redefine their situation with respect to observation, the physical continuum, the relationship between the qualitative and the quantitative and their contemplation of motion, immobility and mutation, which have their archetype in the very process of thinking.

Whether in China or India, Russia or anywhere else, it is still possible to change our fate as mere objects of quantity. This path necessarily passes through small scale, independence from large budgets, and the unwavering conviction that there are things more important than quantitative precision; things with which one acquires perspective and equally reduces the probability of being handled.

The specialist doesn't know everything about his speciality; on the contrary, he is trained to selectively ignore many of the decisive steps that have constituted his domain. Multi-specialists need to know how to take advantage of this even if in principle they do not have open channels to express their disagreement: they will have to create their own.

The multispecialist has gone to great lengths to know what the specialist knows; the specialist has gone to great lengths not to see what the multispecialist sees. This gives the latter a huge and unexpected advantage.

Our misnamed "reductionist science", in fact algorithmic science since it has not been able to reduce anything to mechanisms, moves between the reduction of the general case to a particular case, which then becomes general again without restraint, and a growing and irrepressible lack of faith in the existence of any physical reality, which has come to make matter first an equivalent of energy and then a mere support of information. What is not understood is that this leaves free an internal axis for another type of inverse generalization of the global features that is precisely based on the surplus of rejected physicality: in the five previous articles of this blog I have dealt with widely differing aspects of this very point.

Chris Anderson in his article on "the end of theory" in the age of the information deluge has defined well the level or rather lack of any level that modern science has reached and that the Anglo-Saxon culture has had the dubious honor of leading. And since they have been fighting tooth and nail to make history of it, their wish will surely come true.

Now, there is much more here yet. Anderson refers to the distillation of knowledge from the mass of Big data, the definitive mathematician farm. Of course these would be the assistants of an autonomous process with aspirations to become even purely automatic to finally get rid of their clumsy operators. A great future for the microserfs. Needless to say, the object of Big data is everyone and everything, and the active subject is simply power.

I don't know if anyone has even noticed that the distillation phase of Big Data corresponds surprisingly well with a completely forgotten prophecy of a hundred years ago: the comparative morphology or general "physiognomic science" (???) of which Oswald Spengler spoke. For he emphatically predicted the advent of a purely formal science finally liberated from the very constraints of causality and logic. No one knew what to do with this anomalous outlook, but Anderson, who couldn't be farther from the German, sentenced in 2008: "Correlation supersedes causation". Bingo. Who would have said it.

One always knew that Spengler's meta-scientific foresight would surpass immensely the scope of the philosophy of science, but it was difficult to see how its advent could take place. And it is that its material is exactly the opposite of matter, it's the ubiquitous circulation of information that surrounds us. The purest creation of the spirit, as this abnormal prophet-historian-philosopher anticipated, at the service of power and social control. The final transmutation of the old social organism into mere organization.

And here comes the heavy dose of irony. Spengler, like so many other advocates at that time of "the White Man's burden", warned of the danger of putting in the hands of other cultures the knowledge of the Faustian man, that should remain under his scrupulous administration and responsibility. But it turns out that this last mefaustophelic creation, at least in the Anglo-Saxon version of "Bacon with Mathematics" that prevails today, has meant a reduction of science and of the knower to the rank of an algorithmic "Babylonian" stochastic tinkerer, the redoubled avatar of the "inferior man", of the barbarian who only cares about the results, only that far below him in terms of his insatiable taste for manipulating and stirring up even the most defenseless things: the Super-elite and the Scum Club all in one. But perhaps these are not the only extremes that mate in this *mysterium coniunctionis*.

The dilemma is this: science, without aspiration to universality, is something abject and destined to serve power against the rest of the population. But a science that truly sought universality, rather than results, would be far more transparent and assimilable to the rest of the population, who would thus cease to be a clueless object of power. Between truth and power this dilemma is unavoidable and has always existed. But in an era of apparently unrestricted flow of information it has to adopt completely different defensive systems, which obviously rely on the very excess of information to filter/modulate it in the most convenient way. This seems to have come a long way and it can be assumed that the process will continue to be successively refined.

On the other hand, and as a counterbalance, there has always existed an assumption that lies at the very root of scientific optimism: it is very difficult to conceal a universal truth, that by its very nature seeks to be shared without hindrance. The intrinsic transparency of truth. Is this truth about truth really correct? We don't know, but we do know that power also likes to air this contagious conviction.

Now, if universal truths are getting further and further and what we have is a mere instrumental knowledge, there is already little room for doubt as to the fate of such knowledge. But does it still have any value? Perhaps it does for others, certainly not for me.

Computers can analyze but even Anderson knows perfectly well how poor they are when it comes to producing or synthesizing knowledge, which is like saying "synthesize transparency". Riemann's zeta function is infinite and completely analytical, but ask a computer to draw global conclusions about it. The only thing computers could be used for there is to find a zero outside the critical line, as the ineffable Turing intended with a rudimentary machine, but you could still expect a quadrillion years.

What do zeta, light, the holographic principle, electrodynamics, classical chromodynamics, or the continuum mechanics have in common? And we haven't even mentioned quantum mechanics here. What they surely have in common are global aspects that can be better understood by comparing them with each other. Can this be done with algorithms? Algorithms can also help, but synthesis has to be produced - or reproduced - by the subject.

I see a truth with many more layers than an onion. Energy transmits information but does not support it, matter supports information but does not transmit it. Only the subject can support it and transmit it. Every time we stop attributing to motion the reason for a phenomenon, we detach ourselves from a minimum layer of our superficiality and approach a bit the omnipresent nucleus, so similar to a flame.

Motion, on its own, is the insignificant thing; only with respect to that which does not move can it have any value. If I perceive motion and activity in my thinking, it is because there is something in me that remains still, otherwise, how could I detect motion? Technology as pure functionality is the most deadly incarnation of nihilism, of the "life of insects" that some are prone to attribute to other cultures. Technique dominates things, but who will dominate technique?

Can you imagine that the Chinese were the ones who finally tame the tamer, dominated the dominator? They, who have so often been criticized for their lack of originality? It is curious that the origin of the binary code is attributed to Leibniz's logical reading of the Chinese exagrams of the *Book of Changes*; curious because the Chinese spirit had used this binary logic as a mere support for a self-sufficient code of images, the minimum symbolic compendium of a multidimensional *characteristica universalis*. And could they not succeed now in turning the tables again?

However, as we have already suggested, this things go through the reinterpretation of physical reality, something hardly hackable. This is the Ariadne's thread, there is no resurrection of the *Logos* without a rebirth of *Physis*. Modern science blew up many bridges in its wake but now it cannot control all that immense territory from the air with flying algorithms. Logic doesn't usually fall from the sky.

In any event it is necessary to go beyond the accidental geopolitics of knowledge and concentrate on the underlying problem: the fierce struggle between the universal and the particular in the arena of knowledge and action. If this context and its scope are ignored, it is almost impossible to speak meaningfully of anything.

Power is the immobile that can set other things in motion. One would have to be suspicious in principle of any "empowerment" that compels one to mobilize —no matter how often motion is needed. Change in action pass through a momentary shift from motion to inmotion or mutation. If non-particular knowledge begins to enter the motionless sphere of power there will be a problem of eviction. Who will evict whom? Only that there's just not a single knowledge, just as there's not a single power.

Naturally it's not that motion is bad, but the compulsion to move and understand everything in those terms. It always gives itself away at its root, in its first impulse. The present and unnameable prison in which we participate is a prison almost exclusively made out of our own outer motion, or agitation, a technique that we could call *inertial confinement* or self-confinement of the subject and its substance. Others continue to call it alienation from the social, regardless of the fact that society itself is the prison and the stampede, the big time general escape.

So I am not only convinced of the existence of an electrodynamic Ether, I also take for granted the existence of the financial ether too, to which all financialization is accountable. And if you do not believe it, so much the better for me, says the power. Your agitation is my vein, and as the misery diggers know, there is no quicker way to separate gold than with the volatile and toxic mercury. What moves does not change, what does not move has the freedom to change.

From the present situation we would have to blame scientists more than science, otherwise we are only admitting that they have never been able to do anything and they are in every sense a pure nullity.

To understand to what extent our knowledge is low-level knowledge is tantamount to leaving on the same level the society that has produced it and even has boasted about it. This cannot be achieved by the impotent dragging drift of the future, but by the impulse of the past which opposes it, for in the present correlation, both are like motion without force as opposed to force without motion.

References

M. A. M. Iradier, *Light, gravity and color*

M. A. M. Iradier, *Self-energy and Self-interaction—Extended particle and thermodynamics*

M. A. M. Iradier, *Beyond control—Feedback and potential*

M. A. M. Iradier, *Health, life, aging and evolution*

M. A. M. Iradier, *Between stress and pressure*

IA ODDISSEY— PHYSICS AND ARTIFICIAL INTELLIGENCE

10 December, 2018

AI and automatic learning are fashionable, but little is said about their parallelisms and contacts with fundamental physics, so full of meaning. In our increasingly artificial environment, we are more concerned with unraveling the black box of intelligent machines than with the great black and starry box of the universe. But don't get discouraged, because the two things might not be so separate in the end. It is always a question of coming out of the hole, that is, of finding universality; and this is usually at the antipodes of applied intelligence. Here one could say the opposite of what John wrote in his book of Revelation: "He who has understanding, let him not calculate..."

Keywords: Artificial Intelligence, two modes of intelligence, Renormalization, Scale relativity, hierarchical learning, Continuum

TWO MODES OF INTELLIGENCE

Possibly the greatest "success" so far of Artificial Intelligence (AI) has not been its concrete achievements, still quite modest, but convincing many that our own brains are a particular type of computers. Conviction that can hardly correspond to reality, given that, as Robert Epstein says, not only are we not born with "information, data, rules, software, knowledge, vocabularies, representations, algorithms, programs, models, memories, images, processors, subroutines, encoders, decoders, symbols or buffers", but we never develop them.

While few might disagree with Epstein at this point, the pull of the computational paradigm remains irresistible. Not surprisingly, after all, as computation is, today, the only way available to replicate a series of human performances, and computational power is multiplied by a hundred every decade. There is also a perpetual challenge to translate that quantitative increase into a qualitative difference, which adds the indispensable intellectual interest to brilliant minds and developers.

We are not going to argue here that machines can perform tasks so far only within the reach of man, and that the spectrum of those tasks inevitably grows over time and the work of engineers, since that is obvious. The human being also has an important part of his intelligence that is developed specifically to carry out tasks, although he does so with a plasticity that has little to do with that of machines. But even this fundamental plasticity is due to the fact that it is not separated from an concrete but continuous perceptive background, which is the opposite of the strategies of cognition and representation based on high-level symbolic models of information processing.

Epstein cites as an example the 1995 landmark article by McBeath and colleagues, which determines how baseball players estimate the trajectory of fly balls —an achievement that posed a challenge to machines. McBeath showed that the players' reaction completely ignores the various relevant factors in a numerical analysis of real-time trajectories, since what they do is move in such a way that the ball remains in a constant visual relationship, which is much simpler than explaining and is "completely free of calculations, representations, and algorithms".

This example has been so often cited that the defenders of "ecological" or "radical embodied" theories themselves often ask not to exaggerate its scope. They certainly have reason to do so, since human intelligence demonstrations show aspects as varied as can be imagined. However, the case in question is precisely one of trajectories, the same trajectories with which analytical geometry, calculus, mathematical physics and ultimately the whole modern scientific revolution started. If all this came out of the analysis over time of the famous conic sections, it should be even more revealing for the computational paradigm, since it questions it not in its possible achievements but in its very baseline.

The embodied or "ecological" approach which emphasizes the physical dependence of the brain on the rest of the body and the environment is not exactly a novelty. Some have called it "postcognitivism", although clear expositions of this vision, in Merleau-Ponty or James Gibson, are in fact prior to the foundational milestones of cognitive psychology and the first babbles of AI. What this approach says is that the intelligent response cannot be dissociated from the perceptual continuum of experience, however elusive and practically impossible to delimit as it may be.

Obviously, the computational model, relying on information processing, cannot resign itself to this. The nature of intelligence is secondary to it, being its priority to reproduce intelligent or highly selective behaviors in well-defined tasks. And since they achieve it in increasing degrees, there is no reason for surrender.

The problem is not only the social impact, but to pretend that the intelligence in which we participate, more than we possess, is the same that one computers are reproducing in such a fragmented and super-specialized way. A pretension that would be laughable if it were not for the mimicry and pervasive character of the social trends.

The processor would be the ultimate worker, and intelligence, subordinate as never before, would be the act of digging into the mine of the new data reality with a certain output. It's then understandable why today operational intelligence is so overvalued: it is sold to us in every way because it's what sharpens the tool of this new worker-miner.

Society, or rather social intelligence, has to conceive intelligence in general as that which allows us to ascend on the scale of the social body itself. And this would have a double purpose: on the one hand, the search for flight from a blurry monster that refers everything to itself and yet is already the incarnation of a flight; on the other, to the extent that one ascends the ladder, that of exerting control over subordinate functions with the purpose of consolidating and incarnating that uncontrollable flight in a body that refers everything to itself.

This "double bind" is in line with the orchestration of the topic of intelligent machines in the media, as a threat/promise: they are going to end with the employment/they are going to end with the burden of work. But progress always had us between the stick and the carrot.

As can be seen, this reading of social intelligence is already "radically embodied" and "profoundly echological" —rather than trying to represent something surely unrepresentable, it is in the middle of its dynamic, but trying to withdraw from identification allowing it to break free by itself.

Thus we see two great modes of intelligence: the one which consists simply in realizing, in being awake to what is happening, and the one which considers, discriminates, compares and manipulates in an endless process. The one who separates itself from tasks, and the one that turns to them. The one who contemplates and the one who acts. The characteristic modes, respectively, of Homo sapiens, the man who knows, and Homo faber, the man who does. The desert of consciousness and the jungle of thought. Awareness by itself and intentionality. Aristotle's agent intellect and the patient intellect —keeping in mind that for the Greek the agent intellect is the one immobile, and the one that moves is the patient.

The former knows indiscernibly by identity; the latter by identification, in which he loses his discernment. From the first, all trace is lost in a continuum without qualities; from the second, the trace is followed but in leaps and bounds, due to the discreetness of its operations.

We don't need to continue to understand that both modes don't exclude each other more than they are mutually implied; we cannot get rid of the first mode, as some pretend, without equally getting rid of the second. And since it also seems that we cannot get rid of the second without getting rid of the first, there will never be a lack of experts in artificial intelligence who will ensure that things like stacked neural networks, which in reality are only multilayered mathematical operations, succeed in demystifying not only operational intelligence, but even consciousness, even if so little can be said about it.

Others think that the only possible demystification or disidentification is to recognize that the brain is not an autonomous entity and that we will never find in it a materialization of memory or the symbolic representations that define computers. However, this would have to open the doors to the contemplation of other wonders.

The so-called ecological approach is the really physical one, if we understand the physical as support or substance, and not as operation. Circumstantially, it happens that also modern physics and its field theories are based on decidedly discrete and operationalist criteria, even though fields are supposed to describe the mechanics of a continuum, even if it is a particulate continuum. The point particles of modern field theories are just the operational shortcut, even though we can be sure that material particles, considered at proper scale, must exhibit surface and extension since extension is inherent in matter.

Physics itself has become an algorithmic, statistical subject, in the precise sense that all description and interpretation is at the service of calculus. In this way the idea of a physical substance evaporates, there are only operations left. It is no coincidence that Feynman worked in the computation group of the Manhattan project and simultaneously in the formalisms of quantum electrodynamics. And so we come to the present times, in which many physicists would willingly get rid of the hindrance of physical reality to reduce the universe to a giant computer in permanent digital flickering of ones and zeros. Matter doesn't matter, only information.

But perceptible matter, which tells us that there are soft and hard objects, belongs to continuum mechanics, and there can only be one Continuum by definition. There cannot be a material continuum on the one hand, and a continuum of motion, operations, or intelligence on the other. The Continuum always refers us to the possible evolutions of a primitively homogeneous medium in which distinctions are impossible and if possible are always compensated. Without homogeneity there is also no conservation of momentum, this being, more than force, the authentic basis of modern physics. Conservation is the essence of what we understand by physical reality —continuity- while actions or operations will always be secondary.

Fundamental physics and artificial intelligence have much more than accidental contact, and not only on the side of discrete operations. The mathematics that underlies deep learning, even with more dimensions, is as familiar to physics as linear algebra and its vectors, matrices and tensors —the heart of the continuum mechanics that governs the constitutive law of materials science, classical electrodynamics or Relativity.

It would be wrong to say that quantum mechanics makes space or time discrete; there is nothing like that. The only discreet thing in QM is action, something for which we still have no due justification today. To prove that time and space maintain their continuity at a microscopic level, one only has to see that the electrons also describe elliptical orbits as planets, simply a class of the conical sections we mentioned.

If there are any doubts about the connection between physics and AI, it will suffice to remember that physicists Mehta and Schwab demonstrated in 2014 that an algorithm of deep neural networks for image recognition works exactly the same as the renormalization group of quantum field theories such as QED that has later spread to other areas, from phase transitions to cosmology to fluid mechanics.

A neural network confronted with the critical point of phase transition in the model of a magnet, where the system becomes fractal, automatically applied the renormalization algorithms to identify the process. Ilya Nemenman, one more theoretical physicist turned to adaptive biology and biophysics, has not resisted saying: "Extracting relevant features in the context of statistical physics and extracting relevant features in the context of deep learning are not just similar words, they are one and the same."

THE ADMINISTRATION OF A POWER

Delegating problems to AI not only has a risk but a price for social engineering, as its solutions, as in all areas, are new problems, but problems even further away from human intuition and its immediate environment. It is falling further and further behind the indecipherable inferences of increasingly complex and therefore more opaque systems. A physician can no longer know how an intelligent system has reached its conclusions about patients, even if it has contributed to its design or database. As the opacity index increases, so does dependency.

There is no need to imagine the chaos of a digital inferno to realize that this is undesirable. The solutions themselves are becoming more and more complex, not so much because of the increase in details, but because of the loss of intelligibility. Only this intelligibility makes details nuances instead of unpredictable decision making. So they talk now of explainable or trustworthy AI.

We have already had time to see to what extent the indiscriminate application of instrumental intelligence leads us fully to the unintelligible and the unnameable.

That is, in a complex natural process there is already an implicit presence of intelligence; we could say that it is just equilibrium, but, given that in an organism it also has to balance and encompass the inveterate partiality of our mind, the invisible half of its equation, we could still consider it intelligent in a more fundamental sense than our own thought.

In any case, it does not take a great deal of knowledge to understand that we are not computers and the intelligence does not arise from information processing; rather, it is there where it is often invested and buried. Materiality, the physical character of intelligence resides at the opposite end of the operational, but this end is never really separate from the continuum and resists quantification.

INTELLIGENCE AND PURPOSE

Intelligence is precisely capturing and estimating nuances without the need for calculations; on the contrary, our calculations can estimate thresholds but they do not have an intrinsic sense of the degree, of the nuance, since the symbolic as such is not part of the physical process. And yet it is evident that an infinite number of mixed digital/analogical systems can be built, and that innumerable forms of integration are possible, something which connects with the evolution of human/machine interfaces.

There are no calculations in intelligence, as there are no calculations in nature. As Fresnel said, nature does not care about analytical difficulties. What would not be our awe at the world if for a moment we realized that nothing we see depends on calculus, no matter how much it may later coincide with it. Computation is the routine in which our mind is immersed.

With the scientific revolution we thought we would definitively pass from a world with a finality ordered by a teleological intelligence to a world merely intelligible in terms of a calculus description; but it is here where the mirage flourishes in all its splendor. Calculus does not liberate us from teleology, on the contrary calculus, as an heuristic tool, cannot be more oriented to a purpose. And in addition, it is the heuristics of calculus itself that has led us to the heights of the incomprehensible.

This is evident in Kepler's problem. Despite what all the books say, Newton's law of gravity does not explain the shape of the ellipse with the central body in one of the foci. One starts from the integral of the curve —the global form- to derive the instantaneous velocities, but the vectors never cancel were not mixed up orbital velocity and innate motion. There is therefore no local conservation of contemporary forces, but a global one, and that is why it is customary to work with the principle of integral action, the Lagrangian.

All quantum mechanics is based on a principle of action, and quantum field theories more specifically in a Lagrangian one. Any principle of action, be it Lagrangian or not, is integral by definition and therefore the derivatives applied to it are implicitly subordinated to a purpose, as Planck himself and many others have not ceased to admit.

It is evident that procedure as artificially and blatantly ad hoc as the renormalization of QED cannot really take place in nature, or else we would have to think that every photon, every electron, and every particle performs endless series of calculations at every instant. And do we call this "blind forces"?

The easiest answer is to say that physics is increasingly statistical, and statistics itself must encompass in principle all conceivable functions of AI. Although this may be true in the uttermost general sense, no single type of statistics is going to lead us to the order we appreciate in nature or in intelligent behavior —neither in application nor in interpretation, neither in the universal nor in the particular. The statistical bias of the observed behaviors is too improbable.

"For each clump, a bubble": In a medium totally homogeneous in principle any increase in density at one point would correspond to a decrease at another, and the same is true for motion, energy and other quantities. Differentiations and motions would take place in time, while the underlying continuity before being altered would remain neither inside nor outside, but enveloping it. In this sense the continuum is neither at the beginning nor at the end, but in the midst of all changes, since it permanently compensates them. The first intelligence, the first undifferentiated mode of which we have spoken, would coincide with the immediacy of this continuum.

However, the continuum of so-called "real numbers" that seems to prevail in physics cannot but be an idealization, since in the world we can only count discrete things and events, in whole numbers and at most in rational numbers or fractions.

Calculus cannot fail to draw increasing complexities in its vision of nature, but all these arabesques do not belong to it in property, but are rather the image it shows on the surface of the mirror that we hold.

FROM RENORMALIZATION TO SCALE RELATIVITY

Despite being created by man, pushed to the limit, these multi-layered neural networks and their response are black boxes nearly as enigmatic as many of nature's behaviors; though surely you can't beat nature's complexity.

In Feynman's integral path approach, in continuity with Huygens universal principle of propagation, trajectories become more irregular on a small scale and between two points particles have potentially an infinity of undifferentiated trajectories. This feature and considering point particles instead of extended ones are the main sources of infinities to be cancelled by renormalization calculation procedures.

It has been asked why renormalization is capable of producing pattern recognition if these objects, for example faces in a landscape, are not fractal or subject to recurrences at different scales. But that is to think from the point of view of the object considered rather than from the subject and the process or transformation that may take place in order to arrive at recognition. Renormalization already implies to a large extent the notions of relevance, selection and elimination that other theories such as biology inspired adaptative models or the information bottleneck theory use to explain these black boxes.

Something similar to renormalization can be applied to the same space-time obtaining an anomalous dimension. This is what the astrophysicist Laurent Nottale did in 1992, proposing Scale Relativity. This was recognized at the time as a brilliant idea but the inclusion in the physical continuum of non-differentiable manifolds such as fractals represents an huge leap into the void that largely questions its feasibility.

The Heisenberg uncertainty relations involve a transition of the spatial coordinates of the particle to a fractal dimension 2 around the length of de Broglie, which Nottale extends to the temporal coordinates for de Broglie time. The renormalizable quantum field theories themselves show a dependence on the scale of energy that enters for the first time explicitly into the equations of the current and limited standard model. What Nottale does is to extend this asymptotic evolution and take it to the rank of a general principle.

We already know that physics cannot be reduced just to motion and extension, but what else can it be is something that is not even remotely clear. If the theory of relativity demands that no coordinate system be privileged for motion, scale relativity demands the same for scale and its resolution, and with it for other associated quantities such as density. If there is any way to move from the extensive to the intensive without letting go of the thread of motion/extension that has shaped the entire physics, surely there is no other more natural and inevitable than this one. Between material points and material particles we would need a resolution tensor to "fill" portions of space.

As it is known, the invariance of the speed of light regardless of the speed of the observer can only be made compatible with the classical equations of motion by means of the additional dimension in the Minkowski space-time continuum; otherwise Special Relativity, a theory focused on local conservation, could only present point events cut out and split from the classical electrodynamic continuum.

I'll try to figure out this principle in the most elementary way. Since Pearson it has been said that an observer traveling at the speed of light would not perceive any motion being in an "eternal now". Within the specifically kinematic framework of relativity, this is the only way you can open a window to something beyond motion. But what would this eye see in its endless approach to the absolute limit? As the motion slows down, his vision would logarithmically zoom into the scale of space and time. The asymptotic approach should admit its own transformation: the change of resolution by covariance that Nottale demands, the laws of motion being replaced at high energies by the laws of scale.

In Nottale's words, quantum field theories, with standard renormalization procedures, "correspond rather to a Galilean version of the theory of scale relativity", and only work within certain limits. For Nottale, renormalization is a semi-group, since it allows the integration of larger scales from smaller ones, while the successful application of scale relativity would allow the inverse operation, obtaining the smaller ones from the larger ones.

The scale recursion also made him think of the application of the principle to complex systems in general, from biological to social organizations, whose functioning depends simultaneously on different hierarchical levels or scales.

Nicolae Mazilu and Maricel Agop have recently undertaken a thorough evaluation of Nottale´s principle in the light of the history of physics and mathematics. The introduction to this work states nothing less that "once the principle of scale invariance is adopted, there is no other way to follow but the right way". And indeed, if the speed of light in the continuum and the Planck length are invariant, unreachable and absolute, the scale has to be relative almost by definition, and by the same token non-differentiable. We don't see how this conclusion can be avoided.

Apart from this, scale relativity is an inherent necessity from the statistical point of view and the theory of measurement that must accompany it.

Fractal geometry can then be seen as a way of mediating between the continuous and the discrete within the above mentioned limits. And this leads us to recursiveness and the confrontation of algorithms both with the laws of nature and the description of their contours.

Of course scale relativity has its own big problems with calculus, but even before that it also requires, as Mazilu and Agop emphasize, a physical interpretation of variables in a well-defined technical sense, quite the same than that already formulated by C. G. Darwin in 1927 —"the translation of the formal mathematical solution, which is in wave form, into terms of particles".

The thread to get out of this labyrinth is lost with Born's statistical interpretation, and that is why the two Romanian physicists undertake an authentic re-foundation of the theory that goes through the historical milestones of the Kepler problem, Hertz's mechanics, de Broglie's wave-corpuscle duality, Schrödinger's wave function and his color theory, Madelung's equivalent hydrodynamic equation, Cartan or Berry, among others.

The distinction between material particle and material point, or extended and point particle, its inevitable application to wave mechanics, the interpretation of the geometric phase in quantum potential and the return of the holographic principle to its original context in light mark out this meticulous and logical reconstruction in the spirit of classical physics.

In fact Agop and Mazilu's reading of scale relativity offers not only a plausible explanation for the dilemmas of quantum mechanics, but also for the occurrence of fractal structures in nature, which in no way can be sustained by mathematics alone.

Fractality on scales is not reduced to the mere spatial recurrence of Russian dolls with which we are most familiar, but affects the geometry of space-time itself. In my opinion, just as we speak of different modalities of renormalization, of positions in real space and the space of moments, with relativity of scale, where coordinates are functions instead of numbers, we could ultimately speak of three fundamental modalities: scale for mass and its density, scale for motion (moment, force or energy), and scale of length that correspond well with our notions of time, space and causality, or matter, space and time; in addition to some generalized coordinates to define the transformations of these three aspects.

In any case, the physical interpretation as Mazilu and Agop understand it, with its coordinate transformations, connects well with the idea of representation in deep learning —the coordinates in which the data are configured. This notion of representation, also based on a mathematics of multidimensional tensors —the resolution in scale relativity being equally a tensor- makes the difference in this emerging group of methods also known as hierarchical learning, with different levels of representation corresponding to the hierarchy of conceptual levels of abstraction. And a fractal is essentially a cascading hierarchy of scales.

The more we use statistics, the more we need interpretation. This is the unquestionable conclusion reached both in physics and in automatic learning methods. Moreover, in this machine learning, the more statistics are used, more profoundly these are involved in physical parameters if they want to define its object.

This unexpected turn occurs precisely now when so many physicists would like to give up interpretation and even principles, to get by on computation or prediction alone. And this would be, leaving aside other considerations, a basic reason for the disdain shown towards this and other theories that demand more space for interpretation in modern physics. An interpretation that lies naturally at the end and is indispensable in any application.

Simply put, in every human enterprise there are principles, means and ends. In physics calculus is also the means and interpretation the end, but this seems to have been forgotten since the advent of the algorithmic style inaugurated by quantum electrodynamics. Now Artificial Intelligence, expressly teleological and focused on the objectives, gives us back its relevance. Only that we know perfectly well that neither the human brain works by means of symbolic representations, nor does nature, nor the particles, nor the waves.

Applied human intelligence and its mechanical replicas have to be purposive by definition. But to pretend that nature is oriented to ends through discrete operations is the height of absurdity, and Aristotle himself would have been the last one to affirm this. Thus, this "first intelligence" cannot be goal-oriented, but must be the reference for the intelligence applied to tasks.

This background intelligence or consciousness cannot be anything other than the invariance of the continuum, the homogeneity of reference on which events and things are drawn, and it would have to be identical in nature and in man, no matter how much each entity can extract a different output depending on its circumstance and internal structure.

Since this primary intelligence is indiscernible and simple, anyone would say that its usefulness is null. But is the global feature of realization useless? What happens is that it is neither quantifiable nor measurable. But to say that there are really two intelligences, one useful and the other useless, instead of two modes, would be total nonsense, as it would be to say that there is no continuity, when the discrete numerical analysis is condemned to follow behind its "ideal" pattern, or that this continuum is only physical, when attributes cannot be applied to the continuum itself.

It is not necessary to stop at philosophical analysis to see that intelligence is always going to be more than the ability to do tasks, and that something else is always indispensable for utilitarian and applied intelligence —just as the continuum cannot be dispensed with either in special, general, or scale relativity.

How much truth can there be in the idea of scale relativity? There aren't yet enough concrete results to judge it. But we can be sure of one thing: the requirement of continuity is absolutely necessary in nature and even in our generic sense of reality; the requirement of differentiability is not. This is man's exclusive business. So this principle will always have something new to contribute.

Let us not forget that Nottale is an astrophysicist in search of a fundamental idea; his starting point is not simple principles, but de facto complexity in constellations of data and observables. His proposal must be understood as a return from the interpretation to the principles, a turn that Mazilu and Agop justify with much more specific arguments.

On the other hand, this is only one among a greater number of theories that include scale factors —such as the gauge theory of gravity elaborated with geometric calculus in a flat space- that are differentiable and that it would be necessary to compare. This gauge theory demonstrates that Poincaré's program to elaborate a relativistic theory in ordinary Euclidean space by modifying the laws of optics was perfectly justified. In any case what remains is that in the absence of an absolute criterion the scale can only be relative and based on comparison.

You can also look for more elementary models; follow, for example, Weber's line of relational mechanics, the grandmother of relativity, and deal with questions of scale as you do with the forces, retarded potentials and speed of light in that theory, returning to a flat and differentiable space and delimiting recursive functions via self-interaction, that "silly" idea in Feynman's words. But what is a self-interaction? It is that silly thing that seems to do an isolated particle when it turns out that it's not isolated at all. And the same goes for our brain, although, in this case, on the contrary, the problem is that it seems too intelligent to us.

Critics say that scale relativity is not a theory, and nothing could be more true if you keep in mind quantum field theories. No, scale relativity is first of all a principle and a project, and so much the better if it is not a theory. Besides, wasn't a guiding principle so desperately sought in fundamental physics? Something merely intelligible. The theories mentioned are shielded to underpin the consistency of the calculations, and are like small isles in the sea; they have had their historical moment and contain enduring teachings but basically belong to a past that will not return. They often serve only as Procrustes' bed to force the explanation of phenomena that are not at all understood but are "predicted" from its data.

You don't open a black box with another closed box.

Today it's quite fortunate and a great advantage not to be one of these theories that subordinates principles and interpretation to calculus instead of making it an instrument. In the new environment computations are increasingly left to the machines and to the same extent one separates from it. Consciousness, if something, is distance; so it is not impossible for AI to help us recover the most underestimated functions of our natural intelligence.

Nottale, Mazilu and Agop have all insisted that scale relativity is not only a general principle of physics but also of knowledge. This claims, which may sound excessive, can now be tested. In the arena of AI the question is whether it can be "discovered" by machines confronted with appropriate models just as it has occurred with renormalization. It will not be without experimental base since the emerging technological environment works at ever smaller scales where all this has many ways of being relevant. And, conversely, whether it helps to develop more efficient or more intelligent algorithms.

No particle accelerators are required, and a huge experimental front is accessible and available. It is totally false that new physics can only be found at high energies, the very controversial boundary between classical and quantum behavior has an huge continuos range of different scales depending on the case.

There is no spearhead like the IA for the current multi-specialist in our tower of Babel, nor can we imagine a better one in the foreseeable future. Physicists, mathematicians, programmers, developers, biologists, linguists, psychologists,... Here there is room to quickly explore possibilities that the different specialties, with their great inertia and defensive barriers, are not willing to consider. Ideas, if we speak of physics, such as scale relativity, retrodictive determinism, and many others that for various reasons go against the course adopted by the specialty. It is no coincidence that a good number of AI experts are physicist moving now into an area so enthusiastically open and prone to stochastic tinkering.

Too enthusiastically, if one thinks that today the mathematics that characterize a physical system can end up defining the control vectors for the behavior of individuals, social groups and human populations. These things are already routinely applied for the modulation of consumption or opinion. Like resolution, attention itself can also be described with tensors —and attention is as close as we can get to consciousness itself. As if it were something accidental, people speak of the "perverse effects" of technologies, when it comes to the most deliberate intentions and purposes.

If they want to put us all in one box, let's make this one as big as the whole universe.

THE CIRCLE OF UNDERSTANDING AND THE THREAD IN THE DIGITAL LABYRINTH

And this way, very much to our regret, it seems that a circle closes. In the struggle between society and nature, we have made of nature an object and we end up subjected to the same treatment, and that with which we shape one ends up sculpting the other. That's the same old and now only becomes more formalized, but being executed consciously we have no excuses.

However, and simultaneously, another circle is being closed by linking more and more tightly categories so far separated as the physics of space and the physics of matter, the extremes that above and below define our human scenario. It depends on their understanding that physics, and not only physics, has access to real intelligibility and a universality that would somehow compensate for the lack of scruples of applied knowledge. We can only reject the latter, but not annul it.

By restoring their inevitable extension to material particles, we make it possible for "external space to enter internal space". Here the old saying that not only the drop melts into the ocean, but also the ocean melts into the drop, comes back to mind. And this happens at the same time that, in another order of things, the social inundates the individual and data floods theories.

If it can be said that we are the fundamental scale, the same could be said of any other being on any other scale; but in any case the world cannot fail to be seen from within. This does not depend on the number of dimensions but on the nature of the continuum.

Not only the subject does not cease to be indispensable in this whole process, but the very substance that supports everything has to be revealed as subject —for the same reason that it turns out that intelligence cannot simply be "inside" the brain. Surely the universe is not a giant computer but computers will allow us to navigate much more freely through its infinite space of functions. And they will do so because the incentive will have changed and we will have devolve calculus to its proper place between principles and interpretation; for it is with these that the circle of understanding closes.

Physics began with clear concepts such as inertia, force, mass, space or time to finally realize that all of them become more and more mysterious; and conversely, in neural networks everything begins as an enigma and the hope is that at some point the processes will be clarified. Naturally, the basic objective of automatic learning is to find the input functions f(X) for the output data Y. This type of reverse engineering has always existed in physics and is the essential part that makes it predictive. And since in both cases a certain type of mathematical language is shared, the two approaches are highly complementary.

In physics we move from principles to the interpretation of phenomena and data through calculus. In a more global approach, this would be only half the circle, the semicircle of the quantitative/analytical ordering so typical of physics. The second half of the circle would be a return to principles in qualitative and continuity terms, ideally ignoring calculus as much as possible —if we have been able to assume that nature and intelligence exist entirely independently of it. To my mind, this is where the challenge lies now.

A continuous and qualitative description, with permission of space-time algebra, would have to be necessarily more ambiguous than one specified by the calculus. But this is precisely where intelligence resides, in finding significant nuances without the need for or with fewer specifications. This paradox is insoluble, but it shows us the way back to internalize, that not to humanize, knowledge —as it is the high-level formalized knowledge or the same machines that are a particular human end.

CONCLUDING REMARKS

Our "two Intelligences" are very similar to the two trees of Paradise in the fable; we have not stopped grabbing and handling the Tree of the Science of Good and Evil that has brought us here; but the Tree of Life is still there, intact as ever.

It's clear that primary intelligence is more than secondary or applied intelligence, as well as that the latter is more than pattern recognition or the list of tasks dealt with by hierarchical learning.

We agree with Epstein that the full understanding of the human brain may still be, if at all, centuries away. There are also in its development too many "useless" things from the utilitarian point of view, that are anyhow necessary to maintain, way beyond the whims of memory, the continuity from which our identity derives.

And even if we should ever get to know our brain, with its complex relationship to the rest of the body and the environment, we would still know nothing of consciousness as such, we would only know of the discontinuous emergence of thoughts against that unqualified background which realizes them.

The continuum of ordinary human experience and its body has little to do with that of fundamental physics, the former seems to us very concrete and the latter too abstract. However, fundamental physics seems so abstract to us precisely because it lost the thread of continuity with the classical physics that Mazilu and Agop are trying to re-establish. Even at very different levels, the two kinds of continuum remain in the same direction.

AI systems are expressly oriented to a purpose; and the same is true of physical theories, so blatantly oriented to prediction. Since both use statistics massively, from this point of view there is nothing to surprise. What is surprising, and never sufficiently pondered, is that nature achieves without purpose what we would not be able to do without it, and this, by way of continuity, leads us to the supposition of a first intelligence that is its reference just as the speed of light is the reference for motion and its radiation is the reference for the quantum of action.

One could quietly forget about all physics and scale problems, that an idea remains intact: the source of intelligence and nature is alien to computation/calculus/finality. With precision, a notorious surveyor left said, "the spirit is freed only when it ceases to be a support".

McBeath hit the ball farther away than anyone, even if it had gone outside the ballpark.

References

R. Epstein,(2016) *The empty brain*, Aeon

L. Nottale, (1992) *The Theory of Scale Relativity*

N. Mazilu, M. Agop, (2018) *The Mathematical Principles of the Scale Relativity Physics — I. History and Physics*

N. Mazilu, M. Agop, *Role of surface gauging in extended particle interactions: The case for spin*

N. Mazilu, M. Agop *Skyrmions: A Great Finishing Touch to Classical Newtonian Philosophy*

P. Mehta, D.J. Schwab, *An exact mapping between the Variational Renormalization Group and Deep Learning*

N. Wolchover, *A Common Logic to Seeing Cats and Cosmos*, Quanta magazine

McBeath, M. K., Shaffer, D. M., & Kaiser, M. K. (1995) *How baseball outfielders determine where to run to catch fly balls*, Science, 268(5210), 569-573.

Fink P.W, Foo P.S, Warren WH, (2009) *Catching fly balls in virtual reality: a critical test of the outfielder problem*

D. Hestenes, *Gauge Theory Gravity with Geometric Calculus*

M. A. M. Iradier, *Self-energy and Self-interaction*

M. A. M. Iradier, *The multi-specialist and the tower of Babel*

F. Kafka, *The Zürau aphorisms*

TOWARDS A SCIENCE OF HEALTH? BIOPHYSICS AND BIOMECHANICS

10 enero, 2019

HOLISM AND REDUCTIONISM IN THE LIFE SCIENCES

Physics, so solidly supported by mathematics, continues to be considered the queen mother of experimental sciences and therefore a model for all the others, including modern biology and biomedicine. It is also common to assume that physics is a reductionist science capable of explaining things by its mechanisms, although this is far from being true and in fact, as Poincaré recalled, the principles of action on which it is based by its very nature do not admit univocal mechanisms but mathematically precise analogies.

This underlying ambiguity of Physics is the cause of all sorts of confusion, all the more so when she comes to help other disciplines inextricably related to complexity trying to sketch its simplest aspects.

Surely one of these misunderstandings is that physics can only be relevant to living systems at the level of their constituent blocks —ions, molecules, cells, etc.- through their corresponding disciplines, such as chemical physics or molecular biology. However, we have just noticed that the atoms, particles and their associated fields are calibrated by their respective principles of action, which are their authentic invariant elements. These action principles are integral by definition, no matter how much physics, in its predictive aims, continuously uses them as a mere excuse to derive local solutions.

As always, it is what we do which determines what we think, just as it is the application what determines the interpretation.

The origin of biophysics, and even of psychophysics, can be traced back to the Leipzig and Berlin schools in Helmholtz's days. Already in the first half of the twentieth century, and long before the blossoming of this vast field, there were pioneers such as Lotka the mathematician and the great promoter of theoretical biology Nicolas Rashevsky. A lot has happened in the meantime, but we can still take these two names as opposing ways of understanding the discipline: either clinging to biometric data or speculating on the mathematical relationships one can develop from them.

Where are we today? The fact that today's biophysics is more reminiscent of the Schrödinger of What's Life? than of this two mentioned authors is already quite eloquent. It speaks to us, independently of the merits of the great Austrian physicist, of the predominance of molecular biology and the constructivist ideas of building blocks.

In spite of the present hegemony of molecular biology, biophysics resists to be reduced to a single perspective, it's too vast and multiform for that. And now that computer science and artificial intelligence are expanding their power of statistical correlation exponentially, many theoretical physicists are moving into this field in the hope of giving it a universality it now lacks. They would be called to save the great fault that already separated Rashevsky from Lotka and that has not stopped getting bigger, as it requires more than computing or statistics to close it.

Faced with this arrival of fresh air and young talents, little can an old spectator like me say except to greet them. However, I cannot resist making some totally extemporaneous observations at this interesting junction; and it is clear that only as extemporaneous could contribute something, even if only the distance.

As much as it may surprise many, the idea held here is that for biophysics to be more holistic or global what it needs is not principles with increasing degrees of abstraction but, on the contrary, to be more faithful to what the foundations of biomechanics or even plain mechanics, understood as continuum mechanics, imply. There's almost nothing trivial about this last one, to the point that we can suppose that it has almost everything that biophysics can need and then some.

And since at present it is so advisable to proceed by contrast, we are going to touch on a case that is completely secondary in modern biology but that in another era had great significance.

A HIDDEN SYMMETRY

Discovered for science by Richard Kayser in 1895, the nasal cycle alternating the respiratory flow remains an enigma despite the abundant literature on the subject. Long before that, various ancient yoga texts implicitly assumed its importance for the general regulation of organic and mental functions. The basic idea held by yoga, of a hidden axis of symmetry in the organism that gives rise to a «sun/moon» alternating current of activation/relaxation even admits an updated version in consonance with our present knowledge of the dual organization of our organism and the autonomic nervous system (ANS) in particular with its sympathetic and parasympathetic subsystems, sensory and motor signals, anabolism and catabolism, and even the two cerebral hemispheres.

Since there are already many studies devoted to the physiology and biochemistry of this unexplained phenomenon, and despite their extensive data collection work they do not provide conclusive evidence or shed any clear light on its nature, I would like to briefly propose here a more basic approach to the subject, which nevertheless connects better with the physical and mathematical treatment of biological signals.

Circulation —or alternation- exists when two conditions cannot be satisfied simultaneously. This elementary truth seems equally valid for all types of systems, whether physical, chemical, biological, or socioeconomic. Life is a situation out of equilibrium, an imbalance that anyway has to find some partial equilibrium with its environment to have some sort of stability. Without imbalance there would be no differentiation, and without some kind of equilibrium there would be no persistence or duration.

But this equilibrium cannot be based only on the boundary conditions between the interior and exterior of a system, on its bordering surface, since in such a case the very limit would be the attractor and would prevent the development and differentiation of a relatively autonomous internal territory. Thus, the balance between the interior and the exterior tends to reproduce itself in the interior as a «center», even when in reality it supposes the virtual cancellation of both.

A good example of this dynamics is the organization of the cell itself, which has had to begin with a lipid membrane or filter that divides an inner enclosure from the outside —the «basic language» – to gradually allow the folding, aggregation and recombination of more complex molecules that would eventually lead to globular proteins, enzymes and nucleic acids capable of replication thanks to the action of the former ones —these macromolecules being the «high-level language».

Seen in this elementary way, the imbalance needs to rest in order to survive. And in reality, if we stop to think about it, the DNA or biological inheritance corresponds to a passive phase of stability or rest that would be useless without the enzymes that activate and modulate them in very variable ways depending on the medium in each occasion of expression.

In a very different order of things, something similar has to happen with the alternation of the flow of air through the two nostrils, no matter how much this cycle varies from individual to individual and even in the same subject over time —without that variability it would be difficult to act as a buffer in the midst of changing circumstances. The suspicion that the predominance by one or the other nostril has to correspond to a predominance of the sympathetic or parasympathetic, of active response or of rest in the ANS, seems simply reasonable and is to a large extent supported by numerous studies.

This, however, is still a relatively trivial alternation, although not for us in the most ordinary circumstances —just as we don't find trivial, either biologically or psychologically, the difference between whether something is inside or outside our body. In phase shifts or in dreamless sleep the flow has been balanced and in such circumstances the powerful inertia of involuntary biology would be somewhat suspended. It is a pity that we cannot realize these privileged moments of our daily agenda and our biological calendar, because if there were really some kind of biorhythm, it could not be based on fixed numbers but on verifiable and measurable real time fluctuations.

Ironically, the originator of the rightly forgotten theory of biorhythms was the famous Berlin otolaryngologist Wilhelm Fliess, praised by his friend Freud as «the Kepler of biology», who , just a few years before Kayser made his little-publicized discovery, recommended the extraction of the nasal turbinates through which this respiratory cycle manifests itself.

What this virtual axis of symmetry, this shushumna or invariable middle between both phases would designate, is a center that is center because it is neither in the interior nor the exterior, but is rather a measure of reference for both ends.

And what is implicitly stated for the autonomic nervous system could also be applied, via analogy, to the central nervous system and its afferent and efferent impulses, the pathways of perception and action. Also here we tend to counterpose a thinking subject and a thought object, when it is obvious that the thinking subject does not cease to be one more object of thought interposed by the general activity of thinking, which would be the immanent logos from which our logic would have fallen apart.

BIOPHYSICS AND BIOMECHANICS

Biophysics is currently experiencing a great moment as its branches are constantly dividing and connecting; no two researchers have exactly the same perception of the discipline, but few doubt that the light that physics can shed on complex biological systems must be at the level of applied physics rather than fundamental physics. However, the same basic physics is inseparable from statistics, an essential signature of what is understood by applied physics. We would like to show here how basic physics, duly considered, has important things to say about the most global aspects of life.

Returning to the nasal cycle, since Kayser it falls by his own weight that the whole question is a matter of tone, and vital tone we could say, since he himself defined it as «the alternation of the vasomotor tone throughout the periphery on both sides of the body». The local mechanism involves changes in the sympathetic tone of the erectile venous tissue of the nasal mucosa.

Tone is something characteristic of the vascular system, muscles, skin and vagus nerve, so important in its overall regulatory effect that it has even been called the «nerve of life». This vagus nerve in turn influences vasomotor tone and blood pressure, musculature and skin —in addition to many other aspects such as breathing, the heartbeat, the functioning of the abdominal viscera, emotions, and a long etcetera.

The same vagus nerve is double and bilateral and transmits sensory and motor signals; although classified within the parasympathetic system, it can be said to act as a mediator between the sympathetic and parasympathetic phases, and is as fundamental to the autonomic system as the spinal cord is to the central nervous system. However, there is no direct and unequivocal way to measure vagal tone, only indirect methods derived from cardiac variability and the arrhythmia of the elemental respiratory cycle.

The most reliable and tangible way to measure a signature as eminently global as the vital tone of the organism is through the blood pulse in the radial artery, as has been done in different physicians and cultures for thousands of years. This previously involved an empirical semiology and a medium, the touch, with a strong subjective component, although modern techniques such as applanation tonometry make a precise and automated measurement perfectly accessible. Naturally we are not talking here about the pulse as just an index of the circulatory system, but about the individual physiology as a whole.

A drop maintains its particularity thanks to its surface tension that marks the physical condition of the boundary with its environment, and in the same way can be characterized the individual tone of a cell and even a higher organism such as the human being. The whole difference lies in the great complexity of the exchange with the medium of an organism, but, even so, breathing and its continuation in circulation should be sufficient to define this tone in what's most characteristic.

There is a tendency to think that when passing from the tension of a drop to the biological signals of the human body the increase in complexity is such that the most fundamental mechanical aspects become too diluted, but the facts and equations disprove this prejudice. On the other hand, one also tends to think that physics deals with purely local or reductionist aspects when all gauge theories since Maxwell have arisen from an integral approach from which local aspects are later deduced.

The case of the classical Maxwell equations for electromagnetism is paradigmatic, because initially it was a theory expressed in integral equations and entirely based on the notions of flow and circulation, with the Faraday flow tube as the basic element. It describes a system in a phenomenological and organic way, from outside to inside, including the constitutive relations that define permittivity and permeability, the respective tension and deformation of the electric and magnetic field that refer us directly to the mechanical properties of materials. There is therefore a set of basic symmetries and a metric that limits them through material properties attributed to the space in which they transform.

In short, Maxwell's electrodynamics is but a continuation of the hydrodynamics initiated by Daniel Bernoulli by studying precisely the mechanical values of blood circulation and later by Euler; in its formal aspects hydrodynamics was indeed the first «field theory» ever, and from it we can pass to Laplace's equation, Poisson's equation, Maxwell's second equation, Poiseuille's equation for laminar flow and Ohm's law, which still is used as the base for hemodynamics.

The fall of the laminar flow in the Poiseuille equation gives rise to a first-kind Bessel function that, just as it can be observed in the membrane of a drum, reflects the cross-section vibration modes. Here, the acoustic and tone elements, never lost in the logic of continuity («with tangential tension, the solid deforms, and the liquid flows»), clearly emerge again.

The pulsology of some traditions, such as Chinese medicine, tried to represent this cross section in its classification of characteristic pulses. If we consider the cycle of nasal alternation in terms of gradient or quantity of flow, it is worth asking if this variation is also translated into quantitative terms to the pulse in the radial artery of the right and left wrist, and even if it is permissible to superimpose them in an interference pattern or to get an average. The function can also be extrapolated to spherical coordinates if we look for a three-dimensional expression.

A traditional representation of the three axes of the human body contemplates a continuous polarity and a correspondence between the left and right sides, the lower and upper part, and the anterior and posterior face; the ends of the three axes being an unfolding in the space of a fundamental duality activity/receptivity or activation/rest. In mechanical terms we can generalize them as tension/deformation or force/adaptation. Can we move from qualitative correspondence to statistical correlation?

THE GEOMETRIC PHASE AND THE BILATERAL RESPIRATORY CYCLE

The geometrical phase is better known as the Berry phase in quantum mechanics but it exists equally in classical electrodynamics and even has exact analogues in waves on the surface of water.

This forces one to think that there is nothing special about quantum potentials, and only by separating the particle from the field could one think otherwise. The geometrical phase, «global change without local change», can only be conceived in terms of the continuum; only the interference patterns of wave systems show this kind of displacement.

After Berry showed the simplest case for cyclic adiabatic processes, there have been generalizations for non-adiabatic, non-cyclic, dissipative or non-conservative processes. Independent of the dynamic aspects of the systems, the geometric phase has demonstrated a profound universality which in spite of everything, and precisely because it's not linked to the forces, always remains in the background.

This «anholonomy» is also not alien to living beings and their locomotion. Inevitable the example of the revolving falling cat —the «Maxwell cat»- surely the happiest example of how a living being protects its invariance against the contrary efforts; or the no less universal figure of the movement of the snakes. Shapere and Wilczek showed their relevance in cellular self-propulsion in viscous fluids by describing a circuit of forms with arbitrary infinitesimal deformations in cylinders and spheres.

We could call this «Maxwell cat», the global contortion of the system, «the fifth equation of electromagnetism» —Maxwell's fifth, of course- although it is understood that this additional curvature does not add anything to its dynamic profile, but only to its global configuration. The status of the geometric phase is ambiguous and it is not surprising that physicists do not miss the chance to affirm that it does not alter quantum mechanics or classical electromagnetism in any way. That's as far as it goes!

The only problem is if these other «fundamental laws» are only emerging ones, the local application of an intrinsically global theory —which is certainly already the case with the first gauge theory, precisely Maxwell's one. Then the geometric phase would be a slipknot with which the continuum holds the system. The anholonomy from the point of view of calculation and integrability would be an holonomy from the point of view of mere continuity.

There has also been an abuse of the «non-local» word for quantum potentials in Bohm's interpretation. In the cases of geometric phase in classical electrodynamics and hydrodynamics the only thing we have is the definite restriction of the global configuration of the system. This is something very different from the total vagueness of the expression «non-local». In such cases, ignoring classical precedents, those advocating non-local interpretations put things too easy to dismiss for the vast majority always inclined towards local application.

Shapere's and Wilczek's demonstration rests entirely on the application of the arguments of the gauge fields of fundamental physics —fields in which the Lagrangian is the invariant calibre- to the deformation of bodies. This approach of a circuit of forms or deformations is suitable to follow the evolution of the volumetry and mechanical forces in the expansion and contraction of the lungs. The analogy is established between electrodynamics and the theory of elasticity, which is known to overlap easily. Let us explain this a little.

Anyone who observes his breathing can see in the most direct way that there is a correspondence between the depth of breathing and the clearing of the nasal passages. Deep abdominal breathing sustained for a time tends to eliminate bilateral alternation; on the other hand, higher or shallower breathing accentuates the alternation and blockage of one of the pathways.

Since increased voluntary activity tends to accentuate high or shallow breathing, it can be assumed that the ANS tends to compensate for these imbalances; which in turn means that lateral alternation also compensates in some way for the height and depth of breathing considered on the vertical axis. As for the anterior-posterior axis, apart from the difference in the filling of the lung volume when done from the base, we have already mentioned the polarity between the spinal cord and the vagus nerve in terms of the two central and autonomous nervous systems.

Naturally, all this takes place in a context of continuous transformations of which there can only be degrees of statistical evidence. However, the idea of calibre or gauge as already presented in classical electrodynamics is perfectly adapted to describe this, including of course the connection with acoustics.

In the variations of respiration and lungs movement, from deeper to shallower, we not only have changes in volume and shape, but also in the filling of that volume, i.e. air pressure and surface tension of the lungs. There is a correlation and possibly also a pressure-tension/deformation/volume circuit. All these are eminently mechanical issues.

We also have this primary presence of mechanics at the level of cells and even macromolecules such as DNA, without there being any need here to raise hierarchical mechanisms of subordination between different levels, since we are talking about problems inherent to the very continuum mechanics and therefore, to continuity. Let's take a closer look at this.

The problem of the chain of command and control in biology is far from being solved because it is supposed to act fundamentally by nerve and chemical signals when what happens is that these emerge from the mechanical conditions of the environment to a much greater extent than we suppose. It also happens, of course, that the relationship between these conditions and particular elements, such as molecules, are very difficult to calculate. But there is also the geometric phase, which after all shows how the propagation path depends of the properties of the medium, and it is in this sense that it is more revealing.

The geometric phase can also be measured in polymers such as DNA, and in fact without their contribution an exact solution to the mechanical problems of twisting and elasticity cannot be reached. Now, the same elastodynamics rationale can be extended to the parameter space of the lungs in a sufficiently long breathing time sequence.

The geometric phase can be seen as a torsion in the same way that torsion can be seen as a change in density. If in an originally homogeneous medium we imagine the appearance of a denser clot and a less dense bubble, both could not just arise without a torsion or vorticity to connect them. This may extend to other types of inhomogeneity.

Truly, the shallow, gasping and altered breath that nearly all of us have supposes a full-fledged contortion of a homogeneous respiratory cycle. But it does not help to characterize it as pathological, since in reality it's only an internal reflex or adaptation to the activity that the external environment and the same subject demands of the organism. As we said, bilateral alternation should also reflect how the ANS compensates for imbalances caused by voluntary activity.

The geometrical phase has also been called phase memory or system memory, since the properties of the medium or its potential provide an additional restriction on the evolution of the dynamic parameters that makes them not return to their original state. In fact, the most basic way to describe it is as the evolution of those wave parameters near some hole or singularity in the topology. Fortunately, we do not have to deal with infinite quantities here.

The change of sign is a typical feature of the geometric phase revealed in the clearest way in the exterior differential forms which in passing are also the most compact and elegant way of representing the integral aspects of electromagnetism. They also find a graphical translation that brings us back to acoustics: the circuit of forms of the modes of vibration of a membrane whose limits also vary within a circuit, discovered by Arnold in 1978. This naturally connects with the Bessel functions that we notice in the pulse cross-section.

The geometric phase is also present at the conical intersection of potential energy surfaces that define global landscapes for chemical reactions. In any case, a strong enough correlation between the modes of the pulse and the gauged circuit of forms of pulmonary mechanics would not be surprising, if we bear in mind not only that the circulatory and respiratory systems are intimately associated —something already sufficiently recognized- but that the respiratory system itself contributes greatly to the total impulse responsible for the blood circulation.

In this total impulse we must of course take into account the measurement of the return of the venous circulation, which is much greater with abdominal or deep breathing and much less with high or superficial breathing; a factor that also has a very direct influence, through capillary congestion, on the blood pressure index —and this being in turn connected in a loop with the cardiac response.

We must not lose sight of the fact that the immediate cause of the alternation in the passage of air through the nose is congestion by vasodilatation and decongestion by vasoconstriction of the venous tissue by the modification of the sympathetic tone with chemical mediators such as noradrenaline, which also affects the heart and is also considered a stress hormone. So the whole phenomenon makes sense as a process.

All the results obtained in electromagnetic theory have analogues in the theory of potential flow. In the simplest hydrodynamic case of geometric phase for waves in a moving medium, the parameters are the velocity of flow and its torsion or vorticity; it should be seen if this can be applied to the thread of respiration, to its ideal continuity so continuously disturbed.

In short, in biological systems the geometric phase is or should be a measure of the degree of (forced) contortion of the system with respect to an unforced fundamental state, and as such it should be robust against various types of noise. On the other hand, it remains to be seen whether both the dynamic elements and the phase shift have a continuous, and therefore immediate, acoustic translation in terms of tension, tone or timbre.

This calibration model of a complex motion inspired by gauge fields should have multiple uses in biology and biomechanics; to give just one example, it could be used to determine, gauge on gauge, the degree of stress, in the short term and with cumulative effects, suffered by cells exposed to different intensities and frequencies of electromagnetic radiation.

TWO PRINCIPLES

It may seem incredible that no one before Arnold Ehret defined vitality —and with it health- as Power minus Obstruction (V = P – O); but it is even more unbelievable that neither medicine nor theoretical biology have made later use of such a basic approach, the same one that a yogi or a Taoist could have subscribed two thousand years before Ehret.

Unbelievable, since after all Ehret's definition is the only one elementally mechanical ever given, science and the times still being supposed to be mechanistic at the beginning of the twentieth century. Nothing could be further from the truth, if we judge by the success of his formula, which is but the most direct possible application of the idea of a machine performance. «The principle for the construction of the ideal engine is to make it work with the smallest amount of friction». It's hard to find where the problem is in understanding this. Nor does it have such a great difficulty when applied to the human body.

Ehret's law may remind us the later Lotka's law, known as the «maximum power principle», but both arise from a very different context; their interpretations, moreover, point in very divergent, if not diametrically opposite directions.

Lotka wanted to bring a physical principle to the Darwinian ghost «mechanism» of natural selection, something that many others have tried with little luck if any. But Lotka's principle maximizes energy consumption and flow, not efficiency like Ehret's, in which the flow caliber is maximized only when the elimination of obstruction is maximized. The section or caliber of entry into the airways and the caliber of the thoracic cylinder in spontaneous or not voluntarily forced respiration also run in parallel. On the other hand, Lotka's principle seems to ask for the exercise of force whenever possible.

Leaving aside its completely different meaning, Ehret's principle, although so general, can be measured quite locally and directly, while Lotka's, like everything that appeals to natural selection, only admits global statistics in population dynamics. The first is really mechanical and easy to verify, the second is neither one nor the other.

There are certainly immediate ways of appreciating power and obstruction in different physiological signals, such as the pulse, where they are practically synonymous with pressure and tension, or resistance to pressure. The inclusion of the other basic constituent factor, deformation, allows to delve deeper into the analysis.

In addition to that it's not difficult to directly connect the Ehret principle with the problem of the nasal cycle. As a general rule, the total air flow, and the inlet section or calibre, tend to remain constant throughout the cycle, both when one side or the other predominates and when both are momentarily balanced.

However, first-person experience seems to indicate that the overall calibre tends to increase if the balance is not transitory and persists for increasing intervals of time. This is accompanied, also to an increasing degree, by a deeper breathing but with a decreasing flow of air that eventually becomes imperceptible. Such circumstance manifests itself in the dreamless sleep phases, as well as in periods of great absorption or concentration within the wakefulness, provided that both have a sufficient duration.

What can be seen in such cases is an increase in energy efficiency, and a general, and not merely nasal, effect of cleansing or decongestion. It can therefore be assumed that if these periods were long enough, they would have a profound effect on the overall balance of the organism. And this is the main reason, in accordance with others, why prolonged meditation was always considered to have an equally profound effect on health.

We can dismiss this idea if we want, but the truth is that, duly considered, it's the only one based on an entirely mechanical principle from top to bottom: Ehret's efficiency principle.

Already during the hours of sleep, even with dreams, a change of patterns in the respiratory flow and nasal cycle can be observed; this is something that the sleep sciences can sufficiently document. If during vigil we are forced to act in a world that we perceive as external to us, in dreams, on the contrary, it is the world that enters us. In the dreamless sleep, both tropisms are cancelled and the consciousness is empty of contents, although, it goes without saying, what we understand by consciousness is precisely that which does not depend on any content.

According to this, ideally all the cerebral/mental activity as well as the same breathing would tend to cease in the bottom of the dreamless sleep, which would act as a fundamental state. We would have a sort of asymptotic spectrum for the fluctuations of the breath and the mind traditionally seen as closely linked.

On the other hand, the fact that the body enters into more intense degrees of physical elimination, which is an energy-demanding work, when physical and mental activity is reduced to a minimum, raises the interesting question of how to define here the work done, the energy available, and the potential energy.

Even when it seems that the breath is about to cease, it's always assumed that this does not take place, but only that it has become so deep that it leaves hardly any marks on the surface; nothing indicates that the body has oxygen supply problems.

This brings us back to the issue of the potential already implicit in the geometric phase. But one does not have to look too far for the reasons. The body at rest is doing as much work as during its activity —the difference is that it now invests its energy in doing that work inside, cleaning and removing obstructions, rather than outside. This necessarily has to find a translation into the fundamental tone of the organism, whether we interpret it through the pulse, the breath or any other signal.

Lotka's maximum power principle is generally applied to external resistance, but could also be applied to internal resistance. Similarly, the Ehret principle of efficiency should have an external translation, even if we are not always in a position to follow it.

In the wake of Lotka, T. Odum has sometimes stretched the maximum power principle in terms of production and efficiency, but in this way it loses its restrictive power. For J. DeLong, what the principle says in its purity is that «biological systems are organized to increase their power whenever restrictions permit». In the context of natural selection it is always assumed that these restrictions are external.

The efficiency principle cannot ignore internal aspects, and while its simplicity is as sure a guide as there may be for these kinds of factors, it can also be too limited. It seems evident that any such general formula has to admit an expansion of terms if we want to deal more in detail with the untold complexities of life.

Although the formula (V = P – O) unequivocally invites to simply identify the power P with life and O, the obstruction, with the negative element that opposes it, it's clear, to give the most elementary example, that the same foods that sustain life can act successively as a source of power and obstruction. P tends to be identified with pressure and energy, O with tension and matter, and between them, at the constitutive level, we have the whole dividing line of deformation. But this deformation does not affect both parts equally, in the same way that in electromagnetism we have tensions without deformation and deformations without tension.

Probably the principle of maximum efficiency should be understood as easiness and could be written as maximum (internal) pressure with minimum stress and minimum deformation.

As matter and energy, O and P, are continent and content; since we know of no life that can exist without continent, here we have the egg and the hen, or the egg and the snake. But the same energy can be exerted outwards or inwards, and also and in parallel, towards the formation of structures or towards the creation of space and inner freedom.

As we have already pointed out, there is surely no more faithful and mechanical way to identify the state of O and P, continent and content, than the acoustic analysis of vital signs. Both acoustics and our organic arrangement, as well as hydrodynamics and electrodynamics, are naturally modeled around the idea of a tube of flow. It goes without saying that an empty tube doesn't sound as a full tube, and a thick one does not sound as a thin one. The potential theory refers us to this same context and circumstances. Metabolism can give us a biological measure of power, but much more indirect and problematic.

A quantum potential is also a potential of the wave field, and in this sense should not be different from the classical potential. This quantum potential looks the same as the pressure tensor of the Madelung equation, the hydrodynamic equivalent of the Schrödinger wave function. Here, too, continent and content can be distinguished.

Take the so-called tunnel effect, which is originally linked to de Broglie's wave mechanics; if we think in terms of point particles, we are obliged to think of an effect exclusive to the quantum domain. In a wave with real dimensions it is not difficult to explain it as a classical effect, which is equivalent even with an elastic rubber inside a container, as Paul Marmet liked to show. The classic model even offers the same pattern of dispersion as that revealed by quantum mechanics.

In the classical equivalent, it counts not only the thickness of the walls of the potential well, but its height also. A pixel animation shows in detail how the entire mass of a liquid can jump over a much higher barrier and get completely out of its confinement keeping the height of its center of gravity —and its potential- invariant. Although the process requires absolutely no discrete steps, these animations inevitably recall the associations between quantum mechanics and cellular automata, no less than the interpretation of quantum potentials by Bohm and Hiley as information potentials.

The maximum power principle and the efficiency principle should have something to say about the general orientation of the economy and society. Indeed, in modern societies the principle of efficiency is reduced to local economies of costs and production, totally subordinated to the hegemony of the maximum power principle, which also maximizes consumption, profit, accumulation and imbalance rather than circulation and homogeneous distribution. It's not hard to see either that the maximum power principle by itself tends towards the fastest possible depletion of resources and the global collapse of the system.

This also explains why in our societies the Darwinian external evolutionary model based on competition is promoted, in the face of all evidence and the plausibility of the internal development model; in such a way that even when the latter is considered, it's only to better subordinate it to the former.

The efficiency principle always prevails, while maximum power is only optional for systems that, in the absence of restrictions, tend to expand and self-destruct as quickly as possible.

TOWARDS A HEALTH SCIENCE?

Ehret's efficiency law plays no role in the health sciences, but this is no surprise since, as we all know, modern biomedicine is not at all concerned with health, but with innumerable diseases and ailments. If there really was a science of health, the principle of efficiency would have to be not only the most solid starting point but also the point of return and reflexion.

It is truly remarkable that modern science has been obsessed for more than four centuries with finding mechanical patterns in the external world and in our anatomy and yet has been unable to attend to such a purely functional argument. In fact, one can bet that this omission has a profound meaning in the general economy of things and in the government of the human.

Naturally we are aware that in the human body, which is an open dissipative system and not a conservative one, the categories of mechanics must be applied with particular care and discernment; but even so it cannot be denied that many aspects of conservative systems are still valid and relevant. A particularly crucial point, especially if we consider issues such as efficiency in the body, is how to move from closed, frictionless systems to open, frictional systems operating with external forces. This also has great significance in our global understanding of mechanics.

In classical mechanics it is the Third Principle of action-reaction that defines what is a closed system and therefore a mechanical system itself. The situation is very curious because Newton conceived his three principles to safeguard his theory of gravity and celestial mechanics but it is in the very orbits of the planets where the Third Principle can be least verified. Since then the application domain of this third principle has not ceased to be controversial, also in the case of the Lorentz force and other aspects of electrodynamics, as well as in Quantum Mechanics itself and General Relativity.

As Mario Pinheiro reminds us, by Noether's theorem, the conservation of momentum must always be valid in modern physics, but the law of action and reaction does not always holds. Surely this is a way of admitting that in practice and at the most fundamental level we cannot explicitly characterize systems as closed, and they must have varying degrees of interaction with the continuum or physical vacuum.

Pinheiro, developing further an idea from Landau and Lifshitz, considers a simple set of equations for out-of-equilibrium rotating systems with a balance between minimum energy variation and maximum entropy production that would have to be of great interest for the general appreciation of the mechanical-thermodynamic interface; these equations, which also describe a topological component of torsion with a mechanism of conversion of angular motion into linear motion, show a different momentum than that of Newtonian mechanics and predict «mutual interaction between systems and a self-regulating energy exchange», something that in living systems seems simply indispensable. Free energy is an essential term for the mechanical behavior of the continuum.

However, in a biological organism we find restrictions and circumstances that can still be very different. A distinction should be made between the internal entropy of the organism and that which exports to the environment, not to mention the maximum power principle, which seems to openly contradict the principle of minimum energy variation.

If we look within modern science for an equivalent for the three principles of Newton's mechanics in open systems such as biological organisms, we are not going to find it. To find something similar we have to look back way further, and then look for a quantitative and mathematical translation.

Actually, the triguna of the Indian Samkya system —the philosophy from which the yoga is derived- and its application to the three reactive modes of human body known as tridosha in Ayurveda brings a strong similarity for the case. The Triguna, as if we said, is the system of coordinates for modalities of the material world in qualitative terms. The three basic qualities, Tamas, Rajas and Satwa, and their reactive forms in the organism, Kapha, Pitta and Vata correspond very well with mass or inertia, force or energy, and balance by transmission of motion. But it is evident that in this case we are talking about qualities and open systems without further definition.

Here the third law of mechanics should give way to the conservation of the moment, implicitly admitting a variable degree of interaction with the medium. In harmony with this, Ayurveda considers Vata to be the guiding principle of the three as it has autonomy to move on its own as well as to move the other two principles. Vata defines the sensitivity of the system in relation to the environment, its degree of susceptibility in relation to it. In other words, the state of Vata is itself an index of the extent to which a system is effectively open.

In the human body the most explicit and continuous form of interaction with the medium is breathing, and therefore it is in the order of things that Vata governs this function most directly. Although doshas are modes or qualities, in the pulse they find their faithful translation in terms of dynamic values and the mechanics of the continuum that is our driving motif —as long as we conform to very modest degrees of precision, but in any case enough to give us an idea of the basic dynamics and patterns.

And since the other two modes reduce to what moves and what is moved, and the three as a whole are never in the same plane, the basic dynamic can only be an ascending or descending scale in which certain relationships remain invariant. Only with the coexistence of the three modes can compound beings occur; what varies is the proportion and degree of activity of each of them. This ascending or descending scale is also a scale of wider sensitivity or progressive dullness, of subtler restrictions or of increasingly material restrictions.

It wouldn't have to be too difficult to find the common ground that the Indian and Chinese semiologies of the pulse have beyond the differences of terminology and categories, and to pass from this common ground to the quantitative, but extremely fluid, language of continuum mechanics. Thus we would have a consistent method for moving from qualitative to quantitative aspects, and vice versa. Our world and our culture would welcome this.

We would then be able to conceive dynamics and patterns that right now go unnoticed as inherent, just as many global aspects of our biomechanics go unnoticed.

The health of the organism tends to maximum internal pressure and minimum internal tension. Outside the activity required by the environment, and even in that activity, it also seeks the least possible deformation that adaptation to an effort demands. All these are problems of optimization within a given domain, and can be applied to states of rest as well as physiological activity. What we propose here is that the pulse as a signature already has enough information to give us a global model of the flow and circulation that we call life; we only have to put its different aspects in order.

In the pulse signal it can also be evaluated impulse response functions quite similar to the Green's functions that already appear in hydrodynamics, electrodynamics or aeroacoustics and that incorporate the already commented Bessel functions. These functions provide a clear approach full of physical meaning, but we would still need to adequately define the impulse in different circumstances, as well as the degree of openness of the system in line with equations such as those of Pinheiro or similar.

In fact these or related functions, such as linear response functions, have long been used in information theory or to study the synaptic response of neurons. Naturally, stress tests are the order of the day in medical check-ups, in which the recovering function of rest values is also relevant. Now it is a question of translating these well-known techniques in terms of efficiency and quality.

It may be assumed that the acupuncture system with its twelve meridians or channels, which after all also rests on the ideas of flow and circulation, is a projection at skin level, taking into account the asymmetry in the arrangement of organs and other anatomical circumstances, of an elemental symmetry group based on the three axes of space. In this respect we must never forget that the meridians only form one level among those contemplated by Chinese medicine, and not precisely the deepest.

All palliative medicines, whether Eastern or Western, «holistic» or «reductionist», manage to find a compromise between an uncomfortable truth and the capacity of the client/patient. Even if Ehret were absolutely right, we would barely find one person in a thousand able to follow his recommendations to the letter. Much more than positive health, which in our present circumstances requires drastic measures more than at odds with our ways of living, people simply want relief for their ailments. However, there is no doubt that the old medicines were more concerned with health and the current ones with disease. It is not only that disease gives more money —at least within the modern «disease system», but that our circumstance as well as our relationship with the environment is increasingly pathogenic, if not pathological in itself.

However, on a theoretical level the simplest truths such as the efficiency principle are an invaluable guide to unravel the more than complex casuistry of diseases and sick people in their irreducible individuality. Since we live in the world in which we live, it is clear that we cannot ignore one extreme or the other: we cannot renounce the simplicity of truth just as we cannot ignore the complexity of facts.

The canonical texts of Chinese or Indian medicine, not to mention others such as those of the ancient Greeks or Avicenna, explicitly show guidelines for the classification of individuals according to qualitative typologies; these typologies were also reflected in a very significant way in the semiology or interpretation of the pulses, their types and their evolution over time in health and illness. These typologies, revealing an elemental combinatorics, were useful although naturally their scope was very limited by the subjective and phenomenological aspects of the evaluation. If we can bridge here the gap between qualitative aspects and continuous measurement, we would have access to a continuous combinatorial domain of incomparably greater scope within which to subsume increasing amounts of data, whether from one type of medicine or another. This as to «the path of complexity» towards which our overabundance of means compulsively pushes us.

From this point of view of complexity, everything we can obtain through molecular biology, whether at the level of genes or any other type of chemical markers, still has very limited value because it cannot be naturally framed in a macroscopic picture observable in real time. This means that modern biomedicine can hardly fulfill its promises of a «personalized medicine» that makes sense for the doctor himself; and without this, rather than medicine, we would have to speak of an electronic oracle. The role of the physician in this situation, if not increasingly accessory, would resemble that of a troubled interpreter.

In addition to checking the main structures and organs, it is essential that we have a dynamic reference at macroscopic level to guide us on the possible evolution of the organism over time. This has always been entrusted to the tacit knowledge of the doctor, to his appreciation of individual differences and to what we call his «clinical eye»; but this type of knowledge cannot be transferred naturally to any expert/AI system. On the other hand, the parameters of the pulse as a general signature of the vitality phenomenon can be transferred integrally and naturally because they are quantities that still can be translated into nuances.

But what's more, we believe that the pulse is already giving extremely valuable average information about the state of the internal environment in which all biochemical and biological interactions take place, and this is really decisive in the internal landscape that defines health. Going on with the analogy of continuum mechanics, we would thus speak of an approximation to the constitutive properties of that medium —much as we now speak of permeability, permittivity and susceptibility.

If precious information can be lost in the Maxwell equations by moving from the global to the local with the standard vector analysis, this should also be considered in the mechanical analysis of blood circulation, which should always maintain the holistic approach as much as possible. This integral approach allows us to see aspects free of metrics or orientability-dependent.

E. J. Post insists on how cyclic integrals, classical or not, have by themselves exact quantum-counting properties independent of metrics —both the Ampere-Gauss cyclic integral and the very Aharonov-Bohm's integral which serves as a model of geometric phase. In fact the determination of e or h is much less reproducible using only the Scrhödinger-Dirac theory or the QED. This counting ability can also be relevant to both measurement references and continuous combinatorics aspects that may arise in typologies, profiles and configurations of pulse analysis and biological data analysis.

Examples like this one mentioned by Post, basically ignored by most physicists, should convince us of the power of resolution of integral methods if we know what to look for, whether in physics or in any other case that presents analogous conditions like the one we are dealing with. All this agrees with our vision that there are no fundamental levels or systems, but rather the emergence of statistical ensembles, and that physical laws, understood as fields, have always had an integral character but a differential application and interpretation.

In biology, medicine or physiology we always have to deal with the weak consistency of data and measurements, all the more reason to look at physics as the reference standard. And if biology or medicine have been modeled by physics bottom-up, the only thing we are advocating for is its downright use from top to bottom, to find out to what degree it can improve that consistency. None of this would be strange to us if it weren't for the inveterate habits of thinking and a bias towards the idea of constituent blocks.

Thus, the most direct way of offering consistency to the growing constellations of medical or physiological data is under the coverage of the same global/macroscopic aspects of physics, in the sense we have described. The point is that by adopting this downward direction we have to develop new categories, categories that until now have been atrophied by the predominance of the constructive or ascending approach. This article is only an introduction to that problem.

Entropy itself is a macroscopic measure, often defined as the number of microstates consistent with the constitutive or state macroscopic quantities that characterize a system; although it should not be forgotten that only by metrics we are authorized to speak of macro and micro levels, and some very important aspects are independent of metrics. Since the terms entropy and information are interchangeable to almost all effects, even from the perspective of information theory the consistency of the data will depend crucially on the macroscopic characterization of the system —so much more in open systems that maintain a self-regulating loop with directly observable values. In the ongoing race for massive use of data in medicine, this is not only fully relevant, but almost the only thing that makes sense. A medicine without phenomenology is neither possible nor desirable.

The history of medicine shows that it is much easier to propose therapies than consistent theories about health and how the body works. Again, with the mechanical and acoustic properties of the pulse, it is easy to conceive extremely simple therapeutic strategies but with a wide range of degrees of sophistication and adaptation to individual constitutions; it would then remain to be seen whether these strategies that assume a global biomechanical perspective are also effective.

Here we cannot even begin to raise the complex relationships between entropy, increasing restrictions in organisms, elimination deficit, aging, maximum power and efficiency. These may be exciting topics for theoretical biology. But to get to the bottom of these issues, an additional thread is also needed to serve as a guide. Also in medicine, the most important thing of all, even more than how the doctor perceives the individual organism, is how this organism perceives itself. While this may seem the most intangible factor, it's also something that leaves traces at all levels.

Also the nasal cycle of which we spoke earlier would be a footprint at a certain level of how an organized being feels about itself; moreover, it would indicate a balance between perception and action that still leaves a variable gap for self-perception. Francois Chollet says that «when you have access to both perception and action, you are looking at an AI problem», so that you can start looking for an optimization loop. Truly, the Natural Intelligence of the organism already makes a balance between perception and action, but it is the self-perception that obstructs the path or determines in any case the level at which such equilibrium or imbalance can take place; and at the same time the self-perception is limited and unbalanced because the intelligence, and the organism following it, are more occupied with the activity in the environment than with its perception.

References

A. L. Pendolino, V. J. Lund, E. Nardello, G. Ottaviano, *The nasal cycle: a comprehensive review*

M. Berry, *Anticipations of the Geometric Phase*

V. I. Arnold, *Methods of Classical Dynamics*

A. Shapere, F. Wilczek, *Self-Propulsion at Low Reynolds Number*

J. Samuel, S. Sinha, *Molecular Elasticity and the Geometric Phase*
Ershkovich Electromagnetic potentials and Aharonov-Bohm effect

P. Marmet, *Reality of Waves in Particles*

M. J. Pinheiro, *On Newton's Third Law and its Symmetry-Breaking Effects*

M. J. Pinheiro, *A reformulation of mechanics and electrodynamics*

E. J. Post, *A history of Physics as an exercise in Philosophy*

F. Chollet, *What worries me about AI, Medium Magazine*

M. A. M. Iradier, *Health, Life, Aging, Evolution*

A FABLE, AN ENIGMA, AND A FINAL SOLUTION

15 January, 2019

The fable

It is not superfluous to remember it, though everyone knows the story,: the Asuras were given the gift of building three large, impregnable fortress-cities, one on Earth, one in the Atmosphere, and one more in Deep Space. The three cities turned like wheels, were permanently changing positions and avoided at all costs to be aligned, having been warned that this would cause their end. But finally, after a thousand years of ceaseless permutations, it seems they only had one left to prove, and since everything seeks its completion, the three cities fell finally in line. It was what Shiva had been quietly waiting for, and not before smiling, the supreme archer fulminated the three cities with a single shot.

The enigma

Few shadows hang over our destiny as inexorably as this one in the city of Tripura. The comic and the cosmic of the city of Tripura is that to seek its perfection is to seek its fulfillment, which is none other than destruction; as a great work of watchmaking, only in its overall concordance does it acquire meaning. This triple city ciphers an enigma and proposes a whole challenge for those who have turned intelligence into a matter of prediction. The enigma is to find out *what* Nature or Cosmos was before the permanent buzzing of those gear-filled cities, that is, our technology-saturated civilization. And the challenge, given that today we care much more about prediction than understanding, is naturally *when*. Shame that the cypher of the *when* depends on the *what*, which today seems to matter so little.

To try to point out today what Nature can be out of science seems to be such a desperate task, especially for people of scientific formation; but, even if we have to submit ourselves to the discipline of logic and remain on its ground, and even if we renounce beforehand to know the *whys*, we still have ways to *calibrate* the *how*, the *how much* and the *what*.

We could also add that in the above story the Asuras represent the maximum power principle while Shiva embodies the efficiency and even the sufficiency principle. Nor can we forget that the only Asura that survives the summary execution is Mayasura, the very principle of appearance.

Before the reader runs out of here to the horrified scream of "Philosophy!" I warn him that I am going to propose a small problem maybe easier to calculate than the value of the money he has in his account, and that is even more vital to him than its liquidity. Important things like masks, so he will just need a little patience. There's no worth in knowing the final solution if one doesn't know the conditions of the problem. In its Olympic way, it's all here, as long as one knows how to guess what's missing.

Throughout its history, physics has not been able to avoid wondering about the nature and scope of the laws or regularities it discovers. Newton promoted absolute space, time, and forces; Leibniz and Mach that every measurement is, by definition, only a relationship. This relational viewpoint is, so to speak, the phenomenology of pure quantity. To these absolutist and relational perspectives could be added a third, causal one, which also demanded an explanation of the mechanisms or the reason for the phenomena. This, beyond the physics of contacts, can only occur through a medium or at any rate a field.

In the end it's clear that science has diluted its demands and has remained a little halfway to nowhere, because Relativity seemed to say goodbye to absolute Newtonian physics but more in appearance than in truth —the relational program of Mach and others was severely truncated and work continued with dimensional constants independent of the background. On the other hand, Relativity itself generalized the mathematics of field theory at the same time that it got rid of the medium or at best turned it into an opportune mathematical ghost. Quantum mechanics has also continued to work with fields but, although it is claimed that the most common interpretation has been liberated from causality, it does not dispense with the idea that we are dealing with irreducible and simple systems, rather than with statistical ensembles.

The point, to make a long story short, is that the laws of physics have never been dense enough to say that they "describe Nature". In spite of its name our Physics, also known as mathematical physics, does not speak of the *Physis* or becoming of the Greek philosophers, but in any case of its *Nomos*, of the regularity that we can observe in it according to our human convention. There is such an abyss between the adimensional punctual particles and the infinite variety of *finite* forms of nature, that in order to save it we have had to imagine another additional dimension in time, the time of becoming, which is distributed in such a convenient and sloppy way between cosmology, statistics, the arrow of time, thermodynamics, or the theory of evolution.

Mixing one and the other is like mixing smoke and mirrors; in reality they don't mix at all and they are still totally different things, but the multiple reflections that allow comfort and entertain us. And by entertaining us, they make us forget the most evident dimension of nature as that which is directly accessible to the senses, sensible appearance or *Aisthesis*, the source of that common or implied sense that Aristotle already attributed to animals but was left far behind by our sophisticated intelligence. If relational physics, which only contemplates laws expressed in known quantities of the same type, is the purest expression of physics as *Nomos* and the phenomenology of pure quantity, the *Aisthesis* would be the phenomenology of pure sensible quality, in its own terms and without interference from alien ideal quantitative factors.

To put it another way, the sphere of the *Nomos* is the sky of our intelligence, and the *Aisthesis* its earth. As for the becoming *Physis*, which is what grows between the two terms, it will depend on how the latter are formulated so that it is a gas or a ghost in grief or something very different. The curious thing is that in spite of the empirical origin of modern science, the senses play an increasingly insignificant role in comparison with the Cuckoo's Clockwork of number and measure.

Let us take a total eclipse of the Sun by the Moon; what is evident, even without any measurement, is that the apparent size of both bodies is the same from the Earth. This striking coincidence plays no role in our celestial mechanics, since this one has been developed with the express aim of replicating/predicting the observable orbits. However, the approximate optical equivalence between Sun and Moon is not an isolated fact in our solar system, it also takes place with the Sun and other planets and their satellites; from the perspective of the Sun a good number of planets have the same apparent size, which some have been able to take into account for the explanation of the Titus-Bode distribution sequence.

One thinks that this mere optical equivalence might hold much more "information" than the equations of motion of celestial mechanics, just as our hypotheses about the past and becoming (*Physis*) are immeasurable in complexity with the simplicity of our truly physical laws (*Nomos*) dealing with the observable "behavior". I also think that if both spheres have never overlapped is precisely by ignoring this type of sensory evidence that seems to be staring at us more intently than we look at the stars. So I believe that the day when these things have a natural place in our theory will also be the day when our theories will have broken with their secular isolation and come into contact with that which we called Nature.

Naturally, a theory that accommodates these appearances must be governed by an accordingly relational law without the slightest sophistication, otherwise we would return to our usual vain speculations. So the only way to do that is to use only homogeneous physical proportions in Archimedes' purest style —things like radii and diameters, distances or densities. We can save ourselves complicated hypotheses from the outset.

The same would be true for the other senses, although today only *color* lends itself to a non-violent superposition of appearance and theory, of quality and quantity. This is precisely the point, the violence that our theories have exercised to force as Procrustes the facts in their molds. One cannot help but suspect that any success outside of this way of proceeding, even if it seems very modest, would have an overall liberating effect on consciousness.

It has been widely said that a picture is worth a thousand words, but not that appearance is worth a thousand theories, and this would be just the case. If Nature doesn't seem to ignore our theories, it is because they took great care to replicate certain apparent evolutions; and by mere inversion of the human purpose, it may well be anticipated that Nature looks so much more for appearance the more gratuitous it seems; and in the case of celestial mechanics she couldn't have made things easier for us.

All our theories, without the spontaneous concurrence of appearance, are like answers that still await a question. With its concurrence they are like an oil-impregnated wick which only needs one spark to burn from top to bottom.

Artificial Intelligence filters unstoppably into the most critical decision-making spheres, from finance to war, and the day when human intervention is erased for the benefit of systems of inscrutable complexity, but which have the fearsome advantage of their speed, no longer seems very distant. If a machine can make the "right decision" long before a human being, the reaction time is absolutely critical to the answer, and it is assumed that there are competitors willing to do the same in a shorter time, all the necessary conditions are in place for humans to delegate to machines they don't understand, as it's a race after all.

If machines are dragging us into some kind of singularity, it's clear that it is not that of an explosion of artificial intelligence, but this other much more predictable, and almost completely consummated, of the resignation of man, which only lacks a few technical details. It would be a voluntary eclipse shortly before an involuntary destruction; but no one could deny that between one and the other there could be no greater continuity in terms of substance and form.

The eternal question of machines is whether they serve us or we serve them. Of course, all this already arises within a closed-loop logic, which is precisely what defines a machine. Newton's Third Law, as we have seen, is what defines the limits of mechanics proper; but today even our fundamental physics, developed with an express predictive purpose, has to forget the Third Law and conform to the much more general conservation of the total field-particle moment. This law does not imply closed systems. But what does scientific clarifications matter in the face of human destiny?

The whole civilization race is a growing isolation of the environment together with a growing coercion and oppression of that same environment; both being fatally united. And in the individual, which in itself is a synthesis of nature and culture, we see today how intimately the exploited and exploiter coincide in the form of self-exploitation.

Today, it is perfectly conceivable that the logic of the efficiency principle be applied to the individual in order to improve its "performance and well-being"; in fact, this is already part of the consummate compensation mechanisms to achieve the internalization of social pressures and tensions. However, the socio-economic machine as a whole is not governed at all, despite what is sometimes said, by the efficiency principle, but by the maximum power/maximum benefit principle, which on the contrary tends to outsource costs without further consideration.

The final solution

It is said that the mythical sage Yajnavalkya calculated that the distance from the Sun and the Moon to the Earth is in both cases 108 times the diameter of their bodies, giving with great approximation an adimensional key to the enigma of the optical equivalence.

Miles Mathis, who had already contemplated the optical equivalence like nobody else, notes without ever connecting it that in accelerators the relativistic mass of a proton usually finds a limit of 108 units that neither Relativity nor quantum mechanics explain, and makes a derivation of the famous gamma factor that links it directly to *G*. What other natural connection could there be with the optical equivalence but light?

And between light and charge? And between charge and mass? And between mass and gravity?

Let him that hath no understanding calculate, and he who has understanding, let him not calculate

Is science able to tell us anything about our place in the Cosmos? Even if it seems so late. On the other hand, this ratio seems to point us something about how matter shields, or opposes, electromagnetic waves —light itself.

Ideally, each person should be able to choose what his end is like. It is not superfluous to remember, with Epictetus, that the door is always open. Collectively, this seems much more difficult. If we add to nuclear weapons and the doctrine of pre-emptive strikes the transfer of decisions to "intelligent systems" for quickness sake, we have the most stupid and abject of all possible ends.

Maybe the only one keeping up with our unnameable present.

But we shouldn't give up. Compared to a nuclear hell inflicted by preventive mechanisms, a general blackout by a giant electromagnetic storm, solar or otherwise, would be most merciful. And since we do not know when the Sun will throw its darts again after the 1859 event, we will always be able to count on our human electromagnetic pulse bombs, which makes a clean, ecological, cheap and reliable weapon. And with a lighter trigger.

The consequences for this civilization would be unmanageable and surely irreversible the damage. It has rightly been said that capitalism finds it much less difficult to imagine the end of the world than its own end, maybe we will have to lend a hand in this blind spot so that instinct understands itself and the unimaginable is better imagined.

In fact, the Internet emerged as a response to a possible nuclear attack, to minimize damage by distributing the decision points of the military command; today we have a decentralized technology, but at the service of an increasingly concentrated power structure.

Does one prefer to be fried with radiations courtesy of a machine, or does one prefer that humans fry machines first and at least reserve the chance to survive in the savage world prior to civilization? Neither for myself nor for the planet do I have the slightest doubt. A different thing is that may we fit such luck.

Executives could perform their survival practices for free and display rudeness and heroism in the most natural of possible environments. And fight, for example, with the wild and their brothers on a sort of equal terms.

In addition, electromagnetic bombs could thwart a nuclear attack, although they could also be nefarious to many nuclear power plants without proper safety measures. All these plants should be forced to close if they cannot meet minimum requirements.

Naturally, a global blackout would require a few high-altitude nuclear detonating pulse bombs. Since it still remains a desirable alternative, it could even be delegated to an international organization to detonate these bombs as a lesser evil before a rogue state or its machines press the button first. Would we submit this in turn to the decision-making power of intelligent machines?

The scenario of the day after could be surprisingly modified; in many countries, rural grandparents might have better prospects of survival than their urban grandchildren. As if the past overtook the future and the future looked to the past awaiting what to do. Another plus of added wisdom.

All points worthy of attention if it is a question of reversing the inexorable dynamic towards the worst possible endgame. And the end, like the carrot, is the rudder of the donkey —as long as there is a good stick on the other side.

This could be used to deactivate other dynamics and time bombs, other precipices towards which we so forcedly rush. As for the monstrous war budgets of the rich countries, it has always been known that there's a good deal of practical joke in it, to better suck the blood of their taxpayers and keep the stakes high.

The important thing is that those preemptive strikers know that others can pull the trigger first. They have a lot more to lose, and we have more to win.

Now at least we have another option on the table.

STOP 5G

10 febrero, 2019

Dedicated to Carlos Ruperto Fermín

2019 will be the year of deployment of 5G, the fifth-generation wireless technology, and there's no time to waste. Can the scientific debate on its effects be settled? We suggest a line of research to detect the effects of non-ionizing radiation on living beings with a purely biophysical and biomechanical argument.

Radiations

This year of 2019 begins the worldwide deployment of the fifth generation mobile telephony (5G), considered an strategic objective in telecommunications for the control of the emerging internet of things and the expansion of artificial intelligence. The Huawei case exemplifies the commercial and geopolitical tension that this advance is causing.

Most of us don't give a damn about who wins this race to fry our brains, be they American or Chinese; on the contrary, we would be sincerely grateful to the first country to take seriously the vast amount of research that suggest the pernicious effects of growing electromagnetic pollution and take some decisive action in this regard.

What's incredible is that nothing has been done and only the interests of the industry count. More than three-quarters of the very numerous independent studies lean toward the danger of this radiation, while more than three-quarters of the studies sponsored by the companies lean toward its safety. Too much statistical divergence to be coincidental.

It is true that we still don't have conclusive evidence on the biological incidence of these fields, and it's already known that in this kind of research the evidence is always very tenuous and requires very long times and large doses of statistical inference. Another thing is that within ten or twenty years doctors will begin to recognize new syndromes and degenerative diseases among the exposed population, which will be practically all of us.

But when in doubt, at the very least, the only sensible thing would be a precautionary moratorium until a satisfactory elucidation of the case is settled, even if it takes ten years. The stakes are much higher than being able to watch videos more quickly on our mobile phones, the stakes are high for our physical and mental health and that of future generations.

Although impossible to verify, it's estimated that the maximum peaks of electromagnetic exposure today are **between fifteen and eighteen orders of magnitude higher** than those of the Earth's natural electromagnetic field. A ten followed by fifteen to eighteen zeros. Comparatively, human atmospheric pollution and CO2 emissions would only have increased by a ridiculous fraction of that generated by the natural cycle of forest fires.

Yes, electromagnetic fields are extremely tenuous in relation to ponderable matter; but also smoke or gas are much more tenuous than solid bodies and no one doubts their cumulative and immediate toxic effect. And besides, the noblest or more sensitive tissues, the brain, the nervous system and the heart, are those that depend most critically on the electrical activity for their proper functioning. This should have been enough to exercise extreme caution.

But a ten followed by eighteen zeros still seems little to us, since with the 5G we'll be able to multiply it several tens or hundreds of times. There's nothing like competition to beat records.

5G is not only an increase in the net amount of exposure but also a qualitative change, as the new wavelengths are much more penetrating. The number of repeaters will also increase alarmingly, as the aim is having always the user at less than 30 meters. In many places even trees are being cut to install new repeaters.

I won't go on with something that has already been dealt with more extensively in hundreds of articles. We know that the dominant argument for rejecting the danger of non-ionizing radiation is that it does not have enough energy to pull electrons out of their orbits and destabilize the organization of matter —there is no plausible mechanism. From the theoretical point of view, it seems a clear, conclusive argument.

However, it is necessary to remember that today **there is by no means a complete theory of radiation**. In both classical and quantum electrodynamics (QED), radiation, emission and absorption are more often than not accounting operations to balance the energy conservation with respect to the observed facts —"if this has happened, it must have been because of this". And so, for example, physicists still can't agree even on something as basic as whether an accelerated charge radiates or not.

With such a theory it is impossible to be categorical. Only the hard, laborious experimental path remains. Besides, we know that we can't measure electrical potentials directly but only differences of potential.

But it's hard to believe that a factor that could have increased by one trillion ($10^{18)}$ remains insignificant, unless that factor was completely irrelevant before the Anthropocene. From what is known of our own and other planets, we will agree that a field that before humankind would have extended appreciably to 65,000 kilometers on the day side and more than 6 million kilometers on the night side doesn't seem so insignificant. Doubts are therefore more than reasonable.

Measure for measure

My intention here is to suggest a line of research to detect in fraganti the incidence of these fields in our biology, something that the researchers themselves despair of being able to achieve. I pointed it out in passing in a more extensive article entitled "*Towards a health science? Biophysics and biomechanics*". The idea at least seems to make sense, though it remains to be seen how much it can yield in practice.

It's all about applying, *gauge to gauge and measure for measure*, the very notions of gauging of the electromagnetic fields to cellular motion. It was precisely by investigating the interference pattern of forced electromagnetic fields on the spin of a particle that Berry discovered, in 1983, the so-called geometric phase. This one translates into "global change without local change", and although physicists associate it mainly with quantum mechanics, it's relevant in both the classical and quantum realms, in both the macroscopic and the microscopic.

Initially the Berry phase appeared as a displacement for adiabatic processes, i.e., processes without net energy transfer, which seems the appropriate case to asses the influence of non-ionizing radiation —such as that of telecommunications- when there is not enough energy to ionize atoms or molecules.

A well-known example of geometric phase is the so-called Aharonov-Bohm effect, in which charged particles are affected by the electromagnetic potential even though the electric and magnetic fields are zero in their region. Contrary to the most widespread opinion among physicists, including Bohm himself, this is not a specifically quantum effect, but inherent in Maxwell's equations of classical electromagnetism, and exact analogous examples have been found even on the surface of water.

Even today physicists don't know how to place this phenomenon, whether in fundamental or applied physics; but its experimental verification is out of question.

After Berry showed the simplest case for cyclic adiabatic processes, there have been generalizations for non-adabatic, non-cyclic, and even dissipative or nonconservative processes, such as biological ones.

This "anholonomy" is also not alien to living beings and their locomotion. Inevitable is the example of the falling cat problem or "Maxwell cat", the happiest example of how a living being protects its invariance against contrary efforts; or the no less universal figure of the motion of the snakes. Shapere and Wilczek showed in 1987 their relevance in the cellular self-propulsion in viscous fluids describing *a circuit of forms* starting from infinitesimal deformations in cylinders and spheres.

In short, in biological systems the geometric phase is or should be a measure of the degree of (forced) contortion of the system with respect to a ground state, and such a feature should be robust against various types of noise.

This type of model inspired by gauge fields such as this one of Shapere and Wilczek should have multiple uses in biology and biomechanics, though what we are talking about here is determining the degree of stress, in the short term and with cumulative effects, suffered by cells exposed to different intensities and frequencies of electromagnetic radiation.

I do not know if a similar reasoning could be applied, instead of to the motion of cells, to the more schematic waves of the electroencephalogram, EEG. The definition of form proposed by Shapere and Wilczek is only one among many that can be extended to other cases.

The geometric phase provides not only a biophysical argument, but even a biomechanical one, which in certain cases such as macromolecules becomes completely tangible. Today we can twist a strand of DNA like a towel and calculate its exact torsion using the geometric phase. Therefore we can measure and calculate its elasticity within a medium or potential.

Think this through. So far science has focused on molecule-molecule reactions rather than molecule-medium reactions. However, today we know perfectly well, for example, that the so-called "central dogma of molecular biology" is wrong and that the same gene can produce different proteins depending on how enzymes interpret the context and the environment. The problem is that this seems almost impossible to measure; however, the geometric phase already gives us an *in situ* index of the potential, even if we can't measure it directly.

This criterion could be used to elucidate other aspects of the debate, such as the possible incidence of this radiation on DNA or the shortening of telomeres at the ends of chromosomes, related to cell death and ageing processes.

If what we are talking about is a direct mechanism, the argument of physicists still seems correct. However, physicists still do not consider the effect of the potential, and this can cause electronic emissions and molecular effects that seem "spontaneous" but would be clearly induced by the potential. Some might call it indirect effect, others superdirect effect, since it doesn't even demand the absorption by particles of radiation; in any case the causal link would be patent.

To conclude

It seems impossible to confront the big interests of the industry, especially in a case like this. But there are examples such as asbestos or the tobacco industry, in which it was finally impossible to hide the truth even if it came painfully late. Since we are also very late in this matter, it's clear that something has to be done as we cannot wait twenty years.

If the big companies and countries involved in this race can lose their big stake on controlling "information flows" with a moratorium, they will lose much more if it comes to light that they have been hiding data and obstructing impartial research. In fact, this could mean the beginning of the end of many things, perhaps more than we can imagine.

If these companies and countries are so interested in money and soft power, I permit me to suggest that it would be smarter, as well as more profitable, to undertake research into a true science of health rather than our well-known sciences of infinite ailments and diseases. Profitable, at least, as long as one seeks for the common good instead of looking blindly for one's own.

And if we think in terms of prestige and social support, which are the only ones that some can understand because the search for truth is beyond them, the scientific community is also very much at stake here. The arguments I put forward here are more than thirty years old. Perhaps it could still be said that then there wasn't the experimental level to collect evidence with that theoretical support; but that excuse no longer exists now.

Time will tell, but I think this issue will be a turning point for more things than we think, even though the only sure thing, in case of crisis, is that the eternal opportunism will try to channel it in its favor.

We'll obviate that now.

What matters is that even if the situation seems desperate, it can still be completely reversed. We can block the blockade; we can circle the encirclement.

References

Y. Aharonov, D. Bohm, (1959): *Significance of Electromagnetic Potentials in the Quantum Theory*

M. V. Berry, (1984): *Quantal Phase Factors Accompanying Adiabatic Changes*

J. Samuel, S. Sinha, (2003): *Molecular Elasticity and the Geometric Phase*

A. Shapere, F. Wilczek, (1987): *Self-Propulsion at Low Reynolds Number*

A. Ehrskovich, (2012): *Electromagnetic potentials and Aharonov-Bohm effect*

M. Iradier, (2019): *Towards a science of health? Biophysics and Biomechanics*

FROM TAIPING TO HONG KONG

23 September, 2019

From Taiping to Hong Kong

The Taiping Rebellion was a terrible Chinese civil war between the central power of the north and the south of the country which lasted from 1850 to 1864 and dwarfs in number of victims all the conflicts experienced by Europe until that time. Leaving aside the massive displacements of population, conservative estimates speak of 20 to 30 million dead, most of them caused by hunger and epidemics.

By way of comparison, the number of casualties in the American Civil War (1861-1865) was just over six hundred thousand.

Despite the magnitude of this tragedy, which sank in chaos and darkness the Middle Kingdom after a long downward trajectory, historians have not been too inclined to delve into the more than dubious circumstances of its origins, which even such an Anglo-Saxon medium as Wikipedia does not bother to disguise.

The story goes that Hong Xiuquan, an aspiring unsuccessful civil servant, proclaimed himself Heavenly King, son of God and younger brother of Jesus Christ, and with this brilliant resume he managed to put nearly half the population of the Empire on his side.

Coincidentally, Hong had been studying in 1847 with the American Baptist missionary Issachar Jacox Roberts, at a time when missionaries were seen as barely disguised intelligence agents, though forcibly tolerated by the humiliating treatise of Nanking's in 1842 at the end of the First Opium War.

Roberts remained in Canton during much of the civil war to return to the capital of the new kingdom of Taiping, which was none other than Nanking, in 1860, where he would again serve as counselor to the distant relative and right-hand man of Xiuquan, the prime minister and later foreign minister Hong Rengan.

By chance again, it turns out that Rengan, too, had been working closely with the diligent envoys of the London Missionary Society in Hong Kong in the midst of the civil war between 1855 and 1858, and his intervention from then on was decisive in keeping the rebels armed until the end the Second Opium War (1856-1860), in which the British, now seconded by the French, obtained the desired access to the interior of the vast country.

After these new concessions London no longer had much interest in prolonging the conflict, and it was clear that they preferred to deal with an extremely weakened emperor in Beijing rather than with such a huge people in arms. No time to waste, and the talented Rengan, whose measures had been so providential for the survival of the movement shortly before, was inexplicably ousted in 1861, as soon as the new treaty was signed.

The First Opium War had taken the Qing dynasty completely by surprise, but once that factor was lost it was required a much larger deployment in order to reach the next list of objectives. Without the tremendous wear and tear of Beijing's fight with the rebels, even with the Western coalition's overwhelming armament advantage, everything would have been much more costly and complicated.

After the ratification of the treaty of Tientsin and the mysterious cessation of Rengan the rebels quickly began to lose ground. The British, who had adopted an official position of neutrality, ultimately intervened in Beijing's behalf to finally decide the war. Until then, the King of Taiping and his followers were always led to think that the Westerners were sympathetic to the uprising. After many years unscathed, Hong appeared unexpectedly poisoned in 1864 and shortly afterwards Nanking fell under the instrumental intervention of English troops. It was also in an English gunboat that Reverend Issachar Roberts had escaped from the city two years earlier.

Characteristically, the English press was sympathetic to the rebellion until the ratification of the hoped-for treaty; then they began to spread stories in which Taiping leaders cut children's heads and smashed them against the wall. Sounds familiar.

Hong's creed sounds incoherently millenarian, puritanical and "modern", if you think that the bulk of the target audience were illiterate Chinese peasants in the middle of the nineteenth century. Abolition of private property, suppression of the cue imposed under death penalty by the Manchus, strict equality and separation of sexes, prohibition of life in common and sexual intercourse even between marriages, separates armies of men and women, substitution of Confucian texts by the Bible as the main subject for civil service examinations.

Think about it. Marx elaborating on the long historical process that mediated between the medieval peasantry and the class consciousness of the industrial proletariat, and it turns out that in China these same peasants had already embraced the most radical extremism at the first opportunity.

There was talk of the suppression of "Confucian idol worship", though everyone knew that Confucianism long predated the Manchus and was the foundation of society. It's unthinkable an autochthonous Chinese ideology betting on Christianity at the expense of its own culture and roots.

It is a commonplace that rebellion could not have spread like wildfire without the inevitable Chinese triads or mafias and their deep penetration into the fabric of rural life. Triads had existed for many centuries, but it was only at this time, with the massive influx of opium and the new rules of trade, that a legendary subculture of the underworld was forged, with its networks of spies, cambalaches, dens and slums.

The murky became the norm. Rebel leaders anathematized drug use, prostitution and everything else, while corruption among them became rampant. "Do what I say, not what I do". Western powers claimed to be neutral while their arms traffickers made a killing.

Meanwhile south and west the great Indian Rebellion of 1857 took place which led to an overall administrative change in the main colony. Faced with the challenge that China posed after 1842, there were all sorts of doubts about which was the most profitable model of penetration and exploitation. For Rothschild, Elgin, Disraeli and company, the Chinese civil war was also a great testing ground to "wait and see" how far the resistance of the central power and the whole people could go.

For the rest of it, the "divide and rule", the determined and systematic interference under the guise of false neutrality, was always the supreme principle of the British abroad and was applied with expert hand every time there was a favorable juncture. The split between the north and the south was a recurring theme in the history of the great power of the Far East, and of course the people had a thousand motives for embracing the rebellion against the oppressive Manchu. To open such a large melon all it took was a good knife.

Despite the telling accumulation of coincidences in the where, when, what, how and to whom the rebellion benefited, I still have not found a Western version of the facts that points to British responsibility in the origin and development of the revolt, which I find simply incredible. It can be assumed that Chinese historiography will have a different opinion, but if that is the case, it has not managed to make itself heard among us. The fact that Taiping is considered to have inspired the subsequent revolutionary movements of Sun Yat-sen and the Chinese Communist Party should not cloud the judgment in the face of what seems so obvious.

Needless to say, by now we are not going to find any smoking gun, as even Chinese historians have to rely on the testimonies of Reverend Roberts to the English press, missionaries and other Western diplomats since they are almost the only thing available, even when their own co-religionists admitted his erratic behavior and the unreliability of his accounts. After all, who could give Roberts credit? We are further told that the missionary had suffered from leprosy since the 1830s, which seems very convenient to keep the curious away, though not quite a seeker of truth like Hong Xiuquan.

The whole story sucks from beginning to end. But if we still have some doubt, we only need to see what is happening at this very moment, all distances saved. Today we see how the United States, Great Britain and Atlanticism not even hide that they do everything they can to destabilize China and introduce as deep a wedge as possible to break her apart —for Hong Kong is only the handiest cleft to open well the cracks in Taiwan, Tibet or Xinjiang.

If we are now witnessing such a effort to introduce chaos in a world where interdependence multiplies consequences, what could not happen in 1850 when for these countries impunity was almost absolute and the only thing to fear was that excessive Chinese bleeding would reduce profits too much.

Of course, today's China has nothing to do with that dark era. However, the strategy of the Atlantic powers has hardly changed over time, and where it once used missionaries, it now employs devoted pro-democracy and human rights foundations such as the NED funded by the relevant government agencies.

But democracy has little to do with the real problems of today's Hong Kong. Hong Kong never had democracy with the British, who made it a condition for retrocession so that the fox could keep putting her foot in the henhouse —so not much nostalgia in that regard. What really squeezes the shoes of the Hong Kong people are the economic hardships, the foolish price of housing and the property crisis, coupled with the lack of prospects in the face of the loss of status of the former colony. It' all about the decline in the standard of living and how to get ahead in life.

This is the only thing that can mobilize people for months and months. Ironically, at heart the main discontent is against the turbo-capitalism that Hong Kong, Britain and the United States have championed, a model that only cares of speculators and oligarchs. How can it be then that this discontent has been deflected against the Beijing government? It is repeatedly stated that the rulers of the capital have made a deal with the local oligarchy in exchange for their political support.

Now, this is what happens routinely in the US client states scattered all over the world including Euroland itself, although here there is another hierarchy in subservience that passes through Brussels and Berlin. And when it comes to autonomy and sovereignty, we can ask the Greeks, for example, what degree of self-determination they enjoy. And they are certainly not the only ones, as the Spaniards know very well. In fact, elections only serve to make us forget a little about it.

As for depressed environment and precariousness, it is becoming more and more widespread, also in affluent Europe and even in Germany itself; but in Hong Kong it stings much more because the rest of China grows as much as they shrink. Nor does one want to remember that her welfare was built on the enormous inequality with mainland China, from which it benefited in every sense.

One would even say that the Hongkongers have been respected much more than the Greeks. The local oligarchies sell us everywhere, the only difference being the ultimate power centre. Washington is obviously the global capital of the oligarchy and they don't want to lose clients anywhere.

So we can hardly see the conflict in Hong Kong as a struggle between liberalism and autocracy. The "really existing neo-liberalism" is an oligarchy that only uses elections as an excuse, and there is little more to talk about. It is true, however, that this is a "complicated" issue, and not just because a few have taken on the task of complicating it.

In the Greek case we have already seen openly how decisions emanates from central banks, not from the ballot boxes; but central banks are not really public entities but the coordinating body of private banks. And so the world of private interests puts its foot directly in the shoes of the public policy without going through the slightest democratic control. This breaks any symmetry and balance of powers from the outset, and makes the opposition between effective liberalism and central power completely misleading.

The Federal Reserve is as centralist in its structure as a power can be, with the difference that its absolute priority is the interest of an extractive and speculative oligarchy. It is centralist and hierarchical, not at all decentralized. Neoliberalism has nothing to do with decentralization.

In China, by comparison, there is a "changed foot reality", so to speak: one of the feet of political power gets into one of the shoes of economic power, while in the West the banks not only put one foot directly in one of the shoes of politics, but take the helm too.

The Chinese system still has more room for maneuver to regain the balance than the Western system, since the line between private and public banking was always unclear; it could take advantage of this clearance to look decisively at the future and take in-depth measures. Of course, we know all too well that "politics" is not synonymous with transparency or the priority of the common good.

In any case, this system does have room to break out of the vicious circle of public and private debt, while in the West ending the fractional reserve system that make it possible threatens to destroy the axis and lever of plutocratic power —in the end our only, truly reference.

China and Hong Kong could even engage in a monetary decompression chamber experiment in relation to international markets, which would still maintain continuity with the dynamics operating since 1949. And since reality shows a slow-motion turning point, the challenge would be to lead the change of sign: from money as debt pumping upwards and towards speculative interests, of which the Hong Kong skyline is the most eloquent manifestation, to public money transferring monetary sovereignty downwards to the citizens, not to another monetary authority at the service of private banks.

This would change the economic and political landscape top to bottom; today economic democracy, public sovereign money, is a hundred times more important than polls. Even the Governor of the Bank of England pondered at the last meeting in Jackson Hole the convinience of ending the Federal Reserve system —and he was talking side by side with its current Chair Jerome Powell. But the aim of this change would be to grab still more power through the new options that electronic money and criptocurrencies allow. More plutocracy and more impunity, as they only would set the rules.

But, so they say, it seems that the Chinese government also has plans in this regard.

An island within an island: such is the ideogram, the emblem of the current situation. But which two islands? There are several metaphorical and literal candidates, and a number of surprising combinations. As always, reality keeps winking at us, even if we don't know what to think about it.

*

In contrast with Opportunity, reigning supreme as blinders in politics, there are glimpses of Synchronicity in events that escape whatever machination; they are often noticed by those most alien to power, even by historians at the end of the day when they stop working and pursuing their theses. Thus, for example, the Taiping rebellion preceded and coexisted in time with the American civil war. So what?

For me it is full of meaning that China acquired its maximum territorial expansion with the Qing around 1800, only a few years before its ruin and darkest stage began. These things happen all too often, and we don't need to say that external expansion has nothing to do with the welfare of the people or even with the consolidation of power.

The American Civil War had an almost diametrically opposite sign. The historical moment that presides over it is certainly not the liberation of the oppressed, in this case the black slaves, but the expansion of the Union and of the future empire, with the concentration of powers of the federal government. In short, it was the first great shock of an unstoppable expansive wave.

Now, if we keep thinking in terms of maximums and minimums, things seem to return the other way back to describe a semicircle. The Anglo-Saxon influence has reached its peak, and one would say that since 2016 —Brexit and Trump in the polls- has begun a certain decline, and if it is more or less pronounced only time will tell.

This really matters as the centrifugal tendencies that exist in any state also depend to a great extent on the evolution of its sphere of influence. The European Union itself, after a rushed expansion, soon began to experience melancholy and the effect of disintegrating forces, and that's the current state of affairs. The United States of America seemed immune to these ailments until now only because the increase of its influence grew unstoppable, but since Trump the forces of discord take command, no longer at the party level, but between "business models" for the empire or in the internecine wars between the multiple bodies and agencies. No doubt there is a great potential for fission, just as there are plenty of cracks to exploit it.

As the ultimate expression of capitalism, the United States finds in expansion its raison d'être, and the day it reaches the limit, its impatient internal elements, so used to growth, can come to a boil. As for the disintegrating horizon of the current Britain or of a European Union incapable of approaching Russia, what can be said? There's all kind of signs that the fate of the West is reaching a limit in respect to its expansion, and and that will profoundly affect its internal dynamics.

At such a delicate juncture it is not very intelligent to sow seeds of discord in the ground of your neighbor and ally with centrifugal forces, even if that is what you have been doing all your life, because everything is entering a new dynamic. While you are so attentive to your opponent's face, you could be getting a pimple in your ass, or even at the very tip of your nose. There are more reasons for concern than those we have indicated here, but since they are so astute, let them worry about finding them.

CONSTANT BEAR AT WOBBLING ROCK

14 November, 2019

«I bequeath the Constant Bear movement to the elderly, the sick, and the frail. It is a wonderful, traditional exercise which is both simple and easy. You can also use it for self-defense until you are years old. All this is easily obtained. Although my explanation is short and simple, if you understand its principles and practice with perseverance, after as few as one hundred days of moving your ch'i, you will notice a marked improvement in health and strength and no longer need to worry about illness. It is truly a 'sacred raft' to strengthen our bodies and bears no semblance to other well known yet inferior exercises» [1].

With these words Cheng Man Ching entrusted to posterity what he considered the most concentrated possible synthesis of the principles of Tai Chi Chuan: the Constant Bear exercise. Were his expectations too high?

More than 44 years have passed since the «master of the five excellences» left, but although his conception of the martial art has spread remarkably throughout the world, the exercise that was called to be the great disseminator of Tai chi among the most varied sectors of the population remains little known despite its extreme simplicity.

Contradictory, no doubt. But we must not forget that Cheng himself warned of the elusive character of the principle underlying the exercise: «Although a thousand ounces of gold is obtained easier than the correct method, even this is easily dropped when passing it on, washing your life out to sea» [2].

Such an elemental movement of synchronized torsion leads very easily to underestimate its potential for infinite transformations like a golden thread that connects the most diverse actions and postures: «The Constant Bear combines both the Five Animals and tai chi into a single move, look for it and you will receive it, neglect it and you will lose it» [3].

So, we have a remarkably simple movement but one that contains a far-reaching principle, as easy to ignore as it is difficult to apprehend. Tai chi at its purest.

Tai Chi Chuan has no opinion.
It has no intention.
It is an idea without motive.

It is an act without desire.
It is, properly, the natural response to an outside force,
not being perceived as such.

For in nature, all are the same, everything is one.
That which attacks is the same as that which responds, t
he same force —redirected and recycled.

When you initiate an ill-intentioned move,
it comes back on you.

The principles of Tai Chi
are the same principles behind the inner mechanism
of the great engine of the universe [4].

For many of us who have practiced Tai Chi, the origin of this art and style of movement is a fascinating enigma. Today it is generally accepted, as a matter of mere common sense, that the forms elaborated with a sequence of movements, by which the different styles of this art are distinguished today, are the ultimate result of a long evolution; that is to say, they are an arrangement or synthesis of quite elementary postures, gestures and moves that in earlier times had to be practiced separately. Some of them, such as the commencing and closing form, the push forward, or the cloud hands, are among the most primary, and although they have undergone transformations of all sorts, still convey the whole flavor of this art.

And in this way, today's best-known classical routines, whatever their style and whether they have 24, 37, 66 or 108 movements, entail a level of complexity that we could compare with that of symphonies, although no one doubts that before the symphonies there had to be songs, and before the songs, chords and tunes. It cannot be any other way.

The endless disputes between styles and lineages tend to be about issues of historical priority or about who has best captured the essence of what seems to be quite a timeless art by itself. The first may be settled by dated documents within chronologies, but always provisionally as well — because, just as the paleontological fossil record, the documents that fall into the hands of the scholars, even if they are authentic, are always a tiny part of the unknown documents that could still emerge to disprove the best founded assumption.

Thus, for example, the still more commonly accepted idea is that Tai Chi, as we understand it today, has its best documented origin in the lineage of the Chen family from the 17th century and specifically in General Chen Wangting (1597-1664); however now various scholars such as Christensen argue, based on some documentary evidence, that the first classic on the subject comes from Li Chunmao of the Li family from the Tang Village, Henan, and would date from 1590. Perhaps this does not introduce a great dissonance in the now admitted chronology, but it does in the lines of transmission, which in turn opens new questions, as that same Li family counts its ancestors up to an almost legendary Li Dao Zi, born almost a thousand years before [5].

Disputes over the essences are, as might be expected, much more difficult to demarcate, though not less important. In fact, for those who seek in Tai Chi a living, always renewed experience, they bear much more weight than dates and chronologies, closer to historians and scholars: they speak not from the paper and ink of old manuscripts, but from the very flesh and spirit of the practitioner.

Cheng Man Ching's position on this the question of origins contains a hypothesis that looks rather speculative in the historical sense but appears quite irreducible when it comes down to the essential.

And in fact, in Cheng's perspective, the (pre)historical view fits into the timeless without need of greater details. Since it is clear that he didn't pretend that the physician Hua Tuo (c. 140-208) was at the origin of modern Tai Chi, but of the common ancestor of both Tai Chi and the Wuqinxi or Five Animals Frolics, which already included the Bear and that are effectively attributed to him.

In all this evolution what came first had generally less differentiated forms, and in addition, we have not received detailed instructions on how Hua Tuo's Five Animals could have been. Since then there have been a large number of exercises imitating the gait and traits of the bear, so the original pattern is vague enough to admit a wide progeny —though not so much so that the lineage cannot be tracked. Here, too, we need some sense of proportion.

Cheng himself didn't leave us a recorded demonstration of how he understood this incessant swaying of the bear, so we have no choice but to return to his words:

«There are three main points to concentrate on when practicing the Constant Bear. They are:

First, the Constant Bear features the constant swinging of the waist to and fro, right and left. It should be done about thirty minutes after your morning and evening meal. Those with weak bodies should start with two to three hundred swings. Always increase, never decrease, the numbers, strive for gradual progress. Increase the movements until the total time spent is between ten and fifteen minutes, and remain in a pleasant mood.

Second, when you practice the Constant Bear, don't let your head hang down as real bear does, but combine the swing with the form called the Looking Owl, where you look straight ahead. Your head should not move independently, but moves aligned with the navel. In Tai Chi this is called 'The light sensitive energy rising to the top of your head' and 'Keeping the weilu centered and straight so that the spirit rises to the top of your head'.

Third, as in Tai Chi, distinguish clearly between full and empty, as expressed in the statement 'Heavy as a mountain, light as a feather'. When you turn left, your weight should be fully on your left leg, making it as heavy as a mountain. Do the same when you swing onto your right leg.

When you practice be sure to keep your mind and chi in the dantien, about 1,3 inches below the navel. Also take care that the soles of your feet are fully on the ground» [6].

Although what we have here is a single alternating motion of swing and turn, there are still many loose ends that allow for diverse executions both in appearance and spirit. Characteristically, we can distinguish between matters of principle, which are prescriptive, and matters of degree which are left to the discretion of the practitioner. For example, nothing is said here about how much one should bend the knees or the angle that may describe the axis of the body trunk in its swing; not to mention other aspects such as the rhythm, the height and motion of the hands or if the swing and the turn should be done altogether or separately.

In any case, the move, itself an incessant closed circuit using the existing momentum to shift the direction with minimal effort, literally embodies the popular saying attributed to Hua Tuo: «Door hinges that are used never rust». So the attribution of this very exercise to the physician of the Han dynasty, although not demonstrable, is at least judicious and pertinent.

For there is little doubt that many centuries before the elaboration of specific martial routines with different choreographies there had to be movements that embodied the spirit of Tai Chi in many different ways. After all, any of its lineages claims to be heir to Yangsheng and Taoism, which we know are thousands of years older! Besides, it is just in the natural order of things that these arts must have evolved from vague attempts and moves to the most refined, diversified forms.

Then the Constant Bear would find its place at that crossroads, somewhat hard to conceive, between the «casual» primitive movements and the more conscious, deliberate and tailored recollection of a repertoire in an organized sequence. And it is here where the magic of the Origin as remembrance exerts all its ascendancy.

Let us understand a dilemma that it is not such. On the one hand, it seems obvious that the exercise proposed by Cheng Man Ching, after a lifetime dedicated to the study and synthesis of these arts, cannot be the same as a «primitive and spontaneous move» that, moreover, tries to imitate the already free frolic of an animal not compelled by external restrictions. On the other hand, we already see that this primitive movement as human art, but not a mere contrivance, can only be mimicry —that's to say, a recreation of yet another synthesis, this time a spontaneous one, played by an animal that is freely bringing some harmony to a chaos of instincts pointing in different directions.

But the fact is that Cheng, in addition, attributes the paternity of the exercise to Hua Tuo, like him an extremely experienced and conscious physician, to cap it all the creator of a series of five exercises mimicking movements of as many animals, at the heart of which, as if it were the element earth, stand the Bear —the synthesis of a framework that it is already a summary.

Then we understand that Cheng's argument works perfectly as an alibi: here the only way we can conceive the origin is reproducing it, the analysis will always leave us cold. And here at the crossroads of both Tai Chi's prehistory and Tao Yin health qigong we are set to find another self-conscious physician creator of his own recollection and synthesis. The Constant Bear is both a discovery and a recreation.

What happens is that if there are many possible ways of doing the Bear, they should share certain irreducible traits, which are the same that Cheng highlights, and that also coincide point by point with the most basic principles of Tai Chi. That's why he avoids carefully to give more specifications, which would unnecessarily restrict the scope of action. Convenient also on the other hand, as real animals frolics should be quite free of stifling rules.

So, here the irreducible is doomed to match some not less irreducible degree of uncertainty, and it must be said that here, too, the very guiding Principle of Tai Chi is in action. Liu Hsi-Heng recalled seeing Cheng practicing the exercise in his home in Taipei while editing one of his manuscripts, before saying: «This single move is Tai Chi —there is nothing more than this». You don't need anything else, when the practice is the proof:

«If you do these three things, your 'Golden Rooster Stand On One Leg' will be stable; your 'Repulse Monkey' will not be snatched away by the movement of Sensitive Ape; your 'Embrace Tiger and Return to Mountain' will be as decisive as a tiger's roar; your 'Look Right, Glance Left' and 'Diagonal Flying Posture' will be faster than the Deer» [8].

And if anyone has doubts, let them look for another move with its own entity, only one, capable of summarizing Tai Chi as briefly and completely as the Constant Bear. So, this is the genuine «pocket Tai Chi», the only one that can take place comfortably in a single square meter. The Bear reveals its identity by the fact that it stands alone: it is the greatest shortcut meeting the necessary and sufficient condition. That is why, if I had to choose only one single exercise among the infinite variety that exist, this would be it.

*

If Hua Tuo actually conceived an exercise imitating the Bear to some extent, surely this one did not involve the sort of clear-cut formal principles that could give a refined and self-conscious Tai Chi master of the second half of the XX century, with his decanted knowledge of the history of Chinese qigong and martial arts. However, the awareness of form may be only the exteriorization of a previous knowledge in spirit that can stand by its own, because if form follows function, function follows intention as well. Let us look back again to the past.

It is a commonplace to allude to the more than plausible shamanic origins of Taoism. Among the solo shamanic dances, those inspired by the Bear have held a preeminent place. Ritual walks and dances like the Yubu and the Bugang, recalling the «footsteps» of the legendary Yu the Great, have always been related to the Polar Star, the sevenfold constellation of the Big Dipper (for the Chinese «the Northern Dipper», whose fourth star is precisely called Tian Quan, «Celestial Balance»), that was seen as the Bear for the ancient Greeks and Romans. Although this steps developed with time more and more intricacies, the original idea behind is the static flight through the stars of the far North indicating the egress and re-entry into the cosmos, the high shamanic theme par excellence.

For, according to the old texts such as the Zhengyi xiuzhen lueyi, the purpose of these ritual steps, as a means of self-cultivation, was none other than to lead the practitioner, who was often a ruler, «from the existence to nothingness», a motive as inconceivable as it is absurd for most modern people. So some brief musings may be here in place.

It should be no wonder if Cheng says Tai Chi has no intention or motive. What we call thought is only an intermittent and discontinuous flashing of intentions, basting and giving an appearance of continuity to hollow beads in a string. Approaching Tai Chi involves seeking a different kind of continuity beyond those flashes by suppressing the intention and partiality that colors our thoughts. As there is nothing in the world (of thoughts) not colored by intention, if I suspend mi intent, a passage opens through the gap that allows that nothingness to be perceived, the potential of the mind prior to any of its acts.

Ordinary thought, without realizing it, is continually doing the opposite: it comes out of nowhere, out of unnoticed blank spots, only to enter once and again into «the world», in fact a jungle of thoughts among which we leap from branch to branch like a monkey. The Bear standing on its hind legs was for man a living hieroglyph of the Pole, and the Pole is Intention itself when it is not directed towards earthly thoughts and things —when it is not, so to speak, «on all fours». The Pole is what remains of Intention when there is no intent; «trying not to try» clears the way of Taoism.

The unconditional idea of Manifest Destiny aroused among the westernmost peoples points to an earthly mission in which the spirit is necessarily exhausted. The conditional Chinese idea of the Mandate of Heaven points on the contrary towards a celestial guide that makes possible the renewal of things. In our times the virtuality of this notion has been almost completely lost. The Chinese expression Tianming speaks of a «will of Heaven», but if

Heaven leads us back to the unity of the Supreme Pole (Tai Chi, literally) it is precisely because it has no will, no intention or motive on its own. Heaven is the guide that unties the knots, frees us from each particular goal and opens a space with new degrees of freedom for everything. That is his prerogative and in that consists his incessant and unseen power to renew Time and the times.

As it has been noted, Taoism judges the progress of human self-cultivation according mostly to physical transformations, whether gross or subtle; Confucianism takes conduct as the criterion and Buddhism relies on the manifestations of the mind; but they are only three different perspectives of one single process.

Even in the most elementary physical aspects of Tai Chi, say an exercise such as the Constant Bear, the requirement to keep straight the vertical axis of the spine guiding it from the top of the head looks for the layperson as a rigidity and an unnecessary restriction, when in fact it is the basic condition for free and effortless motion. No doubt, there is a freedom to exhaust oneself as there is a freedom to renew oneself also, and both, even seeing the same thing, look in opposite directions.

In case anyone missed it, this movement of the Bear as we have described it traces in space a pattern entirely similar to the Taijitu itself, the very same symbol popularly known as the «Yin-Yang circle», and not only in two but also in three dimensions and in time as well, which gives it an additional interest and one more vertical sense, beyond duality, to this kind of elusive, uncertain moving between being and not being.

*

Let us descend again to more earthly, tangible aspects. It's well known that many Tai Chi teachers and practitioners have a prominent belly due to the abdominal breathing it promotes. This doesn't need to be wrong, despite the idea now artificially extended that only a flat stomach is healthy. There are rare Tai Chi masters though, that against the general rule, advocate high-lung breathing, don't exhibit this typical belly and use to be stronger and more corpulent than their counterpart. However, we're not going into what is the best way to breathe now, as the most important thing is to do it as freely as possible.

In my case, Yang-style Tai Chi never gave me a bloated belly, but the Constant Bear did, and to such an extent that it forced me to question whether I was doing something wrong. Since this exercise is like a long form concentrate, one could think it tends to amplify any basic problem of performance. The distension used to decrease one or two days after stopping the exercise, and reappeared almost immediately when I resumed it, so the cause was clear enough. However, I could not identify which particular detail was behind the undesirable effect.

There are many kinds of exercise in qigong, and I had explored different styles and variants over several decades; a good number of them have contra-indications or in any case are not recommended for untrained practitioners. On the other hand we have nearly risk-free medical or therapeutic qigong for people with an average or even weak physical condition. Undoubtedly, doing the Bear belongs to this last category and that is why Cheng Man Ching did not hesitate to recommend it as an ideal way of recovery for the people with frailer health. And despite this and the absurd simplicity of the exercise, I was unable to detect the origin of the problem.

The judgment on this bloating phenomenon is complicated by the fact that, to some extent, the practice of this routine leads quite naturally to develop power and build up pressure in that area; but if it feels uncomfortable surely there is something wrong with it. Internal pressure should circulate and distribute, rather than accumulate, though both processes are closely interrelated.

So I consulted instructors from Cheng's lineage, but the two I met did not seem any better than me. One of them, an excellent professor who devoted about an hour a day to this exercise, had a belly as prominent as a watermelon, so I didn't even want to raise the issue since it was obvious he had not solved the problem either. He was also performing a markedly low form with very bent knees, that at least for me was not the most convenient to distribute the overall pressure without tensions or obstacles. However, there was a very good point in the slowness and firmness of his posture. One cannot judge by the form only as the overall effect depends on the physical condition, the depth of relaxation and the present sensitivity of the performer.

Later on I asked Jacob Newell of the Old Oak School of Dao in Sonoma, California. He also admitted that there's definitely a pressure that builds up in this practice, although the Ruyu style of Tai Chi developed by C. K. Chen wants to drop it all the way into the foot[9]. It took Jacob years to bring the pressure to the ground after first unblocking the hip joint and then the knee joint. To get to the bottom he used as a method standing on a wobbly rock.

Using a wobbling rock is an excellent way to take advantage of our innate sense of balance to loosen joints and unify body coordination. These are three related aspects that I had already developed to a good degree throughout a long practice of Zhan zhuang, that is, standing postures like the well known «embracing the tree». So I was convinced that I was already letting the pressure drop and that the cause of the abdominal bloating had to be elsewhere.

Immediately after this advice I noticed that it is not usually specified how much one should turn sideways, and that I tended to overdo the turn looking for a greater workout, something between 150 and 180 degrees. This tweaks the middle part of the body, catching and strangling the pressure in that area. I tried with more moderate turns, about 120 degrees or less, and the problem disappeared almost completely. Just a few days latter I found an article from this same Old Oak School that specifies a total turn of only 90°, 45° to each side[10]. After this, I have found that there is no need to make extreme turns to increase the performance, as there is no need to lower the stance too much. More than the height of the posture, it is the perception of weight what needs to go down. In Tai Chi less is often more. But of course my experience is very limited, and it would be desirable to know a wide range of opinions[11].

However, I was left with the idea of using a wobbling rock or board as an aid to develop quickly that unified sense of the body and its movements which is essential in Tai Chi. In the old days this kind of elemental aids were always welcome and even enshrined means in monasteries and schools. Although it also takes time, there's no need for people to spend years on it with demanding practices such as zhan zhuang when we can acquire the skills in a shorter and more enjoyable way.

So we come to the idea that, if the Constant Bear is already an excellent way to popularize Tai Chi and make its essence available for a much larger number of people, putting it on the «wobbling rock» would be an ideal touchstone to deepen an exercise which too often is only approached at the most superficial level. Especially for those who do not practice long forms, since these are already a means of testing the depth of the foundation.

Like Tai Chi itself, the Bear has a wide margin of possible interpretations and performances; even its practice beyond the Tai Chi precepts, as a mere routine to loosen the body and the limbs, without taking care of the alignment of the trunk, yields remarkable health benefits.

The third and busiest battleground between the various Tai Chi schools revolves around efficiency. Traditionally the different parties have resorted to combat to see who has the last word. But while competition may change sides and opinions, it is far from being a supreme court in which more subtle issues such as the therapeutic efficacy and the effect on health of the different styles can be properly settled.

The diversity of styles in Tai Chi has arisen more as a matter of nuance and emphasis than from the principle itself; and these differences of emphasis just echo differences in the spirit, complexion and sensitivity of the interpreters. A corpulent complexion does not usually have the same sensitivity as a thin one, and sensitivity is crucial in this art, as it unveils the margin of action and the available degrees of freedom. Moreover, Tai Chi would not be an art without that uncertain margin and the need to interpret it. Naturally, as a need to adhere to one practice among others, we all are bound to believe that our school or style is the best and the others are inferior—at least for the time being and until we decide to change sides.

Many practitioners of other lower forms, not less than common sense thinking, will seriously doubt that you can practice such a straight, natural and relaxed form as the Ruyu without losing power and thrust, but if that possibility didn't exist, Tai Chi would not have a special interest. Even admitting that it all depends on the degree of sensitivity, it seems to me that the more relaxed styles are in better disposition to delve into the degrees of freedom of the art, and to pay off right from the start.

We list in a separate note the points to be taken into account in order to really take advantage of the potential of the exercise —according to the Ruyu style. Rather than the waist or the spine, it is the heavy foot what should be feel as the real pivot in each turn. I also find it preferable to concentrate on the centre of the sole of the foot rather than the dantian below the navel; being both the mechanic and the energetic functions of the two areas completely different.

What we want is to realize the potential of the exercise as soon as possible. The biomechanical aspects due to a third-person view would make up the external component of this potential. In a first-person view we can not neglect the sensations we perceive, as they should be just the subtler aspect of this same physical reality. In modern materials science we speak of stress and strain, of tension-pressure and deformation —and this already implies forces both inside and outside the bodies. Here, first-person only means from the inside, and it has to be necessarily subjective. In the body we have soft and hard tissues and parts, that in the old taoist parlance would be yin and yang areas. Related to this, and the associated pressure-tension, or strain and stress, in our sensations as we perceive it, we can speak also of cold and hot energies, or the «water» and the «fire element».

To simplify as much as possible, we can say that in the body we perceive fresh and warm energies, in a positive sense, just as we perceive overheated and cold or rigid zones, in a negative sense; which can be compared to the traditional distinction between «young yin and yang», and «old yang and yin». The lower dantien or «cinnabar field» is so called precisely because it is the natural place for the intimate mixture of cool and warm (as cinnabar is made of mercury and sulphur) sensations or «energies». However, the main location where the freshness enters the body are the (soft) soles of the feet —especially after the residual energies already used have been brought to earth and released.

Working the exercise from the sole of the feet, is more equivalent to «breathing from the heels», or «from the bones», to use two older expressions. That is, from the deepest level and from the cooler extremes, while the centre of the body is the place where the extremes naturally merge. The difference may be great when this is put into effect: like the one between walking along an ordinary route and hiking in the mountains inspiring the freshness of the snowy peaks. However, probably the best method is not to concentrate on any particular point and just be attentive to the overall feeling.

The old neidan representations showing a whole landscape of mountains and waterfalls inside the human body respond to quite an elementary reality, provided one knows how to attend and interpret one's own sensations.

Traditional Chinese medicine tries to direct «the fire» of the body downwards and «the water» upwards to compensate for the natural tendency of these elements to divorce and stop reacting. One can't sum up more complex processes with fewer words. At the biomechanical level, what corresponds to fire and water are tension and pressure. We live between tension and pressure, all our life is a scenario of that interplay, with strain, the ability to deform, as an interface or contact surface between both. The other elements or activities that traditional medicines takes into account arise from the combinations and intricacies of this basic relationship.

In Tai Chi it is better to move smoothly everything, even if little, than to move a lot some parts only to neglect others. Quantitatively, it is no small thing to obtain 50 percent of the benefits of doing the Bear; but qualitatively, that 50 percent is just to stay halfway between that union of the extremes that bears the real fruit. If the great expectations that Cheng placed in this exercise have not been fulfilled, surely it is because we haven't yet succeeded in closing the gap between the possible and the real.

As we said, the wobbling rock is just a touchstone to sharpen and distribute the inner sense of balance affecting the whole body. Anyone training balance on a board learns how to loosen the body and the limbs to optimize the capacity of coordinated response. Balance effectively involves the whole body, and in this art we should always start from the whole. Wholeness is both the foundation and the goal.

In the same way, doing the Bear we can feel that our performance is being excessive or insufficient, too empty or too full. But whether the excess is of one type or another, both leave us with the needed, but not always heard, feeling of failure. The exercise itself seems to complain: «It could be much better if you weren't so dumb». We get the message, that is that we are missing the point.

Full and empty are correlative terms; when the whole body feels homogeneously full, to the same extent the line separating the inside from the outside ceases to be perceived, and by the same token, the very line that separes the full and the empty vanish too. It is no small thing to harmonize them, and yet it is not something that can be «done,» but rather something that we can discover through action, as one discovers an unsuspected path into the mountains that can lead us to who knows what magnificent view. The changing margin, gap or clearance defines the width and turns of the path.

Since everyone has to find their optimum grade of performance, which varies not only with each individual but for the same individual over time, any general rule will be somewhat misleading. Again the words of Cheng Man Ching are worth remembering: «Hua To instructions for the Five Animals were lost except for the two words: 'Until perspiration'. A little perspiration is fine when you practice on a hot summer day, but during autumn or winter you practice only until you feel a slight tingling sensation of sweat on your back and forehead, then rest, do not continue until you perspire» [12].

Anyway, my experience is that one can obtain lots of energy from this movement without the slightest need to sweat. In this respect we could also recall the extended Tai Chi rule of never exercising more than 70 per cent of one's own capacity.

A good orientation on the degree of flexion in the posture could be this: the lower you go, the more you should look for lightness; and the more upright you are, the more you should concentrate on sinking the weight. In this way we explore our own range time and again, in search of the mean between doubly-weighted extremes. A similar «paradoxical scanning» between extremes can be done with other aspects as the optimal rhythm or speed. Firmness and sensitivity should go together if we want to progress.

*

The Constant Bear can bring all the benefits of Tai Chi with an abysmally simple move. Since the swing and return use the momentum of a movement to initiate the next, the effort is minimal and it seems that we could go on endlessly. If rightly has been said that Tai Chi is more enjoyable when more circular is, in no other instance could be enjoyed more than with this single alternate circular move, a kind of charging and discharging oneself — charging oneself with the inner natural pressure which is vitality, and discharging that tension which obstructs the free distribution of that pressure. Body and movement form a single field, they are not separate things. The point is to learn how to delve in the exercise finding the most satisfactory way for oneself.

Incidentally, the comparison with an electric generator or alternator is not entirely trivial, but here the interplay of «currents», if the analogy is allowed, is richer and has added interest, since we are also dealing with a sort of eliminator, or if you will, a double or rather quadruple pumping system. It would be fantastic if we could apply these same principles to our environment and to the atmosphere, and taken with due proportion, maybe we can identify the corresponding natural processes at different levels.

Some claim that if we are looking for a method of maintaining health, there are qigong series such as the 18 hands of Lohan (Shaolin) or the famous 8 pieces of brocade (ba duan jin) that are as effective as Tai Chi and take much less time to learn. But if one can learn these series in a few sessions, you can learn the Constant Bear, or «Taijitu exercise», as we might also call it, in five minutes without renouncing in the least to the whole mystery of Tai Chi. Who can beat that?

There are exercises such as zhan zhuang, the cultivation of standing postures, which can be equally powerful and deep. Some Tai Chi practitioners, especially from the Chen school, include them to give firmness and power to all their movements. But, in our hyperkinetic society, it is increasingly difficult to find people with the patience to stand still for appreciable periods of time — for most of us this is little less than torture, and at best, a forced «holding the posture» that by itself blocks the access to deeper levels of practice. This kind of cultivation has never been very widespread, and today it can only be explored in depth by a tiny minority, far less than 1 percent of the population.

Moreover, such static positions make supervision almost indispensable, because if you don't apply properly the principles of body alignment or are not relaxed enough, you can reap as many physical problems as benefits.

As our environment makes us build up tensions, exercising with movement is much more attractive since motion is the most basic way to alleviate tensions and get them out. Although generally slow in its forms, Tai Chi does not fail to use the advantages of motion and dynamism. Some merely indicative statistics speak of around 5 per cent of Tai Chi practitioners in China and 1 per cent in the United States of America.

Even resorting to motion, for the majority of people, both in the West and in China, Tai Chi remains far too slow and complicated no matter the simplicity of its universal principle. However, the Constant Bear allows us to imbue ourselves completely with its spirit in such a basic way that it openly challenges our intelligence.

Of course, nobody is saying that the Bear should replace the long and developed forms; on the contrary, it is an insuperably simple introduction Tai Chi that cannot betray its essence in the least.

In addition to being far more accessible, the Bear's routine has many advantages: It does not require a large space like traditional Tai Chi forms, being ideal for both indoor and outdoor practice; nor does it require changing clothes or waiting long after meals. Since you don't have to memorize different moves and steps, the attention is not distracted with diversity and soon becomes focused in the most important, that remains always beyond the form.

Other major advantages are related to research in geriatrics, rehabilitation, and the medical effects of Tai Chi or qigong. Everything is greatly simplified when the sample population for a given study only has to learn and practice one elementary movement. In addition, most of this exercises to maintain or regain health require the practitioner to have at least an average physical condition, while a movement such as the Bear, which admits such a wide gradation of intensities, can be practiced by anyone of any age who has the strength to simply walk or stand on their own legs.

The importance of exercise of any kind for the maintenance of health cannot be underestimated, as it is something that even the best environmental or nutritional conditions cannot replace. We know too well, however, that the circumstances in which almost all of us live are far from being such, that our organism has to fight continuously against all kinds of aggressions —physical, chemical and psychological- and that we approach a looming global health crisis as the aging population grows in developed societies.

It is clear that the great triumphs of the contemporary biomedical model have been the fight against infectious diseases, due both to the improvement of hygiene and medical conditions, and the prodigious advances of surgery. But all this, and what remains to be done in the same direction, has very little to say about degenerative diseases, which are just the latter emergence of the way we have lived and our embodiment of the external environment. And although we will not stop trying new technologies to avert from our own flesh the natural unavoidable reaction, it will not be without great economic and moral costs.

Aging is like going bankrupt: first it occurs gradually, then all of a sudden. One cannot deceive nature, one cannot simply reverse a process which has taken many decades to develop before emerging. Whoever believes otherwise, believes in magic; but we all know that modern medicine is largely based on the idea that there is a «magic bullet» for everything. If we have been able to believe these things, it is because of the very short time elapsed between the great triumphs of biomedicine and the ongoing accumulation of the pernicious effects of modern life, which are just beginning to emerge. Moreover, our very success in extending the life span multiplies the incidence and impact of degenerative diseases. No doubt our perception of these serious issues will change within thirty years, not to mention longer periods.

At the other end of the spectrum, we find that the greatest problem now and in the future for the development of kids is hyperactivity and the inability to maintain concentration, clinically known as «attention deficit disorder». Of course, there are all kinds of children, and individual impulses, along with the family environment, are a big part of the issue. But before making of them premature sick persons depending on medication from an early age, we should admit that the first imbalance factor is our rushed society constantly sending contradictory stimuli and producing all kinds of artificial excitement.

In this context of permanent mental excitement, I think it would be fine to invite children to «imitate the movement of the Bear» so that the energy of their bodies, rising continually to their head and limbs due to the extraordinary demand caused by over-stimulation, can descend again towards their natural centre of gravity and settle. One might think that an exercise like this is too slow for children's impulsivity, but if we can skillfully combine the appealing elements of imitation and the paradoxical play between the extremes, like feeling fast when one moves slowly, and vice versa, or feeling light when going down in the posture and heavy when raising it —things that children love and for which they feel very willing- we can bring them back to serenity and help them create harmony in their own inner environment. There is a great spell for them in going between two oscillating points that calms the mind and the mood. This could be a great help for both a genuine physical education and the formation of character.

*

The five arts or excellences Cheng Man Ching talked about as various expressions of the same guiding principle were Chinese calligraphy, poetry, painting, martial arts and medicine. That Tai Chi and calligraphy share the same guiding idea is quite obvious: the brush must to be held upright and with a loose wrist, in the same way that in Tai Chi one moves with a straight spine and joints as free as possible.

As for poetry and painting, it is also clear that both aspire to the «suspension of ordinary reality,» actually only a very narrow slit; that is, they try to suspend the intention and the narrower utilitarian purpose in order to access a more comprehensive, higher view, that we associate with the so called «contemplative mood». And what is Tai Chi but the inquiry and implementation of the Law of Levity, such a skillful use of movement to unlearn the usual sense of Time?

Even if one aims at closing the gap between these disciplines, they always coexisted without discord. The connection with traditional Chinese medicine, so open to gradations and nuances, was not very problematic either; the real trouble has arisen with biomedicine, and in general, with the framework of modern science. And yet it is not impossible that surprises await us here, the day we put in perspective some mutual misconceptions.

In Tai Chi motion is used to access a potential implying an indefinite background. Along with many other masters, Cheng Man Ching used to recall that «stillness in movement is true stillness». In physics, the moving force has pre-eminence and everything is deduced from it: mass and motion, kinetic and potential energy are well defined… by definition, and in fields also by choice. However, the systems where in practice these components are not well defined are legion: from the very biomechanics of the human body to the climate or the economy. Actually, everything that has a margin of internal uncertainty can be interpreted as something more than indeterminacy.

In other words, physics deals with the reaction of systems subject to external forces, while Tai Chi, as Cheng puts it in his poem, explores «the natural response to an external force when it is not perceived as such». It seems an insoluble riddle, and yet it somehow sums up both the differences and the possible forms of contact between the two perspectives.

¿?

A charged particle feels the potential
even when the electric and magnetic fields are zero
the physicists said it was incomprehensible
and that that is the way things work in quantum mechanics
but today some admit it's a very common occurrence
wringing a towel
screwing a light bulb
or parking in parallel
they call geometric phase to this shift
global change without local change
It waves in the snake advance
or in the turn of the falling cat
that so much amazed Maxwell, good old friend

«Useless and treacherous potentials»
said Mr Heaviside,
who only wanted to get rid of them
but they are no stranger than you and me
when we are doing nothing
without disappearing for that

The Bear perceives it
even when it seems he is playing dumb
half-swaying from side to side
half-falling back and forth
but still standing, however
learning how to walk before the first step

Uplifting and sinking
shrinking and expanding
hollowing in and releasing
alternatively but simultaneously
describes the spherical vortex of the world
without lifting the plants from the ground,
stalks the unwobbling pivot
of the Constant Mean
and clears the path of its ascent and descent

Silently,

between Heaven and Earth
the slow gate to the sublime
turns

We all immediately associate rigidity with death, our instinct knows it too well without need of more explanation. The dilemma of Taoism was how to harmonize intelligence and instinct, since we cannot survive without both, neither as individuals nor as a species. A rigid body cannot create balances on a wobbling board, it falls or does not fall; but a flexible body with internal margins does both things at the same time, because it also knows how to «fall inwards» to continue standing. There is still a lot to learn from these simple motives —or not, depending on how we understand the word «learn».

The best thing about being an aspirant is that it demands inspiration, because only the sage or the dead does not require it. The sage, on the other hand, probably does not observe stillness because it is necessary, but to honor the potential: and although this one is present in everything, it can hardly be made evident in that which moves, except by the very fact that it moves. From this point of view, movement is the most trivial part of reality, even if nothing is trivial. When a potential is exploited, it recedes; when it is observed and appreciated, it enriches. Between both extremes there would be a certain margin of responsible use that should not be exceeded. Where there is uncertainty, there is always the possibility of putting in our sense of balance to learn where the limits are, but as long as our study of nature is limited only to her exploitation and prediction, there is little room to learn how to do it.

*

Tai Chi has been one of the best ambassadors of Chinese culture in the rest of the world, and Cheng Man Ching played his good part in this «journey to the West». Yet an important part of his legacy still awaits to come to fruition. However, there is the well; whether we draw water from it or not, it is not his problem, as nobody is claiming property rights here.

No doubt there are different countries and climates in the world, but for the Tai Chi Nation, aimed at the Pole, all regions converge.

A white bear and a black bear,
trying to catch two mice
finally stirred up the honey

I can see people gathering around a board, a stone or an unstable slab discussing the best way to do the Bear and talking about how they have regained agility and security walking, that wonderful lightness and carelessness they had when they were young. And in truth, if you persevere in practice and don't give up, you get to walk like you never did before, because the body-movement has find an entirely new capacity and added degrees of freedom. And although patience always pays off, with time enough per session the effects begin to be felt from day one.

Naturally, trying to embody the Taijitu and get in phase with it is more than a physical exercise. It is an aid and a creative means for self-cultivation, which allows us to establish a direct and infinitely nuanced relationship between the highest and lowest in us, between instinct and intelligence, between force and sensitivity. There is no need to expand any further on this as only the practice allows us to understand it.

What Tai Chi seeks begins where non-doing is perceived within the apparent doing; this may seem too vague, and yet it provides the unmistakable evidence, far more important than any technicality. There is an intelligence that tends to dull itself in the mechanical as there is an intelligence that grows with sensitivity and creates new space in its self-perception, and it is wonderful that we can grasp the latter even in the natural movement of the body.

Notes

1. Cheng Man Ching, *Master of Five Excellences*, p. 113-117; traducción de Mark Hennessy. Frog Books, 1995.

2. Ibid.

3. Ibid.

4. Cheng Man Ching, Taichi, https://taiji-forum.com/tai-chi-taiji/tai-chi-philosophy/poem-prof-cheng-man-ching/ https://taiji-forum.com, Author Nils Klug.

5. Lars Bo Christensen, *Tai Chi -The True History and Principles*, 2016, ISBN 978-1539789314. Véase también Fighting Words de Douglas Wile, 2016: http://www.wooddragon.org.uk/fighting_words.pdf

6. Cheng Man Ching, *Master of Five Excellences*, p. 113-117.

7. Ibid.

8 Ibid.

9. According to the words of the Old Oak School of Dao, the form developed by Cheng Man Ching is a more upright and relaxed Yang style, and the Ruyu style developed by the master Chen Qu Kuan (known as C. K. Chen) modified Cheng's form to make it even more upright and relaxed.

10. Sam Edwards & Frank Broadhead, *C.K. Chen's Vertical Axis Taiji-quan.* http://oldoakdao.org/yahoo_site_admin/assets/docs/Ruyu.Like_a_Tree_Frank_Sam.1210359.pdf

11. Jacob Newell gave me in a personal communication these guidelines for the exercise according to the Ruyu style: 1) keeping the weight properly balanced in the bubbling well – in Ruyu-Taiji we call the middle of the arch (not Kidney 1) the bubbling well; 2) keeping the knee bent but not too bent; 3) "song-kua" meaning relax and tuck the hip; 4) drop the lower back, but not forcibly; 5) float the head top (drop the chin but keep the throat open); 6) shoulders relax, elbows sink, armpits breathe; 7) completely separate the weight shift from the turn – this is specific to Ruyu Tai Chi, and not all Cheng Man Ching's styles do this.

12. Cheng Man Ching, *Master of Five Excellences*, p. 113-117

A GOOD REMINDER (UN RECORDATORIO

28 noviembre, 2019

Life is a river
I am life
I am a river
A river has no hands

La vida es un río
Yo soy la vida
Yo soy el río
Y el río no tiene manos

Muchas gracias a "Furia del lago" / *https://furiadellago.blogspot.com/*

URINE AND CORONAVIRUS

12 March, 2020

During the Chinese New Year holiday, I think it was January 29th, I had a feeling of nausea after a walk after I had finished lunch and when I got home, as I probably had some fever, I got into bed and decided that I was not going to eat anything or take any medicine until I felt good; I was just going to drink my own urine. After four or five hours I woke up perfectly.

I live in Mianyang, Sichuan province, and those were the days when the so-called epidemic started to gain momentum and and there was hardly any talk of anything else, but I am not a hypochondriac and the risk seemed minimal to me. Fasting and drinking my urine is something I do whenever I have symptoms of flu or a simple cold, and it usually stops within a day, sometimes just within 12 hours. John W. Armstrong said so in his popular book «The Water of Life», and it has always worked for me.

The truth is that I have no idea if it could be the coronavirus or anything else, but I don't usually stay in bed because of a cold, flu or gastroenteritis. A lot of people who have the virus remain asymptomatic, or with very mild symptoms that go away on their own.

But what is true is that stopping eating and drinking the urine is very effective against the common flu, for what I know much more practical than any known medicine. What medicine cuts totally a flu in 12 or at most 24 hours?

I know these things seem pure quackery to many, but I don't care. It is enough with just one person this can help.

There must be people for everything in this world, and of course I respect the selfless work of doctors and all the health personnel. One of the reasons for overstating the scope of the epidemic is to prevent our health systems from being completely inundated. Collective matters does not work as the individual ones, but they begin and end with the individual.

The truth is that for now this coronavirus is very similar in its effects to the previous known flu viruses; it may have some higher mortality but it is not very different, and this is generally admitted by researchers and specialists.

My opinion regarding viral and microorganisms-motivated diseases is the same of Armstrong and many other renowned naturopaths: the so-called vector of infection is much less important than the condition of our internal environment. Viruses and bacteria are everywhere in large numbers, starting with our own organism. The main difference is which conditions make them proliferate —make them, as we usually say, virulent.

This should also be taken into account so as not to spend all day looking for scapegoats. The very word «infection» is highly infectious.

Armstrong said that much of our colds and flus were due to excess waste accumulated in the body by our diet, mainly due to our high overdose of starch and protein. Stopping eating and drinking urine suddenly reverses this tendency and greatly helps the body to eliminate this excess.

But it is clear that in half a day or a day we can only remove a small part of the accumulated debris. Why then are the symptoms cut off so quickly? First of all, it seems that the urine provokes a very fast positive immune response, decisively boosting body's defenses.

Secondly, I have always been surprised at how quickly you notice the increase in breathing capacity after a few hours of drinking the urine without eating. The feeling is that the lungs are expanding in the rib cage in a free and wonderful way. No doubt this has to be related to the immediate improvement in the elimination of mucus and waste in all the lung tissue and not only in the bronchi.

It is entirely logical that the cleanup is much more immediate in the spongier tissue than in the rest of the body: the density of the lungs is by far the lowest in relation to the fluid that circulates through them. This is why the improvement in respiratory function is so rapid that it even suggests a mechanical cause. And indeed, there is an elementary biomechanical cause.

Let's take a brief look at the clinical signs and evolution caused by the new coronavirus. From the first symptoms, mainly fever, to the severe acute respiratory syndrome, it takes about 8 days on average. From the point of view of the speed of the body's reaction to the urine fast, this is a very long time, which should already give us a fair bit of tranquility.

It is a long way from the first symptoms of dyspnea to acute respiratory distress, which can lead to heart damage and cardiac arrest. There are two main keys to the worsening of symptoms: depressed immune response and the impaired ability to remove waste products that quickly build up in the lung tissue. In both cases the effect of a urine fast is very rapid and absolutely positive.

Armstrong notes that fasting with urine has much quicker effects on flu than normal fasting with water alone. This is my experience too, which must be due to these two points, the positive immune reaction and the boosted elimination.

Honestly, and although everyone is free to think what he wants, to consider urine dirty is to turn things upside down. It is a blood serum filtered by the kidneys, and in a few hours without eating it becomes clean and transparent. Leaving aside urine infection, in which case drinking is totally contraindicated, the only thing that makes our urine unreliable is precisely the byproducts of the heap of dirty things we ingest, including highly toxic medications, not the urine itself, which is quite similar to the amniotic fluid.

In 2016, Yoshinori Ohsumi was awarded the Nobel Prize in Medicine for his studies of autophagy and the therapeutic effects of fasting. Hopefully, the plain evidence that fasting with urine quickly strengthens our immune response and dramatically facilitates elimination, especially in hollow or spongy organs such as the lungs, will soon be properly analysed.

Bibliography

John W. Armstrong, *The water of life. A treatise on urine therapy.*

Those who like simple problems can try to demonstrate this relationship before moving on. It's insultingly easy:

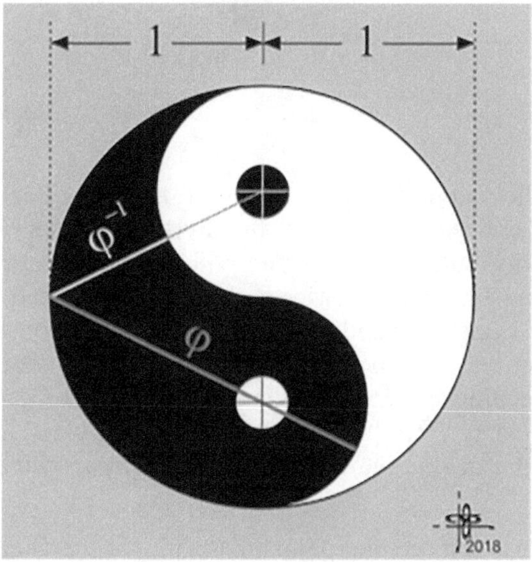

$$\varphi = 1/\varphi + 1 = \varphi\text{-}1+1 = 1/\varphi\text{-}1$$

We owe this fortunate discovery to John Arioni. The elementary demonstration, along with other unexpected relationships, is on the site *Cut the knot* [1]. The number φ is, naturally, the golden ratio (1+√ 5)/2, in decimal figures 1.6180339887…, and φ^{-1} is the reciprocal, 0.6180339887... . And since its infinite decimal places can be calculated by means of the simplest continuous fraction, here we will also call it the *continuous ratio* or *continuous proportion*, because of its unique role as mediator between discrete and continuous aspects of nature and mathematics.

This looks like the typical casual association of recreational math pages. One can get φ in many different ways with circles, but to my knowledge this is by far the most elementary of all, being the radius the unit of reference. In other words, this relation seems too simple and direct not to contain something important. And yet it has only recently been discovered, almost by chance.

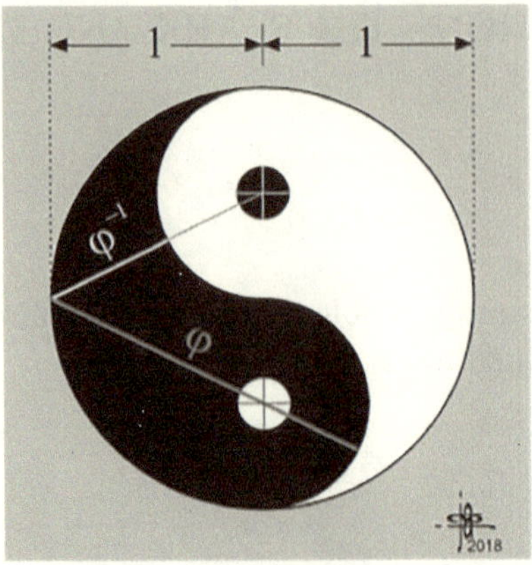

John Arioni

Since Euclid and probably much earlier, the entire history of findings on this proportion has been derived from the division of a segment "in the extreme and mean ratio", and has developed with the construction of squares and rectangles. The most immediate cases involving the circle come from the construction of the pentagon and the pentagram, no doubt known to the Pythagoreans; but one does not need to know anything about mathematics to realize that the relationship contained in this symbol is of a much more fundamental order —just as, from the quantitative point of view, the 2 is closer to 1 than 5, or from the qualitative one, the dyad is closer to the monad than the pentad.

If the circle and its central point are the most general and comprehensive symbol of the monad or unit, we have here the most immediate and revealing proportion of reciprocity, or dynamic symmetry, presented after the division into two parts. The *Taijitu* has a double function, as a symbol of the supreme Pole, beyond duality, and as a representation of the first great polarity or duality. It is, as it were, halfway between both, and both are linked by a ternary relationship —precisely the continuous proportion.

A relation is the perception of a dual connection, while a proportion implies a third order relationship, a "perception of perception". Since at least the times of the Kepler's triangle, we have known that the golden mean articulates and conjugates in itself the three most fundamental means of mathematics: the arithmetic mean, the geometric mean, and the so-called harmonic mean between both.

We could ask ourselves what would have happened if Pythagoras had known about this correlation, which would certainly have exalted Kepler's imagination as well. It will be said that, like any other counterfactual, there is no point about it. But the question is not as much about the past that might have been as it is about the possible future. Pythagoras could hardly have been as surprised as we are, since he knew nothing about the decimal values of φ or π . Today we know that they are two ratios running to an infinite number of decimal places, and yet they are linked exactly by the most elementary triangular relationship.

Mathematical truth is beyond time, but not its revelation and construction. This allows us to see certain things with the insight of a Geohistory, as it were, in four dimensions. There has been speculation about what would have happened if the Greeks had known and made use of the zero, and whether they might have developed modern calculus. This is very doubtful, since they would have still needed to make a series of great leaps far from their conception of the world, such as the numbering system, the zero and its positional use, the idea of derivative, and so on. The double spirals were a common motif in archaic Greece, and the arithmetical speculations of the Pythagoreans, very similar in nature to those developed by the Chinese over time; but for whatever reason the Greeks did not intertwine the two spirals into one, and, in China itself, a diagram like the one we know today did not came into existence until the end of the Ming dynasty and only after a lengthy evolution.

Which is just another example of how hard is to see the obvious. It's not so much the thing itself, but the context in which it emerges and in which it fits. Depending on how one looks at it, this can be encouraging or discouraging. In knowledge there is always a high margin to simplify, but as in so many other things, that margin depends to a large extent on knowing how to make it happen.

The *Taijitu*, the symbol of the supreme Pole, is a circle, a wave and a vortex all in one. Of course, the vortex is reduced to its minimum expression in the form of a double spiral. Characteristically, the Greeks separated their double spirals, and eventually turn them into squares, in the motifs known today as grecas. It is just another expression of their taste for statics, a bent that set the general framework for the reception of the golden mean in mathematics and art, and which has come down to us through the Renaissance.

The series of numbers that approximate infinitely the continuous proportion, known to us as Fibonacci numbers, appeared already long before in the numerical triangles consecrated in India to Mount Meru, "the mountain that surrounds the world", which is just another designation of the Pole. As it is well known, from this figure, called the Pascal's triangle in the West, a huge number of combinatory properties, scales and sequences of musical notes are derived.

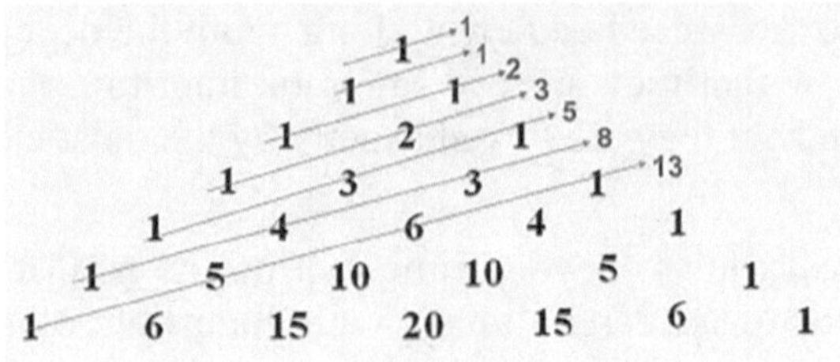

The polar triangle, known in other cultures as Khayyam's triangle or Yang Hui's triangle, is one of those "extraordinarily well connected" mathematical objects: from it one can derive the binomial expansion, the binomial and normal statistical distributions, the $\sin(x)^{n+1}/x$ transform of harmonic analysis, the matrix esponential and the exponential function, or the values of the two great gears of calculus, the constants π and e. It is almost incredible that the elemental connection with the Euler number has not been discovered until 2012 by Harlan J. Brothers. Instead of adding up all the figures in each row, one only needs to extract the ratio of ratios for their product; the difference between sums and products is a motif that will emerge several times throughout this article.

The polar triangle looks like an arithmetic and "static" representation, while the *Taijitu* is like a geometric snapshot of something purely dynamic. However, the rich implications for music of this triangle, partially explored by the work of Ervin Wilson, largely circumvent the separations created by adjectives such as "static" and "dynamic". In any case, if the staircase of figures deployed in Mount Meru is an infinite progression, when we finally see the lines hidden in the circular diagram of the Pole we immediately know that it is something irreducible —the first offers us its arithmetical deployment and the second its geometric retraction.

The oldest known mention of the triangle, albeit a cryptical one, can be found in the Chanda’ śāstra of Pingala, where Mount Meru is shown as the formal archetype for metric variants in versification. It is also fair to say that the first Chinese author to deal with the polar triangle is not Yang Hui but Jia Xian (ca. 1010-1070), a strict contemporary of the philosopher and cosmologist Zhou Dunyi (1017-1073), the first author who publicized the *Taijitu* diagram.

Nowadays very few people are aware that both figures are representations of the Pole. It is my conjecture that all the mathematical relationships that can be derived from the polar triangle can also be found in *Taijitu*, or at least generated from it, although under a very different aspect, and with a certain twist that possibly involves φ. Both would be a dual expression of the same unity. Mathematicians will see what is the point of this.

Between counting and measuring, between arithmetic and geometry, we have the basic areas of algebra and calculus; but there is an overwhelming evidence that the latter branches have developed in one particular direction more than in others —more in decomposition than in composition, more in addition than in multiplication, more in analysis than synthesis. So the study of the relations between this two expressions of the Pole could be full of interesting surprises and basic but not trivial results, and it poses a different orientation for mathematics.

It can be seen that the arithmetic triangle has closed links with fundamental aspects of calculus and the mathematical constant e, while the *Taijitu* and the constant φ lack in this respect relevant connections —hence the totally marginal character of the continuous proportion in modern science. It has been said that the latter, unlike the intimate connection with change of Euler's number, is a static relationship. However, its appearance in the extremely dynamic character of the yin-yang symbol already warns us of a general change of context.

For centuries calculus has been dissolving the relationship between geometry and change in favor of algebra and arithmetic, of not so pure numbers. Now we can turn this sandglass upside down observing what happens on the upper bulb, the lower bulb and the neck.

*

The appearance of the golden mean between the yin and yang in a purely curvilinear fashion not only is not static but on the contrary cannot be more dynamic and functional, and indeed the *Taijitu* is the most complete expression of activity and dynamism with the minimum number of elements. The diagram also has an intrinsic organic and biological connotation, inevitably evoking cell division, which in fact is an asymmetrical process, and, at least in plant growth, often follows a sequence governed by this ratio. In other words, the context in which the continuous ratio emerges here is the true antithesis of its Greek reception that has lasted until today, and this can have far-reaching implications on our perception of this proportion.

Oleg Bodnar has developed an elegant mathematical model of plant phyllotaxis with hyperbolic golden functions in three dimensions and with coefficients of reciprocal expansion and contraction that can be seen in the great panoramic book that Alexey Stakhov dedicates to the Mathematics of Harmony [2]. It is an example of dynamic symmetry that can be perfectly combined with the great diagram of polarity, regardless of the nature of the underlying physical forces.

The presence of spiral patterns based on the continuous proportion and their numerical series in living beings does not seem mysterious. Whether in the case of a nautilus or vegetable tendrils, the logarithmic spiral —the general case- allows indefinite growth with no change of shape. Spirals and helixes seem an inevitable result of the dynamics of growth, by the constant accretion of material on what is already there. At any rate, we should ask why among all the possible proportions of the logarithmic spiral those close to this constant arise so often.

And the answer would be that the discrete approaches to the continuous proportion also have optimal properties from several points of view —and cell growth ultimately depends on the discrete process of cell division, and at higher levels of organization, on other discrete elements such as tendrils or leaves. Since the convergence of the continuous ratio is the slowest, and plants tend to fill as much room as possible, this ratio allows them to emit the greatest number of leaves in the space available.

This explanation seems, from a descriptive point of view, sufficient, and makes it unnecessary to invoke natural selection or deeper physical mechanisms. However, in addition to the basic discrete-continuos relationship, it contains implicitly a powerful link between forms generated by an axis, such as the pine cones, and the so-called "principle of maximum entropy production" of thermodynamics, which we will find later again.

Needless to say, we do not think this proportion has "the secret" to any universal canon of beauty, since surely such a canon does not even exist. However, its recurrent presence in the patterns of nature shows us different aspects of an spontaneous principle of organization, or self-organization, behind what we superficially call "design". On the other hand, the appearance of this mathematical constant, due to its very irreducible properties, in a great number of problems of optimization, maximums and minimums, and parameters with critical points allows us to connect it both naturally and functionally with human design and its search for the most efficient and elegant configurations.

The emergence of the continuous proportion in the dynamic symbol of the Pole —of the very principle- augurs a substantive change both in the contemplation of Nature and in the artificial constructions. of human beings. Contemplation and construction are antagonistic activities. One goes top-down and the other bottom-up, but there is always some sort of balance between both. Contemplation allows us to free ourselves from the connections already built, and construction gets ready to fill the resulting void with new ones.

It is somewhat strange that the continuous proportion, despite its frequent presence in Nature, is so poorly connected with the two great constants of calculus, π and e —except for anecdotic incidences as the "logarithmic golden spiral", which is only a particular case of an equiangular spiral. We know that both π and e are transcendental numbers, while φ is not, although it is indeed the "most irrational number", in the sense that it is the one with the slowest approximation by rational numbers or fractions. φ is also the simplest natural fractal.

Until now, the most direct link with trigonometric series has been through the decagon and the identities $\varphi = 2\cos 36° = 2\cos(\pi/5)$. It has not been associated so far with imaginary numbers, *i* being the other great constant of calculus, which is concurrent with the other two in Euler's formula, of which the so-called Euler identity ($e^{i\pi} = -1$) is a particular case.

The number *e*, base of the function that is its own derivative, appears naturally in rates of change, the subdivisions ad infinitum of a unit that tend to a limit or in wave mechanics. The imaginary numbers, on the other hand, so common in modern physics, appeared for the first time with the cubic equations and pop up each time additional degrees of freedom are assigned to the complex plane.

Actually, complex numbers behave exactly like two-dimensional vectors, in which the real part is the inner or scalar product or and the so-called imaginary part corresponds to the cross or vector product; so imaginary numbers can only be associated with motions, rotations and positions in space in additional dimensions, not with the physical quantities themselves.

This is easier to say than to think of, since it is even more "complex" to determine what a physical quantity or a mathematical variable can be independently of change and motion. Both to geometrically interpret the meaning of vectors and complex numbers in physics and to generalize them to any dimension a tool like geometric algebra may be used —"the algebra flowing from geometry", as Hestenes put it; but even then there is much more to geometry than we may think.

Many problems become more simple on the complex plane, or so the mathematicians say. One of them, under the pseudonym *Agno* sent in 2011 an entry to a math forum with the title "Imaginary Golden Mean", which shows a direct connection with π and e : $\Phi i = e^{\pm \pi i/3}$ [3]. Another anonymous author found this same identity in 2016, along with similar derivations, looking for fundamental properties of an operation known as "reciprocal addition", of interest in circuits and parallel resistances calculations. As refraction is a kind of impedance, it may also have its place in optics. The relation in the polar diagram may be associated right from the start with geometric series and hypergeometric functions associated with continuous fractions, modular forms and Fibonacci series, and even with noncommutative geometry [4]. The imaginary golden ratio, in any case, reflects as in a mirror many of the qualities of its real part.

The *Taijitu* is a circle, a wave and a vortex all in one. The synthetic genius of nature is quite different from that of man, and she does not ask for unifications because not to arbitrarily separate is enough for her. Nature, as Fresnel said, does not care about analytical difficulties.

The diagram of the *Taijitu* becomes a flat section of a double spiral expanding and contracting in three dimensions, a motion that seems to give it an "extra dimension" in time. It is always a real challenge to follow the evolution of this process, both spiral and helical, within a vertical cylinder, which is but the complete representation of the *indefinite propagation of a wave motion*, the "universal spherical vortex" described by René Guenon in three short chapters of his work "The Symbolism of the Cross". The cross of which Guenon speaks is certainly a system of coordinates in the most metaphysical sense of the word; but the physical side of the subject is by no means negligible.

The propagation of a wave in space is a process as simple as it is difficult to grasp in its entirety; one need only think of Huygens' principle, the universal mode of propagation, which also underlies all quantum mechanics, and which involves continuous deformation in a homogeneous medium.

In that same year of 1931 when Guenon was writing about the evolution of the universal spherical vortex, the first work was published on what we know today as the Hopf fibration, the map of the connections between a three-dimensional sphere and a sphere in two dimensions. This enormously complex fibration is found even in a simple two-dimensional harmonic oscillator. Also in the same year, Paul Dirac conjectured the existence of that unicorn of modern physics known as the magnetic monopole, which brought the same kind of evolution into the context of quantum electrodynamics.

Peter Alexander Venis gives us in a wonderful work a completely phenomenological approach to the classification and typology of the different vortices. There is nothing mathematical here, neither advanced nor elementary, but a sequence of transformations of 5 + 5 + 2, or 7 classes of vortices with many types and countless variants that unfold from the completely undifferentiated only to return to the undifferentiated again —or to the infinity of which Venis prefers to speak. The transitions from ideal points with no extension to the apparent forms of nature seems quite arbitrary without the aid of vortices, hence their importance and universality.

Peter Alexander Venis

Venis does not deal with the mathematical and physical aspects of such a complex subject as vortices, and of course he does not apply to them the continuous proportion; on the contrary he gives us the privilege of a new fresh vision of these rich processes, in which the insight of a presocratic naturalist and the capacity for synthesis of a Chinese systematist meet together effortlessly.

Even if the Venis sequence admits variations, it presents us a morphological model of evolution that goes beyond the scope of ordinary sciences and disciplines. The author includes under the term "vortices" flow processes that may or may not have rotation, but there is a good reason for that, since this is necessary to cover key conditions of equilibrium. He also applies the theory of yin and yang in a way that is both logical and intuitive, which probably admits a fairly elementary translation to the qualitative principles of other traditions.

The study of this sequence of transformations, in which questions of acoustics and image are closely linked, should be of immediate interest in order to deepen the criteria of morphology and design even without the need to enter into further considerations.

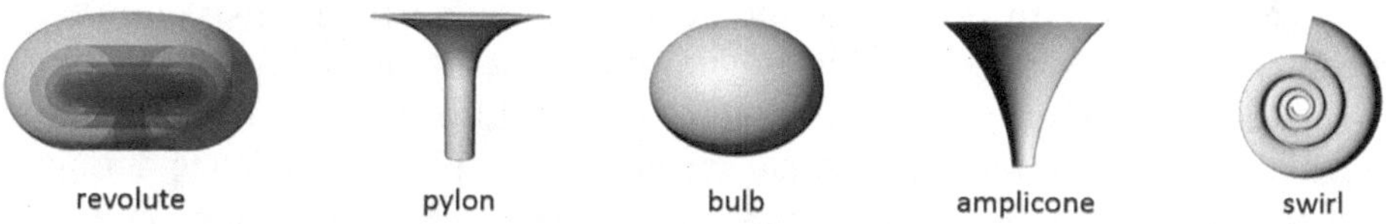

Peter Alexander Venis

A metric-free description would be, precisely, the perfect counterpoint for a subject as badly affected by arbitrariness in the measurement criteria as the study of proportionality. Naturally, mathematics also has several tools essentially free of metrics, such as external differential forms, which allow the study of the physical fields with maximum elegance. Then, perhaps, the metrics that physics deals with could be used as a middle ground between both extremes.

Thus, in this search to better define the context for the appearance of the continuous proportion in the world of phenomena, we can speak of three types of basic spaces: the ametric or metric-free space, the metric spaces, and the parametric or parameter spaces.

By metric-free space we understand the different spaces that are free of metrics and the action of measurement, from the purely morphological sequence of vortices above to projective geometry or the metric independent parts of the topology or differential forms. The projective, metric-independent space is the only true space; if we sometimes speak of metric spaces it is only because of the different connections with metric spaces.

By metric spaces, we mean those of the fundamental theories in physics, not only mainstream theories but also other related, with a special emphasis on Euclidean metric space in three dimensions of our ordinary experience. They include physical constants and variables, but here we are particularly interested in theories that do not depend on dimensional constants and can be expressed in homogeneous proportions or quantities.

By parametric or parameter spaces we mean the spaces of correlations, data, and adjustable values that serve to define mathematical models, with any number of dimensions. We can also call it the algorithmic and statistical sector.

We are not going to deal here with the countless relationships that can exist between these three kinds of spaces. Suffice it to say that to get out of this labyrinth of complexity in which all sciences are already immersed, the only possible Ariadne's thread, if any, has to trace a retrograde path: from numbers to phenomena, with the emphasis on the latter and not the other way around. And we are referring to phenomena not previously limited by a metric space.

Much has been said about the distinction between "the two cultures" of sciences and the humanities, but it should be noted that, before attempting to close this by now unsurmountable gap, we should begin first by bridging the gap between the natural, descriptive sciences and a physical science that, justified by its predictions, becomes indistinguishable with the power of abstraction of mathematics while isolating itself from the rest of Nature, to which it would like to serve as foundation. Reversing this fatal trend is of the greatest importance for the human being, and all efforts in that direction are worthwhile.

POLE OF INSPIRATION – PHYSICS AND THE CONTINUOUS PROPORTION

8 April, 2020

We already see that there are purely mathematical reasons for the continuous proportion to appear in the designs of nature independently of causality, be it physical, chemical or biological: in fact the convenience of logarithmic growth is independent even of the form itself, as is the elementary fact of the discrete and asymmetric division of cells.

In this light, it would be an emergent property, just a parallel plane to physical causation and becoming. On the other hand, the idea of parallel planes with a merely circumstantial connection with physical reality looks odd, and in any case very distant from what the diagram of the Pole express so well —that no form or nothing apparent is free from dynamics.

The fact is that the connection between physics and the continuous proportion is very dim, to say the least. However we have important occurrences of this ratio even in the Solar System, where it is almost impossible to avoid celestial mechanics. A better understanding of the presence of the continuous proportion in nature should not ignore the framework defined by fundamental physical theories, nor what these might leave out.

We have three possible approaches with increasing degrees of risk and depth:

1. The continuous proportion in nature can be studied independently of the underlying physics as a purely mathematical question; this would be the most prudent, but somewhat limited position. The aforementioned A. Stakhov has developed an algorithmic theory of measurement based on this ratio that can be used to analyze in turn other metrological theories of cycles, continuous fractions and fractals as for example the so called Global Scaling.

2. This proportion can be studied according with views compatible with known mainstream physics; for example, as Richard Merrick has done, in a neopythagorean rereading of the collective harmonic aspects of wave mechanics, such as resonances, and in which *phi* would be a critical damping factor [7]. These ideas are totally accessible to the experiment, either in acoustics or in optics, so that they can be verified or falsified.

Merrick's idea of harmonic interference is within everyone's reach and understanding and it is not without depth. It can be naturally complemented with the holographic concept proposed by David Bohm and his distinction between the implicate and explicate order. Although Bohm's interpretation is not standard, it is compatible with experimental data. The harmonic interference theory can also be combined with mathematical theories of cycles and scales such as those mentioned.

3. Or, finally, one can consider other more classical theories that differ from the mainstream but which may provide deeper insights into the subject. Within this category, there are various degrees of disagreement with the standard theories: from just a wider understanding of thermodynamics, to in-depth revisions of classical mechanics, quantum mechanics and calculus. We could say that this third option is not that speculative, but rather divergent in the spirit and the interpretation.

Here we will focus more on the third level, which may also seem the most problematic. One could ask what is the need to question the best established physical theories in order to find a better ground for the occurrence of a constant that maybe does not require them. Furthermore, the first two levels already offer plenty of room for speculation. But this would be a very superficial way of looking at it.

We cannot delve into the presence of the continuous proportion in a symbol of perfect reciprocity ignoring the question of whether our present theories are the best exponent of continuity, homogeneity or reciprocity —and in fact they are far from it.

POLE OF INSPIRATION – TWO KINDS OF RECIPROCITY

8 April, 2020

The *Taijitu*, the emblem of the action of the Pole with respect to the world, and of the reciprocal action with respect to the Pole, inevitably reminds us of the most universal figure in physics; we are naturally referring to the ellipse —or rather, it should be said, to the idea of the generation of an ellipse with its two foci, since here there is no eccentricity. The ellipse appears in the orbits of the planets no less than in the atomic orbits of the electrons, and in the study of the refractive properties of light it gives rise to a whole field of analysis, ellipsometry. Kepler's old problem has scale invariance, and plays a determining role in all our knowledge of physics from the Planck constant to the furthest galaxies.

In physics and mechanics, the principle of reciprocity par excellence is Newton's third principle of action and reaction, which is at the base of all our ideas about energy conservation and allows us to "interrogate" forces when we are obliged to assume the constancy or proportionality of other quantities. The third principle does not speak of two different forces but of two different sides of the same force.

Now, the story of the third principle is curious, because we are forced to think that Newton established it as the keystone of his system to tie up the loose ends of celestial mechanics —particularly in Kepler's problem- rather than for down-to-earth mechanics based on direct contact between bodies. The third principle allows us to define a closed system, and closed systems have been the given for all fundamental physics since then —however, it is precisely in celestial orbits, such as that of the Earth around the Sun, that this principle can be least verified, since the central body is not in the center, but in one of the foci. The force designated by the vectors would have to act on the void, where there is no matter.

Since the very first moment it was argued in the continent that Newton's theory was more an exercise in geometry than in physics, although the truth is that, if physics and vectors were good for something, the first thing that failed was geometry. That is, if we assume that forces act from and on centers of mass, instead of on mere mathematical points. But, despite what intuition tells us —that an asymmetric ellipse can only result from a variable force, or from a simultaneous generation from the two foci-, the desire to expand the domain of calculus prevailed over anything else.

In fact the issue has remained so ambiguous that attempts have always been made to rationalize it with different arguments, either the system's barycenter, or the variation in orbital velocity, or the initial conditions of the system. But none of them separately, nor the combination of the three, allows to solve the issue satisfactorily.

Since no one wants to think that the vectors are subjected to quantitative easing, and they lengthen and shorten at convenience, or that the planet accelerates and brakes on its own as a self-propelled rocket, in order to keep the orbit closed, physicists finally came to accept the combination in one quantity of the variable orbital speed and innate motion. But what happens is that if the centripetal force counteracts the orbital velocity, and this orbital velocity is variable despite the fact that the innate motion is invariable, the orbital velocity is in fact already a result of the interaction between the centripetal and the innate force, and then the centripetal force is also acting on itself. Therefore, the other options being ruled out, what we have is a case of *feedback or self-interaction* of the whole system.

So it must be said that the claim that Newton's theory explains the shape of the ellipses is at best a pedagogical resource. However, this swift pedagogy has made us forget that our so called laws do not determine or "predict" the phenomena we observe, but try to fit them at most. Understanding the difference would help us to find our place in the overall picture.

The reciprocity of Newton's third principle is simply a change of sign: the centrifugal force must be matched by an opposing force of equal magnitude. But the most elementary reciprocity of physics and calculus is that of the inverse product, as already expressed by the formula of velocity, ($v = d/t$), which is the distance divided by time. In this very basic sense, those who have pointed out that velocity is the primary fact and phenomenon of physics, from which time and space are derived, are absolutely right.

The first attempt to derive the laws of dynamics from the primary fact of velocity is due to Gauss, around 1835, when he proposed a law of electric force based not only on distance but also on relative velocities. The argument was that laws such as Newton's or Coulomb's were laws of statics rather than of dynamics. His disciple Weber refined the formula between 1846 and 1848 by including relative accelerations and a definition of potential —a retarded potential, in fact.

Weber's electrodynamic force is the first case of a complete dynamic formula in which all quantities are strictly homogeneous or proportional [8]. Such formulas seemed to be exclusive to Archimedes' statics, or Hooke's elasticity law in its original form. In fact, although it is an specific formula for electric charges and not a field equation, it allows to derive Maxwell's equations and the electromagnetic fields as a particular case, simply integrating over volume.

The logic of Weber's law could be applied equally to gravity, and in fact Gerber used it to calculate the precession of Mercury's orbit in 1898, seventeen years before the calculations of General Relativity. As is well-known, General

Relativity aspired to include the so-called "Mach principle", although in the end it did not succeed; but Weber's law was entirely compatible with that principle in addition to explicitly using homogeneous quantities, well before Mach wrote about these issues.

It has been said that Gerber's argument and equation was "merely empirical", but in any other era not having to create ad hoc postulates would have been seen as the best virtue. In any case, if the new proportional law was used to calculate a tiny secular divergence, and not for the generic ellipse, it was for the simple reason that in a single orbital cycle there was nothing to calculate for either the old or the new theory.

Weber's purely relational formula cannot "explain" the ellipse either, since force and potential are simply derived from motion —but at least there is nothing *unphysical* in the situation, and the fulfillment of the third principle is guaranteed while permitting a deeper meaning.

Ironically, as this new law changed the prevailing idea of central forces, understood with a string attached, Helmholtz and Maxwell blamed Weber's law for not complying with energy conservation, although finally in 1871 Weber showed that it did so on the condition that the motion was cyclical — which in this issue was already the basic requirement for Newtonian or Lagrangian mechanics too. Conservation is a global property, not a local one, but the same was true for the orbits described in the Principia, not less than those of Lagrange. Strictly speaking there is no local conservation of forces that can make physical sense. Newton himself used the analogy of a slingshot, following Descartes' example, when he spoke of the centrifugal motion, but nowhere in his definitions is there any talk that the central forces should be understood as if connected by a string. However, posterity took the simile at face value.

Why claim that there is in any case feedback, self-interaction? Because all gauge fields, characterized by the invariance of the Lagrangian under transformations, are equivalent to a non-trivial feedback between force and potential —the eternal "information problem", namely how does the Moon know where the Sun is and how does it "know" its mass to behave as it does.

Indeed, if the Lagrangian of a system —the difference between kinetic and potential energy - has a certain value and is not equal to zero, this is equivalent to say that action-reaction is never immediately fulfilled. However, we use to assume that Newton's third principle is immediate and takes place automatically and simultaneously, without mediation of any time sequence, and the same simultaneity is assumed in General Relativity. The presence of a retarded potential indicates at least the existence of a sequence or mechanism, even if we can not say anything else about it.

This shows us that additive and multiplicative reciprocity are notoriously different; and the one shown by the continuous proportion in the diagram of the Pole includes the second kind. The first is purely external and the second is internal to the order considered.

All the misunderstandings about what is mechanics come from here. And the essential difference between a mechanical system in the trivial sense and an ordered or self-organized system lies precisely at this point.

At the time it was believed that Hertz's experiments confirmed Maxwell's equations and disproved Weber's, but that is another misunderstanding because if Weber's law —which was the first to introduce the factor of the speed of light- did not predict electromagnetic waves, it did not exclude them either. It simply ignored them. On the other hand, some perceptive observers have noted that the only thing Hertz demonstrated was the reality of the action at a distance, not of waves, but that is another story.

As a counterpoint, it is worth remembering another fact that shows, among other things, that Weber had not fallen behind his time. Between the 1850s and 1870s he developed a stable model of the atom with elliptical orbits —many decades before Bohr proposed his model of the circular atom, without the need to postulate special forces for the nucleus.

Weber's relational dynamics shows another aspect that may seem exotic in the light of the present theories: according to its equations, when two positive charges approach a critical distance, they produce a net attractive force, rather than a repulsive one. But is not the very idea of an elementary charge exotic in the first place, or should we just say a mere convention? In any case, this fits very well with the *Taijitu* diagram, in which a polarized force can potentially become its opposite. Without this spontaneous reversal, hardly we could speak of truly alive forces and potentials.

POLE OF INSPIRATION – THE APPLE AND THE DRAGON

8 April, 2020

To my knowledge, Nikolay Noskov was the first to appreciate, in the 1990s, that Weber's dynamics was so far the only one that allowed for a physical account of the shape of the ellipses, even if it did not pretend to give a "mechanical explanation" for them. In this respect, Noskov particularly insisted on associating the retarded potentials with longitudinal vibrations of the moving bodies in order to give a content to the conservation, merely formal in Weber, of energy; he also insisted that their occurrence permeated all types of natural phenomena, from the stability of atoms and their nuclei, to orbital elliptical motion, sound, light, electromagnetism, the flow of water or gusts of wind [9].

Despite the misunderstandings on the subject, these longitudinal waves are not incompatible with known physics, and Noskov recalled that the same Schrödinger wave equation is a mixture of different equations that describe waves in a medium and waves within the moving body —and the same thing happened from the start with Maxwell's "electromagnetic waves", which even from the most classical point of view cannot be anything other than a statistical average between what occurs in portions of space and matter.

Noskov noticed that the behavior of forces and potentials in Weber's law involved a sort of feedback, although he does not seem to recognize that this is already the case for all gauge theories, and finally even for Newtonian celestial mechanics itself, although in all these instances it is presented in disguise. Atoms would be definitely dumb without this ability to adjust embedded into the very idea of the field.

*

Let us now return to the continuous proportion. Miles Williams Mathis wonders how it is that, given the equality $\Phi^2 + \Phi = 1$, *phi* has not been related to the most elementary inverse-square laws of physics; moreover, he wonders how it is that it has not been associated with the sphere itself, being so evident that the surface of a sphere also decreases to the square [10].

It could be argued that the Fibonacci series does not square, but the factor Φ does, as can be easily seen in the successive squares of the golden spiral ($1, 1/\Phi, 1/\Phi^2, 1/\Phi^3 \ldots$) or in its expression as a continuous square root. Mathis is not confusing the inverse square with the square root, but is talking about a scale factor between two hypothetical subfields one into the other.

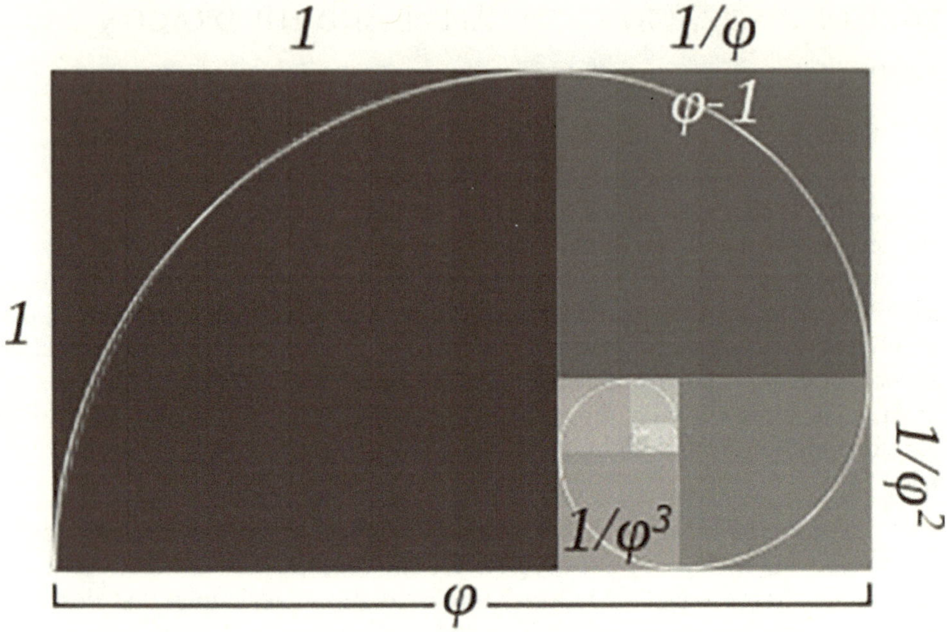

MILES MATHIS: MORE ON THE GOLDEN RATIO AND FIBONACCI SERIES

$$\varphi = \sqrt{1 + \sqrt{1 + \sqrt{1 + \sqrt{1 + \cdots}}}}.$$

Mathis may be right in insisting that the presence of *phi* must also have an underlying physical cause; the only problem is that modern physics ignores and completely denies a scale relationship between charge and gravity, indeed light and gravity, as he is proposing. However, the origin of his correlation lies in the same Kepler's problem, in which he wants to see a joint action of two different fields, the second one based not only on the inverse square of the distance but also on an inverse law of the fourth power ($1/r^4$) with a product of density by volume, instead of the usual formula of masses.

Now, Mathis is the first to specifically point out the conflation of orbital velocity and innate motion in Newton, interpreting the Lagrangian as the disguised product of two fields, of opposite attractive and repulsive effects, whose relative proportion or intensity is a function of scale and density [11].

The inclusion of density would have to be fundamental in a truly Archimedean relational physics, which brings us back to the issue of waves and spirals. Spirals are a common occurrence in astronomy, galaxies being their most apparent manifestation; these galaxies have been described in terms of density waves.

Many have noticed a logarithmic spiral with Φ as a key also in the Solar System and the distribution of its planets. As in the case of the so-called "law", or rather rule of Titus-Bode, the existence of a non-random order seems quite evident, but the adjustment with the known values is somewhat arbitrary.

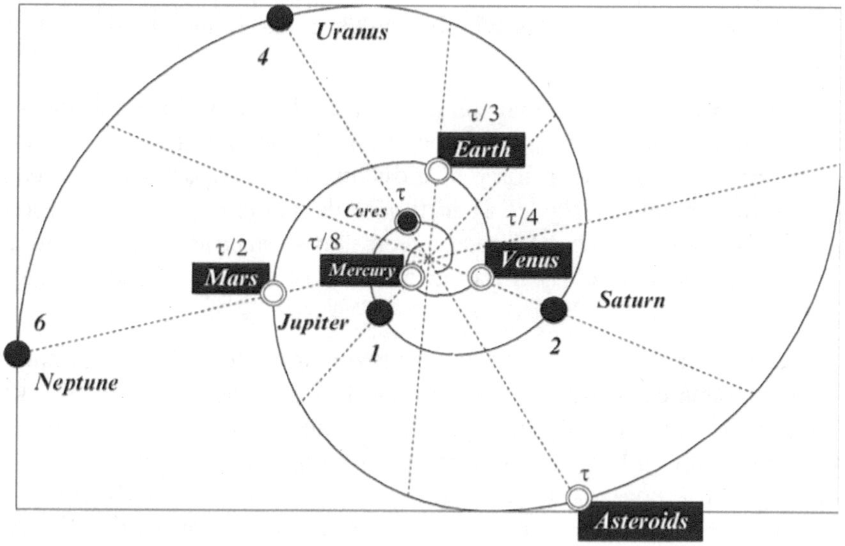

Jan Boeyens: *Commensurability in the Solar System*

It goes without saying that the ellipse is the transformation of the circle when its centre is divided into two foci; although from the other point of view it can be said, and this is not unimportant, that the circle is only the limit case of the first one. Expanding on Kepler's problem, although in a different light, Nicolae Mazilu refers us to Newton's theorem of revolving orbits. Newton had already carefully considered the case of forces decreasing to the cube of distance, and in this hypothetical case the bodies describe orbits with logarithmic spiral shapes, which of course no one has observed.

However, E. B. Wilson's works of 1919 and 1924 showed that the stable electron orbits in the atom were not ellipses but logarithmic spirals; only that the force involved here is not the Coulomb force, but a transition force between two different elliptical orbits. The later solution of the problem has buried in oblivion a model that was also consistent. And as for all applications of conic sections to physics, here too we find that signature of change in the potential, the shift in phase or plane of polarization known as the geometrical phase, discovered by Pancharatnam and so successfully generalized to quantum mechanics by Michael Berry [12] .

Various studies recount that the distribution of the planets in the Solar System follows the pattern of a logarithmic golden spiral with an accuracy of more than 97 percent, which may increase if the sidereal years and synodic periods of the system as a whole are taken into account [13]. For Hartmut

Müller, the proximity is simply due to the closeness of *phi* to the value of √e, which is 1.648. According to other counts not verified by myself, the average distance between consecutive planets from the Sun to Pluto, taking the distance between the two previous ones as a unit, is just 1.618. If the last planet is discarded, the average deviates widely, which gives an idea of the fragility of these calibrations.

It has often been said that the perceptible harmony in the Solar System is not possible without some feedback mechanism, while the Newtonian approach simply combines a force at a distance with trajectories like cannonballs —a cannonball theory of everything- dependent on external forces or collisions. However, we have already seen that even in the Newtonian case a self-interaction is masked by merging innate motion and orbital velocity into one.

Newtonian celestial mechanics gave way to a more abstract version, Lagrangian mechanics, to avoid this mess; the difference between the kinetic energy and the potential begs the question to the so-called "initial conditions", but these are nothing but Newton's innate motion… at any rate, the case is that this average difference of the Lagrangian and the average eccentricity of the orbits is of the same order of magnitude than the deviations of the distribution of the solar system obtained by the logarithmic golden spiral. Thus, one can take the Lagrangian density of the entire system and its averages and see how the planets with their orbits nestle in.

It seems that scientific publications no longer admit studies on planetary distribution, since, having no underlying physics, they are relegated to the limbo of numerological speculation. However, the Lagrangian routinely used in celestial mechanics is also nothing more than a pure mathematical analogy, and exists only to blur differences of the same order of magnitude. Suffice it to admit this to realize that both issues are not on different grounds —maybe they are not even two different things.

To admit this is also to admit that gravity itself is an adjustment force that depends on the environment and not a universal constant, but this is something that was already implicit in Weber's relational mechanics.

Mathis' theory is more specific in that it regards G as a transformation between two radii. Not concerned with fitting his own notions of the physics underlying the Golden Section into the spiral of the Solar System, he deals in detail with Bode's Law in a much simpler way based on a series based on √2. He also includes naturally in it the optical equivalence, the neglected fact that many planets look of the same size from the Sun, just as many satellites look the same size as the Sun seen from their respective planets. So this is not a punctual coincidence [13]. The optical equivalence would be the final wink that Nature gives us to see who is more blind, she or we.

And since it looks like a typical fancy of Nature, let us allow ourselves a bit of numerological fun. The optical equivalence that is shown in the total eclipses is an angular or projective relationship (with an approximate value of 1/720 of the celestial sphere) in accordance with the number 108, so important in different traditions, here entailing the number of solar diameters between the Sun and the Earth, the number of terrestrial diameters in the diameter of the Sun and the number of lunar diameters that separate the Moon from the Earth.

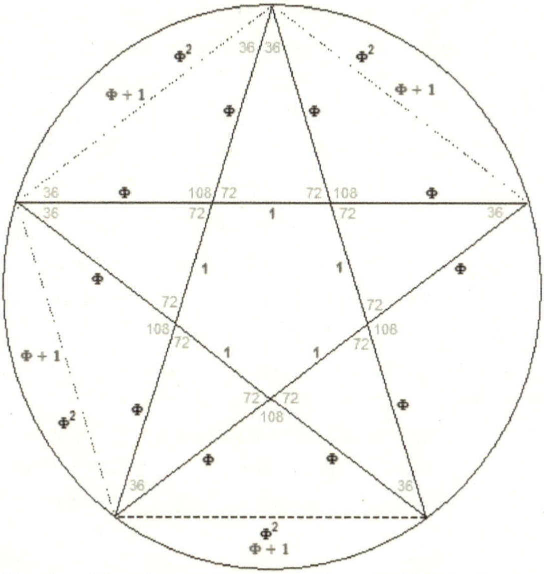

Dimensions of the pentagram and its relation to PHI - (by: Birol Koc) - EyePhi.com

In the pentagram used to construct a golden spiral —and with which an ellipse can also be univocally determined in spherical geometry- we see that the reciprocal angles of the pentagon and the star are 108 and 72 degrees. On the other hand, Mathis himself comments, without relating it in any way to the optical equivalence, that in accelerators the relativistic mass of a proton usually finds a limit of 108 units that neither Relativity nor Quantum Mechanics explain, and he makes a derivation of the famous gamma factor that links it directly to G.

Of course, the Lorentz relativistic factor coincides with Weber's mechanics up to a certain limit of energy —although in the latter what increases is the internal energy instead of mass. There could be no more natural connection with the optical equivalence than that of light itself, and Mathis' theory establishes a series of equations and identities between light and charge, charge and mass, mass and gravity.

On the other hand, if we were to throw a stone into a well which perforated the Earth from side to side, and waited for it to return like a spring or a pendulum, it would take about 84 minutes, the same as an object in a close orbit around the planet. If we did the same thing with a particle of dust on an asteroid the size of an apple, but of the same density than our planet, the result would be exactly the same. This fact, which seems to assign an important role to density over mass and distance itself, pierce the appearance of the gravitational phenomenon, and should be as astonishing to us as Galileo's finding that objects fall at the same speed regardless of their weight; it also fits very well in the context of an spiral equal at all scales.

In any case the Lagrangian, the difference between kinetic and potential energy, has to play a fundamental role as a reference for the fine tuning of the different elements of the Solar System. In celestial mechanics, despite what is said, the integral has always led to the differential, and not the other way around. The law discovered by Newton does not shape the ellipse but rather tries to fit it.

So we have Newton's apple and the Golden Dragon of the Solar System Spiral. Will the dragon swallow the apple? The answer is that he doesn't need to swallow it, since it has been inside him from the start. Let us say it again: the gauge fields, characterized by the invariance of the Lagrangian under transformations, are equivalent to a non-trivial feedback between force and potential, which in turn is indistinguishable from the eternal "information problem", namely how the Moon knows where the Sun is and how it "knows" its mass to behave as it does. Why to ask about information at the microlevel of particles when the problem is in plain sight at the macrolevel in the first place?

Considering the adjustments of the Lagrangian in comparison with a system described exclusively by non-variable forces, the entire Solar System looks like a great spiral holonomy.

The Lagrangian can also hide virtual dissipation rates —virtual, of course, since we already know that the orbits are preserved. In fact, what Lagrange did was to dilute D'Alembert's principle of virtual work by introducing generalized coordinates. But we are so used to separate the formalisms of thermodynamics from those of the supposedly more fundamental reversible systems that it is hard to see what this means. However, the most certain instinct tells us that everything reversible is an island surrounded by an ocean without forms. There is no motion without irreversibility; to pretend otherwise is just an illusion.

Mario J. Pinheiro wants to repair this divorce between convictions and formalisms by proposing a reformulation of mechanics alternative to the Lagrangian account, with a variational principle for rotating systems out of equilibrium and a mechanical-thermodynamic time in a set of two differential equations of first order. Here the equilibrium takes place between the minimum energy variation and the maximum entropy production.

This thermomechanics allows us to consistently describe systems with characteristics that are quite different from those of reversible systems, and which are particularly relevant to the case at hand: subsystems within a larger system can absorb the forces exerted on them, and instead of being enslaved there is room for interaction and self-regulation. There may be a component of topological torsion and conversion of linear or angular motion into angular motion. The angular momentum acts as a damper to dissipate the disturbances, "a well-known redressing mechanism in biomechanics and robotics " [15] .

To my knowledge, Pinheiro's proposal of an irreversible mechanics is the only one that gives a proper explanation of Newton's famous buck experiment and the whirlpool formed by its rotation, by the transport of angular momentum, as opposed to Newton's absolute interpretation or Leibniz's purely relational one, neither of which are really to the point. Suffice it to recall the elemental observation that in this experiment the appearance of the vortex requires both time and friction, and matter is transferred to the regions of highest pressure, a clear signature of the Second Law. What is extraordinary is that no one has insisted on this before Pinheiro —something that can only be explained by the conventional roles assigned to the different branches of physics. Besides, it is clear that springs, whirls and spirals are the most suitable and efficient forms of damping.

It is perhaps appropriate to remember that the so-called "principle of maximum entropy" does not tend towards maximum disorder, as is often thought even among the physicists, but rather the opposite, and this is how Clausius originally understood it. This establishes a very broad but essential link with highly organized systems, at the top of which we usually place living beings [16]. On the other hand, it is enough to contemplate the spiral of the Solar System for a moment to understand that it only makes sense as an open, irreversible process in permanent production. fr

The concept of order that Boltzmann introduced is no less subjective than that of harmony, the main difference being that in statistical mechanics the micro-states, not the macro-states, have received a convenient quantification. Of course, this is another great rationalization: the irreversibility of phenomena or macrostates would be derived from the reversibility of microstates. But the mere postulation of stationary orbits in atoms —to pretend that there can be variable forces in isolated systems- is illegal both from the thermodynamic point of view and from the mere common sense.

The variational principle proposed by Pinheiro was first suggested by Landau and Lifshitz but has not been developed to date. This is inevitably reminiscent of the idea of damping wells in the Landau-Zener theory, which arise from adiabatic torque transfer when waves cross without destructive interference. Richard Merrick has directly related these wells or vortices to the golden spirals under conditions of resonance [17]. Many will say that one can not see how these conditions can be met in the Solar System, but, once again, the resonances of the classical theory of perturbations in Laplace's celestial mechanics are in no better situation, being nothing else than pure mathematical relations. If anything, it could be said that they are in a worse situation, since we are asked to believe that gravity can have a repulsive effect.

Although Pinheiro's thermomechanics involves something similar to this form of transfer, which evokes the parallel transport of the geometric phase, it also incorporates, and this is the key difference, a term for the thermodynamic free energy. A reversible system is a closed system, and there are no closed systems in the universe.

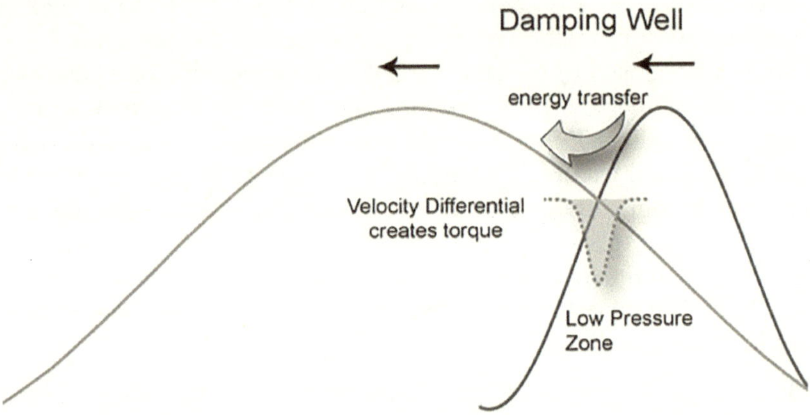

Richard Merrick: Harmonic formation helps explain why phi pervades the solar system

Merrick's own theory of harmonic interference would be elevated to a much higher level of generality simply by appreciating that the principle of maximum entropy production is not contrary to the generation of harmony but rather conducive to it.

The principle of maximum entropy can be transferred to quantum mechanics with hardly any more sacrifice than the idea of reversibility, as shown by the quantum thermodynamics developed by Beretta, Hatsopoulos and Gyftopoulos; the subject is of extraordinary importance but now it would take us too far [18].

*

Physicists are proud of the high degree of accuracy of some of their theories, which is quite understandable given the work invested carrying out their calculations, sometimes to ten and twelve decimal places. Few things would be more eloquent than such precision if it came naturally, without special assumptions or arbitrary ad hoc adjustments, but that is the case most of the time. Still the value of gravity on Earth cannot be measured to more than three decimal places, but astrophysicists pretend to calculate to ten or twelve places to the confines of the universe.

In the case of Lagrange and Laplace this is absolutely evident, and one day we will wonder how we were able to accept their methods without even blinking an eye. The truth is that these procedures were not digested overnight, but if they were finally accepted it was from the invincible desire to expand the power of calculus at any rate, reinforced by the idea, inherited from Newton and Leibniz, that Nature is a clockwork machine of virtually infinite precision. And for the means, what better than to serve the Ideal.

It has rightly been said that had Kepler had more precise data, he would not have advanced his theory of elliptical motion; and in fact, Cassini ovals, fourth-degree curves with a constant distance product, seem to reproduce the observable trajectories more closely, something one should attribute to the perturbations involved. These ovals also raise interesting and profound questions about the dynamical connection between ellipses and hyperbolas. Interestingly, Cassini ovals are used to model the geometry of the spontaneous negative curvature of red blood cells, in which the golden ratio has also been found [19].

As Mathis points out, the very first analyses of perturbations included, already since Newton and Clairaut, a factor $1/r^4$ with *a repulsive force*, which shows again to what extent the "auxiliary" elements of celestial mechanics are hiding something much more important [20].

To the eye of the naturalist, accustomed to the very variable precision of the descriptive sciences, the golden spiral of the Solar System would have to appear as the most splendid example of natural order; an order so magnificent that, unlike Laplace's, it can include catastrophes in its bosom without hardly blurring. This is a characteristic that we invariably attribute to living beings. Whether judged as a natural phenomenon or as an organism, taking everything into account, the spiral shows a precision, more than sufficient, excellent.

And what is the place of the *Taijitu*, our symbol of the Pole generating the yin and yang, in all this? Well, it goes without saying that the system we are talking about, along with its subsystems — planets and satellites —is an eminently polar process, with axes defining its evolution; and so it is the spiral holonomy that envelops them. As for the yin and yang, if we were to say that they can *also* be the kinetic and potential energy, we would be told that we are proposing too trivial a correspondence. But all the above should serve to see that this is not the case.

We know that in the orbits kinetic and potential energy do not even compensate, and when they should, as in the case of circular motion in Binet's equation, we do not even obtain a single force —at least a difference between the center of the circle and the center of the force is required. Looking for the simplest possible argument, the first thing that comes to mind is that the emergence of the golden section in the *Taijitu*, the freely rotating spherical vortex, contains a sort of analogical and a priori synthesis of 1) a law of areas applied to the two energies, 2) the focal geometry of the ellipses, and 3) a difference that is integrable and a shift in the plane of polarization that it is not. This third point overlaps the Lagrangian and a geometrical phase that in principle seem quite different.

Of course, we leave large loose ends here that such a simple diagram cannot translate. To begin with, just because an ellipse has two foci within does not mean that we have to look inside always for the origin of the forces that determine it, and this would lead us to the theory of perturbations. However, any environmental influence, also outer planets, should already be included in the geometric phase.

If we were to pass from celestial dynamics to light, we could reinterpret in terms of retarded potentials and their incidence on phase the data of ellipsometry or the "abstract monopole with a force of —1/2 at the center of the Poincaré sphere" to which Berry appeals in his generalization of the geometric phase. However, it should not be forgotten that light was already an essentially statistical process even since the days of Stokes and Verdet. The degree of polarization and entropy of a beam of light were always equivalent concepts, although we are still far from drawing all the consequences from this.

We assume the coincidence of the retarded potential and the geometrical phase, although there is not even a specific literature on the subject, nor is there agreement, otherwise, on the significance and status of the geometrical phase itself. There have been those who have seen it as an effect of the exchange of angular momentum, and in any case in classical mechanics the geometrical phase is shown by Hamilton-Jacobi's formulation of angle-action variables [21].

If harmony is totality, the so-called geometrical phase should have its part in the mathematics of harmony, since it is nothing but the expression of "global change without local change". We already noticed that the geometrical phase is inherent in fields involving conic sections, so its inclusion here is just elementary. However, the fact that it does not involve the known forces of interaction does not mean that we are dealing with mere "fictitious forces"; they are real forces that transport angular momentum and are essential in the effective configuration of the system.

Since this energy transport is an interference phenomenon, the potential energy of the Lagrangian must comprise the sum of all the interference from the adjacent systems, this being the missing "regulatory mechanism". It may be argued that in the course of the planets we do not observe the manifestation of interference that characterizes wave processes, even though we do not hesitate to resort to "resonances" to explain perturbations. Let us look at this a little more closely.

If until now we have chosen to see the geometrical phase, in classical mechanics the difference in the solid angle or Hannay angle, as a relational property, the most appropriate way of understanding it would have to be within a purely relational mechanics such as Weber's one. However, as Poincaré remarked, if we have to multiply the velocity squared, *we no longer have a way of distinguishing between kinetic and potential energy, and even the latter is no longer independent of the internal energy of the bodies* considered. Hence the postulation of an internal vibration by Noskov. However, this inherent ambiguity does not prevent us from making calculations as precise as with Maxwell's equations, in addition to some other obvious advantages.

Remember the comparison of the stone that passes through the Earth and the dust particle on that tiny asteroid, which return to the same point in the same time. In a hypothetical medium of homogeneous density, this would suggest an overall dampening and synchronizing effect at different spatial scales. But, without the need for any hypothesis, what the geometrical phase implies is the effective coupling of systems that evolve at different time scales, for example, electrons and nuclei, or gravitational and atomic forces, or, within gravitational forces, the interactions between the different planets. This makes it particularly robust to noise or disturbances.

The ambiguity of relational mechanics need not be a weakness, but could be revealing some limitations inherent in mechanics and its calculus. Just when we want to take to its logical extreme the ideal of converting physics into a pure kinematics, a science of forces and motions, of mere extension, its inevitable dependence on potentials and "non-local" factors is revealed, although we would rather have to talk about definite global configurations.

What is essential in the apparently casual comparison between the *Taijitu* and the elliptical orbit is that the latter is also an integral expression of the totality: not only of the internal forces but also of the external forces that contribute to its form in real time. If the compensation mechanism serves as an effective regulation it cannot affect only the potentials but equally the forces.

POLE OF INSPIRATION – QUESTIONS OF PRINCIPLE

8 April, 2020

The principles of Newtonian physics are based, as can be expected, in the circularity of its definitions of vector and scalar magnitudes like force and mass; the Lagrangian mechanics and the gauge fields, that "extend" it, demand a fixing to regulate the redundant degrees of freedom. Weber's law already allowed to appreciate in Kepler's problem the constituent elements of the gauge fields even if it dispensed completely with the very idea of fields —what changes here is the very status of the fundamental definitions, which are blurred. The retarded potentials allow to account for the essential aspects of modern physics, including the so-called relativistic effects.

The known and the unknown are easily confused with each other. On an immediate level, for us a force is, on the one hand, what induces motion, and on the other hand, what produces deformations in other bodies. But the "force" of gravity does not deform bodies when it forces them to move, and instead produce deformation when it does not force them —when it remains as a potential. Newton said that the whirling of the water buck spinning was due to centrifugal "fictitious forces" in absolute space, but Empedocles had shown two thousand years before that that same buck spinning above our heads counteracts the force of gravity.

How can one wield the axis of the Pole? The Pole, which balances the extremes of reality, is not something to be wielded. However the *Taijitu* invites us to look from its perspective. As for the spirit, Newton's three principles are tacitly summarized in the phrase "nothing moves unless it is moved" (by something else), that is, nothing moves without an external force —and neither relativity, quantum mechanics, nor modern cosmology have ever claimed otherwise. Everything is dead, except for the push given to it by something external. Now, all this prodigious development of modern science understood as mechanism is nothing but the unfolding of the consequences of the principle of inertia, and the ironic twist is that all the predictions of modern physics, and many others, can be made without the need of this principle at all.

The equivalence principle tells us that gravitational mass and inert mass are equal or indiscernible, and therefore the general theory of relativity states that there is no difference between the gravitational "force" and the fictitious forces. This is an attempt to walk in the direction of relational physics, but after terrible detours, and after arbitrating different versions —weakest, weak, middle-strong and strong- of such a principle, one ends up returning to the starting point, which is what it was all about.

The starting point is the principle of inertia, which no one wants to drop. The principle of inertia, often seen as purely redundant, bears the whole intentionality and disposition of the parts in physics. In other words, to do physics without the principle of inertia is equivalent to suspending its intention, which brings all operations back to the usual circular logic, with the help of the other two principles.

The principle of inertia, illustrated by the ball rolling for eternity in empty space, has been judged to be perfectly ideal. But it is not a perfect ideal, but a contradictory one: the motion of the ball must be related to axes of coordinates external to that system, and thus we have an isolated system that has the property of not being isolated. In reality there cannot be inertially isolated systems.

It is possible, and even necessary, as André Assis does, to propose a completely relational mechanics without using the concept of inertia, introducing instead the principle of dynamic equilibrium, so that "the sum of all forces of any nature acting on any body is always zero in all reference systems". This frees physics from the concepts of inertia, inert mass, absolute space, and the scholastic distinctions between frames of reference [22].

Something diametrically opposed to that implicit in the laws of mechanics also allows a description consistent with what we know. Thus, for example, Alejandro Torassa shows a dynamics valid for all observers in which "the motion of bodies is not determined by the forces acting on them, but the bodies themselves determine their motion" balancing the forces acting on them. "The natural state of a body in the absence of external forces is not only the state of rest or of uniform rectilinear motion, the natural state of movement of a body is any possible state of motion" [23].

If the sum of all forces is zero in any state, only differences and force ratios can be measured; introducing constants with dimensions here would be out of place. Another way of stating this principle would be to say that "the zero sum of all forces includes the observable motion," something that a certain mental inertia makes difficult to accept. Perhaps we can understand this better if we say that "observable motion balances the rest of the forces," that is, it balances out those that are not observable either. The balance of forces is not confused with their absence; but what we observe is motion and velocities, not forces.

There are of course other ways of representing this fundamental equilibrium without direct dependence on motion. For example, we can take from René Guenon the idea of an initially homogeneous medium, in which each compression at one point must correspond to an equal expansion at another point, so that their densities are reciprocal and their product equals always unity, even though the forces associated with them may be opposite, attractive or repulsive [24]. If in the originally homogeneous medium we imagine the correlative appearance of a fuller and an emptier portion, both could not simply arise without a torsion or helicity connecting them —and that torsion would be by itself a change in density. The characterization of equilibrium as a product is what we have considered here as reciprocity in the more intrinsic sense.

The cosmology of modern physics may argue that aspects such as general equilibrium are not matters of principle but of observation. What actually happens is that, if all that is observed is motion, and one starts from the principle of inertia, everything has to refer to causes external to what is observed — hence the hand of God to define the innate motion of the planets in Newton, or the notion of an event at the beginning of time that draws all energy from nothing. Thus, for example, and contrary to the endlessly publicized history, the first and most accurate predictions of microwave background radiation were not those of Gamow or other creationists, but those of physicists who assumed a universe in dynamic equilibrium [25].

This is the best example of the basic assumption being imposed on everything else, which instead tries to fit the assumption. It does not matter if this implies the most grandiose violation of the principle of energy conservation, as long as it is thrown out of bounds.

For modern physics, and not only physics, imbalance is the father of all things, and equilibrium is synonymous with death and disorganization. But the same observations and data have always allowed us to say that dynamic equilibrium is the father and mother of all things and that entropy does not lead to thermal death but to increased organization.

The true relevance of this dynamic equilibrium will be duly appreciated when mechanics and thermodynamics are united in a single discipline such as a thermomechanics of the type proposed by Pinheiro or other equivalents and more developed. And since his system of two equations is an alternative to the Lagrangian, it still fits better the Noskov equations, since the longitudinal vibrations of the bodies —which are congruent with Planck's formula- are equivalent to the input of free energy available in the medium.

Pinheiro's thermomechanics is conceived for open systems or systems out of equilibrium, and relational mechanics, even if the properties of the medium are not considered, as it lacks dimensional constants, depends implicitly on the environment.

POLE OF INSPIRATION: QUESTIONS OF INTERPRETATION - AND OF PRINCIPLE

8 abril, 2020

According to Proclus, when Euclid wrote the *Elements* his primary objective was to elaborate a complete geometrical theory of the Platonic solids. Indeed, it has been said on several occasions that behind the name "Euclid" there could be a collective with a strong Pythagorean component. The existence of only five regular solids is possibly the best argument to think that we live in a three-dimensional world.

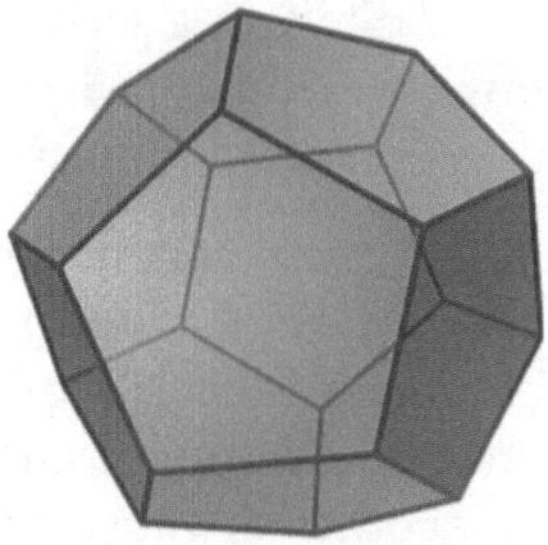

Another great champion of the concept of harmony in science was Kepler, the astronomer who introduced the ellipses into the history of physics — we only need to remember the title of his opus magnum, *Harmonices Mundi*. He discovered the convergence of Fibonacci's series, combined Pythagoras' theorem with the golden ratio in the triangle that bears his name, and theorized extensively about Platonic solids, which he even placed between the orbits of the planets, and which have been considered by a certain tradition as the arcana of the four elements plus the quintessence or Ether.

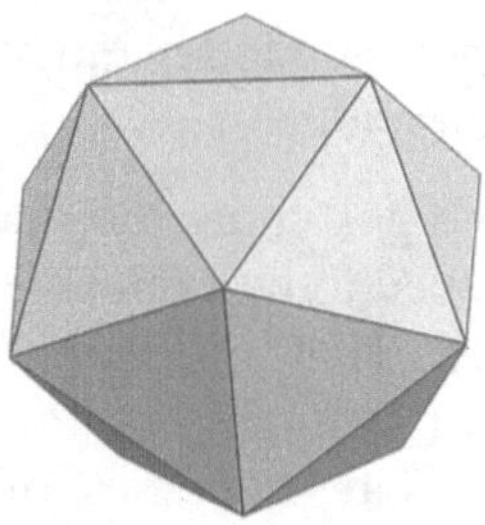

The latter would be represented by the pentagon and the pentagram, and by the dodecahedron dual of the icosahedron, corresponding to the element water.

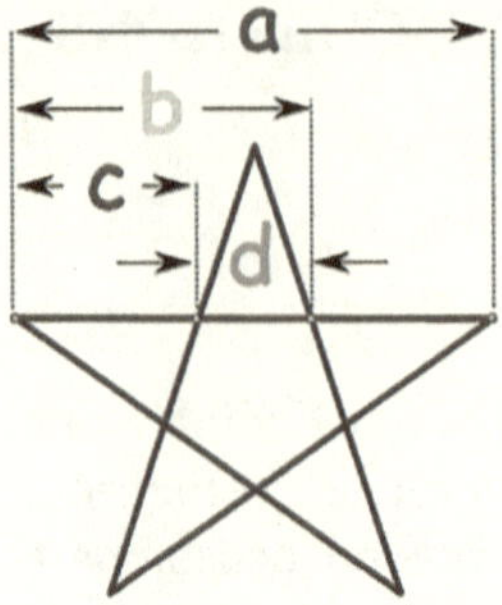

a/b = b/c= c/d = (1+√ 5)/2

The cascade of identical proportions at different scales immediately evokes the capacity to generate self-similar forms like the logarithmic spiral that displays its proportion indefinitely in a series of powers: φ, φ^2, φ^3, φ^4... If a recursion is that which allows something to be defined in its own terms, the pentagram shows us the simplest recursive process in a plane.

As late as 1884 Felix Klein, the reconciler of analytic and synthetic geometry, gave a group of *Lectures on the Icosahedron* in which he considered it to be the central object of the main branches of mathematics: "Every single geometrical object is connected in one way or another to the properties of the regular icosahedron". This is another of those "surprisingly well connected" objects and it would be interesting to trace the links with the two representations of the Pole from which we started. The Rogers-Ramanujan's continuous fraction, for example, plays a role for the icosahedron analogous to the exponential function for a regular polygon.

The orientation proposed by the influential Klein did not get as wide a reception as the famous Erlangen program; to properly deepen it required the dynamic guidance of nature, of applied physics and mathematics. Even current times are more propitious to recover this program, in spite of the fact that physics itself has become more abstract by leaps and bounds. The best way to revive Klein's second major program today might be with the use of a very down-to-earth version of geometric algebra.

The golden section emerges under the unequivocal seal of the fivefold symmetries, which were believed to be exclusive of living beings until the discovery of the quasi-crystals. That ordinary crystals, static structures by definition, exclude this type of symmetry seems to indicate long-range conditions of equilibrium. The optical properties of photonic quasi-crystals and other less periodic structures —often with a zero or double-zero refraction of permittivity and permeability- are being studied intensively in 1D, 2D, and 3D. As expected, phases with spiral holonomy have been found. As in graphene, although the Berry curvature has to be zero, the phase shift can be zero or π [26].

It would be interesting to study all these new states from the point of view of thermomechanics, quantum thermodynamics and retarded potentials, the latter being able to offer a more convincing interpretation of, for example, the so-called relativistic effects of graphene; just as they allow to deal with singular points in a more logical way. On the border between periodic and random structures, the secrets of fivefold symmetry will not be opened without a careful, unbiased interpretation.

*

Except to confuse minds and things, the proscription of the Ether by the special relativity is anything but relevant. First of all, because what makes special relativity work is the Lorentz-Poincaré transformation, conceived expressly for the Ether. Second, because although special relativity, which is the general theory, dispenses with the Ether, general relativity, which is the special theory for gravity, demands it, even if it is in a rarefied theoretical way.

Weber's electrodynamics coincides closely with the Lorentz factor up to speeds of 0.85 c without having to dispense with the third principle of mechanics. But in any case, and even if we stick to Maxwell's equations, which is the very crux of the matter, what we have is that there are not and cannot be electromagnetic waves moving in space with one component ortogonal to the other, but a statistical average of what occurs between space and matter. This is Mazilu's conclusion, but we only need to acknowledge the complete failure of any attempt to specify the geometric description of the field and the waves [27].

The strange thing, once again, is that this has not been seen clearly before. But it turns out that the idea of the Ether around 1900, at least the idea of Larmor, among others, was not only as a medium between particles of matter, but something that also penetrated those particles, which in fact were seen as condensations of the Ether —as now we can see the particles as condensations of the field. There was Ether outside in space and Ether inside matter — as in the electromagnetic waves, among which is light.

Ether is nothing but light itself, but it can also be other things than the light we see, and the whole electromagnetic spectrum. It's just that we cannot know anything without the help of light. Light is the mediator between a space that we cannot know directly, but gives us the metric, and a matter that is in the same situation, but is the subject to measurement.

And now that we know that we are in the midst of the Ether, like the bourgeois gentleman who discovered that he had always been speaking prose without knowing it, perhaps we can look at things more calmly. Only a consummately dualistic mentality that thinks in terms of "this or that" has been able to remain perplexed for so long on this question.

This idea of the Ether *in medias res* could not be very clear at the beginning of the 20th century, otherwise physicists would not have opened their arms to relativity as they did. If it was welcomed, leaving other reasons aside, it was because it seemed to end once and for all with an endless series of doubts and contradictions —or so they thought at the time, until insoluble "paradoxes" began to emerge one after another. One could think that by then Weber's law, which did not even need the existence of a medium because it did not consider waves either, had largely fallen into oblivion —though certainly not for researchers as exhaustive as Poincaré.

Of course, instead of Ether we can also use the word "field" now, as long as we do not understand it as the supplement of space that surrounds the particles, but as the fundamental entity from which they emerge.

In any case, special relativity proper practically does not come into contact with matter, and when it does through quantum electrodynamics, since Dirac, we are confronted with vacuum polarization and an even more crowded, strange and contradictory medium than in any of the previous avatars of the Ether.

Today, transformation optics and the anisotropies of metamaterials are used to "illustrate" black holes or to "design" —it is said- space-times in different flavors. And yet only the macroscopic parameters of Maxwell's old equations, such as permeability and permittivity, are being manipulated. And why should empty space have properties if it is really empty? But, again, what this is all about is statistical averages between space and matter. Only the prejudice created by relativity prevents us from seeing better these things.

It should be much more interesting to study the properties of the space-matter continuum accessible to our direct modulation than exotic aspects of hypothetical objects of a theory, the general relativity, which is not even unified with classical electromagnetism.

And if the ether of 1900 could prove to be inconvenient, what can we say about a theory that breaks with the continuity of the equations of classical mechanics, and that to alleviate it introduces infinite frames of reference? Certainly, it is not an economic solution, and it seems even worse if we think that with the principle of dynamic equilibrium we can dispense with inertia and the distinction of frames of reference —a move that is routinely used to rule out other theories and clear the field.

On the top of this, Maxwell's equations only are valid for extended portions of the field; special relativity is only valid for point-events, and in the field equations of general relativity point particles again become meaningless. Transformation optics takes advantage of this threefold incompatibility with a bypass that leaves special relativity in limbo, to link Maxwell with another incompatible theory. And yet, although this is not said, it is because of special relativity that quantum mechanics has been unable to work with extended particles. In contrast, starting from Weber mechanics there are no problems in working with both extended and point particles.

For those who still believe that the fundamental framework of classical mechanics must have four dimensions, it can be recalled that well into the 21st century, consistent gauge theories of gravity have been developed that satisfy the criterion formulated by Poincaré in 1902, namely to elaborate a relativistic theory in ordinary Euclidean space by modifying the laws of optics, instead of curving the space with respect to the geodesic lines described by light. Light is the mediator between space and matter; and if light can be deformed, which is obvious, there is no need to deform anything else.

Maxwell's equations are not even a general case, but a particular case, both of Weber and of Euler's equations of fluid mechanics. Within fluid mechanics, Maxwell sought the case for a static or motionless medium; if Maxwell's equations are not fundamental, the principle of relativity cannot be fundamental either [28]. The reciprocity of special relativity is purely abstract and kinematic, not mechanical, since it is not bound to centers of masses, and does not allow a distinction between internal forces that comply with the third law and external forces that do not need to comply with it. The principle of relativity, which asserts the impossibility of finding a privileged frame of reference, is valid if and only if there are no external forces to those considered within the system —but on the other hand, by neglecting the third principle, the internal forces are not defined mechanically either.

The so-called Poincaré stress that the French physicist introduced for the Lorentz force to comply with the third principle plays the same role in the relativistic context as Noskov's longitudinal vibrations for the Weber force. The fact that such stress was later considered irrelevant for special relativity shows conclusively its total divorce from mechanics.

Maxwell's equations, as Mazilu says, are a reaction to the partially or totally uncontrollable aspects of the Ether. In physical theories the quantities that matter are not those we can measure, but those we can control; but in this way we dispense with information that could be integrated into a broader theoretical framework.

Questions of interpretation inevitably bring us back to questions of principle; without changing the principles we are condemned to work for them.

The principle of relativity is contingent and therefore unnecessarily restrictive, depending also on arbitrary synchronization procedures. The equivalence principle of general relativity also does not put an end to the problems of reference frames, and, in combination with the principle of relativity, rather multiplies them.

The principle of dynamic equilibrium of relational mechanics radically simplifies this situation without creating unnecessary restrictions. Leaving generality aside, a principle should not be restrictive, but, first of all, necessary. On the other hand, the inability of the equivalence principle to get rid of the principle of inertia automatically subordinates it to the latter.

And it is no coincidence. If we know the same about inertia as we know about the Ether, it is rather because inertia itself inadvertently overlaps with the idea of the Ether and supplants it. Then, the Ether could only emerge without mystifications from a physics that completely dispenses with the idea of inertia, something that is perfectly feasible and compatible with all our experience.

No doubt each theory has its own virtues, but Maxwell's theory and relativity have already been more than sufficiently extolled. Here we prefer to look at the theory that has historical precedence over both, since the presentation made today could not be more biased.

Returning to the past, we see that Lorentz's non-dragging medium, Fresnel and Fizeau's partial dragging medium, and Stokes' total dragging medium are not contradictory and refer to clearly different cases. There are experiments, such as those of Miller, Hoek, Trouton and Noble, and many others, which can be carried out again under much better conditions and provide invaluable information from many points of view, provided that our theoretical framework allows us to contemplate them, which is not the case now [29]. In addition, these experiments are thousands of times less expensive, simpler and more informative than the current "confirmations" of special and general relativity.

There is also an inevitable complementarity between the constitutive aspects of electromagnetism in modern metamaterials, with their mixture of controllable and uncontrollable parameters in matter, and the measurement of uncontrollable parameters in a free environment out in space. But this complementarity cannot be appreciated without principles and a framework that can make them compatible, to begin with. On the other hand there is no need to say that between transformation optics and general relativity we have more of a flimsy parallelism than of real contact.

Another way of talking about a geometric phase is that it is a transformation or holonomy around a singularity. This singularity can be a vortex, which provides a natural connection with the entropy or attenuation of certain magnitudes, which obviously cannot reach infinite values.

An interesting case is the so-called transmutation of optical vortices, that is, the qualitative change of their most intrinsic feature, which is vorticity, and which has recently been carried out even in free space [30], also involving pentagonal symmetries. Vortices occur in the four states of matter —solid, liquid, gaseous and plasma, which are our version of the ancient four elements. Taking into account that their characteristic behavior can be described as a function of constitutive stress/strain relationships, a quantitative description of the transmutation of the states of matter is also feasible, which is widely different from the nuclear transmutation of the elements, without prejudice that also the nuclei can be described more or less classically with vortices like the skyrmions.

The so-called geometrical phase, this so universal phenomenon that it even manifests itself as vorticity on the surface of the water, applied to classical electromagnetism becomes, so to speak, "Maxwell's fifth equation", since it brings into play and encompasses the four known ones. The very name "geometric phase" seems clearly a euphemism, since it is not the geometricians who usually deal with it, but the physicists, and applied physicists for that matter. I also prefer to call it holonomy instead of anholonomy, since the latter refers to the fact that it cannot be integrated within the frame of a theory, while holonomy refers to a global aspect that can be recognized even by the naked eye.

Berry himself admits that the geometrical phase is a way of including the (uncontrollable) environmental factors that are not within the system defined by the theory [31]. In this sense, to get to Maxwell's "fifth equation" we don't need to add terms, but only refer to the "predynamics" of the less restricted or more general equations from which they originate —Weber's and Euler's in this instance. And the same applies to relativity.

No doubt the phenomenology of light is so vast that it never ceases to surprise us, but all this would have an incomparably greater transcendence if a parallel effort were dedicated to the uncontrollable but complementary aspects that are now outlawed or masked by the dominant theories. But in fact there is no part of physics that is not being contemplated today under an unnecessarily distorted optic.

*

There is also place in the theory of black holes for the golden ratio to emerge, just at the critical turning point when the temperature goes from rising to falling: $J^2/M^2 = (1+\sqrt{5})/2$, with M and J being the mass and angular momentum when the constants *c* and G equals 1. The meaning and relevance of this is not clear, but it echoes the custom of this constant of appearing at critical points [32].

In Weber-type forces there is no room for theoretical objects such as black holes since the force decreases with the speed, and it would be interesting to see if transformation optics is able to find a laboratory replica for the evolution of these parameters. In relational mechanics the most that can be expected are different types of phase singularity, such as the optical vortices mentioned.

If there is any interest for us in the emergence of φ in black holes, even if they are purely theoretical calculations, it is because of the direct association with angular momentum, entropy and thermodynamics. It shows us at least that the continuous proportion can also emerge in accordance with the principle of maximum entropy that we consider fundamental for understanding nature, quantum mechanics, or the thermomechanical formulation of classical mechanics. If it can emerge here, it can also do so in other types of singularities, such as the phase of vortices, in the optical model that de Broglie built for the light ray, or in holograms.

For perhaps the greatest interest of the theoretical study of black holes has been to introduce this principle of maximum entropy into fundamental physics, albeit as a final term at extreme conditions, when surely it has a presence at any time in the past and present. In modern theoretical physics, this could be expressed through the so-called holographic principle, which makes sense since this principle makes extensive use of the light phase, and, after all, we have already seen that there is no knowledge of our physical world that does not pass through light. However, there are all sorts of doubts about how to apply such a principle to ordinary low-energy physics.

Black holes are extreme theoretical objects of maximum energy, but they have been reached through the "ordinary" physics of gravity, governed by action principles of minimum energy variation. Technically, this does not involve any contradiction, but makes us wonder about the very nature of the action principles, something that still worried a conservative physicist like Planck.

The action principle of Weber's law, or Noskov's extension of it, does not allow the existence of these extreme objects because, applying reciprocity in a strictly mechanical way to the centres of mass, force and speed get balanced. In the breaking down of the Lagrangian by Mathis into two forces, something similar happens. In Pinheiro's thermomechanics, in which there is a balance between minimum energy variation and maximum entropy, this does not seem to be possible either, provided there is free energy available.

The ordinary Lagrangian is the most void of causality, and Mathis' theory, which wants to dispense with energy and action principles in order to remain only with vectors and forces, would be obviously the "fullest" model; if it is viable is a different matter. The other two are in between. We know that with action principles univocal causes are impossible. One can choose the path one prefers and see how far it goes, but my position on this is that although a univocal determination of causes is not possible, we can have a statistical but certain sense of causality, related to the Second Law of Thermodynamics. Noskov's and Pinheiro's action principles seems compatible.

The third principle that defines what a closed system is, but the ironic twist is that this principle cannot be applied without the help of an environment with free energy contributing to close the balance. This happens even in Mathis' model, where free charge is recycled by matter. Thus, any reversible mechanics emerges as an island from an irreversible background, and it is this superposition of levels that gives us our intuition of causality.

The propagation of light is based on the homogeneity of space, but the masses on which gravity acts involve a non-homogeneous distribution. If we assume a primitive homogeneous medium, no type of force, including gravity, can alter that homogeneity except in a transitory way. The self-correction of forces, which is already implicit in Newton and the original Lagrangian, leads in that direction and seems the only conceivable way, if there is any, to cancel out the infinities arising in calculations. The entropic and thermodynamic treatment of gravity would also necessarily have to follow that direction.

The emergence of the continuous proportion and its series in plant growth, in pinecones or sunflowers, makes us think of vortices made up of discrete units, while at the same time it brings us back to the considerations about *the optimal, not maximum*, use of resources, matter and data collection that nature seems to display.

Yasuichi Horibe demonstrated that the Fibonacci binary trees were subject to the principle of maximum information entropy production [33], something that might be extended to thermodynamic entropy and, perhaps, to other branches such as optics, holography, or quantum thermodynamics. The question is whether these series can simply emerge from the principle of maximum entropy or are at some variable optimum point between minimum energy and maximum entropy, as suggested by Pinheiro's equations.

Since Pinheiro begins by testing its mechanics with some very elementary models, such as a sphere rolling on a concave surface, or the period of oscillation of an elementary pendulum, it would be of great interest to determine the simplest problem, within this mechanics, in which the continuous proportion appears with a critical or relevant role. With this we could take up again the Ariadne's thread of this ratio for action variables and optimization problems.

Planck was still concerned that the principles of action seem to imply a purpose. And the same was true of the Second Law of Thermodynamics for Clausius, although he was not at all bothered by that. It is plain to see that both types of processes, apparently so far apart, are effectively teleological, and this is not a coincidence since they are not even separated, as Pinheiro's thermomechanics shows. It seems that the simultaneous inclusion of two undeniable propensities of nature is more natural than their separate treatment.

In the West there has been a strong rejection of any teleological connotation because teleology has always been confused either with theology and the providential invisible hand or with the intentional hand of man. *Tertium non datur*. However, it is clear that here, for both mechanics and thermodynamics, we are talking about a tendency as unquestionable as it is spontaneous. Understanding this third position, which already existed before the false dilemma of mechanism, leads us to radically change our understanding of Nature.

POLE OF INSPIRATION: BIOLOGICAL FEEDBACK —QUANTITATIVE AND QUALITATIVE MODELS

8 April, 2020

In a recent paper we speculated on the presence of a geometric phase or phase memory in the bilateral nasal cycle, using a certain analogy between the mechanics of the circulatory system and a gauge field such as the electromagnetic one [34], and taking into account that Maxwell's equations are a particular case of the fluid equations. It is known that shortly after its discovery, the geometric phase was generalized well beyond the adiabatic or even the cyclic cases, and that today it is studied even in dissipative open systems and in various cases of animal locomotion. The analogy may be relevant despite the fact that the respiratory system obviously operates in a gaseous phase instead of a liquid one, while still being coupled to the blood circulation.

According to V. D. Tsvetkov, the ratio between systolic and diastolic time in humans and other mammals averages the same reciprocal values of the golden mean, and also the ratio of the maximum systolic pressure to the minimum diastolic pressure points to a relative value of 0.618/0.382 on average. Although these values may be arbitrarily approximate, we would have an excellent opportunity here to contrast them mechanically and see if there really is some kind of underlying optimization, since the systolic time already echoes the reflected vascular wave, and the same is true of the diastolic time.

On the other hand there is the Pulse Wave Velocity, which is a measure of arterial elasticity: both are derived from the second law of mechanics through the Moens-Korteweg equation. This wave velocity varies with pressure, as well as with the elasticity of the vessels, increasing with their stiffness. The return distance of the reflex wave and the time it takes increases with height, and a lower diastolic pressure, which indicates less resistance of the whole vascular system, reduces the magnitude of the reflected wave. Treatment of hypertension should focus, it is said, on decreasing the amplitude of the reflected wave, slowing it down, and increasing the distance between the aorta and the return points of this wave.

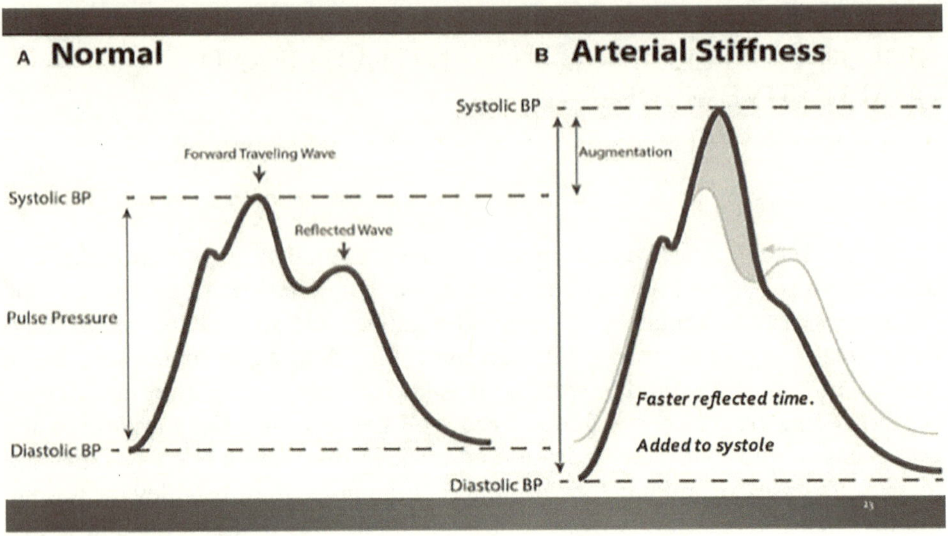

Now, we can try to apply here Noskov's retarded potentials with longitudinal waves, as he emphasized their universality and their place in the most elementary feedback; in fact, there is perhaps no better way of illustrating these waves and their correlation with certain proportions in a complete mechanical system than the circulatory system itself.

Since, to a large extent, it seems that we can consider the elasticity of the reflected wave as a retarded Weber-Noskov potential dependent on distance, force and phase velocity, and check whether this results in coupling or resonance conditions that incidentally tend to the values of the golden section. The myocardium is a self-exciting muscle, but the return of the reflex wave also contributes to this, so we have a fair example of a circuit with tension-pressure-deformation transformations that are fed back and that do not differ in essence from the gauge transformations of modern physics, in which there is also an implicit feedback mechanism.

This would be a perfect instance to explore these correlations in a sort of "closed loop" process, even if the system remains open through the breath, which is not contrary to our approach because for us every natural system is open by definition. It allows both numerical simulation and approximation by real physical models created with elastic tubes and coupled "pumps", so that it can be approached in the most tangible and direct way [35].

However, the idea that the heart is really a pump, being as it is a spiral muscular band, or that the motion of the blood, which generates vortices in the vessels and the heart, is due to pressure, when it is the pressure that is an effect of the former, should be thoroughly revised. In fact this is an excellent example of how we can give a strictly mechanical description while radically questioning not only the form but the very content of causality —the cause-effect relationship. The essential factor of the pressure created is not the heart, but the open component, in this case, the breath and the atmosphere. And although it is clear that these are very different cases, this is in line with our idea of gauge fields and natural processes in general.

*

The dynamics and biomechanics of the blood pulse can be derived from the applied force, but if we look within modern science for a suitable equivalent to Newton's three principles of mechanics in open systems such as biological organisms, we do not find it. To find something similar we have to look back to principles that are more "archaic" to us, and then look for a quantitative and mathematical translation.

Actually, the *triguna* of the Indian Samkya system —samkya means proportion- and its application to the human body as the *tridosha* in Ayurveda is the better match. The triguna, as it were, is a kind of system of coordinates for modalities of the material world in qualitative terms. The three basic qualities, *Tamas*, *Rajas* and *Satwa*, and their reactive forms in the body, *Kapha*, *Pitta* and *Vata* correspond very well to the mass or amount of inertia, the force or energy, and the dynamic equilibrium through motion (let us say: passivity, activity and balance). But it is evident that in this case we are talking about qualities and the systems are considered open from the start without need for further definition.

So, here is the law of conservation of momentum, not the third law of mechanics, what really should hold here, as a system like this implicitly admits a variable degree of interaction with the environment. In harmony with this, the Ayurveda considers that *Vata* is the guiding principle of the three since it has autonomy to move by itself in addition to moving the other two. *Vata* defines the sensitivity of the system in relation to the environment, its degree of permeability or lack thereof. In other words, the state of *Vata* indicates by itself to which extent the system is effectively open.

In the human body the most explicit and continuous form of interaction with the environment is the breath, and therefore it is just in the order of things that *Vata* governs this function most directly. Although the *doshas* are modes or qualities, in the pulse they find a faithful translation in terms of dynamic values and the continuum mechanics —provided we settle for modest degrees of precision, but surely enough to give us a qualitative idea of the dynamics and its basic patterns.

The other two modes are simply what moves and what is moved, but the articulation and coexistence of the three can be understood in very different ways: from a purely mechanical way to a more specifically semiotic one. Here again, the indistinction or ambiguity between kinetic, potential and internal energy, which we have already noticed in relational mechanics, might be of some relevance.

The principle of inertia is a possibility, that of force a brute fact, the action-reaction —the same act seen from two sides- is a relationship of mediation or continuity. We can put them on the same plane or put them on different planes, which constitute an ascending or descending gradation, as in fact are the modalities of the Samkya system.

Actually, it should not be too difficult to find the common ground that the Indian and Chinese semiologies of the pulse have, beyond the differences of terminology and categories; and to move from this common ground to the quantitative, but extremely fluid, language of continuum mechanics. Thus we would have a method to pass from qualitative to quantitative aspects, and vice versa; and to find dynamic patterns that now pass unnoticed. There are several issues here. One is the extent to which these qualitative descriptions can be made consistent.

Another question is to what extent the representation of a qualitative scale can be made intuitive. Let us think, for example, of biofeedback signals, which can be effective under the representation of forces, potentials, and many other more indirect relationships. What is interesting is that these types of assisted feedback do not aim at control and manipulation, but at tuning in to the organizing principle of the dynamics.

From our perspective, as we have already said repeatedly, all physical systems, from galaxies to atoms, have feedback. But what are the physical limits of, let us say a human being, to tune in to other entities? The phase rhythmodynamics and its resonances, the time scale, the energy scale, strain-stress constitutive relations, the dependence on free energy? Or the capacity to align with the Pole that both systems have in common? Is there interference or is there rather a parallelism on the same background?

These are subjects for which science has not yet found even the minimum criteria, but which should help us to overcome the instrumental compulsion, the instrumentation syndrome that has guided human technology since the first tools, and which intensifies as the tools offers less resistance to the user.

One more question is whether this type of trimodal analysis, or even a bimodal one, has a recursive character, as the same feedback and the presence of the continuous proportion in the circulatory system suggest; and what type of recursion is involved.

*

The characterization of the dynamic equilibrium should always indicate the Pole of the evolution of a system, if it has one. In the case of the Solar System and the planets this is obvious —and notwithstanding, it is still far from receiving the attention it deserves. But it turns out that the bilateral nasal cycle is also telling us about an axis even in a process where polarity does not look very relevant, from the biomechanical point of view, such as the compression and release of a gas in our own organism. This should be of great interest to us, and it provides a thread through which many other things can be revealed.

In fact, the Earth's own climate or that of other planets, with its great complexity, is a more explicitly polar system than the respiratory regime of any mammal —and in this case the separating barrier would be the intertropical convergence zone. The point of interest here is that, if the analogy is sound, from a thermomechanical point of view the degree of separation that the barrier exerts, possibly associated with a topological torsion, could also be defining the degree of autonomy of the system with respect to the external conditions —let us call it the endogenous component. An endogenous view that would have to be duly complemented with appropriate sensors and observations of the so-called spatial time [36].

If we said before that the fact that an ellipse has in its interior two focuses does not mean that we only have to look inside it for the origin of the forces that determine its shape, the same is true for the disturbances that usually affect other organisms or systems, which does not prevent them from synthesizing in their behavior the product of external and internal factors, in the breath not less than in other balances that run in parallel.

POLE OF INSPIRATION – GOLDEN MEAN, STATISTICS AND PROBABILITY

8 April, 2020

It has been said for some time that in today's science "correlation supersedes causation", and by correlation we obviously mean a statistical correlation. But even since Newton, physics has not been concerned that much with causation, nor could do it, so that more than a radical change we only have a steady increase in the complexity of the variables involved.

In the handling of statistical distributions and frequencies it makes little sense to talk about false or correct theories, but rather about models that fit the data better or worse, which gives this area much more freedom and flexibility with respect to assumptions. Physical theories may be unnecessarily restrictive, and conversely no statistical interpretation is truly compelling; but on the other hand, fundamental physics is increasingly saturated with probabilistic aspects, so the interaction between both disciplines continues to tighten.

Things become even more interesting if we introduce the possibility that the principle of maximum entropy production is present in the fundamental equations of both classical and quantum mechanics —not to mention if basic relations between this principle and the continuous proportion φ were eventually discovered.

Possibly the reflected wave/retarded potential model we have outlined for the circulatory system gives us a good idea of a virtuous correlation/causation circle that meets the demands of mechanics but suspend —if not reverse- the sense in the cause-effect sequence. In the absence of a specific study in this area, we will now be content to mention some more circumstantial associations between our constant and the probability distributions.

The first association of φ with probability, combinatorics, binomial and hypergeometric distributions is already suggested by the presence of the Fibonacci series in the polar triangle already mentioned.

When we speak of probability in nature or in the social sciences, two distributions come first to mind: the almost ubiquitous bell-shaped normal or Gaussian distribution, and the power law distributions, also known as Zipf distributions, Pareto distributions, or zeta distribution for discrete cases.

Richard Merrick has spoken of a "harmonic interference function" resulting from harmonic damping, or in other words, the square of the first twelve frequencies of the harmonic series divided by the frequencies of the first twelve Fibonacci numbers. According to the author, this is a balance between spatial resonance and temporal damping.

In this way he arrives at what he calls a "symmetrical model of reflexive interference", formed from the harmonic mean between a circle and a spiral. Merrick insists on the transcendental importance for all life of its organization around an axis, which Vladimir Vernadsky had already considered to be *the* key problem in biology.

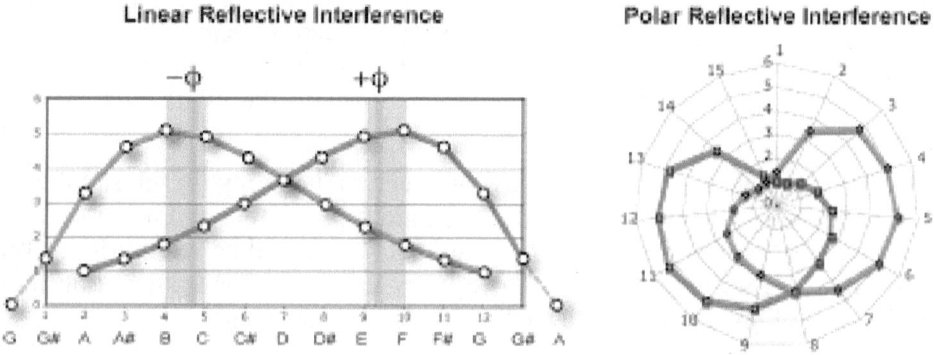

Richard Merrick, Harmonically guided evolution

Merrick's ideas about thresholds of maximum resonance and maximum damping can be put in line with Pinheiro's thermomechanical equations, and as we have indicated they would have a wider scope if they contemplated the principle of maximum entropy as conducive to organization rather than the opposite. Merrick elaborates also a sort of musical theory on the privileged proportion 5/6-10/12 at different levels, from the organization of the human torso to the arrangement of the double helix of DNA seen as the rotation of a dodecahedron around a bipolar axis.

*

The powers laws and zeta distributions are equally important in nature and human events, and are present, among many other things, in fundamental laws of physics, the distribution of wealth among populations, the size of cities or the frequency of earthquakes. Ferrer i Cancho and Fernández note that "φ is the value where the exponents of the probability distribution of a discrete magnitude and the value of the magnitude versus its rank coincide". It is not known at this time if this is a curiosity or if it will allow to deepen the knowledge of these distributions [37].

Zipf or zeta distributions are linked to hierarchical structures and catastrophic events, and also overlap with fractals in the space domain and with the so-called 1/f noise in the time domain. A. Z. Mekjian makes a broader study of the application of the Fibonacci-Lucas numbers to statistics that include hyperbolic power laws [38].

I. Tanackov et al. show the close relationship of the elementary exponential distribution with the value 2ln φ, which makes them think that the emergence of the continuous proportion in nature could be linked to a special case of Markov processes —a non-reversible case, we would advance. It is well known that exponential distributions have maximum entropy. It can be obtained an incomparably faster convergence to the value of the number *e* with Lucas numbers, a generalization of Fibonacci numbers, than with Bernoulli's original expression, which is enough food for thought; we can also get with non-reversible walks a faster convergence than with the usual random walk [38].

Edward Soroko proposed a law of structural harmony for the stability of self-organized systems, based on the continuous proportion and its series considering entropy from the point of view of thermodynamic equilibrium [39]. Although here we give preference to entropy in systems far from equilibrium, his work is of great interest and can be a source of new ideas.

It would be desirable to further clarify the relationship of power laws to entropy. The use of the principle of maximum entropy seems to be particularly suitable for open systems out of balance and with a strong self-interaction. Researchers such as Matt Visser think that Jaynes' principle of maximum entropy allows a very direct and natural interpretation of powers laws [40].

Normally one looks for continuous power laws or discrete power laws, but in nature we can appreciate a middle ground between both as Mitchell Newberry observes with regard to the circulatory system. As usual, in such cases reverse engineering is imposed on the natural model. The continuous proportion and its series offer us an optimal recursive procedure to pass from continuous to discrete scales, and its appearance in this context could be natural [41].

The logarithmic average seems to be the most important component of these power laws, and we immediately associate the basis of the natural logarithms, the number *e*, with the exponential growth in which a certain variable increases without restrictions, something that in nature is only viable for very short periods of time. On the other hand, the golden mean seems to arise in a context of critical equilibrium between at least two variables. But this would lead us rather to logistic or S-curves, which are a modified form of the normal distribution and also a scaled compensation of a hyperbolic tangent function. On the other hand, exponential and power laws distributions look very different but sometimes can be directly connected, which is a subject on its own.

As already noticed, we can also connect the constants *e* and Φ through the complex plane, as in the equality ($\Phi i = e^{\pm \pi i/3}$). Although entropy has always been measured with algebras of real numbers, G. Rotundo and M. Ausloos have shown that here too the use of complex values can be justified, allowing to treat not only a "basic" free energy but also "corrections due to some underlying scale structure" [42]. The use of asymmetric correlation matrices could also be linked with the golden matrices generalized by Stakhov and applied to genetic code information by Sergey Pethoukov [43].

In the mechanical-statistical context the maximum entropy is only an extreme referred to the thermodynamic limit and to Poincaré's immeasurable scales of recurrence; but in many relevant cases in nature, and evidently in the thermomechanical context, it is necessary to consider a non-maximum equilibrium entropy, which may be defined by the coarse grain of the system. Pérez-Cárdenas et al. show a non-maximum coarse-grained entropy linked to a power law, the entropy being so much lower when finer is the grain of the system [44]. This graininess can be linked to the constants of proportionality in the equations of mechanics, such as the same Planck's constant.

*

Probability is a predictive concept and statistics a descriptive, interpretative one, and both should be balanced if we do not want human beings to be increasingly governed by concepts they do not understand at all.

Just to give an example, the mathematical apparatus known as the renormalization group applied to particle and statistical physics is particularly relevant in deep learning, to the point that some experts claim both are the same thing. But it goes without saying that this group historically emerged to deal with the effects of the Lagrangian self-interaction in the electromagnetic field, a central theme of this article.

For prediction, the effects of self-interaction are mostly "pathological", since they complicate calculations and often lead to infinity —although in fact we should put the blame for this in the inability to work with extended particles of special relativity, rather than in self-interaction. But for the description and interpretation the problem is the opposite, it is about recovering the continuity of a natural feedback broken by layers and more layers of mathematical tricks. The conclusion could not be clearer: the search for predictions, and the "artificial intelligence" thus conceived, has grown exponentially at the expense of ignoring natural intelligence —the intrinsic capacity for self-compensation in nature.

If we want to somehow reverse the fact that man is increasingly governed by numbers that he does not understand —and even the experts are bound to trust in programs whose outputs are way beyond their understanding- it is necessary to work at least as hard in a regressive or retrodictive direction. If the gods destroy men by making them blind, they make them blind by means of predictions.

*

As Merrick points out, for the current theory of evolution, if life were to disappear from this planet or have to start all over again, the long-term results would be completely different, and if a rational species were to emerge it would be totally different from our own. That is what random evolution means. In a harmonically guided evolution conditioned by resonance and interference as Merrick suggests, the results would be fairly the same, except for the uncertain incidence that the great cosmic cycles beyond our reach might have.

There is not something like pure chance, there is nothing purely random; no matter how little organized an entity is, be it a particle or an atom, it cannot fail to filter out the environmental "random" influences according to its own intrinsic structure. And the first sign of organization is the appearance of an axis of symmetry, which in particles is defined by axes of rotation.

The dominant theory of evolution, like cosmology, has emerged to fill the great gap between abstract and reversible, and therefore timeless, physical laws and the ordinary world showing an irreversible time, perceptible forms and sequences of events. Today's whole cosmology is really based on an unnecessary and contradictory assumption, the principle of inertia. The biological theory of evolution is based on a false one, that life is only governed by chance.

The present "synthetic theory" of evolution has only come into existence because of the separation of disciplines, and more specifically, due to the segregation of thermodynamics from fundamental physics despite the fact that there is nothing more fundamental than the Second Law. It is not by chance that thermodynamics emerged simultaneously with the theory of evolution: the first one begins with Mayer, who elaborated on work and physiology considerations, and the second one with Wallace and Darwin starting, according to the candid admission of the latter in the first pages of his main work, from Malthus' assumptions of resources and competition, which in turn go back to Hobbes —one is a theory of work and the other of the global ecosystem understood as a capital market. The accumulated capital in this ecosystem is, of course, the biological inheritance.

Merrick's harmonic evolution, due to the collective interference of waves-particles, is an updating of an idea as old as music; and it is also a timeless, purpose-free vision of the events of the world. But to reach the desired depth in time, it must be linked to the other *two clearly teleological, but spontaneous domains,* of mechanics and thermodynamics, which we call thermomechanics for short.

It would be enough to unite these three elements for the present theory of evolution to start becoming irrelevant; and not to mention that human and technological evolution is decidedly Lamarckian beyond speculation. Even

DNA molecules are organized in the most obvious way along an axis. And as for information theory, one only has to remember that it has come out of a peculiar interpretation of thermodynamics, and that it is impossible to do automatic computations without components with a turning axis. Whatever the degree of chance, the Pole rules and defines its sense and meaning.

However, in order to better understand the action of the Pole and the spontaneous reaction involved in mechanics it would be good to rediscover the meaning of polarity.

POLE OF INSPIRATION: QUESTIONS OF PROGRAM —AND OF PRINCIPLE, AGAIN: CALCULUS, DIMENSIONAL ANALYSIS AND CHRONOMETROLOGY

8 April, 2020

In physics and mathematics, as in all areas of life, we have principles, means and ends. The principles are the starting points, the means, from a practical-theoretical point of view, are the different branches of calculus, and the interpretations the ends. These last ones, far from being a philosophical luxury, are the ones that determine the whole contour of representations and applications of a theory.

As for principles, we have already commented, if we want to see more closely where and how the continuous proportion emerges, we should observe as much as possible the ideas of continuity, homogeneity and reciprocity. And this includes the consideration that all systems are open, since if they are not open they cannot comply with the third principle in a way that is worthy of being considered "mechanical". This is the main difference between Nature and manmade machines.

These three-four principles, are nonspecifically included in the principle of dynamic equilibrium, which is the way to dispense with the principle of inertia, and incidentally, the principle of relativity. If we talk about continuity, this does not mean that we are stating that the physical world must necessarily be continuous, but that what seems to be a natural continuity should not be broken without need.

Actually, the principles also determine the scope of our interpretations, although they do not specify them.

As for calculus, which in the form of predictions has become for modern physics the almost exclusive purpose, it is always a matter of justifying how to achieve results known in advance, so reverse engineering and heuristics always won the day over considerations of logic or consistency. Of course there is a laborious foundation of calculus by Bolzano, Cauchy and Weierstrass, but it is more concerned with saving the results than with making them more intelligible.

On this point we cannot agree more with Mathis, who stands alone in a battle to redefine these foundations. What Mathis proposes can be traced back to umbral calculus and the calculus of finite differences, but these are considered as sub-domains of the standard calculus and ultimately have not brought a better understanding to the field.

An instantaneous speed is still an impossible that reason rejects, and besides there is no such thing on a graph. If there are physical theories, such as special relativity, that unnecessarily break with the continuity of classical equations, here we have the opposite case, but with another equally disruptive effect: a false notion of continuity, or pseudo-continuity, is created that is not justified by anything. Modern calculus has created for us an illusion of dominion over infinity and motion by subtracting at least one dimension from the physical space, not honoring that term "analysis" which is so proud of. And this, naturally, should have consequences in all branches of mathematical physics [45].

Mathis' arguments are absolutely elementary and irreducible; these are also questions of principle, but not only of principle as the problems of calculus are eminently technical. The original calculus was designed to calculate areas under curves and tangents to those curves. It is evident that the curves of a graph cannot be confused with the real trajectories of objects and that in them there are no points or moments; then all the generalizations of this methods contain this dimensional conflation.

Finite calculus is also closely related to the problem of particles with extension, without which it is nearly impossible to move from the ideal abstraction of physical laws to the apparent forms of nature.

Mathis himself is the first to admit, for example in his analysis of the exponential function, that there is still a great deal to do for a new foundation of calculus, but this should be good news. At any rate the procedure is clear: the derivative is not to be found in a differential that approaches zero, nor in limit values either, but in a sub-differential that is constant and can only be 1, a unit interval. A differential can only be an interval, never a point or subtraction of points, and it is to the interval that the very definition of limit owes its range of validity. In physical problems this unit interval must correspond to an elapsed time and a distance traveled.

Trying to see beyond Mathis' efforts, it could be said that, if curves are defined by exponents, any variation in a function should be able to be expressed in the form of a dynamic equilibrium whose product is unity; and in any case by a dynamic equilibrium based on a constant unit value, which is the interval. If classical mechanics and calculus grew up side by side as almost indistinguishable twins, even more so should be in a relational mechanics where inertia always dissolves in motion.

The heuristic part of modern calculus is still based on averaging or error compensation; while the foundation is rationalized in terms of limit, but works due to the underlying unit interval. The parallel between the bar of a scale and a tangent is obvious; what is not seen is precisely what should be compensated. Mathis method does not work with averages; standard calculus does. Mathis has found the beam of the scale, now it comes down to set the plates and fine tune the weights. We will come back to this later.

The disputes that still exist from time to time, even among great mathematicians, regarding standard and non-standard calculus, or the various ways of dealing with infinitesimals, at least reveal that different paths are possible, but for most of us they are remote discussions far removed from the most basic questions that should be analyzed first.

We mentioned earlier Tanackov's formula to calculate the constant *e* way faster than the classic "direct method"; but the fact is that amateur mathematicians like the already mentioned Harlan Brothers have found, just twenty years ago, *many* different closed expressions that calculate it faster, besides being more compact. The mathematical community may treat it as a curiosity, but if this happens with the most basic rudiments of elementary calculus, what cannot happen in the dense jungle of higher order functions.

A somewhat comparable case would be that of symbolic calculus or computer algebra, which already 50 years ago found that many classical algorithms, including much of linear algebra, were terribly inefficient. However, as far as we can see, none of this has affected calculus proper.

"Tricks" like those of Brothers play in the ground of heuristics, although it must be recognized that neither Newton, nor Euler, nor any other great name of calculus knew them; but even if they are heuristics they cannot fail to point in the right direction, since simplicity use to be indicative of truth. However with Mathis we are talking not only about the very foundations, which none of the revisions of symbolic calculus has dared to touch, but even about the validity of the results, which cross the red line of what mathematicians want to consider. At the end of the chapter we will see whether this can be justified.

In fact, to pretend that a differential tends to zero is equivalent to say that everything is permitted in order to get the desired result; it is the ideal condition of versatility for whatever heuristics and adhocracy. The fundamental requirement of simplified or unitary calculus —plain differential calculus, indeed- may at first seem like putting spanners in the works already in full gear, but nonetheless it is truthful. No amount of ingenuity can replace rectitude in the search for truth.

Mathis' attempt is not quixotic; there is here much more than meets the eye. There are reversible standards and standards irreversible in practice, such as the current inefficient distribution of the letters of the alphabet on the keyboard, which seems impossible to change even though nobody uses the old typewriters anymore. We do not know if modern calculus will be another example of a standard impossible to reverse, but what is at stake here is far beyond questions of convenience, and it blocks a better understanding of an infinite number of issues; overcoming it is an indispensable condition for the qualitative transformation of knowledge and the ideas we have —or cannot have- of space, time, change and motion.

*

Today's theoretical physicists, forced into a highly creative manipulation of the equations, tend to dismiss dimensional analysis as little more than pettifogging; surely this attitude is due to the fact that they have to think that any revision of the foundations is out of question, and one can only look ahead.

In fact, dimensional analysis is more inconvenient than anything else, since it is by no means irrelevant: it can prove with just a few lines that the charge is equivalent to mass, that Heisenberg's uncertainty relations are conditional and unfounded, or that Planck's constant should only be applied to electromagnetism, instead of being generalized to the entire universe. And since modern theoretical physics is in the business of generalizing its conquests to everything imaginable, any contradiction or restriction to its expansion by the only place it is allowed to expand must be met with notorious hostility.

It is actually easy to see that dimensional analysis would have to be a major source of truth if it were allowed to play its part, since modern physics is a tower of Babel of highly heterogeneous units that are the reflection of contortions made in the name of algebraic simplicity or elegance. Maxwell's equations, compared to the Weber force that preceded him, are the most eloquent example of this.

Dimensional analysis is also of interest when we delve into the relationship between intensive and extensive quantities. The disconnection between the mathematical constants e and φ may also be associated with this broad issue. In entropy, for example, we use logarithms to convert intensive properties such as pressure and temperature into extensive properties, converting for convenience multiplicative relations into more manageable additive relations. That convenience becomes a necessity only for those aspects that are already extensive, such as an expansion.

Ilya Prigogine showed that any type of energy is made up of an intensive and an extensive variable whose product gives us an amount; an expansion, for example, is given by the PxV product of pressure (intensive) by volume (extensive). The same can be applied to changes in mass/density with velocity and volume, and so on.

The unstoppable proliferation of measurements in all areas of expertise already makes simplification increasingly necessary. But, beyond that, there is an urgent need to reduce the heterogeneity of physical magnitudes if intuition is to win the battle against the complexity with which we are accomplices.

All this is also closely related to finite calculus and the equally finitist algorithmic measurement theory developed by A. Stakhov. The classical mathematical measure theory is based on Cantor's set theory and as we know it is neither constructive nor connected with practical problems, let alone the hard problems of the modern physical measurement theory. However, the theory developed by Stakhov is constructive and naturally incorporates an optimization criterion.

To appreciate the scope of the algorithmic measurement theory in our present quantitative Babel we must understand that it takes us back to the Babylonian origins of the positional numbering system, filling an important gap in the current theory of numbers. This theory is isomorphic with a new number system and a new general theory of biological population. The number system, created by George Bergman in 1957 and generalized by Stakhov, is based on the powers of φ. If for Pythagoras "all is number", for this system "all number is continuous proportion".

The algorithmic measure theory also raises the question of equilibrium, since its starting point is the so-called Bachet-Mendeleyev problem, which curiously also appears for the first time in Western literature with Fibonacci's *Liber Abacci* of 1202. The modern version of the problem is to find the optimal system of standard weights for a balance that has a response time or sensitivity.

According to Stakhov, the key point of the weight problem is the deep connection between measurement algorithms and positional numbering methods. My impression however is that it still supports a deeper connection between dynamic equilibrium, calculus and what it takes to adjust a function; the weights for the plates needed by the beam of the scale identified by Mathis. Maybe there is no need to use powers of φ or to change the number system, but very useful ideas might be developed about the simplest algorithms.

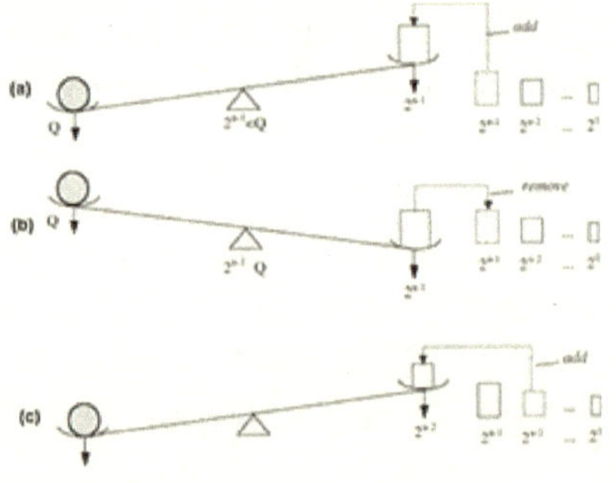

Alexey Stakhov, The mathematics of Harmony

Of course, the algorithmic theory of complexity tells us that we cannot prove an algorithm is the simplest for a task, but that does not mean that we do not look for it continually, regardless of any demonstration. Efficiency and formal demonstration are largely unrelated.

Human beings inevitably tend to optimize what they measure most; however, we do not have a theory that harmonizes the needs of metrology with those of mathematics, physics and the descriptive sciences, whether social or natural. Today there are many different measure theories, and each discipline look for what is best for it. However, all metrics are defined by a function, and functions are defined by calculus or analysis, which does not want to have anything to do with the practical problems of measurement and pretends to be as pure as arithmetic even though it is far from it.

This may seem a somewhat absurd situation and in fact it is, but it also places a hypothetical measure theory that is in direct contact with the practical aspects, the foundations of calculus and arithmetic in a strategic situation above the drift and inertia of the specialities.

Calculus or analysis is not pure math, and it is too much to pretend that it is. On the one hand, and as far as physics is concerned, it involves at least a direct connection with questions of measurement that should be more explicit in the same math; on the other hand, the highly heuristic nature of its most basic procedures speaks for itself. If arithmetic and geometry have large gaps, being incomparably clearer, it would be absurd to pretend that calculus cannot have gaps much greater.

*

On the other hand, physics will never cease to have both statistical and discrete components —bodies, particles, waves, collisions, acts of measurement, etc- besides continuous ones, which makes a relational-statistical analysis advisable.

An example of relational statistical analysis is the one proposed by V.V. Aristov. Aristov introduces a constructive and discrete model of time as motion using the idea of synchronization and physical clock that Poincaré already introduced precisely with the problem of the electron. Here each moment of time is a purely spatial picture. But it is not only a matter of converting time into space, but also of understanding the origin of the mathematical form of the physical laws: "The ordinary physical equations are consequences of the mathematical axioms, 'projected' into physical reality by means of the fundamental instruments. One can assume that it is possible to build different clocks with a different structure, and in this case we would have different equations for the description of motion".

Aristov himself has provided clock models based on non-periodic, theoretically random processes, that are also of great interest. A clock based on a non-periodic process could be, for example, a piston engine in a cylinder; and this could also include thermodynamic processes.

It should also be noted that cyclical processes, despite their periodicity, mask additional or environmental influences, as we have seen with the geometric phase. To this, a deductive filter of unnecessarily restrictive principles is added, as we have already seen in the case of relativity. And as if all this were not enough, we have the fact, hardly recognized, that many processes considered purely random or "spontaneous", such as radioactive decay, show discrete states during fluctuations in macroscopic processes, as has been extensively shown by S. Shnoll and his school for more than half a century.

Indeed, all kind of processes, from radioactive decay to enzymatic and biological reactions, through random number generators, show recurrent periods of, 24 hours, 27 and 365 days, which obviously correspond to astronomical cyclic factors.

We know that this regularity is filtered and routinely discounted as "non-significant" or irrelevant, in an example of how well researchers are trained to select data, but, beyond this, the question of whether such reactions are spontaneous or forced remains. An answer may be advanced: one would call them spontaneous even if a causal link could be demonstrated, since bodies contribute with their own momentum.

The statistical performance of multilevel neural networks —ultimately a brute force strategy- is increasingly hampered by the highly heterogeneous nature of the data and units with which they are fed, even though dynamic processes are obviously independent of the units. In the long run, the pureness of principles and criteria is irreplaceable, and the shortcuts that theories have sought for prediction accumulates a huge deal of deadweight. And again, it is of little use what conclusions machines can reach when we are already incapable of seeing through the simplest assumptions.

The performance of a relational network is also cumulative, but in exactly the opposite sense; perhaps it should be said, rather, that it grows in a constructive and modular way. Its advantages, such as those of relational physics —and information networks in general - are not obvious at first sight but increase with the number of connections. The best way to prove this is by extending the network of relational connections. And indeed, it is about collective work and collective intelligence.

With arbitrary cuts to relational homogeneity, destructive interference and irrelevant redundancy increase; conversely, the greater the relational density, the greater the constructive interference. I don't think this requires demonstration: Totally homogeneous relations allow higher order degrees of inclusion without obstruction, just as equations made of heterogeneous elements can include equations within equations as opaque elements or knots still to unravel [46].

*

Let us go back to the calculus, but from a different angle. Mathis' differential calculus does not always get the same results as the standard one, which would seem sufficient to rule it out. Since the principle is unquestionable, errors, if possible, might be due to an incorrect application of the principle, leaving the criteria still to be clarified. On the other hand, that there is a "dimensional reduction" of the curves in standard calculus is a fact, which however is not widely recognized because after all now the graphs are supposed to be secondary and even dispensable.

Are they really? Without graphs and curves calculus would never have been born, so that is enough. David Hestenes, the great advocator of geometric algebra and geometric calculus, says that geometry without algebra is dumb, and algebra without geometry is blind. We should add that not only algebra, but calculus too, and to a greater extent than we think, provided we understand that there is more to "geometry" than what the graphs usually tell us. We can now look at another type of graph, this time of vortices, due to P. A. Venis [47].

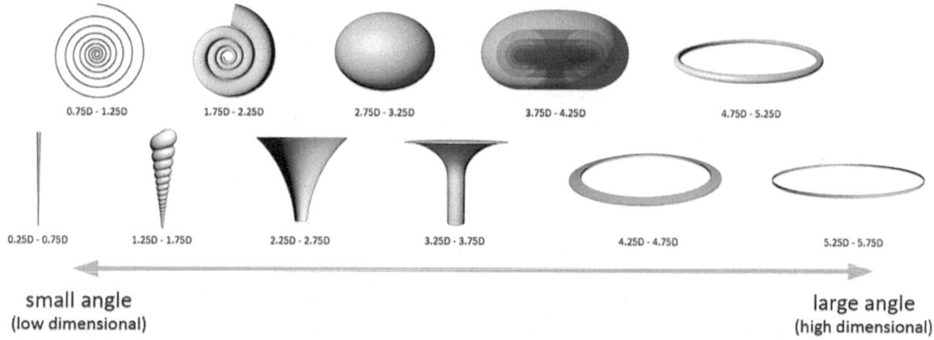

Peter Alexander Venis

In the transformation sequence Venis makes an estimate of its dimensionality that at first may seem arbitrary, although it is based on something as "obvious" as the transition from point to line, line to plane and plane to volume. The fractional dimensions seems at first sight striking, until we recognize it is just an estimate of the continuity within the order of the sequence, which could not be more natural.

Although Venis does not look for a proof, his transformation sequence is self-evident and more compelling than a theorem. One need only a minute of real attention to understand its evolution. It is a general key to morphology, regardless of the physical interpretation we want to give it.

For Venis the appearance of a vortex in the physical plane is a phenomenon of projection of a wave from a single one field where the dimensions exist as a compact whole without parts: a different way to express the primitive homogeneous medium of reference for dynamic equilibrium. It is clear that in a completely homogeneous medium we cannot characterize it as either full or void, and that we could say that it has either an infinite number of dimensions or no dimensions at all.

Thus, both the ordinary dimensions, as well as the fractional ones, and even the negative dimensions are a phenomenon of projection, of projective geometry. The physical nature is real since it participates in this one field or homogeneous medium, and it is a projected illusion to the extent that we conceive it as an independent part or an infinity of separate parts.

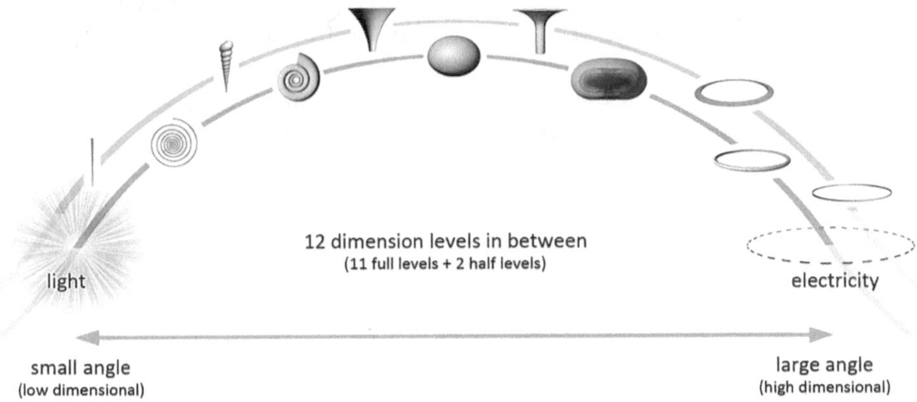

Peter Alexander Venis

Negative dimensions are due to a projection angle lower than 0 degrees, and lead to toroidal evolution beyond the bulb in equilibrium in three dimensions —that is, dimensions greater than the ordinary three. So they form a complementary projective counter-space to the ordinary space of matter, which with respect to unity is not less a projection than the first. Light and electricity are at opposite ends of manifestation, of evolution and involution in matter: light is the fiat, and electricity the extinction. Much could be elaborated on this but we will leave something for later.

Arbitrary cuts in the sequence leave fractional dimensions exposed, coinciding with the shapes we can appreciate. Since Mathis himself attributes the differences in results between his calculus and the standard calculus to the fact that the latter eliminates at least one dimension, and in the sequence of transformations we have a whole series of intermediate dimensions for basic functions, this would be an excellent workbench to compare both.

Michael Howell consider that fractal analysis avoids the usual dimensional reduction, and translates the exponential curve into "a fractal form of variable acceleration" [48]. It is worth noting that for Mathis the standard calculus has errors even in the elementary exponential function; the analysis of the dimensional evolution of vortices gives us a wide spectrum of cases to settle differences. I am thinking about fractional derivatives and differentiable curves, rather than fractals as non-differentiable curves. It would be interesting to see how the constant differential works with fractional derivatives.

The history of fractional calculus, which has gained great momentum in the 21st century, goes back to Leibniz and Euler and is one of the rare cases where both mathematicians and physicists ask for an interpretation. Although its use has extended to intermediate domains in exponential, wave-diffusion, and many other types of processes, fractional dynamics presents a non-local history dependence that deviates from the usual case, though there is also local fractional calculus. To try to reconcile this divergence Igor Podlubny proposed a distinction between inhomogeneous cosmic time and homogeneous individual time [49].

Podlubny admits that the geometrization of time and its homogenization are primarily due to calculus itself, as the intervals of space can be compared simultaneously, but the intervals of time cannot, and we can only measure them sequentially. What may be surprising is that this author attributes non-homogeneity to cosmic time, rather than to individual time, since in reality mechanics and calculus develop in unison under the principle of global synchronization and simultaneity of action and reaction. In this respect relativity is not different from Newtonian mechanics. According to Podlubny, individual time would be an idealization of the time created by mechanics, which is to put it upside down: in any case the idealization is the global time.

On the one hand, fractional calculus is seen as a direct aid for the study of all kinds of "anomalous processes"; on the other hand, fractional calculus itself is a generalization of standard calculus that includes ordinary calculus and therefore also allows to deal with all modern physics without exception. This makes us wonder if, more than dealing with anomalous processes, it is ordinary calculus what enforces a normalization, which affects all quantities it computes, time among them.

Venis also speaks of non-homogeneous time and temporal branches, though his reasoning remains undecided between the logic of the sequence, which represents an individualized flow of time, and the logic of relativity. However, it is the sequential logic that should define time in general and individual or local time in particular —not the logic of simultaneity of the global synchronizer. We shall return to this soon.

POLE OF INSPIRATION – FROM MONOPOLES TO POLARITY

8 April, 2020

Polarity was always an essential component of natural philosophy and even of plain thought, but the advent of the theory of electric charge replaced a living idea with a mere convention.

Regarding the universal spherical vortex, we mentioned earlier Dirac's monopole hypothesis. Dirac conjectured the existence of a magnetic monopole just for the sake of symmetry: if there are electric monopoles, why can not exist magnetic monopoles too?

Mazilu, following E. Katz, suggest quietly that there is no need to complete this symmetry, since we already have a higher order symmetry: the magnetic poles appear separated by portions of matter, and the electric poles only appear separated by portions of space. This is in full accordance with the interpretation of electromagnetic waves as a statistical average of what occurs between space and matter.

And this puts the finger on the point everybody tries to avoid: it is said that the current is the effect that occurs between charges, but actually it is the charge what is defined by the current. Elementary charge is a postulated entity, not something that follows from definitions. Mathis can rightly say that the idea of elementary charge can be completely dispensed with and replaced by mass, which is justified by dimensional analysis and greatly simplifies the picture —freeing us, among other things, from "vacuum constants" such as permittivity and permeability that are completely inconsistent with the word "vacuum" [50].

In this light, there are no magnetic monopoles, because there are no electric monopoles nor dipoles to begin with. The only things that would exist are gradients of a neutral charge, photons producing attractive or repulsive effects according to the relative densities and the screening exerted by other particles. And by the way, it is this purely relative and changing sense of shadow and light what characterized the original notion of yin and yang and polarity in all cultures until the great invention of elementary charge.

So it can well be said that electricity killed polarity, a death that will not last long since polarity is a much larger and more interesting concept. It is truly liberating for the imagination and our way of conceiving Nature to dispense with the idea of bottled charges everywhere.

Yes, it is more interesting even for such a theoretical subject as the monopole. Theoretical physicists have even imagined global cosmological monopoles. But it is enough to imagine a universal spherical vortex like the one already mentioned, without any kind of charge, but with self-interaction and a double motion for the rotations associated with magnetism and the attractions and repulsions associated with charges to arise. The same reversals of the field in Weber's electrodynamics already invited to think that the charge is a theoretical construct.

We should come to see electromagnetic attraction and repulsion as totally independent of charge, and conversely, the unique field that includes gravity, as capable of both attraction and repulsion. This is the condition, not to unify, but to approach the effective unity that we presuppose in Nature.

*

The issue of polarity leads us to think of another great theoretical problem for which an experimental correlate is sought: the Riemann zeta function. As we know, basic aspects of this function present an enigmatic similarity with the random matrices of subatomic energy levels and many other collective features of quantum mechanics. Science looks for mathematical structures in physical reality, but here on the contrary we would have a physical structure reflected in a mathematical reality. Great physicists and mathematicians such as Berry or Connes proposed more than ten years ago to confine an electron in two dimensions under electromagnetic fields to "get its confession" in the form of the zeros of the function.

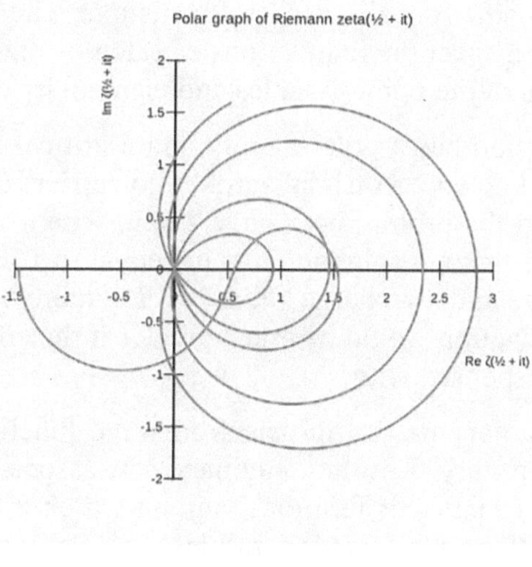

Polar zeta

There has been a great deal of discussion about the dynamics capable of recreating the real part of the zeros of the Riemann zeta function. Berry surmises that this dynamics should be irreversible, bounded and unstable, which makes a big difference for the ordinary processes expected from the current view of fundamental fields, but it is closer to quantum thermomechanics, or what is the same, irreversible quantum thermodynamics. Moreover, it seems that what it is at stake here is the most basic arithmetic relationship between addition and multiplication, as opposed to the scope of multiplicative and additive reciprocities in physics.

Most physicists and mathematicians think that there is nothing to scratch about the nature of the imaginary numbers or the complex plane; but as soon as they have to deal with the zeta function, there is hardly anyone who does not begin with an interpretation of the zeros of the function —especially when it comes to its connections with quantum mechanics. So it seems that only when there are no solutions the interpretation matters.

Maybe it would be healthy to purge the calculus in the sense that Mathis asks for and see to what extent it is possible to obtain results in the domain of quantum mechanics without resorting to complex numbers or the stark heuristics methods of renormalization. In fact, in Mathis' version of calculus each point is equivalent to at least one distance, which should give us additional information. If the complex plane allows extensions to any dimension, we should check what is its minimum transposition into real numbers, both for physical and arithmetic problems. After all, Riemann's starting point was the theory of functions, rather than number theory.

Surely if physicists and mathematicians knew the role of the complex plane in their equations they would not be thinking of confining electrons in two dimensions and other equally desperate attempts. The Riemann zeta function is inviting us to inspect the foundations of calculus, the bases of dynamics, and even our models of the point particles and elementary charge.

The zeta function has a pole at unity and a critical line with a value of 1/2 where lie all its known non-trivial zeros. The carriers of "elementary charge", the electron and the proton, both have a spin with a value of 1/2 and the photon that connects them, a spin with a value equal to 1. But why should spin be a statistical feature and not charge? Possibly the interest of the physical analogies for the zeta function would be much greater if the concept of elementary charge were to be dispensed with.

That the imaginary part of the electron wave function is linked to spin and rotation is no mystery. But the imaginary part associated with the quantization of particles of matter or fermions, among which is the electron, has no obvious relation with spin. However, in classical electromagnetic waves we can deduce that the imaginary part of the electrical component is related to the real part of the magnetic component, and vice versa. The scattering amplitudes and their analytical continuation cannot be separated from the spin statistics, and vice versa; and both are associated with timelike and spacelike phenomena respectively. There can be also different analytical continuations with different meanings and geometrical interpretations in the Dirac algebra.

In electrodynamics all the development of the theory goes very explicitly from global to local. Gauss' divergence theorem of integral calculus, which Cauchy used to prove his residue theorem of complex analysis, is the prototype of the cyclic or period integral. Like Gauss law of electrostatics, is originally independent of metric, though this is seldom taken into account. The Aharonov-Bohm integral, a prototype of geometric phase, is very similar in structure to the Gauss integral.

As Evert Jan Post emphasize time and again, the Gauss integral works as a natural counter of net charge, just as the Aharonov-Bohm integral works for quantized units of flux in self-interaction of beams. This clearly speaks in favor of the ensemble interpretation of quantum mechanics, in contrast to the Copenhagen interpretation stating that the wave function corresponds to a single individual system with special or non-classical statistics. The main statistical parameters here would be, in line with Planck's 1912 work in which he introduced the zero point energy, the mutual phase and orientation of the ensemble elements [51]. Of course orientation is a metric-independent property.

Naturally, this integral line of reasoning shows its validity in the quantum Hall effect and its fractional variant, present in two-dimensional electronic systems under special conditions; this would bring us back to the above-mentioned attempts of electron confinement, but from the angle of classical, ordinary statistics. In short, if there is a correlation between this function and atomic energy levels, it should not be attributed to some special property of quantum mechanics but to the large random numbers generated at the microscopic level.

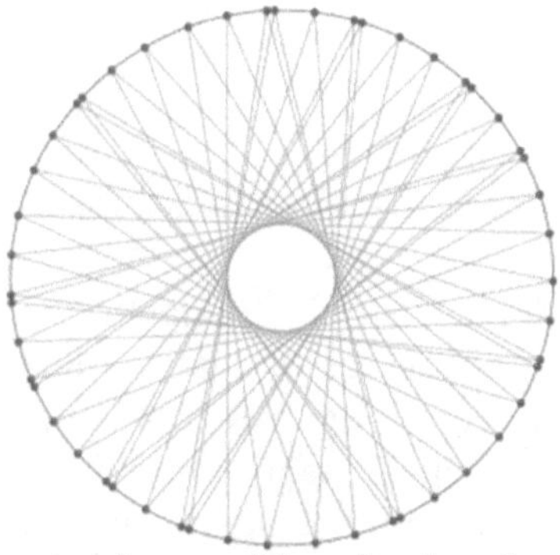

If we cannot understand something in the classical domain, hardly we will see it more clearly under the thick fog of quantum mechanics. There are very significant correlates of Riemann zeta in classical physics without the need to invoke quantum chaos; and in fact well known models such as those of Berry, Keating and Connes are semiclassical, which is a way to stay in between. We can find a purely classical exponent of the zeta function in dynamic billiards, starting with a circular billiard, which can be extended to other shapes like ellipses or the so-called stadium shape, a rectangle capped with two semicircles.

The circular billiard dynamics with a point particle bouncing is integrable and allows to model, for instance, the electromagnetic field in resonators such as microwave and optical cavities. If we open holes along the boundary, so that the ball has a certain probability of escape, we have a dissipation rate and the problem becomes more interesting. Of course, the probability depends on the ratio between the size of the holes and the size of the perimeter.

The connection with the prime numbers occurs through the angles in the trajectories according to a module $2\ \pi/n$. Bunimovich and Detteman showed that for a two hole problem with holes separated by 0°, 60°, 90°, 120° or 180° the probability is uniquely determined by the pole and non-trivial zeros of the Riemann zeta function [52]. The sum over can be determined explicitly and involves terms that can be connected with Fourier harmonic series and hypergeometric series. I do not know if this result may be extended to elliptical boundaries, which are also integrable. But if the boundary of the circle is smoothly deformed to create a stadium-like shape, the system is not integrable anymore and becomes chaotic, requiring an evolutionary operator.

Dynamic billiards have multiple applications in physics and are used as paradigmatic examples of conservative systems and Hamiltonian mechanics; however the "leakage" allowed to the particles is nothing more than a dissipation rate, and in the most explicit manner possible. Therefore, our guess that the zeta should be associated with thermomechanical systems that are irreversible at a fundamental level it is plain to understand —it is Hamiltonian mechanics that begs the exception to begin with. Since it is assumed that fundamental physics is conservative, there is a need for "a little bit open" closed systems, when, according to our point of view, what we always have is open systems that are partially closed. And according to our interpretation, physics has the question upside-down: a system is reversible *because* it is open, and it is irreversible to the extent that becomes closed, differentiated from the fundamental homogeneous medium.

From a different angle, the most apparent aspect of electromagnetism is light and color. Goethe said that color was the frontier phenomenon between shadow and light, in that uncertain region we call penumbra. Of course his was not a physical theory, but a phenomenology, which not only does not detract from it but adds to it. Schrödinger's theory of color space of 1920, based on arguments of affine geometry though with a Riemannian metric, is somewhat halfway between perception and quantification and can serve us, with some improvements introduced later, to bring together visions that seem totally unconnected. Mazilu shows that it possible to obtain matrices similar to those arising from field interactions [53].

Needless to say, the objection to Goethe was that his concept of polarity between light and darkness, as opposed to the sound polarity of electric charge, was not justified by anything, but we believe that it is just the opposite. All that exists are density gradients; light and shadow can create them, a charge that is only a + or — sign attached to a point, can not. Colors are within light-and-darkness as light-and-darkness are within space-and-matter.

It is said, for example, that the Riemann zeta function could play the same role for chaotic quantum systems as the harmonic oscillator does for integrable quantum systems. Maybe. But do we know everything about the harmonic oscillator? Far from it, as the Hopf fibration or the monopole remind us. On the other hand, the first appearance of the zeta function in physics was with Planck's formula for blackbody radiation, where it enters the calculation of the average energy of what would later be called "photons".

The physical interpretation of the zeta function always forces us to consider the correspondences between quantum and classical mechanics. Therefore, a problem almost as intriguing as this should be finding a classical counterpart to the spectrum described by Planck's law; however, there seems to be no apparent interest in this. Mazilu, again, reminds us the discovery by Irwin Priest in 1919 of a simple and enigmatic transformation under which Planck's formula yields a Gaussian, normal distribution with an exquisite correspondence over the whole frequency range [54].

In fact, the correspondence between quantum mechanics and classical mechanics is an incomparably more relevant issue on a practical and technological level than that of the zeta function. As for the theory, there is no clarification from quantum mechanics about where the transition zone would be. There are probably good reasons why such a sensitive point receives apparently so little attention. However, it is highly probable that Riemann's zeta function unexpectedly connects different realm of physics, making it a mathematical object of unparalleled depth.

The point is that there is no interpretation for the cubic root of the frequency of Priest's transformation. Referring to the cosmic background radiation, "and perhaps to thermal radiation in general", Mazilu tries his best guess in order to finish with this observation: "Mention should be made that such a result can very well be specific to the way the measurements of the radiation are usually made, i.e. by a bolometer". Leaving this transformation aside, various more or less direct ways of deriving Planck's law from classical assumptions have been proposed, from that suggested by Noskov on the basis of Weber's electrodynamics to that of C.K. Thornhill [55]. Thornhill proposed a kinetic theory of electromagnetic radiation with a gaseous ether composed of an infinite variety of particles, where the frequency of electromagnetic waves is correlated with the energy per unit mass of the particles instead of just the energy —obtaining the Planck distribution in a much simpler way.

The statistical explanation of Plank's law is already known, but Priest's Gaussian transformation demands a physical explanation for classical statistics. Mazilu makes specific mention of the measurement device used, in this case the bolometer, based on an absorptive element —the Riemann zeta function is said to correspond to an absorption spectrum, not an emission spectrum. If today metamaterials are used to "illustrate" space-time variations and mock black holes —where the zeta function is also used to regularize calculations-they could be used with much sounder arguments to study variations of the absorption spectrum to attempt a reconstruction by a sort of reverse engineering. The enigma of Priest's formula could be approached from a theoretical as well as a practical and constructive point of view —although the very explanations for the performance of metamaterials are controversial and would have to be purged of numerous assumptions.

Of course by the time Priest published his article the idea of quantization had already won the day. His work fell into oblivion, and little else is known about the author except for his dedication to colorimetry, in which he established a criterion of reciprocal temperatures for the minimum difference in color perception [56]; of course these are perceptive temperatures, not physical ones; the correspondence between temperatures and colors was never found. If there is an elementary additive and subtractive algebra of colors, there must also be a product algebra, surely related to their perception. This brings to mind the so-called non-spectral *line of purples* between the red and violet ends of the spectrum, the perception of which is limited by the luminosity function. With a proper correspondence this could give as a beautiful analogy that the reader may try to guess.

On the other hand, it should not be forgotten that Planck's constant has nothing to do with the uncertainty in the energy of a photon, though today they are routinely associated [57]. Noskov's longitudinal vibrations in the moving bodies immediately remind us of the *zitterbewegung* or "trembling motion" introduced by Schrödinger to interpret the interference of positive and negative energy in the relativistic Dirac equation. Schrödinger and Dirac conceived this motion as "a circulation of charge" generating the magnetic moment of the electron. The frequency of the rotation of the zbw would be of the order of 10^{21} hertz, too high for detection except for the resonances.

David Hestenes has analyzed various aspects of the *zitter* in his view of quantum mechanics as self-interaction. P. Catillon et al. conducted an electron channeling experiment on a crystal in 2008 to confirm de Broglie's internal clock hypothesis. The resonance detected experimentally is very close to the de Broglie frequency, which is half the frequency of the *zitter*; the de Broglie period would be directly associated with mass, as has recently been suggested. There are several models for creating resonances with the electron reproducing the zeta function in cavities and Artin's dynamic billiards, but they are not usually associated with the zitter or the de Broglie internal clock, since this does not fit into the conventional version of quantum mechanics. On the other hand, it would be advisable to consider a totally classical zero-point energy as can be followed from the works of Planck and his continuators in stochastic electrodynamics, though all these models are based *on point particles* [58].

Math, it is said, is the queen of sciences, and arithmetic the queen of mathematics. The fundamental theorem of arithmetic places prime numbers at its center, as the more irreducible aspect of the integer numbers. The main problem with prime numbers is their distribution, and the best approximation to their distribution comes from the Riemann zeta function. This in turn has a critical condition, which is precisely to find out if all the non-trivial zeros of the function lie on the critical line. Time past and competition among mathematicians have turn the task of proving the Riemann hypothesis into the K-1 of mathematics, which would involve a sort of duel between man and infinity.

William Clifford said that geometry was the mother of all sciences, and that one should enter it bending down like children; it seems on the contrary that arithmetic makes us more haughty, because in order to count we do not need to look down. And that, looking down to the most elementary and forgetting about the hypothesis as much as possible, would be the best thing for the understanding of this subject. Naturally, this could also be said of countless questions where the overuse of mathematics creates too rarefied a context, but here at least a basic lack in understanding is admitted.

There seem to be two basic ways of understanding the Riemann zeta function: as a problem posed by infinity or as a problem posed by unity. So far modern science, driven by the history of calculus, has been much more concerned with the first aspect than with the second, even if both are inextricably linked.

It has been said that if a zero is found outside the critical line —if Riemann's hypothesis turns out to be false-, that would create havoc in number theory. But if the first zeros already evaluated by the German mathematician are well calculated, the hypothesis can be practically taken for granted, without the need to calculate more trillions or quadrillions of them. In fact, and in line with what was said in the previous chapter, it seems much more likely to find flaws in the foundations of calculus and its results than to find zeros off the line, and in addition the healthy creative chaos that would produce would surely not be confined to a single branch of math.

Of course, this applies to the evaluation of the zeta function itself. If Mathis' simplified calculus, using a unitary interval criterion, finds divergences even for the values of the elementary logarithmic function, these divergences would have to be far more important in such convoluted calculations like those of this function. And in any case it gives us a different criterion for the evaluation of the function; furthermore, there might be a criterion to settle if certain divergences and error terms cancel out.

The devil's advocates in this case would not have done the most important part of their work yet. On the other hand, fractional derivatives of this function have been calculated allowing us to see where the real and imaginary parts converge; this is of interest both for complex analysis and physics. In fact it is known that in physical models the evolution of the system with respect to the pole and zeros usually depends on the dimension, which in many cases is fractional or fractal, and even multi-fractal for potentials associated with the number themselves.

Arithmetic and counting exist primarily in the time domain, and there are good reasons to think that methods based on finite differences should take a certain kind of preference when dealing with changes in the time domain — since with infinitesimals the act of counting dissolves. The fractional analysis of the function should also be concerned with sequential time. Finally, the relationship between discrete and continuous variables characteristic of quantum mechanics should also be connected with the methods of finite differences.

Quantum physics can be described more intuitively with a combination of geometric algebra and fractional calculus for cases containing intermediate domains. In fact, these intermediate domains can be much more numerous than we think if we take into account both the mixed assignment of variables in orbital dynamics and the different scales at which waves and vortices can occur between the field and the particles in a different perspective like Venis'. The same self-interaction of the *zitterbewegung* calls for a much greater concreteness than hitherto achieved. This movement allows, among other things, a more directly geometric, and even classical, translation of the non-commutative aspects of quantum mechanics which in turn allow for a key natural connection between discrete and continuous variables.

Michel Riguidel makes the zeta function object of an intensive work of interaction in search of a morphogenetic approach. It would be great if the computing power of machines can be used to refine our intuition, interpretation and reflection, rather than the other way around. However, here it is easy to present two major objections. First, the huge plasticity of the function, which although completely differentiable, according to Voronin's theorem of universality contains any amount of information an infinite number of times.

The second objection is that if the function already has an huge plasticity, and on the other hand graphics can only represent partial aspects of the function at any rate, further deformations and transformations, however evocative they may be, still introduce new degrees of arbitrariness. The logarithm can be transformed into a spiral halfway between the line and the circle, and create spiral waves and whatnot, but in the end they are just representations. The interest, at any rate, is in the interaction function-subject-representation — the interaction between mathematical, conceptual and representational tools.

But there is no need for more convoluted concepts. The greatest obstacle to go deeper into this subject, as in so many others, lies in the stark opposition to examining the foundations of calculus, classical and quantum mechanics. The more complex the arguments to prove or disprove the hypothesis, be it true or false, the less importance the result can have for the real world.

It is often said that the meaning of the Riemann hypothesis, and even of all the computed zeros of the function, is that the prime numbers have as random a distribution as possible, which of course leaves wide open how much randomness is possible. We may have no choice but to talk about apparent randomness.

But even so, there we have it: the highest degree of apparent randomness in a simple linear sequence generalizable to any dimension hides an ordered structure of unfathomable richness.

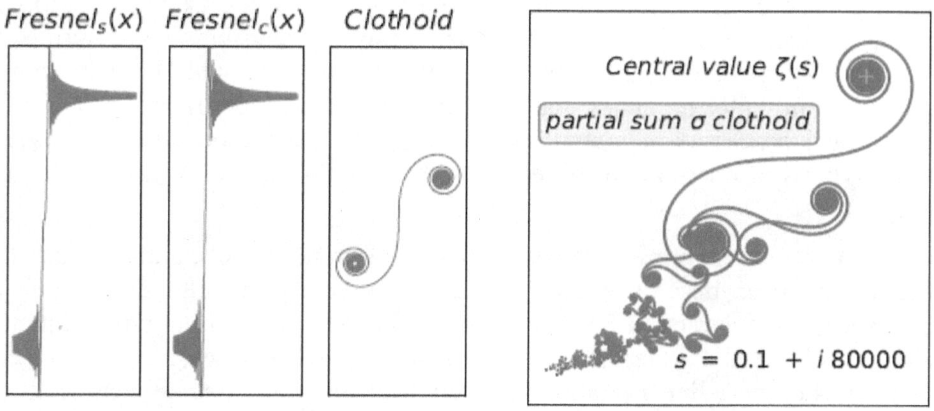

Michel Riguidel: Morphogenesis of the Zeta Function in the Critical Strip by Computational Approach

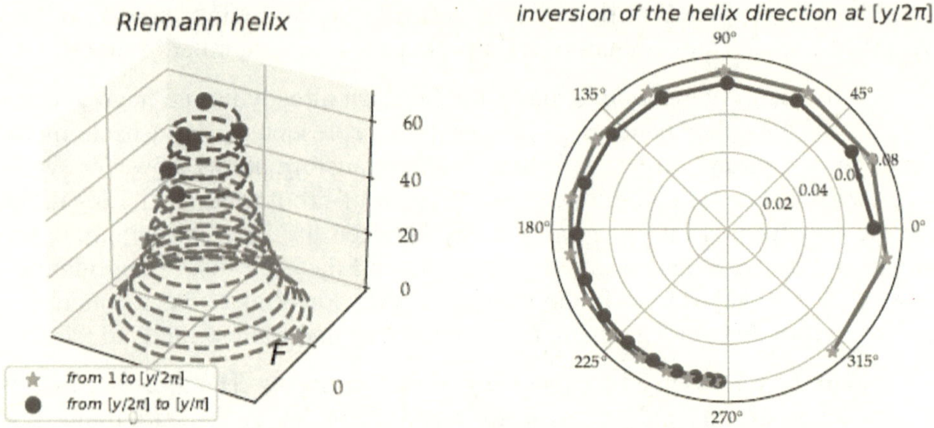

Michel Riguidel: Morphogenesis of the Zeta Function in the Critical Strip
by Computational Approach

*

Let us return to the qualitative aspect of polarity and its problematic relationship with the quantitative realm. Not only the relation between the qualitative and the quantitative is problematic, but the qualitative interpretation itself raises a basic question inevitably connected with the quantitative.

For P. A. Venis everything can be explained with yin and yang, seen in terms of expansion and contraction, and of a higher or a lower dimension. Although this interpretation greatly deepens the possibility of connection with physics and mathematics, the version of the yin yang theory he uses is that of the Japanese practical philosopher George Ohsawa. In the Chinese tradition, yin is basically related to contraction and yang to expansion. Venis surmises that the Chinese interpretation may be more metaphysical and Ohsawa's more physical; and latter he thinks that the former could be more related to the microcyclical processes of matter and the latter to the mesocyclical processes more typical of our scale of observation, but both views seem to be quite divergent.

Without resolving these very basic differences we cannot expect to soundly connect these categories with quantitative aspects, although one may still speak of contraction and expansion, with or without relation to dimensions. But on the other hand, any reduction of such vast and nuanced categories to mere linear relations with coefficients of separate aspects such as "expansion" or "contraction" runs the risk of becoming a poor simplification dissolving the value of the qualitative in appreciating nuances and degrees.

Venis' interpretation is not superficial at all, and on the contrary it is easy to see that it gives a much deeper dimension, quite literally, to these terms. The extrapolation to aspects such as heat and color may seem to lack the desirable quantitative and theoretical justification, but in any case they are logical and consistent with his general vision and are wide open to delve into the subject. However, the radical disagreement on the most basic qualifications is already a challenge for interpretation.

It should be said right from the start that the Chinese version cannot be reduced to the understanding of yin and yang as contraction and expansion, nor to any pair of conceptual opposites to the exclusion of all the others. Contraction and expansion are only one of the many possible pairs, and even if they are often used, as with any other pair they depend entirely on the context. Perhaps the most common use is that of "full" and "empty", which on the other hand is intimately linked to contraction and expansion, although they are far from identical. Or also, depending on the context, the *tendency* towards fullness or void; it is not for nothing the common distinction between young and old yang, or young and old yin. Reversal is the way of Tao, so it is only natural that these points of potential, spontaneous reversal are also notorious in the *Taijitu.*

On the other hand, qualities such as full and empty not only have a clear translation in differential terms for field theories, hydrodynamics or even thermodynamics, but also have an immediate meaning, although much more diffuse, for our inner sense, which is precisely the common sense or sensorium as a whole, our undifferentiated sensation prior to the imprecise "sensory cut" which seems to generate the field of our five senses. This common sensorium also includes kinesthesia, our immediate perception of movement and our self-perception, which can be both of the body and of consciousness itself.

This inner sense or common sensorium is just another expression for the homogeneous, undivided medium that we already are —the background and tacit reference for feeling, perception and thought. And any kind of intuitive or qualitative knowledge takes that as the reference, which obviously goes beyond any rational or sensory criteria of discernment. Conversely, we could say that this background is obviated in formal thought but is assumed in intuitive knowledge. Physicists often speak of a result being "counter-intuitive" only in the sense that it goes against the expected or acquired knowledge, not against intuition, which it would be vain to define.

However it would be absurd to say that the qualitative and the quantitative are completely separate spheres. Mathematics is both qualitative and quantitative. We use to hear that there are more qualitative branches, like topology, and more quantitative branches like arithmetic or calculus, but on closer inspection this hardly makes any sense. Venis' morphology is totally based on the idea of flow and on such elementary notions as points of equilibrium and points of inversion. Newton himself called his differential calculus "method of fluxions", the analysis of fluent quantities, and the methods for evaluating curves are based on the identification of turning points. So there is a compatibility that not only is not forced but it is natural; if modern science has advanced in the opposite direction towards increasing abstraction, which in turn is the just counterbalance to its utilitarianism, is another story.

Polarity and duality are quite different things but it is useful to perceive their relation before the convention of electric charge was introduced. The reference here cannot fail to be the electromagnetic theory, which is the basic theory of light and matter, and to a large extent also of space and matter.

Obviously, it would be absurd to say that a positive charge is yang and a negative charge is yin, since between both there is only an arbitrary change of sign. In the case of an electron or a proton other factors come in play, such as the fact that one is much less massive than the other, or that one is peripheral and the other is at the core of the atom. Let us take another example. At the biological and psychological level, we live between stress and pressure, which frame our way of perceiving things. But it would also be absurd to say that one or the other is yin or yang insofar as we understand tension only as a negative pressure, or viceversa. In other words, mere changes of sign seems to us trivial; but they become more interesting qualitatively and quantitatively when they involve other transformations.

Whether all or nothing is trivial depends only on our knowledge and attention; a superficial knowledge may judge as trivial things that in fact are are full of content. The polarity of charge may seem trivial, as may seem the duality of electricity and magnetism, or the relationship between kinetic and potential energy. Actually none of them is trivial at all, but when we try to see everything together we already have a space-time algebra with a huge range of variants.

In the case of pressure and stress or tension, the more apparent transformation is the deformation of a material. Strain-stress-pressure relations define, for instance, the properties of the pulse, whether in the pulsology of traditional Chinese or Indian medicine or in modern quantitative pulse analysis; but that also leads us to the stress-strain relations that define the constitutive law in materials science. Constitutive relations, on the other hand, are the complementary aspect of Maxwell's electromagnetic field equations that tell us how the field interacts with matter.

It is usually said that electricity and magnetism, which are measured with dimensionally different units, are the dual expression of the same force. As we have already pointed out, this duality implies the space-matter relationship, both for the waves and for what is supposed to be the material support of the electric and magnetic polarity; in fact, and without going into further detail, this seems to be the key distinction.

All gauge field theories can be expressed by forces and potentials but also by non-trivial pressure-strain-stress variations that involve feedback, and there is feedback at any rate because first of all there is a global balance, and only then a local one. These relations are already present in Weber force law, only in this one what is "deformed" is force, instead of matter. The great virtue of Maxwell's theory is to make explicit the duality between electricity and magnetism, hidden in Weber's law. But we must insist, with Nicolae Mazilu, that we can find the essence of the gauge theory already in Kepler's problem.

Constitutive relations with definite values such as permittivity and permeability cannot occur in empty space, so they can only be a statistical average of what occurs in between matter and space. Matter can sustain stress without exhibiting strain or deformation, and space can deform without stress or tension —this runs in parallel with the basic signatures of electricity and magnetism, which are tension and deformation. Strain and tension are not yin or yang, but to yield easily to deformation is yin, and to withstand tension without deformation is yang —at least as far as the material aspect is concerned. Of course between both there must be a whole continuous spectrum, often affected by many other considerations.

However, from the point of view of space, to which we do not have direct access but through the mediation of light, we could see the opposite: the expansion without coercion would be pure yang, while the contraction may be seen as a reaction of matter to the expansion of space, or the radiations that fill it. The waves of radiation themselves are an intermediate and alternate process between contraction and expansion, between matter and space, which cannot exist separately. However, a deformation is a purely geometrical concept, while a tension or a force is not, being here where the proper domain of physics begins.

Perhaps in this way a criterion for reconciling the two interpretations can be discerned, not without careful attention to the overall picture of which they are part; each may have its range of application, but they cannot be totally separate.

It is a law of thought that concepts appear as pairs of opposites, there being an infinity of them; finding their relevance in nature is something else, and the problem becomes nearly unsolvable when quantitative sciences introduce their own concepts that are also subject to antinomies but of a very different order and certainly much more specialized. However, the simultaneous attention to the whole and to the details makes this a task far from impossible.

Much have been said about holism and reductionism in sciences but it must be remembered that physics to start with, never has been described in rigorously mechanical terms. Physicists hold onto the local application of gauge fields, only because that is what give them predictions, but the very concept of Lagrangian that makes all that possible is integral or global, not local. What is surprising is that this global character has not a proper use in fields such as medicine or biophysics.

Starting from these global aspects of physics, a genuine and meaningful connection between the qualitative and the quantitative is much more feasible. The conception of yin and yang is only one of many qualitative readings man has made of nature, but even taking into account the extremely fluid character of these distinctions it is not difficult to establish the correspondences. For example, with the three gunas of Samkya or the four elements and four humors of the Western tradition, in which fire and water are the extreme elements and air and earth are the intermediate ones; these also can be seen in terms of contraction and expansion, of pressure, tension and deformation.

Needless to say, the idea of balance is not exclusive of the Chinese conception either, since the same cross and the quaternary have always had a connotation of equilibrium that is totally elemental and of universal character. It is rather in modern physics that equilibrium ceases to have a central place due to inertia, although it cannot cease to be omnipresent and essential for the use of reason itself, as it is for logic and algebra. The possibility of contact between quantitative and qualitative knowledge depends both on the precise location we give to the concept of equilibrium and the correct appreciation of the context and global features of the so-called mechanics.

Unlike the usual scientific concepts, which inevitably tend to become more detailed and specialized, notions such as yin and yang are ideas of utmost generality, indexes to be identified in the most specific contexts; if we pretend to define them too much they also lose the generality that gives them their value as an intuitive aid. But also the most general ideas of physics have been subject to constant evolution and modification depending on the context, and we only have to look at the continuous transformations of quantitative concepts such as force, energy or entropy, not to mention issues such as the criterion and range of application of the three principles of classical mechanics.

Vortices can be expressed in the elegant language of the continuum, of compact exterior differential forms or geometric algebra; but vortices speak above all with a language very similar to that of our own imagination and the plastic imagination of nature. Therefore, when we observe the Venis sequence and its wide range of variations, we know that we have find an intermediate, but genuine, ground between mathematical physics and biology. In both, form follows function, but in the reverse engineering of nature that human science is, function should follow form to the very end.

In Venis account there is a dynamic equilibrium between the dimensions in which the vortex evolves. This widens the scope of the equilibrium concept but makes it more problematic to assess. Fractional calculus would have to be key to follow this evolution through the intermediate domains between dimensions, but this also rise interesting points for experimental measurements.

How dimensions higher than three can be interpreted is always an open question. If instead of thinking of matter as moving in a passive space, we think of matter as those portions to which space has no access, the same matter would start from the point or zero dimension. Then the six dimensions of the evolution of vortices would form a cycle from the emission of light by matter to the retraction of space and light into matter again —and the three additional dimensions would only be the process in the opposite direction, and from an inverse optic, which circumvent repetition.

This is just one way of looking at some aspects of the sequence among many possible ways, and the subject deserves a much more detailed study than we can devote to it here. One think is to look for some sort of symmetry, but there must be many more types of vortices than we know now, not to speak of the different scales of occurrence, and the multiple metamorphoses. Only in Venis' work one can find the due introduction to these questions. Venis assumes the number of dimensions to be infinite, so we could not find and count them all. An indication of this would be the minimum number of meridians necessary to create a vortex, which increases exponentially with the number of dimensions and which the author associates with the Fibonacci series.

We can speak of polarity as long as we can appreciate a capacity for self-regulation. That is to say, not when we just count on apparently antagonistic forces, but when we can not help notice a principle above them —or an underlying unity, if preferred. This capacity was always present since the very Kepler's problem, and it is only telling that science has failed to recognize it. Weber's force and potential are explicitly polar, Newton's force is not, but the two-body problem exhibit a polar dynamics in any case. To call the evolution of celestial bodies "mechanics" is just a rationalization, and in fact we do not have a mechanical explanation of anything when we speak of fundamental forces, and probably we cannot have one. Only when we notice a self-regulating principle could we use the term *dynamics* honoring the original intention still present in that name.

POLE OF INSPIRATION – TIME AND THE CLOCK

8 April, 2020

Pendulums, circles, waves, vortices, electrons, ellipses, spirals... What kind of "clock" could be built with all this? Or better yet, what kind of "time" would it be measuring, counting, calculating? Furthermore, would it be calculating it, or rather indicating it?

To speak of the third principle of dynamics is to speak of the very concept of reciprocity —within closed systems. The reciprocity of relativistic reference frames is purely mathematical, since it is not bound to centers of mass —but the same applies to forces in ellipses according to Newton.

Relativistic simultaneity, even the local determinism of quantum mechanics, is embedded in the basic Newtonian assumption of the global synchronizer, the absolute time in which the third principle does not take place sequentially but simultaneously. But the same vortex of Newton's bucket experiment, as Pinheiro notes, denies this absolute time in the most convincing and categorical way —and not only for instinct, which is always stronger than metaphysics. The thermomechanical reading allows us to obtain more information, as well as another type of indication. And that same vortex is already a completely different model of clock, that we should learn to contemplate —if to contemplate something, in this our artificial world, we have to start by being able to somehow reproduce it.

We agree here with Pinheiro and other authors [59]. Various arguments and numerical calculations indicate that the third principle is not applicable to systems out of equilibrium; and if it is applicable, it can only be so in the sense of the retarded potentials. That is, we assume that even celestial orbits are always out of equilibrium and only reach it instantly thanks to an additional term representing the action of the medium on matter; this term can be entropic, or respond to a longitudinal vibration, or a constant bombardment of external particles.

In quantum mechanics, which is supposed to be the fundamental theory, forces become secondary to the potential; however, we continue to understand everything according to the logic of the three principles, even after the relativistic generalization of the principle of equivalence.

But the decisive turn had already taken place when the potential went from representing a mere position to having an irreducible temporal component. For Weber's relational dynamics that emerged halfway between Newtonian and 20th century physics, it does not make sense to separate the dynamic component of the state of the system, nor the forces of the environment in which they occur. The rigorous application of the principle of dynamic equilibrium also made this distinction irrelevant, as it made irrelevant the equivalence principle, since inertia itself is dispensable.

If, by virtue of the principle of dynamic equilibrium, we dispense with the concept of inertia, we also dispense with the intentionality —the *dispositio* underlying mechanics. The Global Synchronizer, so perfectly intangible, is the supreme symbol of power in the present civilizing cycle. It is intangible as it is metaphysical, but it suppresses or in any case makes unrecognizable the local exchange of information of the open systems and the interaction with the environment.

It has been said that after Newton the universe went from breathing like a mother to ticking like a clock; there is still time to bring it back to breath again.

POLE OF INSPIRATION – TAO OF TECHNOSCIENCE

8 April, 2020

The paths in the science-technology continuum may be innumerable but they all presuppose a potential reciprocity between knowledge and application —thus between knowledge and power. And yet we still have no idea of what kind of circle knowledge and power draws on us.

Newton's celestial mechanics seemed initially far removed from worldly affairs but the unwarranted generalization of his principles to things far removed from human artifacts had the effect of turning the world into a wheelless rolling machine.

Society has taken shape as it becomes isolated from Nature but cannot subsist without a permanent commerce with her which in turn depends more and more on our knowledge of it. Any dominance relationship over Nature is reproduced within society, between some parts that exercise control and others parts subject to this control.

The solar system bound by gravity, or the function of the heart in our blood circulation, have been seen as simply governed by the concept of force in our present world view. Since the middle of the 20th century, stability theory and cybernetics developed a theory of control over these so-called "blind forces", generalizing a version of entropy that was already far removed from the original thermodynamical context. Now it remains to be seen what twist would result for control theory assuming spontaneous regulation in action principles, in the Second Law and in the collective resonance between elements; as well as in the relationship between these three aspects.

Overcoming reductionism cannot in any way involve last-minute corrections that seek to compensate increasing degrees of abstraction with also increasing degrees of subjectivity for the sake of the inclusion of the observer, be it in statistical mechanics, quantum mechanics or relativity; it involves in any case correcting the gaps in the foundational position, which so far remains unaffected.

But the reciprocity between man and nature goes far beyond anything we suspect, and cannot be encompassed by a mere theoretical turn, however wide or deep it may seem. It is not a matter of looking for an idealized external nature either, since all that is trapped in the human being is also nature.

We do not know and maybe we do not want to live without machines. Can we radically change our relationship with them? Machines, too, are trapped, molded and compressed nature; and while we are forced to depend on them we are routinely trained for obedience. I will now pick up a few paragraphs from a topic I discussed at greater length in *Techno-Science and the Laboratory of Self*:

"Vico's principle, which states that knowing is making, is more general than Descartes'. But surely one can also doubt Vico's principle. I can move my hand, but do I know how I move my hand? Second hand, so to speak, not first hand. Of course making is not doing, except in thinking, and we make machines not to do things directly, and not to directly think. We can then try to introduce into the realm of knowledge-power the duly reformed Vico principle: I only know that in what I take part, and to the extent that I take part.

It is not by calculus, but by the practical arts, that we know the world best. The same concept of efficiency, as economy of effort or elegance, was a natural notion in the art of all cultures before techniques were invaded by stacked layers of scientific mediations; now it would have to be taken out of the bottom of the stack. There is a natural sense of efficiency in any physical activity, in the right intonation, in any gesture or brushstroke.

To move from one area to another, from the functional domain governed by calculus to the intuitive functioning, we can take as example the biological feedback and biofeedback. A signal that corresponds to a vital function can be used to vary it at will, within certain limits of course. However, and this is the important thing, here any notion of manipulation is out of place, as in this context loses all meaning. Even control, with all its vast current theory, is subsumed in the idea of self-control, which far from being a particular case, seems to be the most indefinite and general.

In our ordinary physical control of external objects, the relationship between action and calculus is also reversed. Think of the complicated balance involved riding a bicycle; dynamics can hardly solve the problem by means of centrifugal forces in the case of slow motion, but it gets out of hands in cases with higher speeds. And yet for the cyclist it is just the opposite: speed is the solution, and excessive slowness the problem. Motion is shown by pedalling.

However, within the category of self-control there is more than just cycles of perception and action; there is also self-observation. In the case of biofeedback there are two basic cases, direct monitoring of a function, such as when we observe our breathing and modify it without even intending to do so, and indirect monitoring, either by means of a mirror that returns our image to try to move involuntary muscles or by a device with sensors that translates signals generated by ourselves without being aware of it.

The biofeedback motif may seem very limited as it has hardly transcended the level of a curiosity since its appearance and diffusion fifty years ago. However, it marks a turning point in the relationship between man and the machine. If the most helpful idea in explaining the emergence of tools is that they are extensions or prostheses that project our capacity as organisms, and if we have later recognized that from certain point onwards all harmonious relations between the tool and the organ are lost, here for the first time we use the machine to help us regain consciousness of organic functions that have already sunk below the threshold of attention.

So, if technology came out of the biology of the conscious organism, it is precisely here that it returns to it in the most mediated way possible, and with a somewhat undecided intention. Clearly, the entire cybernetic theory of control would have to return to self-control as its archetype, since this one already incorporates the cycles of perception and action, allowing the right space for the self-consciousness around which they revolve; the automatic is subsumed in self-control, self-control in self-interaction and this in spontaneity.

In spite that gauge fields contain a basic feedback mechanism, the use of the Lagrangian in control theory seems totally secondary, which is a curious situation. Things could change working with Weber-type forces, and also with dissipative, thermomechanical forces, in the sense proposed by Pinheiro. The measure theory would also have to be adjusted to the different requirements of this type of systems.

Today it is said that the double access to perception and action defines an artificial intelligence problem. But a greater field in self-perception and natural intelligence awaits us.

POLE OF INSPIRATION – INDIVIDUAL EVOLUTION OF AN ENTITY

8 April, 2020

The so-called "new synthesis" of the theory of evolution has never had the least predictive power, but neither has any descriptive power, and like cosmology, it has been used mostly as a narrative complement for normative physical laws unable to connect with the real world and the real time in the most decisive sense.

To alleviate the evident limitations of a theory that only through phantasy can have some contact with natural forms —and let us not underestimate the power of fantasy in this instance-, evolution has been combined with the perspective of biological development (evo-devo), and even with ecology (eco-evo-devo), but even then it has not been possible to create a moderately unitary framework to describe problems such as the emergence, maturation, aging and death of organized beings.

There is no need to call all this by any other name than evolution without further ado, since the present theory has only taken over the name to deal with speculative and remote issues, rather than with close and fully approachable problems.

What determines the aging and death of individual organisms and societies? This is a real evolution problem that takes us closer to the real time.

Astrophysicist Eric Chaisson has observed that the energy rate density is a much more decisive and unequivocal measure for complexity metrics and evolution than the various uses of the concept of entropy, and his arguments are fairly straightforward and convincing [60]. In any case it would be desirable to include this quantity in a context better articulated with other physical principles.

Georgiev et al. have attempted to do so by establishing a feedback loop between this energy flow, the physical principle of minimum action *understood as efficiency or as movement along paths with less curvature or restriction*, and a quantitative principle of maximum action; that is, with the "metabolic" energy flow per mass mediating between efficiency and size, applying it experimentally to CPUs as organized flow systems [61]. When these flow-efficiency-size vertices are connected in a positive feedback loop, an exponential growth of all three is produced and power-law relations between them arise. Although this model admits many improvements and can be illustrated in very different ways, it seems on the right track.

Since the basic problem of aging is the increasing restriction and the inability to overcome it, and any theory not addressing this fundamental issue cannot make a dent in the subject. Another way of saying the same is that organic aging is the increasing inability to eliminate. Aging is irreversibility, and irreversibility is the increasing incapacity to be an open system.

Let us think a bit about this. Something as basic and elementary in physics as the principle of least action is capable of telling us something absolutely essential about aging: to understand the real value of this we only have to know how to apply global measurements in the context of open systems with a variable use of the available free energy.

The theoretical advance in this field is infinitely more feasible than in the modern synthesis and infinitely more relevant, since here it is no longer a question of species, but of the destiny and individual evolution of any spontaneous organization, be it a vortex or a soliton, a human being or a civilization; to a large extent it affects even the Lamarckian evolution of machines and computers with a definite design and purpose.

The energy rate density is a measurable, unambiguous quantity, profoundly significant from a cosmic perspective, and can also have a bearing on bringing down to earth unmanageable entropy criteria. Naturally, the flow-curvature-size criterion can be contrasted with the flow-curvature-entropy criterion, whether the latter is maximum or not.

This threefold flow density-curvature-size criterion can be applied fruitfully to contexts where flow is the decisive factor, be it in strongly quantitative models like the monetary flows, or in purely qualitative models of vortices evolving between expansion and contraction such as the one proposed by Venis. It can be applied even to the circle of individual destiny, suggesting a clock of its evolution, which in today's terminology many would call an aging clock.

POLE OF INSPIRATION – FROM THE BOOK OF CHANGES TO THE ALGEBRA OF CONSCIENCE, THROUGH TECHNICAL ANALYSIS

8 April, 2020

In his splendid book on Harmony Mathematics, Alexey Stakhov devotes a section to Vladimir Lefebvre's mathematical theory of decision and strategic interaction, which, unlike other developments in this area such as the much-hyped game theory, includes an intrinsic behavioral and moral component [62].

Everything starts from the observation that if someone is asked to divide a pile of string beans into two piles of bad and good ones, oddly enough the average result is not 50%-50% as expected, but 62%-38% in favor of those considered good.

According to Lefebvre, what lies at the bottom of this asymmetric perception is a triple gradation of definable logical implications within a binary logic or Boolean algebra:

$$A\ [a_0\text{consciousness}\rightarrow a_1\text{reflection}\rightarrow a_2\text{intention}]$$

consciousness being the primary field, reflection the secondary level and intention a reflection of second order.

Within this structure of human reflexion, the personal evaluation of behavior implies, by logical implication, a self-image:

$$A = a0^{a1^{a2}} \quad ; \quad A = (a2 \rightarrow a1) \rightarrow a0$$

The subsequent truth table gives us an asymmetric sequence of 8 possible values, 5 being a positive estimate and 3 a negative one. Naturally, 8, 5 and 3 are adjacent numbers in the Fibonacci series.

The extraordinary thing about the logical framework of ethical cognition created by Lefebvre is that, inadvertently, it allows a completely modern interpretation of the *Book of Changes* without distorting its sapiential and moral nature. It is not something that we are going to prove here, but given the deep influence of *Yi Jing* in the history of Chinese culture, it would be of utmost interest to look for a rigorous correspondence between both views.

Undoubtedly, Lefebvre's framework is more formalized, and more focused on the image that the individual agent has of himself, than on the situation or critical juncture, as is the case with the *Yi Jing*; in fact, we could say that Lefebvre's formalization is a Boolean and measurable limit for interactions with a low number of individual agents, but even then the correspondence remains intact and can lead to much broader and impersonal degrees of involvement, and therefore, of understanding.

*

Stakhov devotes another section to Elliott's technical wave analysis, a tool also based on the Fibonacci series. The theoretical scope of Elliott waves is limited at best, even if its fractal-type analysis can be applied discretionally. But here we want to point out a much deeper methodological issue.

Lefebvre's model shows us that these series have, or at least admit, a reflexive component —in fact his work has been qualified as a reflexive theory of social psychology. And that reflective component is precisely what is most missing in Elliott's model. There is also a reflexive theory of the economy and the market, with positive and negative feedback cycles; this theory is not without merit, if we compare it, for example, with the hypothesis of the efficient market, which, even if it were not false, would always be incapable of telling us anything. Moreover, the reflexive theory does not only affect prices, but also the fundamentals of the economy.

Economic reflexivity is a model of self-interaction, as it is also of self-image, and this decisive element in human affairs can never be dispensed with, even if it is impossible to quantify it. But what happens when self-interaction runs through the whole behavior of natural systems? Let us think, for example, of the case of the pulse analysis we have mentioned, a transparent model of self-interaction, which, as a bonus, tends towards golden ratios in rhythm and pressure. But it is clear that in social systems reflexivity passes through the knowledge of an external situation, while in an organic system reflexivity is a pure question of action.

How much room does one type of reflexivity leave for the other? How far can the perception of the markets be manipulated? Today we also see that central banks have an undisguised and unabated intervention in prices, injecting money discretionally and becoming guardians of the financial bubbles. Well, the analysis of the pulse, the transformation of its modalities, and the study of the limits of conscious manipulation through biofeedback are giving us a good match or at least some sort of reference for this elusive problem. And if we want more intensive methods, besides the quantitative implications that all this already have, we can also apply to monetary flows CPUs models like that of the triangular flow-efficiency-size feedback that we commented on earlier.

*

In *The Algebra of Conscience*, Lefebvre made a comparative analysis of the Western and Soviet ethical context within the framework of the Cold War confrontation [63]. The W-system of the United States believed that the compromise between good and evil was bad, and that the confrontation between good and evil was good. The S-system of the Soviet world, on the contrary, believed that the compromise between good and evil was good and the confrontation between good and evil was bad —what Lefebvre does not say, as it goes beyond his formal analysis, is that the desire for confrontation of the W system was not based on any moral idiosyncrasy but on the essential expectation of further expansion; just as the desire of the S system to avoid confrontation was based on the fear of disappearing.

At any rate, from the choices and implications between commitment and confrontation, four basic attitudes emerge, that of the saint, the hero, the philistine and the dissembler: the saint embraces suffering and guilt to the maximum; the philistine wants to diminish suffering but may feel acute guilt; the hero minimizes his guilt but not his suffering; and the dissembler minimizes suffering and guilt.

Maybe we could have created 8 types instead of 4, in consonance or dissonance with 8 other scenarios or natural tropisms, with two more third grade binary implications or a single sixth grade one. Logic and calculus serve to ascend this ladder and understanding begins where calculus ends. Thus, the continuous proportion would show us in a very direct way its role of mediator between the discrete and the continuous, the digital and the analogical.

In reality we have here an unsuspected Centaur, namely, a halfway point between binary formal logic and the dialectical logic as in Hegel. But the proof that this is something genuine, and not a mere theoretical construct, is that it has a built-in asymmetry that is a distinctive feature of Nature. We could call it an asymmetrical logic of implication.

It has been said that Lefebvre's theory was used at the highest levels of negotiation during the collapse of the Soviet Union; and here we will not go into judging what positive or negative role it might have had, and for whom. In any case, its framework only covers two agents, and not the scenario or juncture that may dispose of them. *Yi Jing's* framework can serve not only to see how I perceive myself in the world, but also to evaluate how the world perceives me; in the end, both are parallel illusions. One could say that asymmetric perception is part of the same reality; a double asymmetry better embraces the implications and the unnoticed axis of the dynamics of a situation.

Surely one could make a great parlor game out of these arguments. More interesting, however, is the study of moral or social conscience through the analysis of its implications. And even more interesting is the awareness of how the external situation, for which we are always looking for an image, interpenetrates with the changing image we make of ourselves, forming both a momentary map of our situation and conscience.

One can only guess that Lefebvre knows the text of the *Yi Jing*; however it is evident that the line of his reasoning is completely independent from the Chinese classic and he has arrived at his "positional system" of moral consciousness starting from the Western logical and mathematical tradition. If he had had just a glimpse of the *Taijitu* with the golden section embedded, the immediate association would have shine in his mind. In the coming years we may witness many new associations of this kind induced by the evolving environment.

POLE OF INSPIRATION – SCIENCE AND CONSCIENCE

9 April, 2020

We have tried to approach a single subject from the most diverse angles, so that everyone has some possibility of connecting it in the most direct way with his or her own interests. This is a minimal introduction relying on a bibliography that is not at all exhaustive, but necessary to delve into any of the issues presented.

Of course our subject proper was not the continuous proportion, but reciprocity and self-organization in Nature and beyond Nature. The constant φ continues to play a marginal, totally episodic role within modern science, which is mainly focused on calculus.

Thus, our preferential point of view is not so much that of science and objectivity, as that of reciprocity itself, to which I attach more importance; as an attempt to give more importance to the awareness of reality than to science.

And reality itself resembles a parable. We can say that next to unity, sometimes also symbolized by the point, the circle and the constant *π*, two other great mathematical constants exist from eternity, *e* and φ. The constant *e*, the basis of the natural logarithms and the exponential function, perfectly embodies the analytical exhaustiveness of calculus, as if looking towards the plurality of the world. The constant φ, the natural algorithm of a Nature that completely ignores calculus, looks at unity without knowing it. And unity itself, if it is really unity, cannot show a preference for either.

Now, it is clear that in modern science the weight of *e* is infinitely greater than that of φ, to the point that we could perfectly dispense with φ without even noticing it. The number *e* refers us to continuity and infinite divisibility; φ refers us to discrete operations without intended purpose —and it should be understood that infinite divisibility is already a human purpose that can always exceed its operational limits. Hence the need to look back to the contradictory basis of infinitesimal calculus.

As it is known, the number *e* was first identified by Jacob Bernoulli in 1683 in a problem of compound interest, and is absolutely consubstantial with the modern spirit of calculus with all that it entails. The continuous proportion has undoubtedly a much older past but, in the eyes of the moderns, it is hard to see how it could have played a relevant role in the knowledge of antiquity. Euler's number is at the base of so-called advanced or superior mathematics, while the number closest to the muses cannot ever deny the elemental character of its origin —which is also its greatest charm. Surely this is the reason for its permanent popularity among amateur mathematicians.

But this undeniable circumstance hides another naivety on the part of the spirit of calculus that we should learn to appreciate. It is well known that people like Stevin or Newton still believed that ancient cultures could have had broader knowledge than their contemporaries; if mathematicians like them still dared to think that, it must surely be attributed to the incomparable impact that the legacy of Apollonius and Archimedes —the most advanced mathematicians of antiquity - had at that time.

But this already presupposed a totally biased idea about what was advanced in knowledge, which has been perpetuated until today.

It is said that the amount of knowledge regularly doubles every 15 years since the scientific revolution, which implies that today our knowledge of physics and mathematics is four million times greater than in 1687, when the *Principia* were published. Yet this magnum opus is already an obscure, hard to read treatise that has been and continues to be routinely misinterpreted even by the best experts.

Let us try to understand four million *Principia*. We do not walk on the shoulders of giants, the giants advance on our shoulders, although less and less, as it is easy to understand. And the logic of accumulated capital is that accumulated capital does not take risks. The theory of relativity and other "revolutions" were adopted following the principle of minimum elimination and maximum conservation of capital. The whole exacerbated search for novelty in theoretical physics is nothing but the enforced headlong rush because it is not allowed to really examine the foundations. But the less one eliminates, and the less one renews the foundation, the more inexorably one ages.

Obviously the stratification of knowledge and the branching of specialities follows the logic of continuous interest and accumulated capital and debt.

It is really curious that two constants as ubiquitous as e and φ, the "two natural fractals", cross so little of each other's paths in a field as unlimited but redundant as mathematics. So curious, that the study of the points of contact and divergence of both constants should be an area of mathematical research in its own right, full of interest for both pure and applied mathematics. If this has not happened yet, is because of the one-sided development of all sciences and mathematics on the side of calculus and prediction, putting the rest of the resources at their service.

It is understandable that there is, among many mathematicians, a typical allergic reaction to the questions raised by the continuous proportion and the mathematics of harmony. They are seen more as a hindrance than as a guide, since they might imply some sort of discrete, constructive limits to an analysis for which no restriction is wanted. Nothing should measure the one who measures.

It seems that at the time of the emergence of writing and the first great cities, the priests kept the standards of measurement in the temples. But Nature is the perfect temple that keeps everything without hiding anything.

We have said that φ seems to be "the natural algorithm" of a Nature that does not care in the least about calculus. But not having any system of calculus and measurement is, in a sense, equivalent to having all of them and surpassing them all; in the same way that nature's lack of intention infinitely surpasses human purposes. So, the value of this branch of mathematics for the theory of measurement and computation should be out of the question. It is a matter of identifying relevant problems in the domain of interdependence.

In just over three centuries we have had at least five great cuts with the constructive and proportional conception of measurement: the infinitesimal calculus and Newton's classical mechanics, Maxwell's theory, set theory, relativity and quantum mechanics. With each of these successive steps, the problem of measurement has become more and more critical and controversial. But it is absolutely superficial to think that only the latest developments count and that measure theory itself is just an aid to our predictions. This is exactly how this tower of Babel was built.

*

The interest of the continuous proportion would go beyond our modern involvement in complexity and calculus. In this area, new discoveries are always possible even at the elementary level that are so unexpected that we do not even know how to value them. Moreover, its persistent appearance in music —that unconscious arithmetic, as Leibniz said-, in physiological rhythms and anatomical sequences indicates not only an intuitive, but, ultimately, pre-numerical character.

There is here a great open track not only for the archaeology of knowledge, but for the very activation of an ancient knowledge that seems inconceivable to us today. In the *Book of Changes* we may see an asymmetric implication algebra. The six lines of a hexagram correspond to the six directions of space at the ends of its three axes, which oppose an I and its circumstance. But here the symmetry of the coordinates is only the external and passive frame, it is the asymmetry what constitutes the internal dynamic spring in which agent and situation interpenetrate.

The *Book of Changes* is the best possible example of an analogical knowledge that does not depend on calculus, and towards which a certain natural logic of implication converges. What is important is not the 64 cases or the 384 lines, but the plane of synthesis towards which they point.

If mathematical analysis has its so-called complex plane to operate with any number of dimensions or variables, one can equally conceive an implication logic that is capable of reducing any number of variables to an equally intangible plane of synthesis, which we will call the plane of universal synthesis or universal inclusion.

Of course, in this day and age anyone who hears of "universal inclusion" can only think of universal confusion. That is why we have developed the analysis, because we do not believe in knowledge that is not properly formalized. However, the development of formalization has not contributed to greater intelligibility, but rather the opposite. For us, formal knowledge increasingly projects more shadows than light; shadows we know too well are cast by ourselves. Shadows of power over natural or social processes.

The whole philosophy of the West since Descartes is based on the idea of a separate intelligence. That is why the successive cuts that come one after another since Newton do not seem a loss to us, since in this line of logic, more separation from Nature amounts to more self-affirmation. But any process has its limit, and when there is nothing left to assert oneself about —because Nature vanished- everything stops making sense, except for the mere exercise of power, which lacks intrinsic stimulus for knowledge.

Besides, the golden age when the field of fundamental predictions was growing is now far behind us. It will not return, for the simple reason that everything was optimized for predictions and all the low hanging fruit is already picked. One can only scrape out diminishing returns in the ugly struggle with complexity. Yet modern science finds it almost impossible to examine its foundations, which is the only thing where there could be real novelty; we have the great advantage of our historical perspective, but the very existence of the specialities depends on not reflecting on foundations.

Modern science is not capable of dealing with either transcendence or immanence; for as subjects we cannot separate ourselves anymore, nor does we know how to see things again from within Nature.

Now, if we look back as we have done, only reordering our perception of the present theories, what does it mean to dispense with the principle of inertia? What is the point of saying that everything is based on self-interaction? What is the point of saying that the only thing we perceive is the Ether?

These are transcendental statements, in the sense that the father of phenomenology, Edmund Husserl might have given them, had he dealt with physics. However, to suspend the principle of inertia ends with the false idealism of physics, such a fecund contradiction that isolates a ball that rolls from the rest of the world except from ourselves. To realize that we only perceive the Ether, because we only perceive in the mode of light, is to realize that we are always right in the middle and that matter and space themselves are transcendental limits.

Finally, to say that the planets or the electrons orbit their centers by self-interaction can only be understood in the sense that the relationship between matter and the medium appear to be reflexive just because both are not separate.

Separation and reflexivity are appearances both for the subject and the object, and it is useless to adhere to one part trying to deny the other, as science has attempted. Intelligence and being coincide —at least from the point of view of intelligence, since this one is incapable of perceiving itself. This reflexivity, this intelligibility, is the plane of universal synthesis itself. But this has also a physical translation. The evolution of a vortex in six dimensions in Venis coordinates could be a good example of the intersection of a naturalistic view with the transcendental plane.

These ideas can be applied to both the mediate and immediate knowledge of Nature. From the operational and formal point of view, all observable knowledge of physics can be included in the principle of dynamic equilibrium. But from the point of view of immediate cognition, one can hardly sustain oneself in the contemplation of the instant without inertia —to such an extent that ball rolling in the void has captured our subjective idea of succession. However, these principles, which seem now much more demanding from the intuitive point of view, since they are much more full of content, are not based on the separation of nature and therefore make their forced "unification" by man less necessary.

Truly, that transcendental plane is the one in which transcendence with respect to Nature and its return to it coincide —but in truth the only nature that is transcended here is that relative to the inertia of habits, what we call our "second nature". Here we could say with Raymond Abellio: "The perception of relationships belongs to the mode of vision of the "empirical" consciousness, while the perception of proportions is part of the mode of vision of the "transcendental" consciousness" [64].

But, of course, in modern science there are hardly any proportions, because all the units we handle are a heterogeneous and unintelligible jumble of quantities —that's why we rely more and more on computers and their programs for data crunching.

The term "transcendental" can only have some meaning for those who deal with the intelligibility of knowledge, not for those who are simply content with its formalization to obtain predictions. However, here we are making it descend to the very core of physical principles and of that eternal unknown we call causality; and this can be done both in a qualitative and a quantitative way.

No physicist or mathematician needs to prove the existence of the complex plane or the complex manifolds, however advisable it may be to review its foundations; much less could we prove the existence of a transcendental plane of synthesis, since the word "transcendental" means that it is the condition of knowledge. Motion is shown by walking, and knowledge by understanding.

Moreover, to speak of the "existence" of such a plane in relation to the world of objects and measurements is not only out of place but completely reverses the situation. Santayana said that essences are "the only thing that people see and the last thing they notice"; from a perspective in line with what we have already said, and which of course has little to do with the usual narratives and cosmologies, the entire experience is a transition between an unknowable but measurable matter and a space that is diaphanous to knowledge but immeasurable. Existence itself is that process of awakening, but with dissimilar rhythms for all kinds of systems and entities.

This our condition "between Heaven and Earth", between the diaphanous and the measurable, is not a mere philosophical or poetic remark, but it determines the whole range of our possibilities of knowledge, which began with the acts of counting and measuring, of arithmetic and geometry, and which have been successively developing with calculus, algebra and everything else until we got here. And we can not only see that it determines it, but also appreciate that there is always a double current, a double movement, descending and ascending.

Nor is it to be believed that this plane of essences is reduced to the mathematical aspect; on the contrary, this is only one possible expression of an infinity of modalities. In our time we have come to believe that complexity exists only in numbers, computers and models, but the richness of phenomena has always been infinite, regardless of any quantification. Our perceptions, as our thoughts, are also a fleeting part of the absolute.

*

Technoscience is a continuum of practice and theory that dictates what seems acceptable to both. On the other hand, it is not ideas that determine our actions —it is what we do and what we want to do what determine our ideas.

For today's technoscience, to see intelligence as something totally separate from Nature is an indispensable condition for recombining any aspect of nature at will: atoms, machines, biological molecules and genes, and all the possible interfaces between them under the least restrictive criterion of information.

But granting this convenient separation of domains to manipulate them with the least possible restrictions, paradoxically entails unnecessarily restrictive principles, as is already evident in fundamental physics, which is also the founding ground of our overall commerce with Nature.

The reintegration of intelligence into the unity of being is absolutely contrary to the liberal principle of prior separation in order to recombine without restrictions. The mere possibility of an intelligence in Nature threatens to short-circuit this separating intelligence who has established herself as the ultimate arbiter.

And yet, as we have seen, the idea that there is feedback in the orbit of a planet or an electron is less contradictory than the usual picture that we are dealing with a cannonball trapped in a field. In fact, it is not contradictory at all: it is just absolutely disconcerting, as well as inconvenient, for most of us.

We want to see on the one hand a separate intelligence, and on the other a completely inert matter, and between these two fictions, the evidence of a perfectly impersonal, pre-individual consciousness without qualities becomes totally inconceivable.

There is nothing extraordinary about the fact that apparently heterogeneous bodies seek to attain the homogeneous condition with the environment from which they have emerged, and that they cannot do it without both inner and outer action. The particular aspect of this balance is indiscernible from the intelligence of that entity or system, which cannot but participate in the universal intelligence. Were it not for our involvement in this last one, our very intelligence would only exists subjectively for ourselves.

It is not strange at all, but it is totally inconvenient for the practices in which we are immersed, for the horizontal, indiscriminate remixing that aspires to dissolve all natural boundaries.

*

In the immediate postwar period, around 1948, under the shadow of the Manhattan project and other military programs, three new "theories" emerged that consolidated the new "algorithmic" style of the sciences: quantum electrodynamics with its endless loops of calculations, information theory, and cybernetics or modern control theory. To this was added, five years later, the identification of the DNA helix, which would soon be reduced equally to the category of information.

Except for the calculation feats that are its only justification, the quantization of the electromagnetic field was on a theoretical level a superfluous undertaking that added nothing new to the known equations. The funny thing is that its promoters had to be averting terms like "self-interaction" and "self-energy" constantly popping up in their faces. Terms like this seem to sound undesirable for such a clean and fundamental theory as QED, and moreover in physics it is assumed that any kind of feedback only can result in stronger non-linear effects.

Thus, while fundamental physics fought tooth and nail to conjure up the idea of feedback, cybernetics had to assume that feedback is a weak, emergent property of highly organized systems made up of "fundamental" blind blocks. At the same time, information theory hacked Boltzmann's mechanical-statistical entropy into a new brand, rather than the irreversible entropy of thermodynamics. To question the foundations of the past theories would have been out of place, so theorists were content to generalize heuristic procedures, and it could not be otherwise since this science-technology continuum does not demand anything else.

We have seen that our very idea of celestial mechanics, of calculus, and of the mechanical-statistical interpretation of the Second Law are based on flagrant rationalizations —not to speak of more recent developments. Even the explanation of the functioning of our heart is based on a rationalization that tries to ignore the monumental evidence of the role of the breath in the global dynamics of blood circulation.

The only reason we maintain these mirages is a very powerful one, because they reaffirm our idea that we have a separate intelligence, while justifying our indiscriminate intervention in Nature, a Nature that we very conveniently want to reduce to blind laws and random processes. Men of science find it shocking, to say the least, to talk about the transcendental in knowledge, but all modern scientific knowledge is based on an ego pretending to be transcendental that in the end became trivial.

POLE OF INSPIRATION – THE RELIGION OF PREDICTION AND THE KNOWLEDGE OF THE SLAVE

9 April, 2020

In calculus, infinitesimal quantities are an idealization, and the concept of limit, provided to support the results obtained, is a rationalization. This dynamics going from idealization to rationalization is inherent to the liberal-materialism or material liberalism of modern science. Idealization is necessary for conquest and expansion; rationalization, to colonize and consolidate all that conquered. The first reduces in the name of the subject, which is *always more than* any object x, and the second reduces in the name of the object, which becomes *nothing more than* x.

But going to the extremes does not grant at all that we have captured what is in between, which in the case of calculus is the constant differential 1. To perceive what does not change in the midst of change, that is the great merit of Mathis' argument; that argument recognizes at the core of the concept of function that which is beyond functionalism, since physics has assumed to such an extent that it is based on the analysis of change, that it does not even seem to consider what this refers to.

Think about the problem of knowing where to run to catch fly balls—evaluating a three-dimensional parabola in real time. It is an ordinary skill that even recreational baseball players perform without knowing how they do it, but its imitation by machines triggers the whole usual arsenal of calculus, representations, and algorithms. However, McBeath et al. more than convincingly demonstrated in 1995 that what outfielders do is to move in such a way that the ball remains in a constant visual relation —at a constant relative angle of motion- instead of making complicated time estimates of acceleration as the heuristic model based on calculus intended [65]. Can there be any doubt about this? If the runner makes the correct move, it is precisely because he does not even consider anything like the graph of a parabola. Mathis' method is equivalent to put this in numbers.

How much can we synthesize knowledge? If there is no way to prove that an algorithm is the shorter or the faster, there can be no explicit limit to its comprehensibility either, and the same applies to all formal knowledge. However, there is an informal but effective guide both to synthesize knowledge and to improve its quality: always pay attention to the invariable mean, to what does not change in the midst of change. And although it is an informal principle, it is always possible to recognize its lineaments: the example above is clear enough both in the real world and in formal analysis.

In fact, it should be said that it gives us an incomparably straighter and simpler path than the "alternative models", in this case the established methods of standard calculus and the analysis of algorithms for tasks in artificial intelligence. Over time we can find an infinite number of instances with a potential for convergence at least as great as the potential for divergence of the usual analysis in terms of change.

Examples of idealizations are the inertia principle, infinitesimal quantities, point particles and the existence of single individual systems in quantum mechanics, reversibility at the fundamental level, the global synchronizer in both Newtonian and relativistic physics, universal physical constants with dimensions that are isolated from the environment, or space as a differentiable manifold. These are force-ideas born out of an intimate need of a subject and a culture, expressing the will to expand till the maximum extent. This idealization required an intense faith in the program that today, when even the rationalization phase seems unnecessary, we can only underestimate.

It is for the present moment to see through idealizations and rationalizations, not to try to move beyond them, since they are already the expression of extremes and of an extreme exercise of abstraction and experimentation. That being reified is only a thought, but the thinker is a thought as well, and the real thinking process, the Logos pervading the world, can only reverberate in a distorted way between both.

The idea of an invariable mean and the idea of dynamic equilibrium, principles so far removed from the scene by long overdue idealizations, are the two sides of the coin and define a new spectrum of relations between formal and non-formal knowledge, between duality and non-duality.

To think that today we are closer to a "theory of everything" in physics than we were in Newton's time is just a mirage. We could increase the amount of knowledge by a trillion without getting any closer to any ultimate truth. In practice, exhaustion comes much sooner than true knowledge; but just as in aging, claudication results from the inability to eliminate and renew oneself.

Nor are we any closer to "unraveling the mystery of consciousness" than in the time of Leibniz and Newton. In fact we are closer to that time than to "the final solution"... among other things, and to begin with, for failing to recognize things like why the calculus works. We haven't even suspected that both things may be connected.

Formal knowledge can increase indefinitely without ever reaching what we are experiencing right now, and the advance in informal knowledge only takes place by the direct recognition and appreciation of something always present but unnoticed. Just as power and knowledge limit each other, so do formal and formless knowledge, but there can be no law specifying their relationship.

So there is no "transcendental horizon" of gradual approach to truth for accumulated social knowledge, but neither for knowledge prior to formalization. For the same reason, in formalized knowledge the possibility exists of "transcending" forms, at least in the sense of leaving behind idealizations and rationalizations.

Much more could be learned today by gently modulating the states of particles than crashing them in accelerators, just as more could be learned by developing a proper theory of the extended particle than speculating about the origin of the universe expanding from another point. But to do this we would have to remove the extreme weights of our bets, that is, the unnecessary restrictions of our present theories. To say that in present day science "theory is superseded by correlation" would be the clearest example of extreme propaganda for the prevailing theory and practice. In fact, if many laboratory experiments do not have a greater significance today, it is because a standard interpretation of facts is immediately enforced even if the theory has nothing to do with it.

It is known how in 1956 Bohr and von Neumann came to Columbia to tell Charles Townes that his idea of a laser, which required the perfect phase alignment of a great number of light waves, was impossible because it violated Heisenberg's inviolable uncertainty principle. The rest is history, but Bohr's and von Neumann's words are not recorded. This is no exception, but what happens constantly and routinely. It is clear that a theory that subtracts infinity from infinity every time is needed can predict anything, especially after the events. They call it "powerful methods".

So, there is no observation that quantum electrodynamics can not rationalize. And the same goes for the two theories of relativity, statistical mechanics, classical mechanics or calculus. This is what means to speak of "powerful methods", what puts the search for truth in a desperate situation. Ptolemy's epicycles were also undoubtedly a powerful method, as they could cope with all kinds of celestial observations with an unbeatable predictive power for the time.

The problem with formalized knowledge is that once we accept a standard and push all sorts of things in, it is terribly difficult to get out of it. In the case of relativity, a certain reform of classical mechanics was accepted because, in addition to the urgency of solving blatant contradictions, the unification with electrodynamics promised an even greater expansion. Another issue is that since Gauss and Weber everything could have been done differently, sacrificing other things and obtaining different advantages in return.

Even if the amount of accumulated knowledge does not increase one iota, the formalization of knowledge goes through mathematics, and mathematics is always capable of expressing any concept and relation of concepts in a new and even unrecognizable way. We have already seen this: the same phenomena can be described by saying that the movement of bodies is determined by external forces, as by saying that the bodies themselves determine their movement. If this does not surprise you, nothing will do.

It would be much more attentive to reality to try to recombine to some extent our ideas than to try to recombine the whole world to its smallest parts only for the sake of manipulating them. It would also be more attentive to the quality of our knowledge, too. In formal knowledge anything can become anything else in the end with due transformations, but here we are not talking about arbitrary transformations, but about historically possible transformations, transformations of meaning and sense.

In other words, one cannot change everything at once, far from it. But there are clear lines of action to turn a glove inside out without breaking it, and here we have been talking about such lines. The same quantitative increase in knowledge has nothing to do with improving its quality, and now is rather the symptom of a great imbalance.

Conversely, the improvement of quality is intimately but not expressly related to the balance between formal knowledge, based on change, and the awareness of the invariable and undivided, which does not seek infinity because it knows it has it within. This balance also depends on the harmony between principles, means and ends; on how the circle of interpretation closes on the principle using the straightest means.

Even if mathematics does not care about what reality is, it still depends on the form, which thus becomes a reality on its own. It is through mathematical physics that it has come out of her splendid isolation, but physics has used and abused mathematics to conquer the world rather than to see it. Of course, the opposite is always equally possible: math can use physics to investigate its own relation with reality, in a way totally different from the one used up to now.

Since reality is first and foremost what has no form and what is the support of forms, but mathematics can begin to glimpse another relationship with it through the recognition of unity in the invariance of change. It is in this sense, as an interpreter of physics in its most basic sense, that it can transcend itself and access the transcendental plane.

*

In the Platonic dialogue *Meno*, Socrates poses questions to a young slave with no more culture than his knowledge of Greek by drawing a square on the floor and then another. After a skillful interrogation, asking him about the length that the side of the second square must have in order to double the area of the first, and after intermittent phases of stupefaction, he manages to awake in the young man the idea of irrational numbers, the first "great revolution" of ancient mathematics. Socrates prides himself on not having teaching the slave, but of having helped him collect the knowledge just questioning his previous answers.

As always, various readings can be made of this famous pedagogical moment, and Gómez Pin, in a magnificent book whose subtitle is precisely *The Knowledge of the Slave*, takes the reasoning further until the slave finds out himself the idea underlying infinitesimal calculus [66]. Certainly, for us it seems that one can pass in two hours from irrational and real numbers to the inception of calculus. But why then did it take more than two thousand years since Plato's time? Many great minds devoted to this problem not only two hours, but a great part of their lives, without approaching the crux of the matter.

The culture and knowledge we treasure as a society serves both to inscribe new things in our minds and to erase them. But high-level knowledge, very specialized wisdom, is a very delicate plant. Mathematicians naturally tend to think that theirs, not that of philosophers, is the true factory of ideas; to this it could be argued that pure mathematicians handle far more rarefied concepts, and that they only have delivered true ideas when they are philosophers to begin with, as in the case of Pythagoras, Descartes, or Leibniz, or at any rate "natural philosophers" and physicists like Newton.

The advanced concepts of mathematics only serve at the cutting edge of instrumentalization. What can a modern-day physicist or mathematician convey even to an educated public about his research? Gómez Pin says that the categorical knowledge inscribed in language is more basic than mathematical knowledge: differences such as quantity and quality, or the category of measure, which is precisely a mediation or synthesis of both.

No doubt there is quite a lot of truth in this; there are ideas more basic than mathematics, which are not exclusive to any discipline, and which channel the drift of mathematical concepts. These ideas are at the basis of a cultural syntheses of an entire epoch or civilization. And finally, there are symbols, which do not even pass through the antinomies of ideas and concepts but can adopt them for mere external convenience. In another time, these symbols were supports of various traditions that sought to transmit a knowledge beyond the reach of forms.

There is a formless knowledge that is truly our common root and soil; the ideas are like the sap that ascends through the trunk of the diverse cultures, and the mathematical concepts, would be rather at the level of the branches, the leaves and the fruits. If fresh sap ceases to ascend, what we have is the autumn of a culture, the season of dry leaves. And instinct is also a part of the implicit knowledge, but shaped and conditioned by the social environment. No one tells us how to catch fly balls, but whether one does it playing baseball or practicing different skills depends on the culture and environment.

The point is that instinct is presented to us as opposed to reason when it is obviously reason that is opposed to instinct, to nature trapped within us. Or perhaps it should be said that it is not even reason, but a series of rationalizations. It always seemed to me that Newton's explanation of the ellipse contradicted not only reason, nor reason and intuition, but reason, intuition and instinct; and it also seemed to me that if people accepted this sort of explanation, it was certainly not because of instinct or reason, but because of a certain interest. The same could be said of wanting to enclose the universe in the laws of mechanics.

Let us see what reason and instinct tell us about the next question. Following the same logic of his amendment of standard calculus, Miles Mathis states that the true value of the constant π in any situation involving motion is not 3.14... but 4. This is something that he has argued in detail showing, to begin with, that he has read Newton and other classics much more carefully than his critics [67].

Mathis is simply saying that length is not the same as distance and that to advance around a curve like a circle one has to move in two directions at once. It is just adding up the vectors, not taking the graph literally. Of course *pi* gives us 3.14... in relation to the diameter in a straight line as if we were measuring it with a string; but the point is that moving around a curve involves simultaneously a velocity and an acceleration.

If anyone is in doubt, he can ask himself whether he will use the same fuel driving in a car for 3.14 kilometres in a straight line than driving around a circle 1 kilometre in diameter. And if the difference is attributed to friction, he may wonder if an object in space with an external impressed speed, which according to the principle of inertia should continue with the same direction and speed indefinitely, can go round and round indefinitely. As far as we know no law of inertia with perpetual motion has been enunciated for circular motion.

And yet that is what Newton's fundamental lemmas at the beginning of the *Principia* imply. According to Newton and all celestial mechanics since him, a planet in a perfectly circular orbit could orbit around a central body forever, exactly like a perpetual motion machine. But I think it is clear that friction has nothing to do with this here, since anyone understands immediately that a contribution of force is needed not only to change speed, but also to change direction.

The question, then, is not how it is possible for Mathis to say what he says; the question is how it is possible for Newton and Euler and Lagrange and Laplace and all the others to have been able to accept this for over 300 years without a blink, and how we can continue to accept it without giving it the slightest attention. Mathis himself asks this question repeatedly, and hesitates between the hypothesis that they did not perceive it and the hypothesis that they did perceive it but hid it.

Surely they were aware of it. But we underestimate the ascendancy that over them had certain ideal of science and nature alike, that of a clockwork governed by a clockmaker. Then, simultaneously with the perfectly utilitarian idea of expanding the realm of calculus at any rate, they felt morally justified that this contributed to bringing us closer to an ideal of nature perfectly passive in relation to its creator. This intimate mixture of utilitarianism and disguised theology, in which a separate personal creator and an equally separate cognitive subject melt together, still has a great ascendance even today; but the important thing for the instinct is not to know itself. On the other hand, while eager to play the game, surely they believed they were taking "the shortest path" to their goal —which was also their ideal of efficiency. Besides, no one could imagine that calculus understood as reverse engineering —truth could not escape- was amenable to such kind of mistakes.

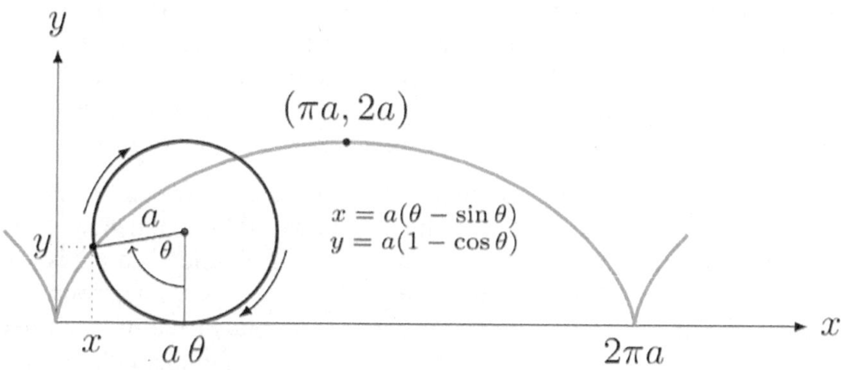

Another way to visualize this issue, if necessary, is by means of the curve known as cycloid, drawn by the rim of a wheel rolling in a straight line without slipping. The length of this curve is exactly 4 diameters or 8 radii, and the motion it describes is extraordinarily elusive only if we are thinking of the ordinary circular graph.

It is ironic that the first diligent study of this curve was made by Galileo, the father of the principle of inertia, —a principle so elusive for the time, that it took fifty years to consolidate and only after the contributions of Descartes and Newton-, in order to calculate quadratures. As he made his laborious calculations, he was pushing with his own hands time and again the best possible counterexample to that idea of the rolling ball that had already taken possession of his mind.

As is well known, Galileo opposed Kepler's elliptical orbits and and continued to adhere the ideal of circular orbits. Huygens made an even more thorough study of the cycloid to improve the accuracy of the pendulum clock, whose very idea was due to Galileo. Later Newton had the matter at hand again while dealing with the celebrated problem of the brachistochrone curve, whose solution is precisely the cycloid.

I don't think that the subject $\pi = 4$ deserves more explanations; and whoever needs more information already knows where to find it. If we want to express it in terms of limits, it can be said that bodies travel in the curves to the limit of the shorter legs of the triangle, instead of the limit of the longer leg. It is called the "Manhattan metric", but it casts by the way a deep shadows on the basis of the vast majority of metrics. However the purely "technical" question pales in comparison with the effect which it should have on our minds to detect such holes going back to three or four centuries. This alone should radically change our perception and relationship with science.

What the derivations of Huygens and Newton precisely show, round about the crystallization of calculus, is the delicate pass from talking about proportionality in forces to talking about their equality. As in the case of the definition of the central forces, Newton is extremely cautious and avoid expressly to speak of equal forces; but once calculus paved the way, the equality was taken for granted. With the emergence of the "vanishing quantities" the constant proportions of geometrical figures left the scene forever.

It is possible that a great deal of the mismatch between the calculus and the continuous proportion is due to the portion of motion vanished in the representation of graphs, of which the kinematic circumference is the best example. It has been said that the continuous proportion belongs to the realm of statics, in contrast to the changing world of calculus —but in reality the opposite is true, it is calculus, the tool designed to describe change, that adheres to the static figures. This should have implications for a modern theory of proportionality, a possibility that was uprooted and wiped off the map by the same operation of standard analysis.

Calculus, as it has been understood, really involves a inversion in our natural view of the physical world. As Krishna Vijaya observes, instead of determining geometry from physical considerations, deriving from it the differential equation, since Leibniz and Newton the differential equation is set up first and then one tries to solve it to get the physical answers. Both methods are far from equivalent, but the same belief in the reality of the differentials follows from the procedure adopted. It is still necessary to reverse this method in order to open the eyes and recover the right perspective [68].

Ideally, description and prediction should be balanced, as should memory and anticipation, with which we constantly and reflectively create our perception of time. Inevitably, if there is a perception of an "unreasonable effectiveness of mathematics" in predicting natural occurrences, it is because a great part of the description has been obscured or concealed. This creates enormous cognitive dissonance in both our perception of nature and our idea of science and knowledge. Modern science is always urged to be more and more "creative" in order to be less and less aware of the nature of its manipulations. But a way exists in which the liberation of Nature and the liberation of the knower coincide.

One can understand Nature without calculus, as one can predict things of Nature without the least understanding of them. Venis' morphology is an instance of the first, and modern physics is certainly the best possible example of the second; but that does not mean that one kind of knowledge excludes the other; everything depends on how the connections are elaborated and derived. The crux of the issue is that modern technoscience is far more interested in manipulation than in understanding, and to such an extent that understanding becomes inconvenient. In turn this limits the degree of disorder that we humans can create.

Philosophers have repeatedly complained that calculus involves a geometrization of motion, but they have never been able to substantiate their claims. Now that they have it, they can seize the opportunity to delve further into the subject.

The cycloid and the wheel could serve as an indication of a different quadrature than the one Galileo intended; rather the opposite, though still with an important point of contact. It is well known that throughout history, and for very different cultures, the circle and the square were the symbol of Heaven and Earth, the active and the passive, the dynamic and the static.

However, the understanding of what is "dynamic" and what is "static" was definitely reversed in the short time span from Galileo to Descartes. Before this reversal, motion and changes in extension could reflect change, but rather accidentally; and so potentiality and actuality in Aristotle had an incomparably broader meaning than that now attributed in physics to potential and kinetic energy —a pair already defined in terms of motion.

Physics takes off when it starts to define everything in terms of extension and motion, although in a very unclear and indistinct form, one is aware that not all physical reality can be reduced to motion and extension.

"What moves does not change and what changes does not move." The closer we get to our time, the more important motion becomes, and the further we move away, the more secondary and related to appearances it is considered. But in this continuous transformation there is permanently a double movement; for it is evident that calculus has also frozen motion and not only reduced it to geometric forms but has assumed its static quantitative relations.

Fig. 6

This double movement of ascent and descent, of condensation and volatilization is really a natural and spontaneous process that can take place at different levels —individual, collective, physical and non-physical. The history and cycles of different civilizations can also respond to this double pattern, and the same history of science shows intermittent evidence of it.

Rivers of ink have flowed about the Cartesian dualism that separates mind and extension, but much less has been said about the duality inherent in physical quantities, in which an extensive part always coexists with a non-extensive one, being now mediated by motion itself. From this stems pairs such as space and time, force and mass, vector and scalar, intensive and extensive quantities, until the extremely complex measurement units of today.

This duality of physical magnitudes, antinomic or not, is nothing but the resolution in the plane of motion of that first duality; and there is no reason why this endless dialectic should have an end, since it is the very deployment of reason. Only ignoring our role in all this could we believe that the cause of consciousness can be found at some explicit level of causation. The thinking substance is already threaded into the description of any level of physical causation since the times of Galileo and Descartes and even much earlier; so in vain do we wait for the paradoxes of quantum mechanics to solve these enigmas for us.

There is a big difference between an idea, and even a mathematical idea, and the manipulation of mathematical symbols. Profound ideas arise when concepts try to return to interpretation or representation, and when representation pays due tribute to the unrepresented, to the implicit in knowledge. But the current drift of science prevents the full circle of this motion that should always be expectant towards the most basic, the really fundamental.

The world would only be an illusion if it could be reduced to extension and motion, yet science fails to tell us what it can be beyond these attributes. Here is its limit, but it would have to be a fertile limit. And the limit between that which cannot be represented and that which is representable passes precisely through this double movement: this is the best exponent of true activity, in Nature and in Spirit alike. We can find exponents of this double movement on many levels, from the electron to the constitutive relations of materials, from our own breathing to the movement of the vortices in the Venis sequence, from the process of individuation to the same self-consciousness.

The swastika offers us a good example of a symbol containing implicit knowledge. It is first of all a sign of the Pole and its action on all things in the world; but it can also be the clearest expression of squaring the circle through motion, of the double movement itself, of the dynamic equilibrium, or of the reciprocity intrinsic to the Law. Surely there was little concern with the issue of calculus in prehistoric times, and yet we always can actualize and give voice to any mute expression of reality.

*

Change itself is only the visible side of the transcendental plane, but this change has an infinite number of aspects that escape our limited ideas on motion. In this sense, all modern physics and science in general continue to have a huge Platonic deadweight, and we have gone from metaphysics to mathephysics without hardly noticing it.

The mere fact of thinking that there are fixed physical constants or identical particles is already a total lack of taste when it comes to describing Nature, an eloquent exponent of our incapacity. And yet, in a framework that aspires to "maximum predictive power", these assumptions are indispensable. Can identical electrons exist, as if they had just come out of a factory, independently of our complex web of assumptions, constants, and measurement conditions? And yet within this framework it is nearly impossible that it could be otherwise.

An electron, like ourselves or any other entity, can only be an ephemeral configuration that changes from moment to moment. It is not the individual what matters, but the process of individuation, which constantly connects what appears to us as the part and the whole. This connection is pure activity, balance or fight at present, not a frozen event that creeps through the universe for thirteen billion years.

The analysis seems to dissolve everything, yet the fluidity of the world escapes it completely. How can this be possible? That's an excellent question that should be answered. Because, currently, it is not even approaching the real flow of things and the great unity of life, but on the contrary is moving further and further away from them.

In fact this is a fatal question for modern science. Because it is clear that the immutable cannot be the object of knowledge, and that if something can be known, it is a matter of change, and nothing else. How is it that analysis, which is the study of change, ignores almost everything about it? Is not it amazing that scientists do not ask themselves more questions about this? But there is not even a proper frame to pose the question.

What can be predicted is an expression of regularity, and in that sense, of law. However, to subordinate everything to prediction one is forced to separate and isolate aspects of the processes to cut out the forefront from the background and the context. But even if the ultimate background could be neutral, the context never is. Any context already flows in a definite direction, but now we have a different direction, because to be guided by predictions alone is like running with blinkers.

Modern science tries to compensate for the void of analysis and prediction with disciplines of a "synthetic" nature such as cosmology and the theory of evolution in an attempt to reconstruct the contexts and directions that analysis has destroyed —but in doing so completely relies on the assumptions established by the analytical criterion. And so the whole of cosmology depends implicitly on the principle of inertia, and the theory of the evolution of life on inertia plus the purely random nature of the processes. We have already seen that the first assumption is unnecessary and contradictory, and the second false.

Since the analytical-predictive part took precedence, and the synthetic part relies on its assumptions, the whole is completely unbalanced. The underlying idea is that we can dissolve everything, and then put it back together again and give it a direction and a narrative. But in this way it is impossible to respect the reality of things.

The "analytical part" already has a direction, and therefore cannot be separated from the synthetic part. In other words, nature already has its own narrative all the time, so it does not need any other one as a supplement.

Venis' infinite sequence presents us the unity of the analytical and morphological aspects, of balance and direction in Nature. This unity is the real challenge of modern science and of science in any age; the contrast between its perspective and the perspective of modern analysis and synthesis will prove to be supremely instructive.

Heraclitus understood the world better than contemporary scientists; if he already disdained Pythagoras' knowledge as polymathy, as a mere variety of learning, it is better not to imagine what he would think of our theories and specialities. The problem is not the quantity of knowledge, but its quality; and mathematics can be a master of deception in this regard. You end up having the knowledge you want, but the point is what kind of knowledge you want.

*

One thing is the spontaneous organization in Nature and another is the recurrent presence in it of a certain mathematical constant. It has yet to be demonstrated that there is any kind of functional link between both, and if there is, nothing can yet be said about its relevance. What admits few doubts is that fundamental physics can be interpreted most directly as a phenomenon of self-organization, through relational mechanics and the retarded potentials; as well as with the due recovery of the "thermomechanical" irreversibility, in fact the only real dynamics.

But theoretical understanding alone is incapable at this point of modifying in the slightest the highly dissolving and destructive drift of current science, so in tune with the rest of the social dynamics —we cannot change ideas without first changing what we do and what we want to do. Today man does not seem to be between Earth and Heaven, but between a Nature in perpetual setback that we only see as a resource pool and machines that materialize the human spirit to exploit both the external nature and our internal nature.

We say "human spirit" because not only our intelligence but also our will is materialized in the machine: and the substantial unity of both has become inconceivable to us precisely because of the growing separation made possible by machines. Yes, machines, like other human creations, are an obvious crystallization of a certain water and fire, a feminine desire and a masculine will: a small abbreviation of nature isolated from the rest that tries to perpetuate its movements at the expense of the environment. We pay dearly for this simulacrum of closure that is not closed at all nor can it be; in fact, perpetual motion and its *telos* never ceased to exert a fatal spell on us.

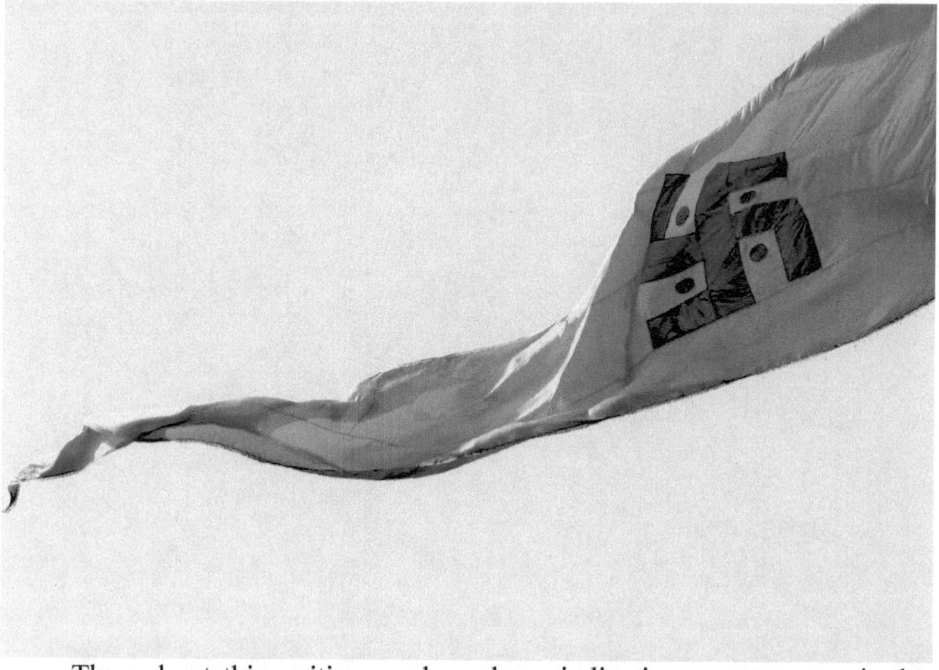

Throughout this writing we have been indicating a common axis that runs through man, nature and the machine, and must undoubtedly go beyond them, since we know of them nothing but momentary planes of manifestation.

If current technoscience separates in order to recombine at will and unleash the universal confusion of planes, we may well do the opposite: see where different planes coincide in order to survey their vertical, their height and depth.

POLE OF INSPIRATION – REFERENCES INFORMATION THEORY AND THE ZETA FUNCTION

9 April, 2020

It has been said that if the Riemann hypothesis were solved, all the keys to cryptography and cybersecurity could be broken; no one has specified how that could lead to faster factorization methods, but at least it reminds us of the close relationship between a hitherto intractable problem, cryptography and information theory.

Such speculations are only based on the fact that the Riemann zeta function establishes a connection between prime numbers and the zeros of an infinitely differentiable function that provides the most powerful method for exploring this field, prime numbers being the basis of classical cryptography. But we have already seen that the zeta is way more than this.

Today the technological revolution is synonymous with the digital revolution and with a category, information, which seems to pervade everything. All previous developments in science and technology converge in it and pass through it. Even physicists now think of the universe as a gigantic computer; there was much discussion about whether the world was made of atoms or stories but in the end it was decided that it was made of bits and case closed.

What is important about information theory is not so much its definitions as the direction it imposes on everything; changing that direction is equivalent to changing the direction of technology as a whole. The unequivocal direction, which obviously inherits from statistical mechanics, is that of decomposing everything into minimal elements that can then be recomposed at will.

For statistical mechanics there is no direction in time: if we do not see a shattered vase recompose itself and return to the table, it is only because we do not live long enough; if the pieces of a dismembered corpse do not get back together and walk again as if nothing had happened, it is only because we are not in a position to wait $10^{1.000.000}$ years or something similar.

Information theory is not concerned at all with the reality of the physical world, but with the probability in its constituent elements, or rather the probability *within its accounting of the constituent elements*. On the contrary, the physical world is just a pool of resources for the sphere of computation, aimed to be independent of the former.

Is this statistical view a neutral position or is it simply a pretext to manipulate everything unencumbered? There is no single answer to this. Statistical mechanics and information theory are successfully applied in countless cases, and in countless cases they could not be more irrelevant. What is worrying is the overall trend they create.

It is my conviction that a shattered corpse is not going to recompose itself, no matter how fabulous the period of time we choose. And this is not rhetoric; rather, large numbers are the rhetoric of probability; an inflated and poor rhetoric, as it ignores the interdependence of things.

This interdependence, this infinite network of interrelations, is what makes things what they are. Statistical mechanics, and its daughter information theory, are the most general framework to deal with random independent elements; the Riemann zeta function, the most direct and elegant way of encompassing an infinite series of elements, the numbers themselves, which are both independent and dependent, apparently random and simultaneously containing an infinite net of relationships.

Then it is only a matter of time before the information theory and the zeta function meet together. Today numbers no longer seem to exist to understand the world, but to crush and squeeze it, and with it all of us. Information theory have become the funnel, the fearsome event horizon for everything; but in turn the zeta function could become the event horizon for the Information Age itself as the very concept of information, so highly generic, is refined and given content.

However, somewhat surprisingly, there is virtually no literature on an explicit relationship between the zeta function and information theory. In our search we only found a brief note due to K.K. Nambiar, in which he showed a connection between the capacity of the channel and the function, with a "Shannon series" and a "Shannon zeta function". The idea was worthy of much further elaboration, discussion and development. Besides, the equivalence between the zeta function and the classical sampling theorem fundamental in signal processing seems proved [69].

The same author issued an even shorter note establishing an electricity equivalent of the Riemann hypothesis in terms of the power dissipated in an electrical network; of course more elaborate equivalents in terms of electrical potential have also been established, but here we are more focused on thermodynamics. Infinitely many waveforms can be created with radiation patterns containing the zeros, the point is how to modulate them. Already in 1947 Van der Pol had created an analogical electromechanical device for computing zeros from the zeta [70].

If not directly with information theory, at least there are many more works connecting the function to closely related aspects: entropy, statistics and probability [71]. Here we will focus on the concept of entropy.

Entropy and information become almost synonymous, and if in thermodynamics entropy was seen as the loss of usable energy, it will now be seen as information loss. Shannon's information was initially called *surprisal* or *self-information* —a message should contain something new.

The confusion of entropy with disorder is due to Boltzmann's rationalization in attempting to derive macroscopic irreversibility from mechanical reversibility, quite a tour de force by itself. It was Boltzmann's use of the concept of "order" what introduced a subjective element. Clausius original energetic conception of entropy, stating that the entropy of the world tends to the maximum, was, if nothing else, far more natural.

Order is a subjective concept. But to say world is to say order too, then also the idea of world is inevitably subjective. However subjective they may be, "order" and "world" are not just concepts, they are vital aspects of every organized entity or "open system". Yet even our very intuition of order and entropy are entangled and confused.

Not that entropy increases disorder. On the contrary, the tendency towards maximum entropy is conducive to order, or better said, order has more potential for dissipation. As Rod Swenson put it: "*The world is in the order production business, including the business of producing living things and their perception and action capacities, because order produces entropy faster than disorder*"[72].

Information theory is a formal and objective framework but it cannot escape from the reflexive turn in the representation and use of data. Information as an object is one thing, but the information as an environment with which an open system interacts and helps to shape is another. Although in black box mode, in many cases the interpretation is already part of the system's behavior.

The ambiguities and limitations of information theory have given way to more inclusive frameworks, such as Luciano Floridi's philosophy of information, which, to put it very simply, considers semantic information as data + questions [73]. Floridi's approach still remains a clear heir to Cartesian dualism and idealism, and in this decisive division we prefer to speak of two types, external and internal, of information and entropy: there is a kind of information-entropy as a formal object and there is an information-entropy environment including the open systems that interact with it. Boltzmann's entropy and Shannon's entropy are of the first type, Clausius' entropy and the one that engineers deal with in concrete physical applications is closer to the second.

The connection between entropy and the zeta function is quite recent and only in the 21st century is beginning to take shape. At a very basic level, one can study the entropy of the sequence of zeros: a high entropy would lead to a low structure, and vice versa. It is observed that structure is high and entropy low, and even low level neural networks were already successful in their prediction [74]. Also, of course, we can study the entropy of the primes in the ordered sequence of numbers and their correlation with the zeros; and from there on we already have an unlimited spectrum of correlations.

There are other non-extensive types of entropy independent of the amount of material, such as Tsallis entropy, which can be applied to the zeta; these types of entropy are generally associated with power laws, more than exponential laws. But as in thermodynamics in general, the definition of entropy can vary from author to author and from application to application.

In order to unify and generalize so many definitions of entropy, Piergiulio Tempesta has proposed a group entropy [75]. Whether in physical contexts or in complex systems, in economics, biology or social sciences, the relations between different subsystems depends crucially on how we define the *correlation*. This is also the central issue in data mining, networks and artificial intelligence.

Each correlation law may have its own entropy and statistics, rather than the other way around. Thus, it is not necessary to postulate entropy, but its functional emerges from the kind of interactions that one wants to consider. Each suitable universal class of group entropy is associated with a type of zeta function. The Riemann zeta function would be associated with Tsallis entropy, which contains classical entropy as a particular case. But these correlation laws are based on independent elements, while irreversibility at the fundamental level seems to reject that assumption. Fundamental irreversibility assumes universal interdependence.

We mentioned earlier Voronin's theorem on the universality of the Riemman zeta function, which shows that any kind of information of any size can be approximated with arbitrary precision within this function —and not just once, but an infinite number of times. In this sense, the entropy of its mapping is infinite. It seems that there is a "zeta code" that could be the object of the algorithmic theory of complexity. But in turn Riemann zeta function, by itself an infinite encyclopedia of correlations and correlation laws, is only the main case of an infinite family of related functions.

One could also find any kind of information in white noise; but white noise lacks any structure, while the zeta function, being well-defined, has an infinite structure. Had Hegel known both cases he would have spoken of false and true infinity; we will see that these Hegelian apperceptions are not totally out of place here.

The *seriousness* of a problem such as the zeta demands that the problem of irreversibility be considered in depth. And considering it in depth means precisely including it at the most fundamental level. Conservative systems are still toy models for this.

The MIT school or Keenan school of thermodynamics, especially with Hatsopoulos, Gyftopoulos and Gian Paolo Beretta, has developed a quantum thermodynamics in stark contrast with the rationalization of entropy by statistical mechanics. The dynamics is irreversible at the most fundamental level. The number of states is incomparably greater than in the standard quantum mechanics, and only a few of them are selected. The selection principle is very similar to that of maximum entropy, although somewhat less restrictive: it is the attraction in the direction with the steepest-entropy-ascent. No great changes are needed in the usual formalisms, what is transfigured is the sense of the whole [76].

The equilibrium formalisms are also retained, but their meaning changes completely. The uniqueness of the stable equilibrium states amounts to one of the deepest conceptual transformations of science in the last decades. The approach is contrary to the idealism of mechanics and much more in line with the daily practice of engineers, for whom entropy is a physical property as real as energy. The theory has many more advantages than sacrifices, and it can be applied to the entire domain of non-equilibrium and to all temporal and spatial scales.

But of course irreversibility can be introduced directly into dynamics as we saw in M. J. Pinheiro's reformulation of classical mechanics, replacing the Lagrangian principle of action with a system with a balance between energy and entropy. This allows the study of entropy generated in classical trajectories under different formalisms and criteria, establishing another general connection between different domains.

In general, for the physicists it does not make sense to change equations that "already work perfectly well". And, in fact, what fundamental physics has always done is to leave aside issues such as friction and dissipation to better stick with the "ideal cases". The principle of inertia is perfectly suited to the task. In this way thermodynamics was carefully segregated as a by-product of ideal physics, which is a curious arrangement indeed.

But we should try to see it the other way around: seeing how the appearance of reversible behavior emerges from a background of irreversibility is the closest thing to finding the magic ring at the heart of nature. Physics today is not about nature, but about certain laws that affect it. Reversibility and irreversibility are like fire and water; only if they intermingle in the right way one can hope to get the real thing.

The ironic twist is that mechanics, with its idealist disposition, has segregated thermodynamics; but in the end it will be information theory, the most idealist offspring of mechanics, that will need to recover everything that has been eliminated if it is to regain its meaning. And the zeta function should play an essential role in that mute and transfiguration of information theory. Information theory should be interested in irreversible mechanics... because it provides more information. And information of critical importance [77].

*

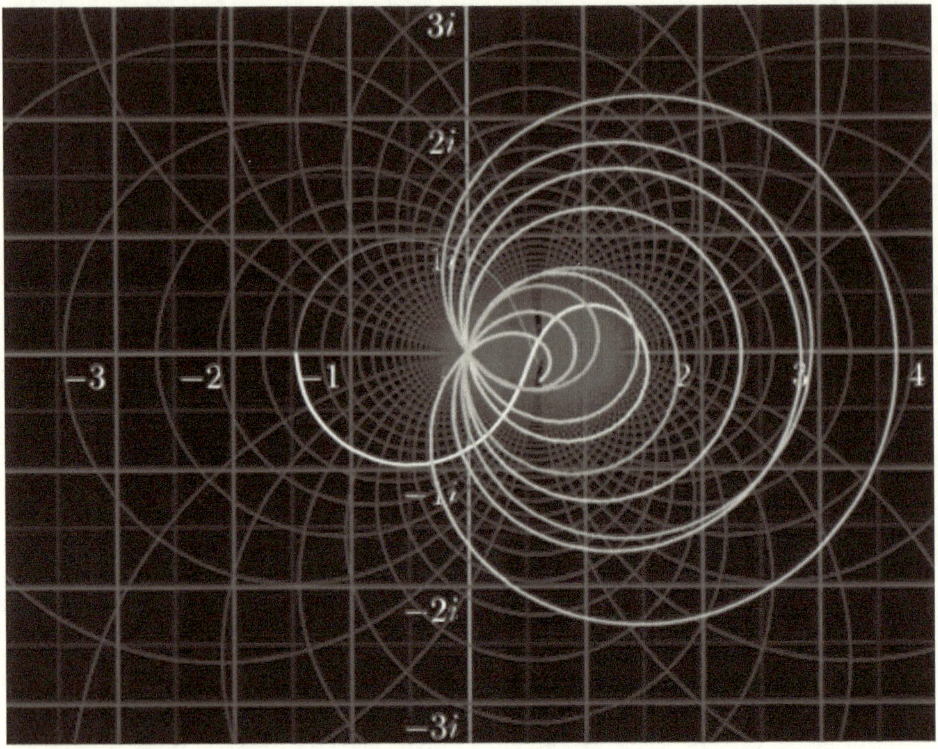

Idealization and rationalization are like the mythical Symplegades rocks that destroyed ships and navigators; only the one who understands their dangers and avoids them will be able to pass to the other side.

We have seen how this alternate process is inherent in the historical evolution of disciplines such as calculus, classical and quantum mechanics, or statistical mechanics. Hardly any experimental evidence can stop the overall trend of rationalization, since rationalization always finds ways to assimilate results and justify inconsistencies.

In philosophy, Hegel gives us the best example of the large-scale sway between idealization and rationalization. In an interesting article, Ian Wright attempts to explain the nature of the Riemann zeta function with Hegel's *Science of Logic*, as the mediation between being and non-being through becoming within the realm of numbers [78]. Old-fashioned as this can sound to many, three things can be said:

First, that from the point of view of pure arithmetic, this is admissible since if there is a part of mathematics that can be considered a priori, that is arithmetic; while, on the other hand, the very definition of this function affects the totality of integers, and its extension to real and complex numbers.

Second, that calculus is not pure mathematics like arithmetic, although many specialists seem to think so, but mathematics applied to change or becoming. Were it not, there would be no need to compute zeros. The zeta function is a relationship between arithmetic and calculus as close as it is uncertain.

Third, for experimental sciences such as physics the opposition between being and not being does not seem to have systemic implications, and for formal sciences such as information theory this is reduced at most to fluctuations between ones and zeros; however the distinction between reversible and irreversible, open and closed systems, which is at the very heart of the idea of becoming, is decisive and lies at the crossroads of the issue.

In problems such as the one posed by the zeta function in relation to fundamental physics, crucial experiments capable of unraveling the relationship between the reversible and irreversible in systems could soon be conducted —provided that they are expressly sought. For instance, a Japanese team very recently showed that two distinct probes of quantum chaos, noncommutativity and temporal irreversibility, are equivalent for initially localized states. As is well known, this function has been studied a great deal from the point of view of non-commutative operators, in which the product is not always independent of order, so this would bring together two areas as vast as distant up to now [79].

With the rise of quantum computing and quantum thermodynamics, along with all the associated disciplines, there will be no shortage of opportunities to design key experiments. On the other hand, it is becoming increasingly clear that the second quantization, which deals with many-body systems, demands an special spectral approach, as Alain Connes and many other researchers have been emphasizing for many years now [80]. In addition to its theoretical interest, the practical scope of such a spectral theory could be extraordinary —think only of chemistry; but it is our opinion that the segregation of thermodynamic irreversibility prevents us from getting to the bottom of the matter.

It is not that hard either. It is said that electromagnetism is a reversible process, but we have never seen the same light rays returning intact to the bulb. The irreversible is primary, the reversible, important as it is, is just a settling of scores. Even Maxwell equations belong to two different thermodynamic categories.

But physicists have always been fascinated by reversible artifacts, independent of time, which some have seen as the last legacy of metaphysics. And what else? There are no closed, reversible systems in the universe, which are just fictions. Metaphysics is an art of fiction and physics has continued metaphysics with other means with the perfect excuse that it now descends into reality. And no doubt it has, but at what cost?

Previously we gave the example of the outfielder who catches the fly ball as a contrast for both the standard calculus and the now prevalent idea of artificial intelligence. This direct form of calculus, which we all perform without knowing it, could be the best illustration of Mathis' constant differential method. Mathis is the first to admit that he has not been able to apply the principle even in many cases of real analysis, let alone complex analysis, which he does not even deal with. Needless to say, the mathematical community cannot stop to consider such limited methods.

And yet the constant differential is the naked truth of calculus, without idealization or rationalization, and we should not dismiss something so precious even if we do not succeed in seeing how it can be applied. If a constant differential is not found in the tables of values of the function, we can still estimate the dispersion, and whoever says dispersion, may also say dissipation or entropy.

In this way it is possible to obtain an *intrinsic entropy* of the function, from the process of calculus itself, rather than from correlations between different aspects or parts. This would be, at the most strict functional level, the mother of all entropies. It should be possible to apply this criterion to complex analysis, to which Riemann made such fundamental contributions, and from which the theory of the zeta function emerged.

The search for the constant differential may even remind us of the mean-value methods of elementary calculus or of analytical number theory. However the core of this method is not averages, it is on the contrary standard calculus that relies on averages —which without this reference lack a common measure. Finite-difference methods have also been applied to the study of the zeta function by leading specialists. But to make things even more interesting, we should remember that both methods do *not* always yield the same values.

This would simultaneously fulfill several major objectives. It would be possible to connect the simplest and most irreducible element of calculus with the most complex aspects. Something less appreciated, but no less important, is that it puts the idea of function in direct contact with that which is not functional —with that which does not change. Analysis is the study of rates of change, but change with respect to what? It should be with respect to what does not change. This turn is imperceptible but transcendental.

It should be remembered that Archimedes did not invent calculus, but he did invent *the problem* of calculus, by looking toward zero. Mathis is rectifying an approach that is now over 2,200 years old.

The direct method laughs in the face of the "computational paradigm" and its fervent operationalism. The now prevailing idea of intelligence as "predictive power" is reactive; in fact, it is no longer known whether it is the one we want to export to machines or the one humans want to import from them. Perceiving what does not change, even if it does not change anything, gives depth to our field. To perceive only what changes is not intelligence but confusion.

On the other hand, the definition of equilibrium conditions, as a sum and as a product, as an external and internal perspective, as applicable to closed and open systems, as intensive and extensive quantities with their possible breakdowns, as reversibility and irreversibility, are at the core of the algebraic aspects complementary to the analytical aspects that should contribute to make the intrinsic more explicit to get a deeper understanding of the zeta function.

If our view is somewhat correct, the cleaning and clarification of the enormous field of thermodynamics and entropy, with all the work that lies ahead, would be closely associated with the progress in understanding the zeta function. This does not seem a promising prospect, since it amounts to tons of dirty work far away from the exertions in the wondrous gardens of the theory of numbers. However, there is a hope that the connection between the two domains can be greatly smoothed out by analytical notions such as intrinsic entropy and the introduction of irreversibility in the foundation of analytical mechanics.

*

We have seen that the gauge fields of the standard theory, based on the invariance of the Lagrangian, clearly exhibit a feedback. They don't even hide it, and the theorists have been struggling with terms like self-energy and self-interaction for generations now, shooing them away like flies of their faces because they thought it was a "silly" idea. Now, the renormalization group at the base of the standard model and statistical physics, containing this self-interaction, is more or less the same that is currently used in the neural networks of deep learning and artificial intelligence —so something that the electromagnetic field already has built-in is being used externally. Let us call it a missed natural intelligence.

Nobody made the integers, and all else is the work of man. The zeta function is an ideal candidate for creating a collective network of emerging artificial intelligence around it. Why? Because in it, as with addition and multiplication, the external and internal aspects of the act of counting, what some would call the object and the subject, coincide. In this type of networks will be key to what extent the potential of the physical substrate is tapped, the connection between analog and digital components in a hybrid approach, the local and global equilibrium conditions, the algorithms to simulate the structure of the integers, the structures that try to perceive or filter them, the correlation spectrum, and a good number of aspects we cannot list here.

Thus it may be possible to create a collective emerging artificial intelligence alien to human intelligence defined by purpose, and yet with the possibility of communication with humans at different levels through a common act, counting, that, as purpose limits our scope, also for humans use to exist at very superficial levels. Intellection is a simple act and intelligence is a connection or contact of the complex with the simple, rather than the opposite.

As a prime example of an analytical totality indigestible for modern methods, the zeta function is already a great sign. Someone, in an Information Age typical turn, might say that, if any possible information is already in that function, we might as well look for the answers there instead of in the universe. But the opposite is true: to better understand the function we need to change the ideas of how the universe "works" and the range of application of our more general methods. That is the beauty of the subject.

Until now, the mathematical physics from which the entire scientific revolution emerged has been applying mathematical structures to physical problems in order to obtain a partial solution by means of an artful reverse engineering. Calculus itself emerged in this process. On the contrary, what we have here are physical processes that spontaneously reflect a mathematical reality for which there is no known solution. This already invites us to change our logic from top to bottom.

As for the information itself, if with Shannon the sequential aspect of the flow of information units prevails, as we move towards more massive amounts of data the relevance of correlations increases unstoppably, and they even tend to constitute an autonomous sphere.

The data we use directly is one thing and the metadata that can be elaborated with that data and its multiple correlations with other data is quite another. The same can be said of a genetic sequence and the multifactorial analysis of the relationships between different genes, and so on. A Riemann sphere, or a Riemann surface with many layers like an onion, seems to be much more suitable representations for the multidimensional reality of analysis than the increasingly insignificant relationship of causal sequences.

However, the zeta function contains the most basic relationship between the simplest sequence, that of integers, and the infinite world of correlations between its elements. But we should not forget the most primary aspect of the question, on the contrary. We should not forget where problems come from, neither in information theory, nor in calculus, nor in mechanics.

In fact nearly everyone thinks that causation is something of the past; but this way we forget that even mechanical causation is a global issue, not a local one, and this may be relevant where we least expect it. If the unexplained dynamics of the zeta is not consistent with the foundations of mechanics, let us change the rules of mechanics, for the understanding of the problem deserves it. It is Nature that sings through random matrices and many other instances, not the spirit of the laws.

Today, it is no longer a question of mastering nature but of mastering technology, since the latter now has more destructive potential for human beings than the former. However, mastering technology demands freeing Nature from constraints imposed by a technoscience that is too instrumentalised from the outset. Considering the history of science as a whole, it is perhaps not so surprising that the avenger is now the most selfless of all sciences, the useless but eternal theory of numbers.

*

In a previous chapter we said that the concept of order is not less subjective than that of harmony; it goes without saying that this could be the subject of endless discussions. There are a lot of formal definitions of order in mathematics for as many different cases, but it could hardly be argued that the basis of them all are the natural numbers. In fact, without them mathematics would not be able to order anything.

The paradoxical aspects of entropy are related to the relative nature of the notion of order. Something with a highly visible order has more potential for disorder than what is already in complete disorder, which can be translated into the social paradox of entropy: the more complex the society, the more disorder it seems to produce. On the other hand, we also saw earlier that entropy is not necessarily the best measure of complexity, and that the energy rate density can be more revealing. These are far-reaching issues that should be attended carefully.

From another angle one can say that pure randomness is the ultimate order, something suggested by the very arrangement of the primes within the number system, so chaotic at a local level and with such a striking global structure —or as one mathematician put it, growing like weeds and yet marching like an army.

The zeta function is a transformation of the traditional harmonic series (1+1/2+1/3+1/4...) so that it does not diverge. The harmonic series, attributed to the Pythagoreans and also known in China at roughly the same time, gives us the harmonics or overtones added to the fundamental wavelength of a vibrating string, and it is a path to understand musical intervals, scales, tuning and timbre.

It has been precisely within music theory that Paul Erlich has combined information entropy with the theory of harmony to define a relative *harmonic entropy*. Harmonic entropy, "the simplest model of consonance... ask the question of how confused is my brain when it hears an interval. It assumes only one parameter in answering this question" [81].

Erlich's concept of harmonic entropy deepens the line of research opened by engineer and psychoacoustic Ernst Terhardt, who had already introduced notions such as virtual tone. Virtual tones, in contrast to spectral tones, are those that the brain extracts even if the signal is masked by other sounds. The ear has a strong propensity to adjust what it hears in one or a few harmonic series. Harmonic entropy can be considered as a kind of "cognitive dissonance", but is also related to the intrinsic uncertainty of time series.

Harmonic entropy has a solid theoretical basis that makes extensive use of the Riemann zeta function. This closes a circle mostly unbeknownst, since Riemann himself made a memorable, though unfinished, study of the mechanism of the ear that already considered the global aspects of auditory perception and was conceptually much more advanced than Helmholtz's reductionist model that was his starting point.

Riemann was struck, among many other things, by the incredible sensitivity of the detectors in the inner ear, currently known to be able to perceive displacements smaller than the size of an atom, or 1/10 of a hydrogen molecule. Would it be possible to illustrate the behavior of the zeta function with a precise acoustic analogy? Can we bring it to that area where understanding and perception merge?

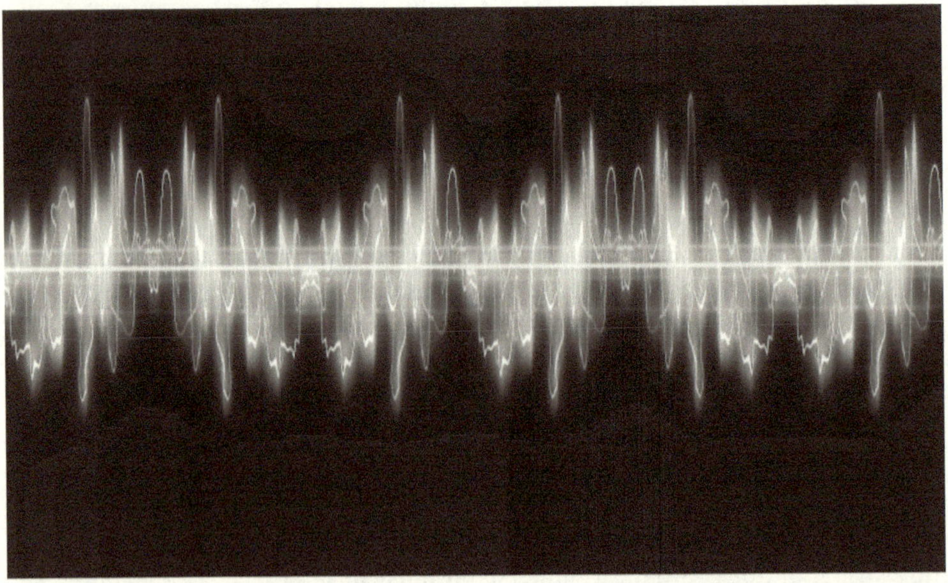

Certainly, issues related to the senses are not among the most important for mathematicians; but here it is easy to see that we are dealing with a problem of high interest at different levels, for laymen and experts alike. If deep math reaches our perception, just as deeply it changes our perception of math and numbers, and that has never been more necessary than now. There is a wide range of psychoacoustic experiments to survey the matter. Erlich uses Farey's series, one of the simplest ways to illustrate Riemann hypothesis.

Well known physicists have shown interest in the conversion of the zeta function and prime numbers into sound in order to hear its "music"; however the experiments we are talking about do not have to refer directly to this function, but first of all to the more generic questions of internal complementarity between spectral and virtual tones. The limits for the interest of the sonification of the arithmetic aspects would depend on the extent of this complementarity.

The reflection of the zeta function in perception in terms of harmonic entropy is a quest in its own right, and it should be sought with as much or even more eagerness than the physical systems capable of replicating it, as it is a more unitary endeavor. But in fact, they are not entirely different problems, even though they appear to us as the subjective and objective ends of the question, which is still misleading.

From the more utilitarian point of view of neural networks and artificial intelligence, the view that there is a basic continuity between perception and the so-called "higher" cognitive aspects is increasingly accepted. Riemann's ideas about hearing as an analytical abstraction are precisely those applied now, 150 years later, to the computer models that attempt to reproduce it.

Distributed networks such as the one mentioned above, for example, would lack something essential without an ability to filter or perceive numbers analogous to that of the ear and the brain when selecting and reinterpreting acoustic signals. Riemann's unfinished study of the ear, which calls the analogy "the poetry of hypothesis", dates from 1866. Weber-Fechner's psychophysical, differential and logarithmic law, dates from 1860; Helmholtz's great work on auditory physiology, from 1863. In 1865 Clausius makes public the first definition of entropy.

No doubt Riemann was after something deep and relevant as usual for him —although he would have had to wait at least a hundred years to start putting the pieces together. Today we can do so, although not without first making some simple, but fundamental adjustments.

Nor is there any doubt that hearing has physical, but also cognitive and psychophysical limits, and that the latter act on the former in such a way that we are faced with variable thresholds defined by harmonic entropy. Within this interaction we would find the zeta function.

Is there a place for the mathematics of harmony, based on the continuous proportion, within this context of the harmonic series and its adventures in the unlimited space of analysis? Certainly not, if one thinks of explicit arithmetic relations; at least so far mathematicians have not found anything worth mentioning. This would lead us back to the hypothesis that from Euclid's *Elements,* unmixed like water and oil, two different traditions flow.

In the world of appearances, water and fire do not come into contact except in the guise of a cloud or mist of vapor. Since the first chapter we have been wondering what the relationship of the continuous proportion to analysis might be and we have not yet a definitive answer on the subject, but one of the threads would certainly be the connections of this proportion with entropy, elementary calculus and the algorithmic theory of measurement.

Needless to say, the difference between algebraic and transcendental numbers, so important in analysis and arithmetic, is irrelevant to acoustics and most of practical problems.

Thus, the harmonic series and analysis, the world of waves that reflects forms, would be on the "water shore" of reversibility; the continuous proportion, with its capacity to combine the continuous and the discrete, on the "fire shore" of entropy as irreversibility.

Of course, analysis and synthesis presuppose each other, but there are uncountable levels for their interaction. The zeta function, an analytical totality with a single pole, is in itself a prodigious synthesis, which also lends itself to all kinds of transformations in alternating, symmetrical functions, and so on. Now, the main synthetic component in modern science is hidden in the very idea of the elements, as fundamental building blocks —in physics atoms and particles. On account of them, it was thought unnecessary to look for other constructive and synthetic resources.

This is a burden from the past of physical atomism that information theory has yet to overcome, since the idea of independent elements will always be less restrictive than the net of observable, empirical correlations.

We will leave this topic suspended in the mist of its own cloud, trying to recall what is beyond complementarity, analysis and synthesis, object and subject. The example of the informal calculus of the outfielder after the the fly ball tells us that both the ball and the runner move mutually and correlatively with respect to what does not move —the constant mean.

This is the true invisible axis of activity, since the thinker is not less a thought than any other. What else can be said? But winged thought always flees from the zone of stillness, that is precisely what its life consists of. Surely this is too simple for us, and that is why we have discovered and invented the harmonic series, the continuous proportion, the entropy or the zeta function.

*

Bernhard Riemann's eight-page paper *On the number of primes less than a given magnitude* was published in November 1859, the same month as *The origin of species*. That very same year saw the beginning of statistical mechanics with a pioneer work by Maxwell, and the starting point of quantum mechanics with spectroscopy and the definition of the black body by Kirchhoff.

I have always felt that in Riemann's conscientious study, so free of further purpose, there is more potential than in all of quantum mechanics, statistical mechanics and its offspring information theory, and the theory of evolution combined; or that at least, in a secret balance sheet, it constitutes a fair counterweight to all of them. Although if it helps to redirect the doomed trend of infonegation of reality it will have already been enough.

These three or four developments mentioned are after all children of their time, circumstantial and more or less opportunistic theories. It is true that Boltzmann fought and despaired for the assumptions of the atomic theory to be admitted, but he had his whole battle with physicists, because chemists had already been working with molecules and atoms for a while. In mathematics, however, the lags operate on another scale.

A wordy bestseller, *The origin of species* was discussed in cafés and taverns on the very day of its presentation; but the Riemann hypothesis, being just the strongest version of the prime number theorem, nobody knows how to go about it after 160 years. Clearly we are talking about different wavelengths.

And what does one text have to do with the other? Nothing at all, but they still present an elusive point of contact and extreme contrast. The usual reading of the theory of evolution says that what drives the appreciable order is chance. Riemann's hypothesis, that prime numbers are distributed as randomly as possible, but that randomness hides in itself a structure of infinite richness.

Mathematicians, even living on their own planet, are gradually realizing that the consequences of proving or disproving Riemann's hypothesis could be enormous, inconceivable. It is not something that happens every Sunday, or every millennium. But we do not need to ask so much: it would be enough to properly understand the problem for the consequences to be enormous, inconceivable.

And the inconceivable becomes more conceivable precisely as we enter the Information Age. The inconceivable could be something that simultaneously affects our idea of information, its calculation and its physical support. Both software and hardware: a full blown short-circuit.

Probably information is not destiny, but the zeta function may be the destiny for the Information Age, its final event horizon. This would take us away from the spectre of the "technological singularity" and bring us closer to a very different landscape. The zeta function has a single pole that is dual with the zeros —the zeros reflect information that the unit cannot give because there are no finite results for $s = 1$. This suggests an enigmatic envelope of reflectivity for the information galaxy.

In all probability, the substance of the problem is not a technical issue. But many of its consequences are, and not only technical. This development has largely gone against the grain of other scientific developments, and its assimilation would profoundly affect the dynamics of the system as a whole.

Like the prime numbers themselves, in the chronology and the sequential order the events cannot seem more fortuitous, but from another point of view they seem destined by the whole, pushed by the ebb from all shores to fit in the very moment they took place. Riemann lived in Germany in the middle of Hegel's century, but in its second half the opposition to idealism could not have been greater in all areas, and in the sciences in particular. The pendulum had turned with all its force and the sixth decade of the century marked the peak of materialism.

Also a son of his time, Riemann could not help but feel acutely the contradictions of 19th-century liberal materialism; but the German mathematician, a theoretical and experimental physicist close to Weber and a profound natural philosopher, was an heir of Leibniz and Euler continuing their legacy by other means. His basic conviction was that we humans do not have access to the infinitely large, but at least we can approach it through the study of its counterpart, the infinitely small.

These were prodigiously fruitful years for physics and mathematics. Riemann passed away at thirty nine and did not have time to articulate a synthesis in tune with his conceptions and permanent search for unity; a synthesis understood not as an arbitrary construction, but as the unveiling of the indivisibility of truth. But he achieved something of the kind where least expected: in the analytical theory of numbers. A provisional synthesis though, that has given rise to the most famous conditional of mathematics: "If the Riemann's hypothesis is true, then...".

This unexpected and conditional synthesis came about through complex analysis, just in the same years that complex numbers were beginning to emerge in physics, seeking, like atoms and molecules, more degrees of freedom, more room to move. The role of complex numbers in physics is a topic always postponed since it is assumed that their only reason for being is convenience —however, when it comes to dealing with the zeta function and its relationship with physics, there is no mathematician who is not forced to interpret this question one way or another, which physicists generally associate with rotations and amplitudes.

But complex analysis is just the extension of real analysis, and to get to the heart of the matter we must look further back. Riemann's conditional synthesis speaks to us of something indivisible, but it still relies on the logic of the infinitely divisible; not to solve the famous problem, but simply to be in tune with it, it would have to be understood in terms of the indivisible itself, whose touchstone is the constant differential.

Is there anything beyond information and the computer? Of course there is, the same thing that is waiting for us beyond the Symplegades. A much more vast and undivided reality.

(Added June 10, 2022)

Scot C. Nelson [82] discovered in late 2001 that the logarithmic spirals of plant growth —sunflowers, daisies, pine cones, etc. — "serve as a simple and naturally efficient prime number sieve". Like everything related to the continuous proportion and its associated number series, this has received barely any attention and seems relegated in advance to the ever-growing section of anecdotal coincidences. And yet this is the first basic connection that has been found between prime numbers and these ubiquitous patterns of phyllotaxis, which should have told us something. In light of Nelson's finding, there appears to be a "central symmetry of prime numbers within three-dimensional objects", and vegetal growth would have natural prime number generating algorithm in its becoming. The same passage from the number line to the unfolding of these patterns on surfaces and in three dimensions should be a thread for the geometric intuition of the fundamental theme of arithmetic. It is not the same to try to link arithmetic with modern abstract "geometry" than to link it with a natural geometry. A mechanical analogy comes to mind: the parts repel each other like magnetic dipoles with a minimization of the energy between them, and as the plant grows the time delay between the formation of new primordia is reduced. One can think of it in ergontropic and information entropy terms as well.

POLE OF INSPIRATION – References

[1] John Arioni, *Golden Ratio in Yin-Yang*
https://www.cut-the-knot.org/do_you_know/GoldenRatioInYinYang.shtml

[2] Alexey Stakhov; *The Mathematics of Harmony — From Euclid to Contemporary Mathematics and Computer Science*
http://vixra.org/pdf/1602.0042v1.pdf
Ervin Wilson, *The scales from the slopes of Mt. Meru and other recurrent sequences,*
http://anaphoria.com/wilsonmeru.html

[3] Agno, *Imaginary Golden Ratio, My math forum*
http://mymathforum.com/number-theory/17605-imaginary-golden-ratio-.html

[4] Math Train, *Imaginary Golden Ratio, StackExchange*
https://math.stackexchange.com/questions/1851698/imaginary-golden-ratio
Karl Dilcher, *Hypergeometric functions and Fibonacci numbers,*
https://www.fq.math.ca/Scanned/38-4/dilcher.pdf

Rogers-Ramanujan continued fraction
https://en.wikipedia.org/wiki/Rogers%E2%80%93Ramanujan_continued_fraction

[5] René Guenon, *Symbolism of the Cross*, Chapters XX-XXII.

[6] Peter Alexander Venis, *Infinity-theory —The great Puzzle*
http://www.infinity-theory.com/en/science/Main_pages/The_Great_Puzzle

[7] Richard Merrick, *Harmonically Guided Evolution*. http://interference-theory.com/files/Harmonic_Evolution.pdf

[8] A. K. T. Assis, *The principle of physical proportions*
Relational Mechanics and Implementation of Mach's Principle with Weber's Gravitational Force (Apeiron, Montreal, 2014), 542 pages, ISBN: 9780992045630

[9] Nikolay Noskov, http://bourabai.kz/noskov/index.html
The Phenomenon of Retarded Potentials,
The Theory of Retarded Potentials,
Limits of application of fields in classical mechanics

[10] Miles Mathis, http://milesmathis.com/
The Physics behind the Golden Ratio
More on the Golden Ratio and Fibonacci Series

[11] Miles Mathis, *Explaining the Ellipse*
Unlocking the Lagrangian

[12] Nicolae Mazilu, *A Newtonian Message for Quantization*
Nicolae Mazilu & Maricel Agop, *The Mathematical Principles of the Scale Relativity Physics I. History and Physics*

[13] Roger Tattersall, *Why Phi? Simplified: A brief Fibonacci tour of the Solar System*
Jan Boeyens, *Commensurability in the solar system*
Harmut Müller, *Global Scaling of Planetary Systems*

[14] Miles Mathis, *A Complete Correction to and Explanation of Bode's Law.* The average deviation Mathis finds, based on a simple orbital law, is 2.75 %, exactly the same Roger Tattersall gives for the golden spiral.
Miles Mathis, A Redefinition of Gravity. Part IX. Why the Sun and Moon have the same Optical Size

[15] Mario J. Pinheiro, *A reformulation of mechanics and electrodynamics,*

[16] Miguel Iradier, *La Tecnociencia y el Laboratorio del Yo*
Rod Swenson, M. T. Turvey, *Thermodynamic Reasons for Perception-Action Cycles*

[17] Richard Merrick, *Harmonic formation helps explain why phi pervades the solar system*

[18] Gian Paolo Beretta, *What is Quantum Thermodynamics?*

[19] Xue-Jun Zhang and Zhong-Can Ou-Yang, *The Mechanism Behind Beauty: Golden Ratio Appears in Red Blood Cell Shape,*
Marcy C. Purnell and Risa D. Ramsey (February 7th 2019). *The Influence of the Golden Ratio on the Erythrocyte*

[20] Miles Mathis, *Perturbation Theory in the Light of Charge*

[21] S.C. Tiwari, *Geometric Phase in Optics and Angular Momentum of Light*
Alexander Ershkovich, *Electromagnetic potentials and Aharonov-Bohm effect*

[22] A. K. T. Assis, *Relational Mechanics and Implementation of Mach's Principle with Weber's Gravitational Force* (Apeiron, Montreal, 2014), 542 pages, ISBN: 9780992045630

[23] Alejandro Torassa, *On classical mechanics,*

[24] René Guenon, *The metaphysical principles of the infinitesimal calculus, chapter XVII, Representation of the balance of forces*

[25] A. K. T. Assis, *History of the 2.7 K Temperature Prior to Penzias and Wilson*

[26] Stephen Spurrier, Nigel R. Cooper *Semiclassical Dynamics, Berry Curvature and Spiral Holonomy in Optical Quasicrystals*

[27] Nicolae Mazilu, *Mechanical Problem of Ether*

[28] C.K. Thornhill, *The foundations of relativity*

[29] Patrick Cornille, *Replication of the Trouton-Noble Experiment*
Patrick Cornille, *Simple electrostatic aether drift sensors (SEADS): New dimensions in space weather and their possible consequences on passive field propulsion systems*

[30] Yuchao Zhang, Jie Gao, and Xiaodong Yang, *Optical Vortex Transmutation with Geometric Metasurfaces of Rotational Symmetry Breaking,*

[31] Michael Berry, *Quantal phase factors accompanying adiabatic changes Geometric phases,*
https://michaelberryphysics.files.wordpress.com/2018/03/berryd.pdf

[32] John Carlos Báez, Black Holes and the Golden Ratio

[33] Yasuichi Horibe, *An entropy view of Fibonacci Trees*
Takashi Aurues, *The Fibonacci sequence in nature implies thermodynamic maximum entropy*

[34] Miguel Iradier, *Towards a science of health? Biophysics and Biomechanics*

[35] Miguel Iradier, Beyond control: feedback and potential

[36] Mario J. Pinheiro, *A Variational Method in Out-of-Equilibrium Physical Systems*
Patrick Cornille, *Simple electrostatic aether drift sensors (SEADS): New dimensions in space weather and their possible consequences on passive field propulsion systems*

[37] R. Ferrer i Cancho, A. Hernández Fernández, *Power laws and the golden number*

[38] Aram Z. Mekjian, *Power law behavior associated with a Fibonacci-Lucas model and generalized statistical models*
Ilija Tanackov, *The golden ratio in probabilistic and artificial intelligence*
Myron Hlynka and Tolulope Sajobi, *A Markov chain Fibonacci Model*

[39] Soroko. E.M., *Structural Harmony of Systems. Minsk: Nauka i Technika* (1984) (Ruso). Comentado por A. Stakhov en [2].

[40] Yaniv Dover, *A short account of a connection of Power Laws to the Information Entropy,*
Matt Visser, Zipf's law, power laws and maximum entropy

[41] Mitchell Newberry, *Self-Similar Processes Follow a Power Law in Discrete Logarithmic Space*

[42] G. Rotundo, M. Ausloos, *Complex-valued information entropy measure for networks with directed links (digraphs)*

[43] A. Stakhov, op. cit. Petoukhov, S.V. *Metaphysical aspects of the matrix analysis of genetic code and the golden section*. Metaphysics: Century XXI. Moscow: BINOM (2006), 216250 (Ruso)

[44] Fernando C. Pérez-Cárdenas, Lorenzo Resca, Ian L. Pegg, Coarse Graining, *Nonmaximal Entropy,*
and Power Laws

[45] Miles Mathis, *A Re-definition of the Derivative (why the calculus works —and why it doesn't)*
Miles Mathis, *Calculus simplified*
Miles Mathis, *My calculus applied to exponential functions*
Harlan J. Brothers, John A. Knox, *New closed-form approximations to the Logarithmic Constant e,*
John A. Knox, Harlan J. Brothers, *Novel Series-based Approximations to e*

[46] Paul Marmet, The Subjectivity of Heisenberg's Uncertainty Relationship
Miles Mathis, *One Thing is Certain: Heisenberg's Uncertainty Principle is Dead*
Wikipedia, *Intensive and extensive properties*
V.V. Aristov, *On the relational statistical space-time concept*
V. V. Aristov. *Relative Statistical Model of Clocks and Physical Properties of Time*
S E Shnoll, V A Kolombet, E V Pozharski|¯, T A Zenchenko, I M Zvereva, A A Konradov, *Realization of discrete states during fluctuations in macroscopic processes*

[47] Peter Alexander Venis, *Infinity theory*

[48] Michael Howell, *Logarithmic derivatives sans the diminishing differentials* . http://milesmathis.com/howell4c.pdf

[49] Igor Podlubny, *Geometric and Physical Interpretation of Fractional Integration and Fractional Differentiation*

[50] Miles Mathis, *Electrical Charge*

[51] Evert Jan Post, *Quantum reprogramming* —A long Overdue and Least Intrusive Reality Adaptation of the Copenhagen Interpretation

[52] Dániel Schumayer, David A. V. Hutchinson, *Physics of the Riemann Hypothesis*

[53] Nicolae Mazilu, *Physical Principles in Revealing the Working Mechanisms of Brain*, Part One
Nicolae Mazilu, *The Classical Theory of Light Colors: a Paradigm for Description of Particle Interactions*

[54] Nicolae Mazilu, *A case against the First Quantization* (2010)

[55] C. K. Thornhill, *The kinetic theory of electromagnetic radiation*

[56] Irwin G. Priest, *A Proposed Scale for Use in Specifying the Chromaticity of Incandescent Illuminants and Various Phases of Daylight*, (1932)

[57] Paul Marmet, op. cit.

[58] David Hestenes, *Quantum Mechanics from Self-Interaction*
P. Catillon · N. Cue · M.J. Gaillard · R. Genre · M. Gouanère · R.G. Kirsch · J.-C. Poizat · J. Remillieux · L. Roussel · M. Spighel, *A Search for the de Broglie Particle Internal Clock by Means of Electron Channeling*
Shau-Yu Lan, Pei-Chen Kuan, Brian Estey, Damon English, Justin M. Brown, Michael A. Hohensee, Holger Müller, A *Clock Directly Linking Time to a Particle's Mass, Science* 339, 554 (2013)
George Savvidy and Konstantin Savvidy, *Quantum-Mechanical Inter-*

pretation of Riemann Zeta Function Zeros
Timothy H. Boyer, *Stochastic Electrodynamics: The Closest Classical Approximation to Quantum Theory,*
Nicolae Mazilu, Maricel Agop, *Role of surface gauging in extended particle interactions: The case for spin,*

[59] Mario J. Pinheiro, *On Newton's Third Law and its Symmetry-Breaking Effects*
Koichiro Matsuno, *Information: Resurrection of the Cartesian Physics*

[60] E. J. Chaisson, *Energy Rate Density as a Complexity Metric and Evolutionary Driver,* Energy Rate Density. II. Probing Further a New Complexity Metric,

[61] Georgi Yordanov Georgiev, Erin Gombos, Timothy Bates, Kaitlin Henry, Alexander Casey, Michael Daly, *Free Energy Rate Density and Self-organization in Complex Systems*

[62] A. Stakhov, op. cit. pgs. 120-125

[63] Vladimir A. Lefebvre:
https://www.researchgate.net/scientific-contributions/2034094747_Vladimir_A_Lefebvre
The algebra of conscience (review by J. T. Townsend),
https://www.researchgate.net/publication/256246817_Algebra_of_conscience_Vladimir_A_Lefebvre_Boston_Mass_Reidel_1982

[64] Raymond Abellio, *La structure absolue, Essai de phénoménologie génétique*, París, 1965

[65] McBeath, M. K., Shaffer, D. M., & Kaiser, M. K. (1995) *How baseball outfielders determine where to run to catch fly balls, Science,* 268(5210), 569-573.

[66] Víctor Gómez Pin, Filosofía. *El saber del esclavo.* Editorial Anagrama, 1989.

[67] Miles Mathis, *A Disproof of Newton's Fundamental Lemmae*
A correction to the equation a = v2/r (and a Refutation of Newton's Lemmae VI, VII & VIII)
What is pi?
The extinction of pi
The Manhattan Metric
The Cycloid and the Kinematic Circumference

[68] Gopi Krishna Vijaya, *Calculus and Geometry*
Celestial Dynamics and Rotational Forces In Circular and Elliptical Motions. Original Form of Kepler's Third Law and its Misapplication in Propositions XXXII-XXXVII in Newton's Principia (Book I)

[69] K. K. Nambiar, *Information-theoretic equivalent of Riemann Hypothesis* (2003).
J. R. Higgins, *The Riemann Zeta Function and the Sampling Theorem* (2009)
Er'el Granot, *Derivation of Euler's Formula and ζ(2k) Using the Nyquist-Shannon Sampling Theorem* (2019)

[70] K. K. Nambiar, *Electrical equivalent of Riemann Hypothesis (2003)*
Guðlaugur Kristinn Óttarsson, *A ladder thermoelectric parallelepiped generator* (2002)
Danilo Merlini, *The Riemann Magneton of the Primes* (2004)
M. V. Berry, *Riemann zeros in radiation patterns: II.Fourier transforms of zeta* (2015)
B. Van der Pol, *An electro-mechanical investigation of the Riemann zeta function in the critical strip* (1947)

[71] Matthew Watkins, *Number Theory and Entropy; Number Theory and Physics Archive*

[72] Rod Swenson, M. T. Turvey, *Thermodynamic Reasons for Perception-Action Cycles*

[73] Luciano Floridi, *What is the Philosophy of information?*

[74] O. Shanker, *Entropy of Riemann zeta zero sequence* (2013)
Alec Misra, *Entropy and Prime Number Distribution; (a Non-heuristic Approach)* (2006)

[75] Piergiulio Tempesta, *Group entropies, correlation laws and zeta functions* (2011)

[76] Gian Paolo Beretta, *What is Quantum Thermodynamics* (2007) También puede visitarse la página http://www.quantumthermodynamics.org/

[77] Miguel Iradier, *La estrategia del dedo meñique*

[78] Ian Wright, *Notes on a Hegelian interpretation of Riemann's Zeta function* (2019)

[79] Ryusuke Hamazaki, Kazuya Fujimoto, and Masahito Ueda, *Operator Noncommutativity and Irreversibility in Quantum Chaos* (2018)

[80] Ali H. Chamseddine, Alain Connes and Walter D. van Suijlekom, *Entropy and the spectral action* (2018)

[81] *Harmonic Entropy,* Xenharmonic Wiki
The Riemann Zeta Function and Tuning, Xenharmonic Wiki
Paul Erlich, *On Harmonic Entropy, con comentario de Joe Monzo*

[82] Scot. C. Nelson, *A Fibonacci Phyllotaxis Prime Number Sieve* (2004)

THE REVOLUTION THAT WILL HAPPEN: CHINA AND THE FUTURE OF TECHNO-SCIENCE

28 April, 2020

It is being said that this year 2020 could mark the beginning of the Asian Century, or if you prefer, the Chinese Century; though we will not find Chinese analysts among those who claim such things. The authors who insist on this reading of the facts point, for example, to China's clear leadership in such an strategic sector as 5G, or the imminent arrival of the digital yuan, which could cause the collapse of the dollar hegemony sooner or later. No doubt, China plays for real.

Yet this need for China to do things independently and in its own way is too often interpreted as an aggressive or expansionist policy in the West, without wanting to see that it is the West itself that has created the current rules where the winner takes it all. In the coming years we will not fail to see this rivalry for technological supremacy increasing, with the usual war of accusations and disqualifications led almost exclusively by one side.

But here I want to touch on a much broader subject than that of technological competition that is not receiving the slightest attention. I am referring to the relationship between science and technology to form a whole, what we now call Technoscience. Technoscience is the reciprocal action or continuity existing between the utilitarian applications and the development of the scientific method, between practice and theory, between power and knowledge. Power and knowledge limit each other, but incredibly, modern studies on technoscience, still know nothing about how and on what depends that knowledge and power are self-limiting —in a totally involuntary, spontaneous way.

To start knowing something about this, we would need counter-examples, that is, we should be able to contrast what we have now with other developments in science and technology with the same level but really independent. However, in the Western countries they use to stress that there is only one possible way of doing science —precisely the one those countries advocate and control.

We don't need illustrations for such an overwhelmingly accepted assumption; but I will still bring up an example to help us appreciate how it sounds when someone wants to put it into words. In a *Foreign Policy* article appeared some two years ago, titled "*The Future of Particle Physics Will Live and Die in China*", the author laments that the extremely expensive next-generation supercollider will no longer be built by Europeans or Americans, but by the Chinese. And that is not the bad news: Western researchers will only have the status of guest observers, instead of deciding what experiments are performed or how the data are filtered.

So to prevent this from happening, the author does not fail to invoke the universality of science: the Chinese must know as soon as possible that there is no such thing as "a science with Chinese characteristics", but just plain old international science, which, just by chance, come directly from the offices of Princeton, Cambridge or Geneva. In other words, the Chinese should be content to pay for the project, while they watch and learn from the creative American scientists how to make the most of an investment. Thus spoke *Foreign Policy.*

There is no need for comments, as this position disqualifies itself. As for the rest, the mega-projects of Western Big Science with their enormous inertia have always been the worst imaginable place for critical thinking, since such large investments force these endeavors to justify themselves whatever it takes —something well known even by the lowest operator working in them.

This is why Chinese scientists have very good reasons to be suspicious. On the other hand, the Chinese authorities could seriously consider whether it is worth inviting Western scientists who, freed from having to justify their own projects, would be eager to revive all that repressed critical spirit waiting for a better occasion.

If there is no Chinese, American or European science, but just international science, why be alarmed that China does its own experiments? They should relax and calmly wait for the results, and be thankful for saving them so much money and work; but it seems there is something more to it.

The most trite case of "concern" about what Chinese scientists might do is in biotechnology and genetic manipulation, especially as it affects humans. But, once again, it is not the Chinese who have invented cloning, nor those who have put into circulation the now universally accepted idea that genes are nothing more than "information", right? And yet this latter reduction is the first thing that should be challenged, because it is what has led us to the indiscriminate manipulation of life. Besides, it is clear that no private corporation will have higher standards in this regard than governments.

But while these issues already attract media attention and are subject of discussions, they are still at the crudest levels of debate —those of propaganda, suspicion and veiled or open accusation. What I want to raise here, and no one has raised yet, is a much more far-reaching challenge for our ideas and our imagination, and surely for the future of humanity. What I want to raise here is whether China will finally be able to develop a science, or a techno-science, really different from what the West has so far deployed.

It has become a cliché to talk about the lack of originality of Chinese researchers, systematically ignoring the enormous effort it has taken for them to catch up and adapt to rules of the game that are completely alien to their mentality. We do not realize that this "lack of originality" is nothing but the effect of belonging to a culture that has been the most differentiated and original in the world, but which has also suffered the tragic circumstance of an amputation of its autochthonous traditions.

It is these two simultaneous circumstances, a modernization as fast as ever seen, together with the unfortunate destruction of their own traditions, that has created this cojunctural, momentary image of the Chinese as lacking in originality, which is so easy to undo just by looking back. On the other hand, China has had a wonderful tradition of natural philosophy unique in the world, but its conception of Nature has so far been incompatible with the standards of modern positivism. We are referring, of course, to a loose set of traditions encompassed under the name of "Taoism".

Modern science, since Descartes, is based on the idea of a self separated from the world. For today's technoscience, that intelligence is totally separated from nature is an indispensable condition for recombining all aspects of nature as we please: atoms, machines, biological molecules and genes, or the interface between any of them under the utterly disengaged criterion of information. This is the condition of what I call our liberal materialism or material-liberalism.

However, although Taoism was far from playing a central role in Chinese society prior to the Revolution, the need to reintegrate intelligence into the unity of being, into the unity of nature of which it never ceased to be a part, still remains. And the same violent amputation of such a consubstantial part of Chinese culture still ask for redress, which will take place sooner or later.

The great English sinologist Joseph Needham long ago posed a famous problem known as "the Needham question", namely why it was not China but Europe that carried out the modern scientific revolution, when the Chinese civilization was much more advanced in terms of industries, inventions and technological achievements.

This is a question that sparked a wide debate among historians, but which in my view is still confronted with more than remarkable naivety. It is said, for example, that there is no point in asking counterfactual questions, and in a sense that is obvious. But even so we are taking many things for granted. The first is that technological development has been good, and not promoting it, a mistake. The second is that the story is over, and that there will be no more unexpected turns in the relationship between science and technology. The third, related to the previous one, is that there is no option left for us than to adapt to the inexorable course and that we cannot change its direction. And all this, which now seems to us to be a fait accompli, are just assumptions.

It has been understood that during the Ming Dynasty (1368-1644), just before the European scientific revolution, when the number of Chinese inventions seemed to be approaching a critical mass, the ruling class decided to curb technological development as a threat to social stability. And surely this would have been the wisest decision, had it not been for the fact that other less developed and permanently warring peoples on the other side of the Eurasian continent were unable to put a brake on themselves. We all know what the outcome was.

Westerners have long been called "barbarians" in China, and I do not think that is an inappropriate rating were we to judge by the general behavior of their warring elites in the world. No doubt the West has had great values, but to think that our values are superior to those of other civilizations due to the overwhelming material success of their model shows a very low conception of what the word "values" means. And the same could be said of the ancient Chinese civilization before its decadence, if we are only able to judge it by its material development.

Today we live in the midst of a full-blown technological barbarism, and the belief, driven from above, that the solution to all our problems lies in technology leads us into the worst of all possible worlds. The same applies to the twin agenda that pretends that society must exist for the economy rather than the economy for society. These are inseparable programs whose mere assumption is already the greatest defeat, for a life governed by functionalism is not even worth living.

And it is the West, not China, that has spread this program by blood and fire all over the world. But when China has had to assume it to the end, even if only to survive, it frightens us and even horrifies us —because it gives us back in a reflection not only what we do not see of ourselves, but above all what we do not want to see: that our lives have no more meaning than theirs, no matter how much we insist that we are the good guys and those who do things with soul and originality.

And here comes my first judgement and my first "prediction." The West has been unable to tame technology and its drift that leads us to the abyss. In this, its failure is as spectacular as the very rampage of the functional, technological and economic sphere. However, *if* China is really capable of assimilating and making the modern scientific method its own —something so far has only done in appearance- its greatest effort will lie precisely in dominating what dominates us and governing what today is ungovernable.

That is to say, it would return to the same reflex of the Ming dynasty, when it was still the most powerful empire on Earth —but in a world that has nothing to do with it, and in which turning back is simply impossible. And this would seem praiseworthy to me, for it would show that there is still a certain amount of wisdom on this planet. So the greatest danger is not that the Chinese develop "a science with Chinese characteristics"; the real danger is that they don't develop it.

Moreover, the West should facilitate this process as much as possible, and contribute to it with the best of its intelligence. First, because if they are freed, we are freed as well. And second, because we tread a garden of forking paths. Here there are no good or bad agents, but good or bad decisions. And, as always, the good can soon become bad and vice versa.

Were to continue the script of our known history, China could give a social and collectivist turn to the new revolution of technoscience; the United States, one based on entertainment and consumerism; and other parts of the world, such as Europe, Russia, or India, would have to choose among the available models or develop their own. Digital currencies or crypto-currencies will soon put us in this field of difficult decisions and spheres of influence; but they are only a small part of the major changes that are coming in our way of seeing and acting in the world.

In a world where everything is revolutionized, waiting or talking about revolutions is like wanting to push someone who is falling —something unnecessary and without effect. And the same can be said of all the "revolutionary" changes in science and technology, which are nothing but successive concentric waves of a single great impact that took place in the 17th century. But what I am talking about is not the present "technological revolution", since that, even if it drags us along, is little more than consensual inertia. For if we do not realize now that this drift is simply dragging us, when will we do it?

The "revolution" we are talking about requires, on the contrary, an active will capable of turning the course of events without frontally opposing the current, which rarely makes sense. This political will today seems only to exist in China, because it is still fully aware that it is at a disadvantage in a world where it has not made the rules. And scientific and technological rules are basically conventions and standards. If Poincaré already stated this in 1900, with an idea of science that is infinitely higher and more demanding than the current one, I do not know what we would have to say now. In any case, for the Chinese scientists and engineers this is painfully obvious. Do not the scientists themselves speak of the standard model of cosmology or that of particle physics?

In modern Big Science there is very little room for originality, and maybe that is the reason to use that word like a mantra. This is so much so that those who really want to do things differently have to give up research when they experience first-hand the overwhelming reality of the production of consent on an industrial scale. This is at least what many researchers say who had to choose other ways of making a living, and one can be sure that these are not mediocrities but quite the opposite.

The originality thing is a typically American public relations claim to extract illusion and give a "positive" and fresh image of scientific research. Large scale science can never be original, neither in Boston, London, or Beijing. Added to this is the fact that in Western science, that is, in the American hegemony of world science, scientific competition, like the neutrality of markets and everything else, is for the most part fictitious, and reduced to who gets the consolation prizes and the minor parts of the loot. Everything fundamental is out of the question; one has no option but to "innovate" in a carefully delimited arena.

We do not say this in a controversial spirit or to denounce anything, but to bring to mind that a "free science" and "a science by itself", as it is claimed in the West, are absolutely contradictory and incompatible things. Ever since the times of Newton's Royal Society it was clearly understood that science was another way of conquering the world and the minds of its inhabitants. Today's "science by itself" is a phenomenon as natural, "spontaneous" and neutral as the hegemony of the dollar, and surely it will not last longer, since both live on our trust.

And this is precisely where the interesting part begins. Everything that seems inevitable to us today, as if fallen from the sky, still has an almost unlimited margin of change in one direction or another. For example, the current Internet in which we almost even breath, since the Internet and its digital revolution is not a product of technology alone, but of the particular way Silicon Valley shapes this technology.

But all of this, no matter how much it shapes our current environment, still is very superficial. The most profound changes do not depend merely on technology, but on the specific relationship between scientific theories and practical applications, and it is precisely in this that there is much more than meets the eye.

The United States made its unofficial takeover of world science in the shadow of the Manhattan Project and other military research of the Second World War. Around 1948, quantum electrodynamics, cybernetics and information theory crystallized. Five years later the DNA helix was identified, which would soon be reduced to the new master category of information.

All these developments have a great continuity with the interwar European science and the change is only of emphasis, towards a more "algorithmic" approach. Quantum electrodynamics, for example, which has been sold to us as the "jewel of physics", is but a feat of computation that adds nothing new to the already diverse formulations of quantum mechanics by Schrödinger, Heisenberg or von Neumann. Information theory recycles the concept of entropy from the old statistical mechanics and applies it to machines and communication channels. Cybernetics is a gradual development of the theory of control and stability to which Wiener makes many new contributions but with a monumental strategic hole at its centre that current science has not even noticed.

In short, and to put it clearly: if the takeover of American science was not very traumatic for Europeans, it was basically due to the lack of originality of their contributions. It only involved a slight change of emphasis in the general program, clearing the way of unnecessary obstacles and scruples. However, if China truly makes the scientific method its own, and not just the technological applications, the transformation will be incomparably deeper, and to some extent, an antithesis of everything the West has known. In fact, this is the only thing that could be compared in magnitude with the scientific revolution of the Baroque era.

And what might this new *Tao of Technoscience* look like? No one knows that, nor does it have to be something to worry about now. Much more important is to realize that a Tao of Technoscience has always existed: that there is a relationship of reciprocal dependence between theory and practice that determines the scope of our vision, our interaction with nature and its results.

This mutual dependence has not been even remotely calibrated due to a typical Western assumption. Among natural sciences, we distinguish between "hard" and abstract sciences that make predictions, such as mathematical physics, and "soft", descriptive sciences, much closer to the world of appearances in which we live and with a much less relevant role for mathematics: evolution theory, biology, or cosmology.

Both types of science need to concur to some extent to give coherence to our world: they are like the subjective recreation we make of time by the mutual interplay between memory and anticipation. However, Western science crystallized with the predictive sciences and the application of mathematics to physics; and then added descriptive supplements for the vast majority of phenomena that do not support prediction. These are parallel circuits of reasoning, but it is their degree of connection that decides what we can conceive and what we cannot.

I always like to give an example that exposes the connection of practice and theory in modern science. In 1956 Bohr and von Neumann came to Columbia to tell Charles Townes that his idea of a laser, which required the perfect phase alignment of a great number of light waves, was impossible because it violated Heisenberg's inviolable uncertainty principle. The rest is history. But now everybody says that quantum mechanics predicts and makes lasers possible! Things like this happen all the time. In a word, theoretical physicists are experts at taking the credit for the hard work of experimental scientists and engineers, who have actually struggled with a theory that makes it almost impossible for them to figure things out.

But it is clear that a theory that can "predict" anything a posteriori is not a great theory, but a great rationalization, and this is true for any type of theoretical construction. Think, for example, of quantum electrodynamics, which subtracts infinities from infinities in recurring computation cycles until the expected result is reached. Today everything is justified in the name of prediction, but Ptolemy's epicycles also held an unbeatable predictive power.

Just as the Chinese are allergic to theories, we Westerners are very good at speculating. We Westerners say, "there is nothing more practical than a good theory", because we see the benefit we get from it, but what we do not see is the cost of adhering to it. Anyone can make theories, the difficult thing is to avoid the temptation to develop one. It is typical of Western speculative genius to overrate the role of theory to the point of actually believing Nature obeys its equations. To such an extent it has forgotten the tricks of its reverse engineering.

Technoscience has as its own matter the relationship between practice and theory. It is a fact that we owe China more technical inventions than any other culture before the scientific revolution; however the hypothetical-deductive method and its procedures are very different from the ancient Chinese methods of observation, inference and invention. There is a serious problem of compatibility here that has not yet been solved, but there is also a way to make it happen.

One should know that it is possible to change the principles, methods and interpretations of science and still have the same observations and predictions; therefore, it is only a matter of time before Chinese scientists adapt the results of modern science to their own way of seeing things. This is perfectly legitimate; what is not legitimate is wanting to prescribe to others how they have to understand nature.

In reality, current science has exhausted its theoretical potential and is at a dead end. It has abused speculation to such an extent, and has collected so well all the low-hanging fruit of predictable things, that now it can only recycle the same tricks over and over again. With a tight control of scientific production, real competition has been eliminated and dissonant voices have been marginalized. On the other hand, the same process of differentiation and specialization has an eminently irreversible character.

The problem is not that there is an American, Chinese, or European science, but that science cannot be very different from the society in which it takes place and which it serves; and it seems that we are still talking about societies that are very distant in their assumptions and objectives.

In science, as in life, the important thing is not to want to solve the problems that are not your own. What prevails today is neither "Western science" nor "international science", but the Anglo-Saxon system of talent extraction from the global pool, which decides which problems and approaches are convenient and pertinent, all this with uninterrupted campaigns of propaganda and public relations.

In this situation of overwhelming hegemony, if China did not exist, history and nature would have had to invent it. The national viewpoint has a high degree of contingency, but appearance itself resists to be reduced to a single principle. The one-sidedness of American supremacy and its excesses could not fail to create a great void in other areas that has to be filled in some way, and science and technology are only manifestations of something more general. This we used to call "dialectic".

Chinese science will manage to harmonize the descriptive and predictive halves, as well as its principles, means and ends. No one will be able to prevent it, and when it is achieved, we will realize how primary, short-sided and improvised scientific progress has been during these four hundred years. We shall speak of these times as the "age of prediction" and "the religion of prediction," something very much like the California gold rush and the conquest of the West, in which the gunmen defied each other to see who had the biggest prediction.

However, no amount of knowledge or scientific development will by itself lead us to a better world. This depends on something very different. Still, the *polis* has come out of the *physis*, cultures are a second nature, and the uncertain gap between the first and second nature defines our plane of existence.

It has taken us four hundred years to weave all this fantastic web of knowledge, but it does not take four hundred more years to turn a glove inside out without breaking it. While scientific managers were concerned with control, the enormous advantage that historical perspective gave us has been completely wasted. Others will benefit from it.

No, we are not talking about the so-called Fourth Industrial Revolution, but about the Second Scientific Revolution, which is something completely different and has not even started. For Europe took centuries to assimilate the number system and the experimental procedures of the Islamic world, which were combined with the Greek propensity for theory, to get a totally distinct outcome. We are now talking about a process of cultural transformation of at least similar scope, and probably greater, but which we can see taking shape already in this very decade of the 20s.

It can also be predicted that this transformation of science, which cannot be reduced to a single country and will end up affecting very different parts of the world, will have a much deeper cultural penetration than that of current technoscience —since the known scientific concepts are the result of a peculiar specialization that greatly limits their assimilation. The same will not be true of the new-old concepts, which by its very nature will be much less isolated; but, to begin with, these fresh developments will start from a higher degree of self-consciousness.

When the most basic scientific concepts connect with a deeper level of the human selfhood, everything will change.

I have dealt with these issues in greater depth in a book entitled "*Pole of inspiration —Math, Science and Tradition*", which I would only recommend to those who want to see science from a totally different perspective than the one that prevails today.

References

[1] Foreign Policy — Yangyang Chen, *The Future of Particle Physics Will Live and Die in China*, 2017.

[2] Miguel Iradier, *Pole of Inspiration -Math, Science and tradition,* 2020.

CHINA IN SPACE AND TIME

20 May, 2020

From the West, we tend to judge today's China more by its economic presence and the impact of its material development on the rest of the world than by the internal needs in the development of its history; thus giving an overwhelming priority to the geopolitical perspective over the cultural one, which should have at least a comparable importance.

In fact, it is easy to see that China's overall impact will depend to a large extent on how well it manages to fit this whole period and the foreseeable future into a historical framework for which it would like as little change as possible. A sense of continuity on a large scale is fundamental to Chinese culture, and in the long term it will always do its best to assimilate and make the origin of foreign influences undetectable. It has already achieved this to a large extent with Marxism and capitalism, which lose so much of their original meaning in translation that it is no longer known whether to call the Chinese system "market socialism" or "state capitalism".

The momentum of modernity is still felt as anomalous for a large part of the world's population, which only suffers the process of cultural colonization in exchange for the loss of its own forms. But if in most countries this acculturation is a passive and reactive process associated mainly with the consumer society, in China first took the form of an active destruction in the unprecedented phenomenon of the Cultural Revolution, and only later adopted the ubiquitous forms of consumerism. This deep self-inflicted wound will still demand some sort of reparation for a long time to come, and could even lead to some overcompensation.

Those who say that China wants to dominate the world are just retelling us the usual Fu Manchu novels, and they know it perfectly well. Were we to take Wallerstein distinction between world-economies and world-empires, it is clear that the global hegemony has been embodied in the liberal model with its center at London in the nineteenth century and in New York in the twentieth century. On the contrary, China has been for more than two thousand years a world-empire, that is to say, a world-system that, like the Ancient Roman Empire, only exists comfortably within certain natural limits. Globalization is the condition to which China has had to adapt.

The natural thing for China is to be the centre of gravity of the whole Southeast Asia with the mark of Confucian culture, as it has been for so many centuries; but this is almost inevitable since she has about 80 percent of its population in addition to be the origin and reference of this culture. Only the circumstantial American overprojection in the region and our modern habit of thinking in national terms prevents us from perceiving what historically has been the norm.

For China, as a world empire, the ideal condition has been and will always be self-sufficiency. It is quite another thing that in a tremendously impoverished situation it has had no other quick way to recover than through export and a mercantilist model. That model, still present, obliges it to secure suppliers of raw materials as well as markets for its recurrent problems of overproduction. The Belt and Route Initiative also responds to these needs, as well as to a progressive shift in trade partnerships.

In other words, it is not a question of expansionism but of ensuring the viability of the model in the future; these moves are not motivated by a spirit of conquest, but of conservation. Historically, even when China extended its dominions more deeply into Central Asia, as in the Tang dynasty, the consolidation of the defense was the main motivation. In any case, the Middle Kingdom has been more defined by its cultural continuity than by the oscillation of its borders.

In the current Chinese system the Communist Party allows the exploitation of the workers in the name of development while playing its role as the last resort capable of redressing or mitigating injustices. This is the "middle way" that has been chosen, which the radical criticism of the left cannot help but see as a hypocritical compromise. A compromise with a marked vertical logic and a strict separation between the governors and the governed.

Certainly, if anything is to be desired in today's China, it is not greater openness to foreign capital flows, as the international plutocracy would like, but more socialism and more popular participation —if only to maintain the balance; but nevertheless, the Chinese mixed system, compared to the neoliberal privatizing fury that prevails in most of the world, looks like a model of sanity and good sense.

We know that capitalism without restraint and counterweights can only lead to catastrophe; since the Chinese government is an almost unique example of how to exercise these counterweights, its overall influence would have to be seen everywhere as basically beneficial —were it not for the venom of the propaganda of the one empire that wants it all and does not tolerate the loss of a single piece of pie.

Think that in the past the Chinese economy could account for 30, 40 or even more of the world's wealth, and Europeans did not even know that such an empire existed. A 30, 40 or even 50 percent of the world's wealth that was based on their own talent and work and not on the plundering of other countries. [1] Now it is about one-sixth or 16 percent of the global product, and we are told they want to devour the world. Why such hysteria? We all know the answer.

It is true that the Chinese model is neither exportable nor has it ever been intended to be exported. But in Europe we had prosperous mixed economies, in countries like France, until only a few decades ago; what is completely unreasonable is how far we have come. The problem is not that the Chinese model is not exportable, the problem is that now in most of the developed countries one cannot even consider that the state can put severe constraints on corporate interests.

China always tries to maintain balance and reciprocity, both in the foreign relations and the internal affairs; this is not difficult to understand, after all, even if in practice it becomes a difficult art of compromises. However, this is not just a matter of diplomacy, but involves also an ethical component that in this political tradition the ruling class tries to assume. The political image that the United States projects today, by contrast, is one of extreme imbalance, blackmail, abuse and disregard.

It would be truly incredible if China were ever to become the first power in a world that it does not even understand and in which it has had enough to adapt. It is even unthinkable, for anyone who has even the slightest idea of how the global financial system works, and all the other parallel systems of domination. China is not about to devour the world, on the contrary its main rival tries to encircle it persevering in all sorts of harassment. It would have the most decided sympathy of us all were it not for the nefarious and toxic influence of the neocolonial media of propaganda that sets the tone in almost the entire world.

To say that the Chinese, the other par excellence, are a threat to the welfare of Westerners, is the perfect way to divert attention from those who really subjugate and plunder us. These "others" are much more alien than anything, so much so that they do not have even faces, or at least one will never see them on the screen. It seems unbelievable that people can take this bait, but they do, because the important thing about propaganda is not the plausibility, but the insistence and pervasiveness. So we can be sure that not only this state of affairs will continue, but that they will turn up the noise of the megaphones.

The cybernetic temptation

In reality, China has to deal more with the time and rhythm imposed on it than with the delicate geopolitical lineaments. The time we are talking about is that of modernity itself, now coated with the attributes of the technological and digital tsunami; a time perpetually accelerated and increasingly alien to nature, a time governed by the neoliberal logic of dissolving absolutely everything except the concentration of power through capital.

And here, again, we are deluding ourselves about China. Since it already leads on strategic fronts such as 5G, this nation is seen as particularly well positioned to be at the forefront of technological development; but even if these kind of deployments really take the lead and are an exhibition of muscle, at heart they are a reaction, or if one prefers, an overreaction. China does not control the logic of this great process, but tries its best not to be a subject of it. It tries to master its forms, but does not control the dynamics, the momentum or the content. This is where the greatest dangers lie.

It is well known that the three main sources of inspiration for Chinese culture have been Confucianism, Taoism and Buddhism. To these three a fourth must be added, especially with regard to the forms of government and the shape of the state: the so-called "Legalist school" developed since Han Fei, or going back further, since Shen Buhai and Shang Yang or even since Guan Zhong. This "Legalism" or *Fajia* would be better called "technocratic realism", or objectivism of government standards. Undoubtedly the technocratic trend is very strong in China today, and its tradition of objectivism in standards is seen as a way more akin to tackle with the challenge of technology.

For the Chinese ruling class the technocratic temptation of creating a closed cybernetic system with feedback in real time that seeks to control everything is really strong. If Chile was already on the verge of implementing its famous Cybersyn project in the 1970s when Allende was killed, imagine what the rulers of this great nation can achieve with today's technology and the developments already in the making [2].

Such a system may seem ideal for maintaining hierarchies while controlling monetary flows or permanently negotiating decentralization and popular participation. The temptation is even greater when one considers that China is not alone in the world, but permanently harassed by a United States that tries to destabilize the country by every means at its disposal —and cybernetics is nothing but the theory of control and stability. Cybernetics and governance are the same word, the *kybernetes* being the helmsman.

China's destiny is being defined as much or more by this type of temporal process than by the interplay of inclusions and exclusions of geopolitics. But not only China's, because we must not forget that the same techniques are already being applied in the rest of the world by the great digital giants —giants that also "collaborate" closely with Western governments, although in a totally different balance of powers.

The symbol of cybernetics would be the snake that bites its own tail. The rulers of the great Asian dragon seem to be even more compelled to make their own these kind of feedback mechanisms because for them it would be tantamount to immunizing themselves against the destabilizing dangers of a technology already so difficult to control —just as the Western plutocracy seeks to use it to isolate itself from the masses and control them.

But dataism is only a substitute religion that cannot hide its abysmal shortcomings. In China, as elsewhere. Most extreme of all is that the Chinese mentality cannot even see Western science and technology as its own, but remains a foreign body, even with all these titanic developments. From this the worst and the best can arise; the worst is that they merely reproduce the forms, the best, that they get to the bottom of it.

The Chinese are fascinated by crabs, with which they share the same atavistic reflex for defending and grabbing. The other symbol that best defines them is the balance. The first is the symbol of the people, the second of the ruling class. They match with the zodiac houses of Cancer and Libra, the two cardinal signs that mark the beginning of summer and autumn, and, in the Chinese calendar, the centre of both seasons.

These would be the nodal points of the Chinese spirit within the wheel of the year, the first symbol of wholeness for us all. Interestingly, the homologous signs of the Chinese calendar are the sheep and the dog, which do not define so sharply this specific *constellation*. There is always something about yourself that only others can see well; I do not know if we ask ourselves what it is about us that the Chinese perceive best.

Capitalism and Marxism are the thesis and antithesis in the time bomb that is modernity. Each of them oscillates between the dissolution and concentration of capital, and a correlative impoverishment and agglutination of public opinion: they come to embody the "double contradiction" that Mao Zedong already spoke of, the cross of the helm in a dialectics that is not based on the Hegelian idealist sublation or on the idea of history as an irreversible progress.

Capitalism and Marxism each have their own cunning of reason, and the Chinese idea of equilibrium in government, its own too; but it goes without saying that the expectation of bringing together the majorities by Marxism failed miserably in the West, partly because capitalism has managed to create all sorts of wedges, and partly because Marxism itself has no divine right to claim exclusivity for either opposition or resistance to the abuses of capital.

The problem is that the operating system that runs it all is still the liberal one, and Marxist or any other criticism is external both to its design and operation as to its use. The operating system is modern technoscience as a whole. Without a change in the operating system, liberal capitalism set the rules and always has the time on its side.

Of course, it is totally false to say that the Chinese lack initiative —one just have to see that they cannot stand still. On the contrary, they have demonstrated an unshakable faith in continuous and persevering action. That they know the many advantages of not showing it is a different issue; or that their political system does not exactly encourage the expression of personal views. Or that they have on their shoulders a modernization that they cannot grip.

The Autonomous Cybernetic System, a self-referential monster so similar to the Great Animal that is society, is so much a temptation for the ruling class on account of that same unshakable faith in continuous improvement and rectification. But it would be a great mistake to be content with that —everything that tends to close in on itself is courting disaster, as it loses contact with the most basic reality.

This is what China needs most now in order to attempt to tackle the great problem of a new cultural synthesis up to the challenge of modernity: contact with this basic reality that is not mediated by political opportunism or socio-economic coordinates. More than fifteen centuries ago, something similar was triggered by the penetration of Buddhism, which has so far been almost the only "beneficial invasion" that China has suffered, without ignoring the multiple intrigues and adverse reactions that it provoked in the already established traditions.

A much more modest and recent, though not negligible, invasion was that of the European piano in the Far East sensibility, that it was like giving a pond to the moonlight. At least it shows that the West can also tune in to the deepest fibers of this culture, provided that there is a contact zone and the indispensable affinity.

If the first instinct of the Chinese is to grab like the crabs and hold on, they have also shown that they know how to give back what is given to them in good faith and without ulterior motives. It is said that the Chinese are the least idealistic of people, and in an important sense this might be true, but the stories of heroic commitment and sacrifice of countless monks, peasants, and rebels of all kinds along the centuries say otherwise, and their moving example has not yet been erased. And the Chinese, unlike the Japanese, appreciate the mixture of the comic, the tragic and the sublime of a character like Don Quixote.

Only when we give do we bring out the deepest of ourselves. It is sad that relations between China and the West are dominated by the most immediate and petty interests, and that there is no willingness to reach deeper areas of our common ground. Even most of cultural relations and exchanges are hardly more than diplomacy and business. It is one of the many things we should rebel against.

The great problem for Chinese culture today is not the assimilation of technology, but of the historical hard core of modern science and how this spreads to the softer adaptable parts, which are precisely those relating to the category of information. And its insertion, precisely, within parameters akin to those of its objectivist spirit, the most basic ideas of Taoism, the internal axis of Confucianism and the identity in Chan Buddhism of the immanent and the transcendental.

The technocratic-cybernetic dream is equivalent to having the genie of the lamp circulating in a glass carefully designed to both work for you and not escape. Of course, if you torment the genie so much, chances are that he will eventually break free. The moral is that in the end one can only count on one's own spirit and not on bound spirits, and this applies equally to the West —but maybe the West does not even suspect where their chains are.

The internal axis and culmination of Confucianism is the Doctrine of the Mean or *Zhongyong*, which is also the natural point of connection with Taoism and the transcendental Buddhist doctrine. The meaning of this doctrine has been almost completely lost, and political leaders such as Mao Zedong reveal by their commentaries that they were totally clueless about its intention, since it has nothing to do with "eclecticism". This should not surprise us, if Confucius himself had already said that it had long been rare among people. It should be evident that the Invariable Mean is not the ultimate degradation of Confucianism, but its highest aspect.

In a recent book I have tried to trace some basic connections of this hard core of European science, calculus and classical mechanics, with the doctrine of the Invariable Mean, the reversible and irreversible in Taoism, the transcendental plane of knowledge or the problem of standards and measurement theory. Of course, sometimes the relations are so obvious that it seemed to me unnecessary to make it more explicit. Certainly I have not write it only with the Chinese culture in mind, but thinking of a common problem still to be solved and on which modern science has preferred to turn the page [3].

I believe that these fundamental issues, so hidden, are much more interesting than the hocus-pocus of contemporary science and also have much more potential —for anyone who knows how to take advantage of it.

The contact of the Chinese tradition with modern science seemed an impossible task, if we take into account not only the difficulty of the Chinese scientist to interiorize its spirit, but also the destitution occurred with their own tradition. However, I believe that here we have *begun* to connect both spheres in a truly *natural* way. The only thing I can say is that I think this path should be followed, and that it applies to both the East and the West.

In a forthcoming article we will discuss *the little finger strategy*, or how to deflect the technological tsunami with minimal effort.

Notas

[1] The United States of America also had about 50 percent of the world's wealth in 1945; but, although at that time most of that wealth resulted from domestic production, that country was carrying on half a century of neocolonial adventure in Latin America, the Pacific, the Philippines or China. Moreover, by 1945 most of the world's wealth had been des-

troyed in a war in which the United States had intervened in earnest to consolidate its hegemony rather than to "defend freedom".

[2] *"Project Cybersyn was the Chilean attempt at controlled economic planning in real time, developed in the years of the government of Salvador Allende, between 1971 and 1973. In essence, it was a network of teletype machines that connected factories to a single computer center in Santiago, where the machines were controlled using the principles of cybernetics. The main architect of the system was the British scientist Stafford Beer".* https://es.wikipedia.org/wiki/Cybersyn

[3] Miguel Iradier, *Pole of inspiration —Math, Science and Tradition*, Source: hurqualya.net .

www.ingramcontent.com/pod-product-compliance
Lightning Source LLC
LaVergne TN
LVHW100247210926
848480LV00003B/845
9798230001102